FORTSCHRITTE DER CHEMIE ORGANISCHER NATURSTOFFE

PROGRESS IN THE CHEMISTRY OF ORGANIC NATURAL PRODUCTS

PROGRÈS DANS LA CHIMIE DES SUBSTANCES ORGANIQUES NATURELLES

HERAUSGEGEBEN VON   EDITED BY   RÉDIGÉ PAR

L. ZECHMEISTER
CALIFORNIA INSTITUTE OF TECHNOLOGY, PASADENA

FÜNFUNDZWANZIGSTER BAND
TWENTY-FIFTH VOLUME   VINGT-CINQUIÈME VOLUME

VERFASSER   AUTHORS   AUTEURS
P. R. ASHURST · F. BOHLMANN · L. FARKAS · Y. GAONI · H. KLING
R. MECHOULAM · G. A. MORRISON · L. PALLOS · J. ROMO
A. ROMO DE VIVAR · J. K. SUTHERLAND · E. WALDSCHMIDT-LEITZ
TH. WIELAND

MIT 25 ABBILDUNGEN   WITH 25 FIGURES   AVEC 25 ILLUSTRATIONS

1967

WIEN · SPRINGER-VERLAG · NEW YORK

ISBN-13: 978-3-7091-8166-9   e-ISBN-13: 978-3-7091-8164-5
DOI: 10.1007/978-3-7091-8164-5

Softcover reprint of the hardcover 1st edition 1967

Titel Nr. 8232

Inhaltsverzeichnis.

Contents. — Table des matières.

Inhaltsverzeichnis. — Contents. — Table des matières. V

Recent Advances in the Chemistry of Hashish. By R. MECHOULAM,
   School of Pharmacy, Laboratory of Natural Products, The Hebrew
   University, Jerusalem and Y. GAONI, Institute of Organic Chemistry,
   Weizmann Institute of Science, Rehovoth......................... 175

Biogenetische Beziehungen der natürlichen Acetylenverbindungen.

Von **F. Bohlmann**, Berlin.

I. Einleitung.

Vor etwa 10 Jahren waren zirka 50 natürliche Acetylenverbindungen bekannt (70). In der Zwischenzeit ist die Zahl derartiger Verbindungen auf etwa 440 angewachsen. Vor allem aus höheren Pflanzen, insbesondere aus Compositen und Umbelliferen, sind sehr viele neue Typen von Verbindungen isoliert worden, die die große Variationsbreite dieser Naturstoffklasse erkennen lassen. Mit der Anwendung moderner physikalischer Methoden, wie Kernresonanz- und Massenspektroskopie, ist es gelungen, auch die Strukturen von Substanzen aufzuklären, die in äußerst geringer Konzentration neben den Hauptinhaltsstoffen zu isolieren sind. Gerade diese Verbindungen haben wertvolle Einblicke in die wahrscheinlichen biogenetischen Beziehungen dieser Substanzklasse ergeben, so daß sich jetzt ein erster Überblick über die Hauptbiogeneseschritte bei den natürlichen Acetylenen abzeichnet. Gestützt werden diese Überlegungen durch zahlreiche Untersuchungen mit ^{14}C- oder tritium-markierten Polyinen, deren Umwandlungen in den Pflanzen studiert wurden.

Darüber hinaus haben sich in der letzten Zeit sichere Anhaltspunkte über die Biogenese der natürlichen Acetylenverbindungen ergeben. Die enge Verknüpfung ihrer Bildung mit der Fettsäurebiogenese ist gesichert. Neben der Bildung der Dreifachbindung sind es im wesentlichen bekannte Biogeneseschritte, die zu der großen Mannigfaltigkeit der schon jetzt bekannten Verbindungen führen. Das Studium der enzymatischen Vorgänge steht allerdings noch sehr in den Anfängen, ebenso wie die Klärung der Frage, welche Bedeutung diesen reaktionsfähigen Verbindungen, die sehr rasch metabolisiert werden, im gesamten Pflanzenstoffwechsel zukommt.

Über die Verbreitung im Pflanzenreich können noch keine eindeutigen Angaben gemacht werden, da bisher nur wenige Familien auf Acetylenverbindungen eingehender untersucht wurden. Bis heute hat man aus etwa 15 verschiedenen Familien derartige Substanzen isoliert. Genauer untersucht ist jedoch nur die Familie Compositae, hier sind etwa 1000 Arten bearbeitet worden, die fast alle Acetylenverbindungen enthalten. Lediglich die Arten des Tribus Cichoriae sowie einige Gattungen des Tribus Senecioneae scheinen eine Sonderstellung einzunehmen, hier sind Acetylene nicht charakteristisch.

Die große Mannigfaltigkeit der isolierten Verbindungen gibt interessante Ansatzpunkte für die Pflanzensystematik, da schon jetzt klar zu erkennen ist, daß offenbar bestimmte Gattungen oder Sektionen besondere Enzymsysteme besitzen, die bei sehr ähnlichen Arten fehlen, so daß klare Abgrenzungen möglich sind. Dieses scheint besonders bei einigen großen Gattungen der Compositen interessant zu sein, da hier die anatomischen Merkmale oftmals kaum ausreichende Differenzierungen ermöglichen. Auch phylogenetisch sind die Acetylenverbindungen eventuell von Interesse.

II. Zur Biogenese der natürlichen Acetylenverbindungen.

Die Frage, wie die natürlichen Acetylenverbindungen gebildet werden, ist in den letzten Jahren teilweise geklärt worden. An vielen Beispielen konnte zunächst einmal gezeigt werden, daß die stets unverzweigten Verbindungen aus Acetat- bzw. Malonat-Einheiten aufgebaut werden.

Bei Verfütterung von $1\text{-}^{14}\text{C}$-Acetat an *Polyporus anthracophilus* erhält man *trans,trans*-Matricariaester, der alternierend markiert ist *(86)*:

$$CH_3\overset{*}{C}OO^{\ominus} \rightarrow CH_3\overset{*}{C}H\underset{t}{=}CH\overset{*}{C}\equiv C-\overset{*}{C}\equiv C-\overset{*}{C}H\underset{t}{=}CH\overset{*}{C}OOCH_3$$

(t = trans und c = cis).

Entsprechend liefert $1\text{-}^{14}\text{C}$-Acetat Odyssinsäure, die durch Eliminierung der endständigen Methylgruppe in Nemotinsäure übergeht *(87)*, die wiederum alternierende Aktivitätsverteilung aufweist:

Literaturverzeichnis: SS. 54—62.

$$CH_3\overset{*}{C}OO^{\ominus} \rightarrow CH_3C\equiv C-C\equiv C-CH=C=CHCHCH_2CH_2COOH$$

$$\downarrow \quad | \quad OH$$

$$H\overset{*}{C}\equiv C-\overset{*}{C}\equiv C-\overset{*}{C}H=C=\overset{*}{C}HCH\overset{*}{C}H_2CH_2\overset{*}{C}OOH$$

$$| \quad OH$$

Durch Verfütterung von 2-^{14}C-Malonat an *Tricholoma grammopodium* kann weiter gezeigt werden, daß als Startmolekül Acetyl-Coenzym A benutzt wird, das dann wie bei der Fettsäurebiogenese mit Malonyl-Coenzym A kondensiert wird (94):

$$COO^{\ominus}$$

$$CH_3COSCoA + 4\ \overset{*}{C}H_2COSCoA \rightarrow CH_3C\equiv\overset{*}{C}-C\equiv\overset{*}{C}-C\equiv\overset{*}{C}-CH=\overset{*}{C}HCH_2OH$$

Bei höheren Pflanzen liegen die Verhältnisse ganz analog, wie z. B. durch Verfütterung von 1- bzw. 2-^{14}C-Acetat an *Chrysanthemum frutescens* L. und *Coreopsis lanceolata* L. gezeigt werden kann (42). Der Abbau der erhaltenen markierten Polyine ergibt wiederum eine alternierende Aktivitätsverteilung (42) (*Schema 1*).

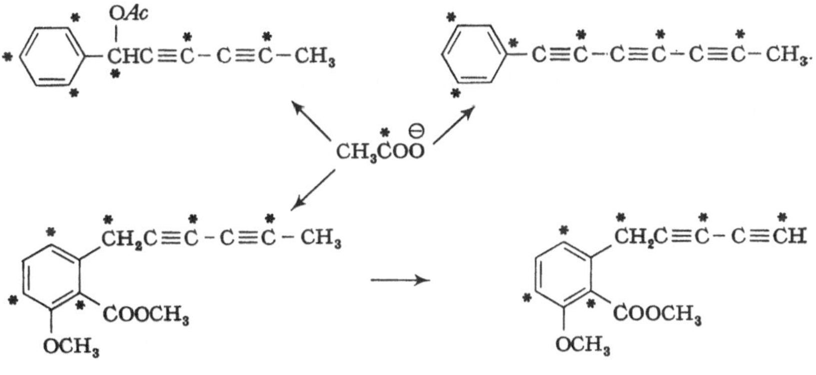

Schema 1.

Diese Ergebnisse sowie die Tatsache, daß alle bisher isolierten Acetylenverbindungen, bis auf das Sesquiterpen (526, S. 50) unverzweigte C-Ketten besitzen, lassen die enge Verwandtschaft mit der Fettsäurebiogenese erkennen, wobei prinzipiell zwei Möglichkeiten bestehen: Entweder erfolgt die Bildung der Dreifachbindung aus den β-Carbonylzwischenstufen (A) oder es findet eine direkte Dehydrierung von Doppelbindungen statt (B):

$$A: \quad R-COCHCOSCoA \longrightarrow R-C=C-COSCoA$$

with the enol structure:
$$\substack{\text{CO}_2^{\ominus}\\|}$$
$$\begin{array}{c} {}^{O^\ominus}\!\!\diagdown C=O \\ R-C=C-COSCoA \\ {}^{O-\textcircled{P}-\textcircled{P}} \end{array}$$

$$R-C\equiv C-COSCoA$$

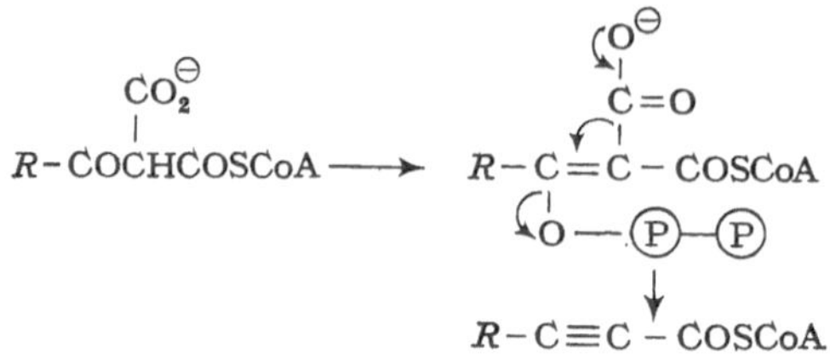

$$B: \quad -CH_2-CH_2- \rightleftharpoons -CH=CH- \rightleftharpoons -\dot{C}\equiv C-$$

Der Weg A ist mechanistisch einleuchtend und hat seine Parallele durch die in vitro sehr leicht verlaufende Fragmentierung entsprechender Enol-brosylate (105):

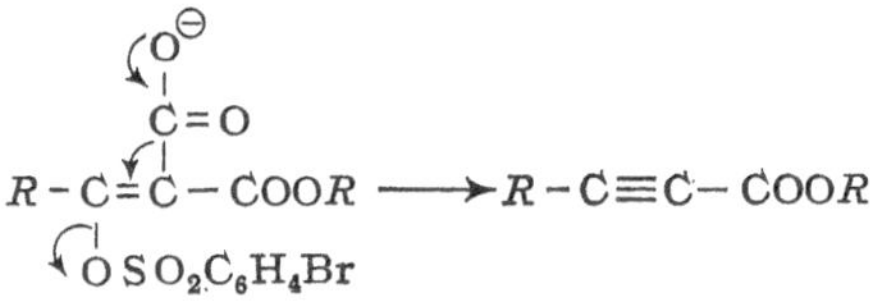

$$\begin{array}{c} {}^{O^\ominus}\!\!\diagdown C=O \\ R-C=C-COOR \longrightarrow R-C\equiv C-COOR \\ {}^{O\,SO_2C_6H_4Br} \end{array}$$

Beim Weg B ist die Dehydrierung gesättigter Fettsäuren zu ungesättigten grundsätzlich als Biogeneseschritt bekannt, lediglich die Dehydrierung von Doppelbindungen zur Dreifachbindung wäre neu. Energetisch sollten sich diese beiden Dehydrierungen nicht sehr wesentlich unterscheiden. Denkbar wäre sogar eine Kombination mit der normalen Fettsäure-Synthese:

$$2-CH=CH-COSCoA \longrightarrow \begin{array}{l} -CH_2CH_2COSCoA \\ -C\equiv C-COSCoA \end{array}$$

Dabei würden jeweils unter Mitwirkung spezieller Enzyme zwei Wasserstoffatome von einer Äthylen-Bindung auf eine andere übertragen.

In der letzten Zeit sind aus höheren Pflanzen zahlreiche Acetylenverbindungen isoliert worden, die die cis-9,10-Doppelbindung der Ölsäure enthalten (27, 47, 53, 55, 56, 68, 71, 72), so z. B. der Kohlenwasserstoff (5) (71) mit 17 C-Atomen, dessen nahe Verwandtschaft zur Ölsäure offensichtlich ist. Folgendes Schema wäre für die Biogenese von (5) wahrscheinlich, zumal die Säure (2) aus Compositen isoliert wurde (130).

Durch Verfütterung von markierter Ölsäure an verschiedene Mikroorganismen konnte inzwischen gezeigt werden, daß hier in der Tat eine entsprechende Umwandlung erfolgt.

Literaturverzeichnis: SS. 54—62.

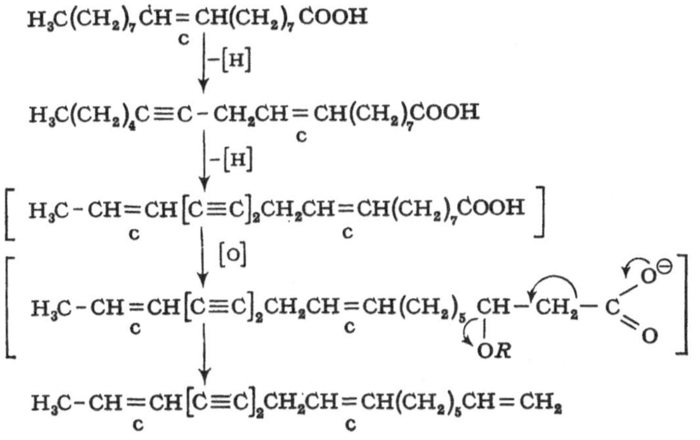

$$H_3C(CH_2)_7\overset{c}{C}H = CH(CH_2)_7 COOH \qquad (1)$$

$$\downarrow -[H]$$

$$H_3C(CH_2)_4C \equiv C - CH_2\overset{c}{C}H = CH(CH_2)_7COOH \qquad (2)$$

$$\downarrow -[H]$$

$$\left[\; H_3C - \overset{c}{C}H = CH[C \equiv C]_2 CH_2\overset{c}{C}H = CH(CH_2)_7COOH \;\right] \qquad (3)$$

$$\downarrow [O]$$

$$\left[\; H_3C - \overset{c}{C}H = CH[C \equiv C]_2 CH_2\overset{c}{C}H = CH(CH_2)_5 \; CH - CH_2 - C \overset{\ominus}{\underset{O}{\overset{O}{\diagdown}}} \atop OR \;\right] \qquad (4)$$

$$\downarrow$$

$$H_3C - \overset{c}{C}H = CH[C \equiv C]_2 CH_2\overset{c}{C}H = CH(CH_2)_5 CH = CH_2 \qquad (5)$$

So wird z. B. 10-^{14}C-Ölsäure (6) von *Tricholoma grammopodium* über Linolsäure (7) und Crepissäure (8) in Dehydromatricarianol (9) (95) umgewandelt:

$$H_3C(CH_2)_7\overset{c}{C}H = CH(CH_2)_7COOH \qquad (6)$$

$$\downarrow$$

$$H_3C(CH_2)_4\overset{c}{C}H = CHCH_2\overset{c}{C}H = CH(CH_2)_7COOH \qquad (7)$$

$$\downarrow$$

$$H_3C(CH_2)_4C \equiv C - CH_2\overset{c}{C}H = CH(CH_2)_7COOH \qquad (8)$$

$$\downarrow$$

$$H_3C[C \equiv C]_3\overset{t}{C}H = CHCH_2OH \qquad (9)$$

Entsprechend erhält man bei Verfütterung von markierter Ölsäure an *Coprinus quadrifidus* das aktive Triol (10), dessen Aktivität an C_1 lokalisiert ist (*112*):

$$HC \equiv C - C \equiv C - C \equiv C - \underset{OH}{CH} - \underset{OH}{CH} - \overset{*}{\underset{OH}{CH_2}} \qquad (10)$$

Fütterungen höherer Pflanzen mit markierter Ölsäure zeigen ebenfalls die glatte Umwandlung in die verschiedensten Acetylenverbindungen (*84*). Triinsäuren mit 14 bis 18 C-Atomen werden sehr glatt in Dehydromatricariaester (11) umgewandelt (*45*), während entsprechende Diinsäuren praktisch nicht in (11) übergeführt werden (*84*). Lediglich die Diinsäure mit isolierter *cis*-Doppelbindung wird relativ glatt in Dehydromatricariaester umgewandelt (*84*). Das bestätigt gleichzeitig die angenommene Bedeutung der Ölsäure als eigentliche Vorstufe.

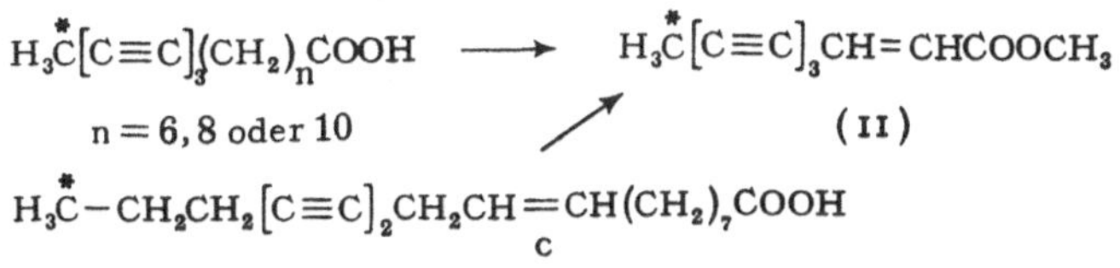

$$H_3\overset{*}{C}[C \equiv C]_3(CH_2)_nCOOH \longrightarrow H_3\overset{*}{C}[C \equiv C]_3CH = CHCOOCH_3$$

$$n = 6, 8 \text{ oder } 10 \qquad\qquad \nearrow \qquad (11)$$

$$H_3\overset{*}{C} - CH_2CH_2[C \equiv C]_2 CH_2\overset{c}{C}H = CH(CH_2)_7COOH$$

Auch die C_{10}-Diinsäure sowie n-Decansäure und die C_{18}-Diinsäure werden nicht in (11) umgewandelt (84):

$$H_3\overset{*}{C}-CH_2CH_2[C\equiv C]_2\underset{t}{CH}=CHCO_2H$$

$$H_3\overset{*}{C}(CH_2)_8CO_2H \qquad\qquad \longrightarrow\!\!\!/\!\!\longrightarrow \qquad (11)$$

$$H_3\overset{*}{C}-CH_2CH_2[C\equiv C]_2\underset{t}{CH}=CH(CH_2)_8CO_2H$$

Alle diese Ergebnisse zeigen, daß auch bei den höheren Pflanzen die notwendigen Polyinsäuren aus Ölsäure gebildet werden.

Damit ist die Frage der Bildung der Dreifachbindung in höheren Pflanzen beantwortet, und es spricht alles dafür, daß auch hier die Dehydrierung von Doppelbindungen zur Dreifachbindung der wahrscheinlichste Biogeneseweg ist. Es ist jedoch nicht auszuschließen, daß eventuell auch Weg A (siehe S. 4) in Betracht zu ziehen ist. Endgültig werden sich diese Fragen erst nach Isolierung der Enzymsysteme klären lassen.

III. Biogenetische Beziehungen.

Überblickt man die bereits jetzt bekannten natürlichen Acetylenverbindungen, so ergeben sich zahlreiche wahrscheinliche biogenetische Beziehungen, die zum Teil auch bereits durch Fütterungsversuche mit markierten Verbindungen sichergestellt werden konnten.

Für derartige Untersuchungen müssen zunächst die entsprechenden Acetylene radioaktiv-markiert aufgebaut werden. Als Isotope kommen ^{14}C oder Tritium in Betracht. Als Beispiel für derartige Synthesen sei der Aufbau von 1-^{14}C-Dehydromatricariaester (46) und von 1.2-^{3}H-Trideca-tetrain-(3.5.7.9)-dien-(1.11) (78) angeführt (Schema 2).

Die *Verfütterung* der markierten Verbindungen an intakte Pflanzen gelingt am besten auf folgendem Wege: Die praktisch wasserunlöslichen natürlichen Acetylenverbindungen, die mit ^{14}C oder Tritium markiert sind, werden in Baumwollsaatöl gelöst und unter Zuhilfenahme eines Emulgators in Wasser emulgiert. In diese Emulsion stellt man die intakten Pflanzen für 24—48 Stunden ein, wobei die Emulsion fast vollständig aufgesogen wird. Die zu untersuchenden Umwandlungen sind praktisch immer in dieser Zeit abgelaufen, so daß man die Pflanzenteile anschließend sofort extrahieren kann. Die Inhaltsstoffe werden dann wie üblich chromatographisch aufgetrennt und die Einzelstoffe bis zur konstanten spez. Aktivität gereinigt. Anschließend müssen die isolierten Substanzen definiert abgebaut werden, um die Aktivität zu lokalisieren, da stets mit der Möglichkeit von Abbau und Resynthese aus Acetat-Einheiten zu rechnen ist.

Literaturverzeichnis: SS. 54—62.

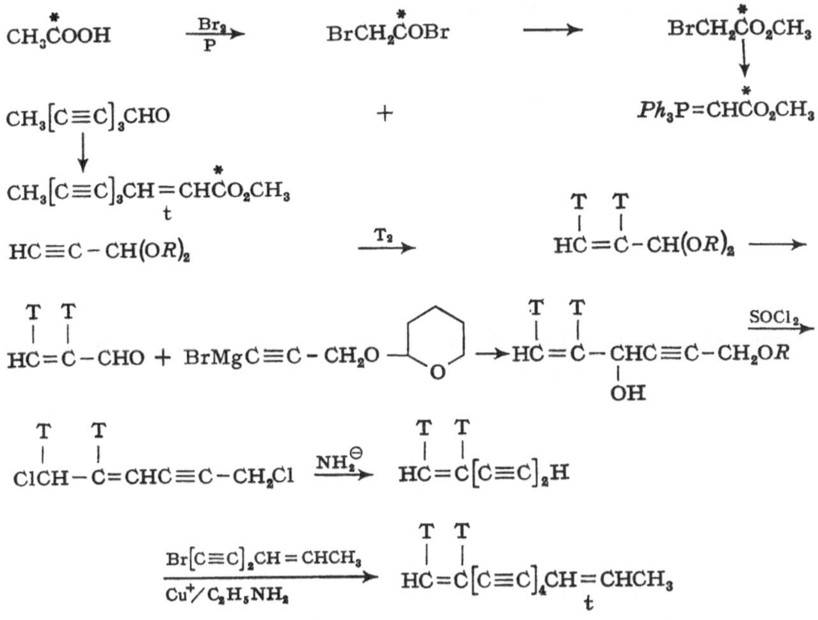

$$CH_3\overset{*}{C}OOH \xrightarrow[P]{Br_2} BrCH_2\overset{*}{C}OBr \longrightarrow BrCH_2\overset{*}{C}O_2CH_3$$

$$CH_3[C\equiv C]_3CHO \qquad\qquad + \qquad\qquad \downarrow$$

$$\downarrow \qquad\qquad\qquad\qquad Ph_3P=CH\overset{*}{C}O_2CH_3$$

$$CH_3[C\equiv C]_3\underset{t}{CH}=CH\overset{*}{C}O_2CH_3$$

$$HC\equiv C-CH(OR)_2 \xrightarrow{\;T_2\;} \overset{T}{HC}=\overset{T}{C}-CH(OR)_2 \longrightarrow$$

$$\overset{T}{HC}=\overset{T}{C}-CHO + BrMgC\equiv C-CH_2O\!-\!\!\langle\;\text{(THP ring)}\;\rangle \rightarrow \overset{T}{HC}=\overset{T}{C}-\underset{OH}{CH}C\equiv C-CH_2OR \xrightarrow{SOCl_2}$$

$$\overset{T}{ClCH}-\overset{T}{C}=CHC\equiv C-CH_2Cl \xrightarrow{NH_2^\ominus} \overset{T}{HC}=\overset{T}{C}[C\equiv C]_2H$$

$$\xrightarrow[Cu^+/C_2H_5NH_2]{Br[C\equiv C]_2CH=CHCH_3} \overset{T}{HC}=\overset{T}{C}[C\equiv C]_4\underset{t}{CH}=CHCH_3$$

Schema 2. Synthese markierter Polyine.

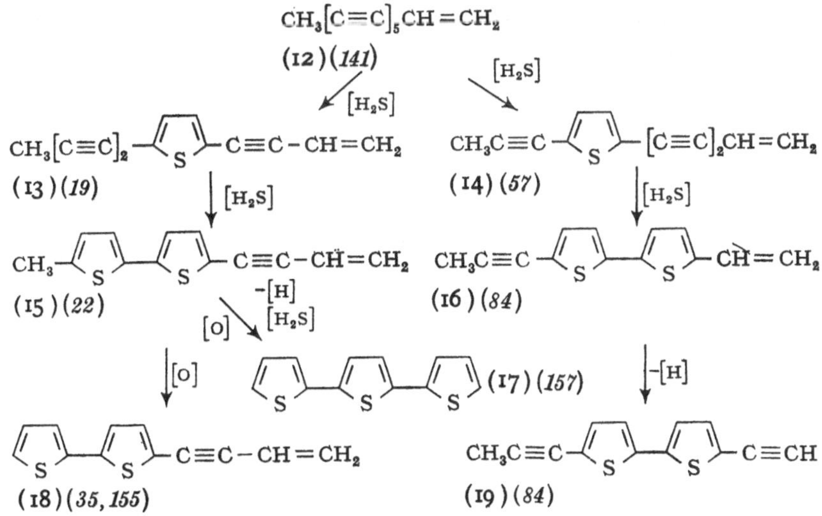

$$\overset{*}{C}H_3[C\equiv C]_5CH=CH_2$$

(12)(*141*)

Schema 3. Umwandlungsprodukte des Pentainens.

In den folgenden Abschnitten sind die einzelnen Gruppen im Hinblick auf ihre biogenetischen Beziehungen zusammengestellt, die zum Teil durch Fütterungsversuche bewiesen sind, während in anderen Fällen die aus Analogieschlüssen wahrscheinlichen Übergänge aufgezeigt werden.

1. C_{13}-Polyinen-Kohlenwasserstoffe als Vorstufen.

a) Pentainen.

In der Familie Compositae ist das Pentainen (12) besonders weit verbreitet. Die Konzentration dieser Verbindung ist jedoch stets sehr gering. Häufig findet man aber neben (12) andere Verbindungen in etwas höherer Konzentration. Vielfach sind diese Substanzen formal sehr nahe verwandt mit (12), so daß offensichtlich enge biogenetische Beziehungen vorliegen. Besonders häufig findet man Thiophenverbindungen, die sich von (12) ableiten, wie *Schema 3* (S. 7) zeigen möge.

Abgeleitet von (13), (15) und (18) findet man weiterhin die folgenden Derivate, die durch sehr häufig zu beobachtende Umwandlungen der endständigen Vinylgruppe sowie durch oxydativen Angriff der Methylgruppe gekennzeichnet sind:

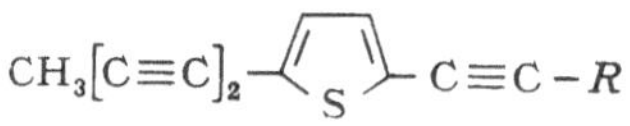

$$CH_3[C \equiv C]_2 - \langle S \rangle - C \equiv C - R$$

(20) $R = -CH_2CH_2OH$ (19);   (21) $R = CH_2CH_2OAc$ (19);
(22) $R = CH(OAc)CH_2OAc$ (19);   (23) $R = CH(OH)CH_2OH$ (19);
(24) $R = CHClCH_2OAc$ (19);   (25) $R = CHClCH_2OH$ (19);
(26) $R = CH - CH_2$ (19);
        $\diagdown O \diagup$

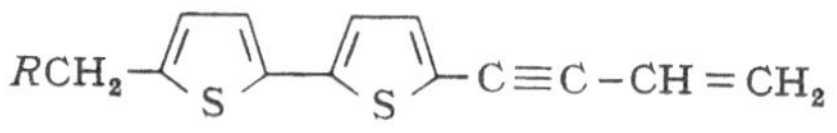

$$RCH_2 - \langle S \rangle - \langle S \rangle - C \equiv C - CH = CH_2$$

(27) $R = OH$ (19)
(28) $R = OAc$ (51)

$$\langle S \rangle - \langle S \rangle - C \equiv C - R$$

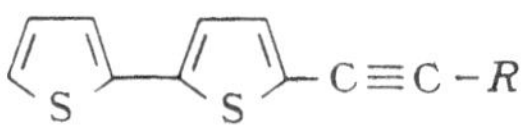

(29) $R = CH_2CH_2OAc$ (57);   (30) $R = CH_2CH_2OH$ (19);
(31) $R = CH(OH)CH_2OH$ (19);   (32) $R = CH(OAc)CH_2OAc$ (19);
(33) $R = CH(OAc)CH_2OH$ (19);   (34) $R = CH(OH)CH_2OAc$ (19);
(35) $R = CH(OH)CH_2Cl$ (4);   (36) $R = C = CHOAc$ (84):

Literaturverzeichnis: SS. 54—62.

C1

Durch Verfütterung von 1.2-^{3}H-(12) an *Echinops sphaerocephalus* L. konnte gezeigt werden, daß (18) über (13), (15) und (27) aus (12) gebildet wird. (31) entsteht ebenfalls aus (12) über (18). (24) wird über (13) und (26) aus (12) gebildet, während (29) aus (18) entsteht (39). Der Abbau zur Lokalisierung der Aktivität sei am Beispiel von (18) gezeigt *(Schema 4)*.

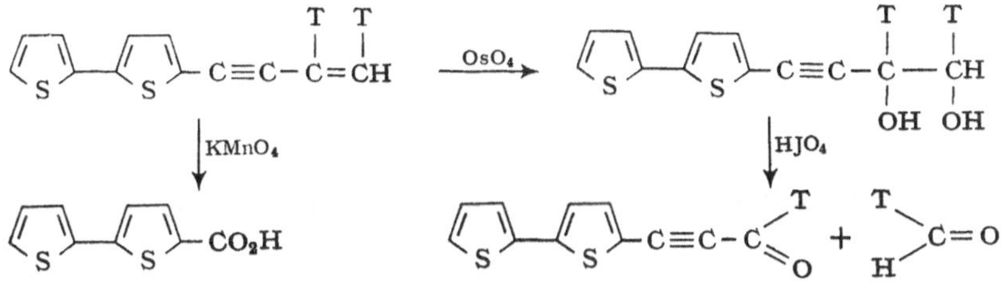

Schema 4.

Neben (13) und (14) findet man häufig auch die rotgefärbten, wahrscheinlich valenztautomeren Dithioverbindungen (37) und (38), die außerordentlich leicht unter Abspaltung von elementarem Schwefel in (13) bzw. (14) übergehen (54) *(Schema 5)*.

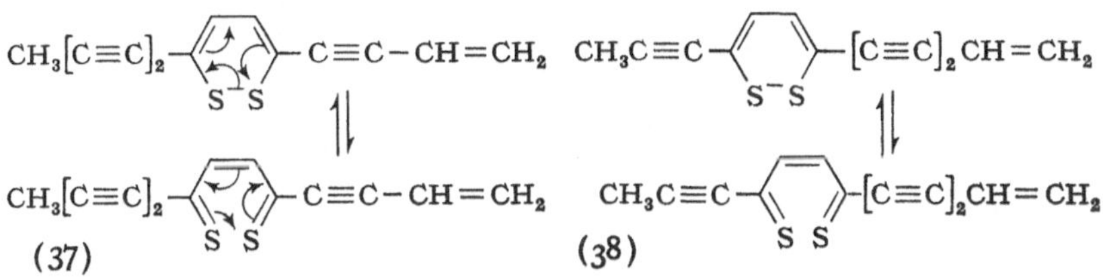

(37)   (38)

Schema 5.

Dieser leichte Übergang läßt vermuten, daß derartige Verbindungen möglicherweise Zwischenprodukte der Thiophenbiogenese darstellen:

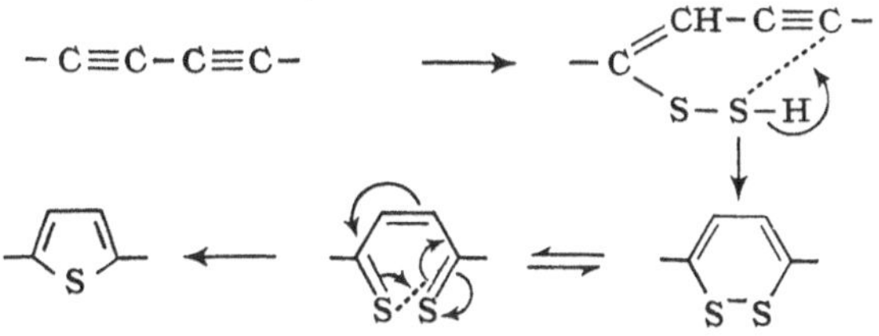

$$-C\equiv C-C\equiv C- \longrightarrow$$

Die Umsetzung von (39) mit Natriumdisulfid liefert jedoch (40) (31), so daß die obige Annahme nicht sehr wahrscheinlich ist.

Wie der Übergang einer Diingruppierung in ein Thiophen in der Pflanze erfolgt, ist ungeklärt. Formal ist eine H_2S-Addition notwendig, eine Reaktion, die in vitro mit Natriumsulfid relativ glatt abläuft (*135*):

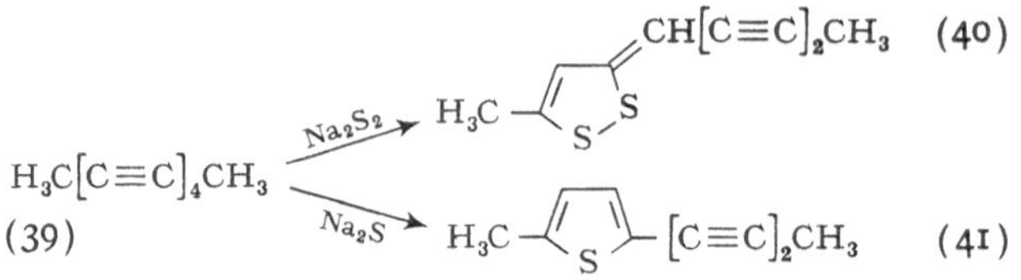

Durch Fütterungsexperimente konnte gezeigt werden, daß Schwefel in Form von Sulfat und Cystein eingebaut wird (*136*).

Das Terthienyl (**17**, S. 7) wird ebenfalls aus (**12**) gebildet. $1.2\text{-}^3H\text{-}$(**12**) liefert ein Terthienyl, das nur in α-Stellung Tritium enthält (*78*). Daraus muß man schließen, daß auch hier ein Diin als Vorstufe für den dritten Thiophenring erforderlich ist. Würde die formale H_2S-Addition an der Vinylacetylenkette erfolgen, so sollte zunächst ein Dihydro-terthienyl entstehen, das nach Dehydrierung in α- und β-Stellung Tritium enthält. Interessant ist in diesem Zusammenhang das Vorkommen von (**19**, S. 7), das offenbar durch Dehydrierung von (**16**) gebildet wird (*84*). Weiterhin ist bemerkenswert, daß (**18**) nicht in Terthienyl übergeht, wie durch Fütterungsversuche gezeigt werden konnte (*134*). Offenbar ist das noch nicht isolierte α-Methylterthienyl Zwischenprodukt. Das entsprechende Carbinol ist allerdings kürzlich aus *Eclipta alba*, die auch (**12**), (**14**), (**17**), (**18**) und (**27**) enthält (*84*), isoliert worden (*126a*). Demnach ist offensichtlich nur die Dehydrierung von (**15**) und nicht von (**18**) möglich (*Schema 6*).

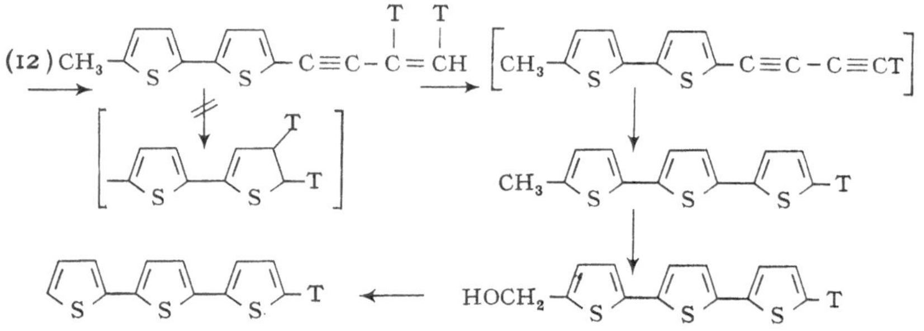

Schema 6.

Neben der Bildung von Thiophenderivaten wird (**12**) in verschiedene andere Verbindungen umgewandelt. Wie *Schema 7* zeigt, ist eine formale Methylmercaptan-Anlagerung mit oxydativen Folgereaktionen und die Umwandlung der Vinylgruppe ein weiterer Biogeneseweg.

Literaturverzeichnis: SS. 54—62.

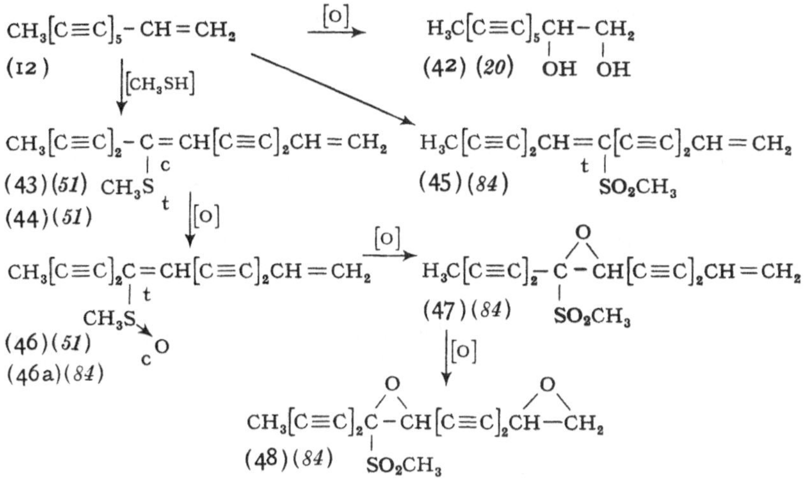

$$CH_3[C\equiv C]_5 - CH = CH_2 \xrightarrow{[O]} H_3C[C\equiv C]_5CH - CH_2$$

(12)

(42)(20)   OH   OH

$\downarrow [CH_3SH]$

$$CH_3[C\equiv C]_2 - \underset{\underset{CH_3S}{|\ c}}{C} = CH[C\equiv C]_2CH = CH_2 \qquad H_3C[C\equiv C]_2CH = \underset{\underset{SO_2CH_3}{t\ |}}{C}[C\equiv C]_2CH = CH_2$$

(43)(51)   CH₃S
(44)(51)   t $\downarrow [O]$

(45)(84)

$$CH_3[C\equiv C]_2\underset{\underset{CH_3S}{|\ t}}{C} = CH[C\equiv C]_2CH = CH_2 \xrightarrow{[O]} H_3C[C\equiv C]_2 - \overset{O}{\overset{\diagup\diagdown}{C}} - \underset{\underset{SO_2CH_3}{}}{CH}[C\equiv C]_2CH = CH_2$$

(46)(51)   CH₃S
(46a)(84)   c O

(47)(84)   SO₂CH₃

$\downarrow [O]$

$$CH_3[C\equiv C]_2\overset{O}{\overset{\diagup\diagdown}{C}} - \underset{\underset{SO_2CH_3}{}}{CH}[C\equiv C]_2\overset{O}{\overset{\diagup\diagdown}{CH}-CH_2}$$

(48)(84)   SO₂CH₃

Schema 7.

$$HOH_2C - CH = CH[C\equiv C]_4CH - CH_2$$

(49)(20)   t $\uparrow [O]$ OH   OH

$$H_3C - CH = CH[C\equiv C]_4CH - CH_2 \qquad\qquad R - CH = CH[C\equiv C]_4CH = CH_2$$

(50)(20)   t   OH   OH $[O]$ t

(51) $R = CH_2OH$ (16)
(52) $R = CH_2OAc$ (16)
(53) $R = CHO$ (16)

[O]

$$H_3C - CH = CH[C\equiv C]_4CH = CH_2$$

(54)(142) t
(55)(148) c $\downarrow [H_2S]$

$$H_3C - CH = CH - \!\!\!\underset{S}{\diagup\!\!\!\diagdown}\!\!\!\underset{S}{\diagup\!\!\!\diagdown} - CH = CH_2$$

(57)(20) t

$[H_2S]$ $[H_2S]$
$[O]$

$$CH_3CH = CH[C\equiv C]_2 - \!\!\!\underset{S}{\diagup\!\!\!\diagdown}\!\!\!- R$$

t

(61) $R = CH(OH)CH_2OH$ (84)
(62) $R = CH(OAc)CH_2OAc$ (84)
(63) $R = CH(OH)CH_2OAc$ (84)

$$H_3C - CH = CHC\equiv C - \!\!\!\underset{S}{\diagup\!\!\!\diagdown}\!\!\!- C\equiv C - CH = CH_2$$

(56)(57) t

$\downarrow [O]$

$$R - CH = CHC\equiv C - \!\!\!\underset{S}{\diagup\!\!\!\diagdown}\!\!\!- C\equiv C - CH = CH_2$$

t

(58) $R = CH_2OAc$ (57)
(59) $R = CH_2OH$ (57)
(60) $R = CHO$ (57)

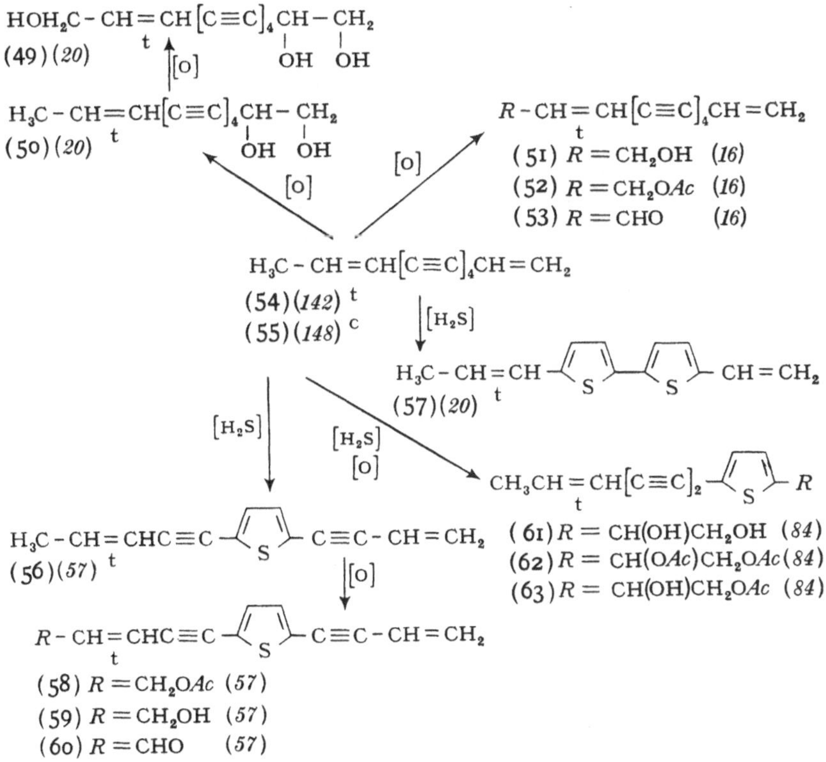

Schema 8. Umwandlungsprodukte des Entetrainens.

Die Umwandlung von (12) in (43) bzw. (44) wurde mit markiertem (12) bewiesen (39).

b) Entetrainen.

Neben (12) ist auch das Entetrainen (54) relativ weit verbreitet bei den Compositen. Auch dieser Kohlenwasserstoff wird in verschiedener Weise in zahlreiche andere Acetylenverbindungen umgewandelt. Wiederum sind die meisten derartigen Biogeneseschritte, neben dem Übergang in Thiophene, durch eine Umwandlung der Vinylendgruppe und der endständigen Methylgruppe gekennzeichnet, wie aus *Schema 8* (S. 11) zu ersehen ist.

Der Übergang von (54) in (56) bzw. (57) und in (51) konnte mit 1.2-^{3}H-(54) sichergestellt werden (78).

c) Tetraindien.

Der Tetraindien-Kohlenwasserstoff (67) und die isomere *cis*-Verbindung (68) kommen relativ selten vor, aber einige Umwandlungsprodukte sind bei den verschiedensten Vertretern der Compositen häufiger zu finden, wie *Schema 9* zeigt.

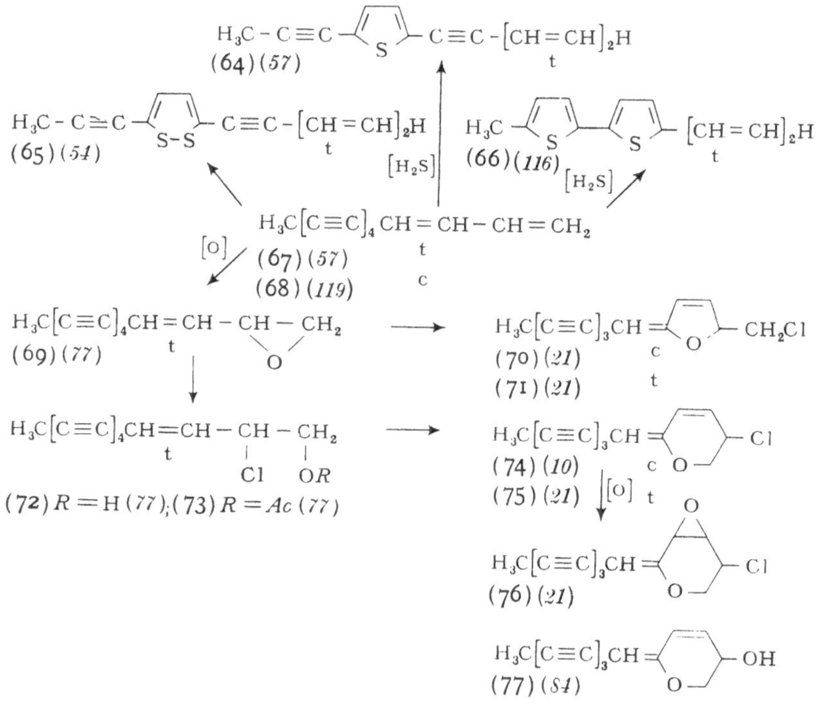

Schema 9. Folgeprodukte des Tetraindiens.
Literaturverzeichnis: SS. 54—62.

Der Übergang von (67) über (69) in (72) konnte durch Verfütterung von $1\text{-}^{14}C\text{-}(67)$ an *Centaurea ruthenica* Lam. bewiesen werden (40). Die Fünfring- und Sechsring-Sauerstoffheterocyclen dürften aus den entsprechenden isomeren Chlorhydrinen mit *cis*-Doppelbindung gebildet werden:

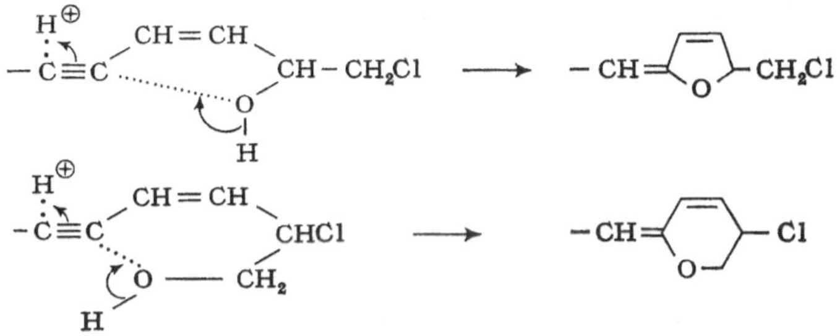

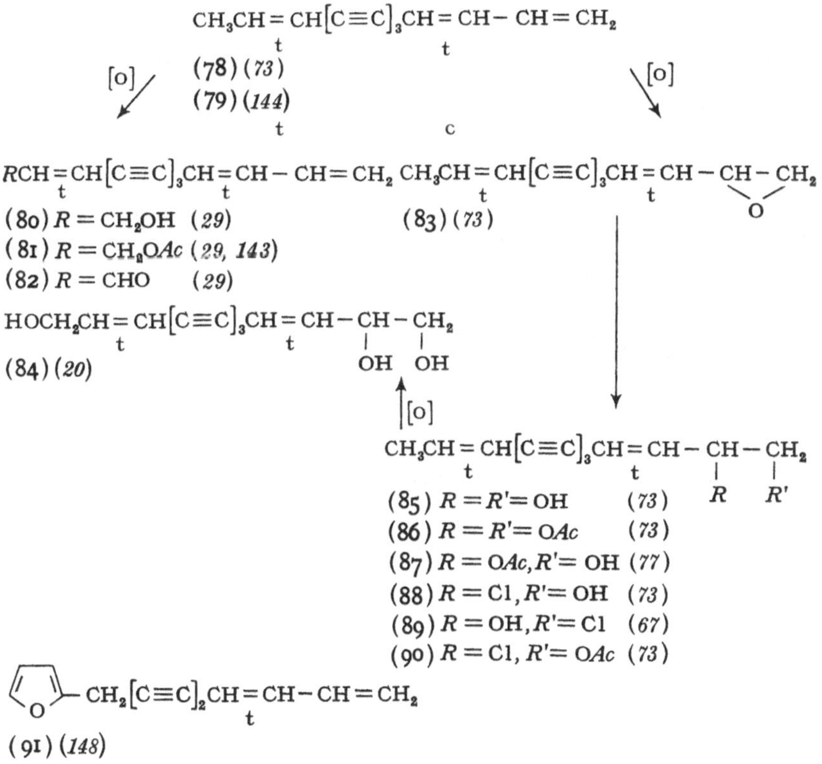

Schema 10. Umwandlungsprodukte des Entriindiens.

d) Entriindien.

Der Kohlenwasserstoff (78) kommt relativ häufig im Tribus Heliantheae und Cynareae vor. Auch hier sind wiederum die üblichen Umwandlungsprodukte zu finden (*Schema 10*, S. 13).

Die Bildung von (83) und (88) aus (78) konnte mit $1\text{-}^{14}C\text{-}(78)$ bewiesen
werden (*38*). (91) dürfte aus dem *cis*-Isomeren von (80) gebildet werden.

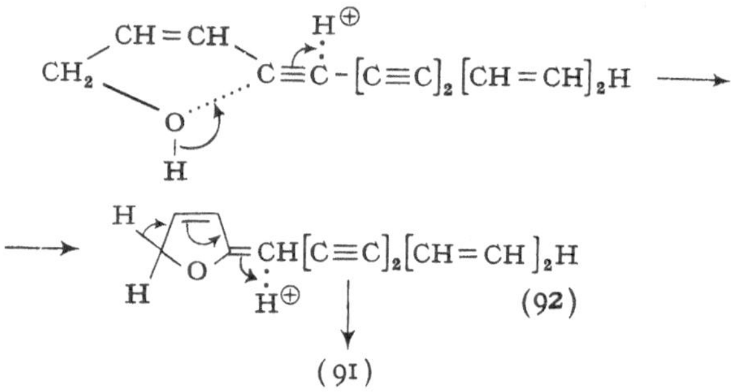

(92) kommt ebenfalls in der Natur vor (*84*). Es geht durch Protonen-
katalyse sehr leicht in (91) über.

e) Triintrien.

Von dem im Tribus Anthemideae relativ weitverbreiteten Kohlenwasserstoff (93) leiten sich nur wenige Verbindungen ab, von denen jedoch
(98) sehr häufig vorkommt:

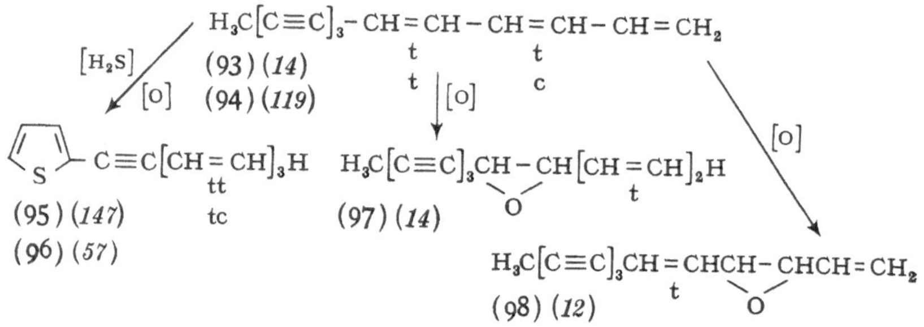

Die Bildung von (95) aus (93) muß wiederum mit einer Eliminierung
der endständigen Methylgruppe verknüpft sein. Mit $1\text{-}^{14}C\text{-}(93)$ kann
gezeigt werden, daß (98) direkt aus dem Kohlenwasserstoff (93) gebildet
wird (*40*).

Literaturverzeichnis: SS. 54—62.

f) Endiintrien.

Von dem im Tribus Heliantheae und Cynareae relativ häufigen Kohlenwasserstoff (99) kommen ebenfalls einige abgeleitete Verbindungen vor *(Schema 11)*.

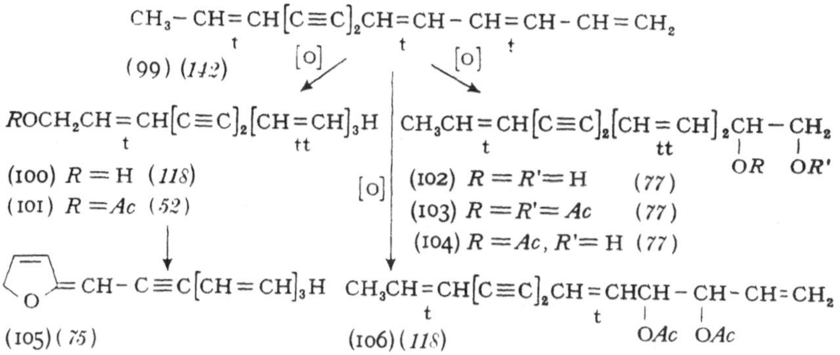

$$\text{CH}_3-\text{CH}=\text{CH}[\text{C}\equiv\text{C}]_2\text{CH}=\text{CH}-\text{CH}=\text{CH}-\text{CH}=\text{CH}_2$$

(99) (*112*)   [O]   [O]

$R\text{OCH}_2\text{CH}=\text{CH}[\text{C}\equiv\text{C}]_2[\text{CH}=\text{CH}]_3\text{H}$ | $\text{CH}_3\text{CH}=\text{CH}[\text{C}\equiv\text{C}]_2[\text{CH}=\text{CH}]_2\text{CH}-\text{CH}_2$

(100) R = H (*118*) | (102) R = R' = H   (77)   OR   OR'

(101) R = Ac (*52*)   [O]   (103) R = R' = Ac   (77)

(104) R = Ac, R' = H   (77)

(105) (*75*) : $\text{=CH}-\text{C}\equiv\text{C}[\text{CH}=\text{CH}]_3\text{H}$ $\text{CH}_3\text{CH}=\text{CH}[\text{C}\equiv\text{C}]_2\text{CH}=\text{CHCH}-\text{CH}-\text{CH}=\text{CH}_2$

(106) (*118*)   OAc   OAc

Schema 11.

Für die Bildung von (105) ist offensichtlich das *cis*-Isomere von (100) als Vorstufe anzunehmen (vgl. S. 14).

g) Endiindien.

Der bisher nur in Umbelliferen aufgefundene Kohlenwasserstoff (107) ist offensichtlich die Vorstufe für die Derivate (108) bis (113), die ebenfalls nur aus Umbelliferen isoliert wurden *(Schema 12)*.

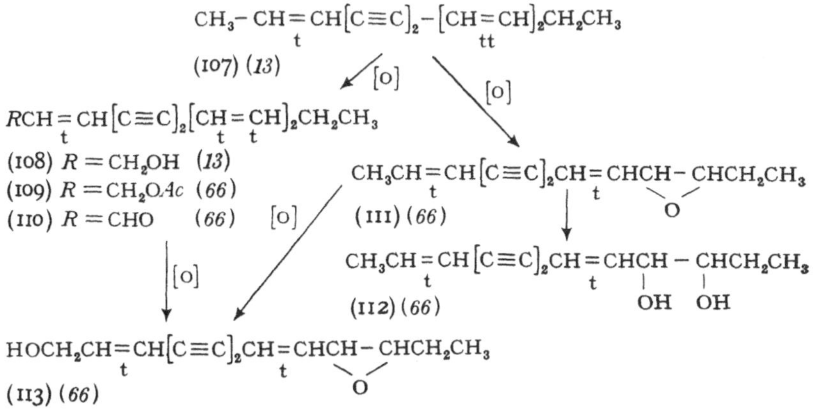

$$\text{CH}_3-\text{CH}=\text{CH}[\text{C}\equiv\text{C}]_2-[\text{CH}=\text{CH}]_2\text{CH}_2\text{CH}_3$$

(107) (*13*)   [O]   [O]

$R\text{CH}=\text{CH}[\text{C}\equiv\text{C}]_2[\text{CH}=\text{CH}]_2\text{CH}_2\text{CH}_3$

(108) R = CH$_2$OH (*13*)

(109) R = CH$_2$OAc (*66*)

(110) R = CHO (*66*)   [O] $\text{CH}_3\text{CH}=\text{CH}[\text{C}\equiv\text{C}]_2\text{CH}=\text{CHCH}-\text{CHCH}_2\text{CH}_3$

(111) (*66*)   O

$\text{CH}_3\text{CH}=\text{CH}[\text{C}\equiv\text{C}]_2\text{CH}=\text{CHCH}-\text{CHCH}_2\text{CH}_3$

(112) (*66*)   OH   OH

[O]

$\text{HOCH}_2\text{CH}=\text{CH}[\text{C}\equiv\text{C}]_2\text{CH}=\text{CHCH}-\text{CHCH}_2\text{CH}_3$

(113) (*66*)   O

Schema 12.

Wahrscheinlich entsteht (107) über (115) aus (114), einem Kohlen-wasserstoff, der neben (107—113) und (115—117) in *Aethusa cynapium* L. vorkommt:

$$CH_3CH=CH[C\equiv C]_2CH_2CH=CHCH_2CH_2CH_3$$
$$\qquad\quad {}_t \qquad\qquad\qquad\qquad {}_c$$

(**114**) *(84)* $\downarrow$ [O]

$$CH_3CH=CH[C\equiv C]_2CH=CHCHCH_2CH_2CH_3 \xrightarrow{-H_2O}$$ (**107**)
$$\qquad\quad {}_t \qquad\qquad\qquad {}_t \;\;|$$
$$\qquad\qquad\qquad\qquad\qquad\qquad OR$$

(**115**) $R = H$ *(13)*; (**116**) $R = Ac$ *(66)*

$$HOCH_2CH_2CH_2[C\equiv C]_2[CH=CH]_2CH_2CH_3$$
$$\qquad\qquad\qquad\qquad\qquad\quad {}_t \;\;\; {}_t$$
(**117**) *(66)*

(**117**) entsteht möglicherweise durch Hydrierung von (108).

2. Verbindungen mit mehr als 14 C-Atomen.

Unter den natürlich vorkommenden Acetylenverbindungen gibt es inzwischen eine große Anzahl von Verbindungen, bei denen es sich um ungesättigte C_{18}-Fettsäuren handelt. Daneben findet man jedoch bei Compositen und Umbelliferen zahlreiche C_{17}-Verbindungen sowie auch einige C_{16}- und C_{15}-Verbindungen, die wahrscheinlich in enger biogeneti-scher Beziehung zu den C_{18}-Säuren stehen. In *Tabelle 1* sind die C_{18}-Acetylensäuren sowie einige C_{17}-Säuren zusammengestellt.

Tabelle 1. Natürliche Fettsäuren mit Dreifachbindungen.

$CH_3(CH_2)_{10}C\equiv C(CH_2)_4COOH$ .... (**118**)   *(70)*

$CH_3(CH_2)_7C\equiv C(CH_2)_7COOH$ .... (**119**)   *(114)*

$CH_3(CH_2)_5CH=CHC\equiv C(CH_2)_7COOH$ .... (**120**)   *(70)*

$CH_3(CH_2)_5[C\equiv C]_2CH(OH)(CH_2)_6COOH$ .... (**121**)   *(109, 131)*

$CH_3(CH_2)_5CH=CHC\equiv C-CH(OH)(CH_2)_6COOH$ .... (**122**)   *(70)*

$CH_3(CH_2)_5[C\equiv C]_2(CH_2)_7COOH$ .... (**123**)   *(109, 131)*

$CH_3(CH_2)_3[CH=CH]_2C\equiv C(CH_2)_7COOH$ .... (**124**)   *(85)*

$CH_3(CH_2)_3CH=CH[C\equiv C]_2(CH_2)_7COOH$ .... (**125**)   *(85)*

$CH_3CH_2[CH=CH]_2[C\equiv C]_2(CH_2)_7COOH$ .... (**126**)   *(85)*

$CH_3CH_2CH=CH[C\equiv C]_3(CH_2)_7COOH$ .... (**127**)   *(85)*

$CH_3CH_2CH=CH[C\equiv C]_3CH(OH)(CH_2)_6COOH$ .... (**128**)   *(85)*

$CH_2=CHCH=CH[C\equiv C]_3(CH_2)_7COOH$ .... (**129**)   *(85)*

$CH_2=CH(CH_2)_4[C\equiv C]_2(CH_2)_7COOH$ .... (**130**)   *(70)*

$CH_2=CH(CH_2)_2CH=CH[C\equiv C]_2(CH_2)_7COOH$ .... (**131**)   *(85)*
$\qquad\qquad\qquad {}_c$

$CH_3(CH_2)_2CH=CH[C\equiv C]_2CH_2CH(OH)(CH_2)_6COOH$ .... (**132**)   *(70)*
Literaturverzeichnis: SS. 54—62.

$CH_2=CH(CH_2)_4[C\equiv C]_2CH(OH)(CH_2)_6COOH$ .... (133)   (85)

$CH_2=CH(CH_2)_2CH=CH[C\equiv C]_2CH(OH)(CH_2)_6COOH$ .... (134)   (85)

$CH_3(CH_2)_4C\equiv C-CH_2CH=CH(CH_2)_7COOH$ .... (135)   (130)
${}_c$

$CH_3(CH_2)_4C\equiv C-CH=CHCH(OH)(CH_2)_7COOH$ .... (136)   (132)
${}_t$

$CH_3(CH_2)_5CH=CHC\equiv C(CH_2)_7COOH$ .... (137)   (133)

$H_2C=CH(CH_2)_4CH=CHC\equiv C(CH_2)_7COOH$ .... (138)   (133)

$H_2C=CH(CH_2)_6C\equiv C(CH_2)_7COOH$ .... (139)   (133)

$H_2C=CH(CH_2)_4CH=CHC\equiv C-CH(OH)(CH_2)_6COOH$ .... (140)   (133)

$CH_3(CH_2)_5CH=CHC\equiv C-CH(OH)(CH_2)_6COOH$ .... (141)   (133)

$CH_3(CH_2)_5CH=CHC\equiv C(CH_2)_6COOH$ .... (142)   (133)

$H_2C=CH(CH_2)_4CH=CHC\equiv C(CH_2)_6COOH$ .... (143)   (133)

$CH_3(CH_2)_5CH=CHC\equiv C-CH(OH)(CH_2)_5COOH$ .... (144)   (133)

$H_2C=CH(CH_2)_4CH=CHC\equiv C-CH(OH)(CH_2)_5COOH$ .... (145)   (133)

$H_3C(CH_2)_{10}CH=C=CH(CH_2)_3COOH$ .... (146)   (133)

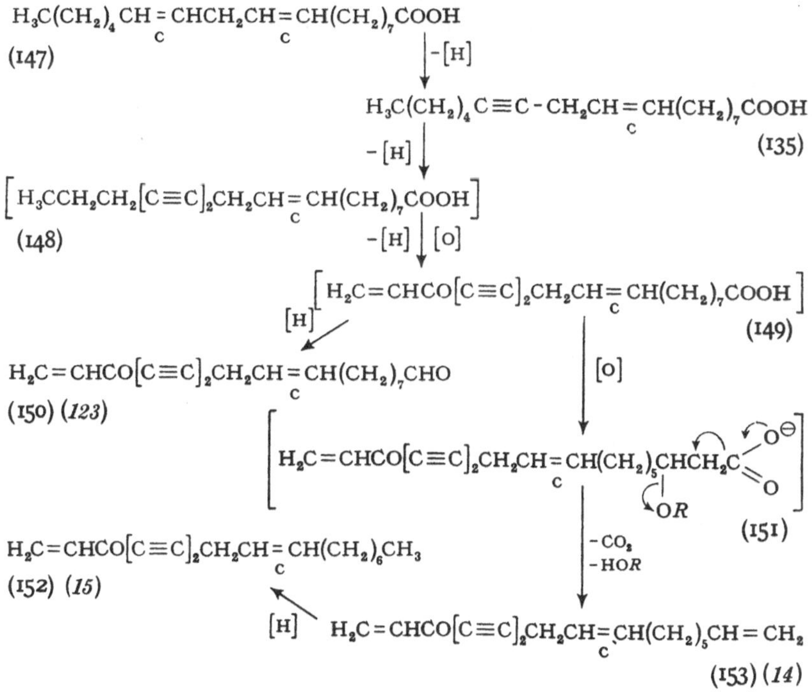

Schema 13. Biogeneseschema der C_{17}-Verbindungen.

Während die meisten Fettsäuren aus den Samenfetten von Santalaceen, Olacaceen und Simarubaceen isoliert wurden, kommen die Enin-säuren (135) und (136) in Compositen vor. Wie schon eingangs ausgeführt (siehe S. 5) ist die Crepissäure (135) biogenetisch sehr interessant, sie kommt auch in Mikroorganismen vor und wird dort aus Linolsäure gebildet (95). Linolsäure ist ebenfalls sehr verbreitet bei den Compositen. Sie wird als charakteristisch für diese Familie angesehen (138).

Wenn man annimmt, daß die analoge enzymatische Dehydrierung (147 → 135) auch in höheren Pflanzen stattfindet, läßt sich die Biogenese der natürlichen Acetylenverbindungen recht zwanglos deuten, wie im folgenden gezeigt werden soll.

Eine Gruppe von C_{17}-Polyinen, die sowohl bei Compositen als auch bei Umbelliferen vorkommt, hat die typische cis-Doppelbindung der Ölsäure bzw. Linolsäure. Neben diesen C_{17}-Verbindungen ist kürzlich auch der C_{18}-Aldehyd (150) isoliert worden (123), der den biogenetischen Übergang von C_{18}-Fettsäuren zu den C_{17}-Verbindungen weiterhin stützt. Schema 13 (S. 17) wäre durchaus denkbar.

Neben Dehydrierungen zur Dreifachbindung wären also Eliminierung der Carboxylgruppe unter Ausbildung einer Vinylgruppe und Allyloxydationen typische Biogeneseschritte.

Von (152) und (153) sind mehrere Umwandlungsprodukte bekannt, die entweder aus (152) bzw. (153) oder aus gemeinsamen Vorstufen entstehen:

$$(\mathbf{152})$$
$$\downarrow$$

$$H_2C=CHCO[C\equiv C]_2CHCH \overset{c}{=} CHC_7H_{15} \qquad (\mathbf{154}) \quad (15)$$
$$\underset{OH}{|}$$

$$H_2C=CHCO[C\equiv C]_2COCH \overset{c}{=} CHC_7H_{15} \qquad (\mathbf{155}) \quad (15)$$

$$H_2C=CHCH[C\equiv C]_2CH_2CH \overset{c}{=} CHC_7H_{15} \qquad (\mathbf{156}) \quad (72)$$
$$\underset{OH}{|}$$

$$H_2C=CHCH[C\equiv C]_2CHCH \overset{c}{=} CHC_7H_{15} \qquad (\mathbf{157}) \quad (72)$$
$$\underset{OH}{|} \qquad \underset{OH}{|}$$

$$HOCH_2CH_2CO[C\equiv C]_2CH_2CH \overset{c}{=} CHC_7H_{15} \qquad (\mathbf{158}) \quad (72)$$

$$HOCH_2CH_2CH_2[C\equiv C]_2CH_2CH \overset{c}{=} CHC_7H_{15} \qquad (\mathbf{159}) \quad (72)$$

$$HOCH_2CH_2CO[C\equiv C]_2CH \overset{c}{=} CHC_8H_{17} \qquad (\mathbf{160}) \quad (72)$$

$$H_2C=CHCOC\equiv C[CH \overset{t}{=} CH]_3C_6H_{13} \qquad (\mathbf{161}) \quad (72)$$

Literaturverzeichnis: SS. 54—62.

$$(\mathbf{153})$$
$$\downarrow$$

$$H_2C\!=\!CHCO[C\!\equiv\!C]_2\underset{\underset{OH}{|}}{CH}CH\!=\!CH(CH_2)_5CH\!=\!CH_2 \qquad (\mathbf{162}) \quad (72)$$

$$H_2C\!=\!CHCO[C\!\equiv\!C]_2COCH\!=\!CH(CH_2)_5CH\!=\!CH_2 \qquad (\mathbf{163}) \quad (72)$$

$$H_2C\!=\!CHCH[C\!\equiv\!C]_2CH_2CH\!=\!CH(CH_2)_5CH\!=\!CH_2 \qquad (\mathbf{164}) \quad (72)$$
$$\underset{OH}{|}$$

Bemerkenswert ist, daß man nur in Compositen die Verbindungen mit endständiger, isolierter Vinylgruppe findet, während diese in den Verbindungen aus Umbelliferen stets fehlt.

Eine weitere Gruppe von C_{17}-Verbindungen kommt in Umbelliferen vor. Es sind dieses die teilweise sehr toxischen Substanzen aus *Cicuta-* und *Oenanthe-*Arten (2). Auch diese Verbindungen dürften aus C_{18}-Fettsäuren gebildet werden. Die Lage der Diingruppierung entspricht völlig der des Falcarinons (152).

$$HO(CH_2)_3[C\!\equiv\!C]_2[CH\!=\!CH]_3\underset{\underset{OH}{|}}{CH}C_3H_7 \qquad (\mathbf{165})$$

$$HO(CH_2)_3[C\!\equiv\!C]_2[CH\!=\!CH]_3CH_2C_3H_7 \qquad (\mathbf{166})$$

$$CH_3CH\!=\!CH[C\!\equiv\!C]_2[CH\!=\!CH]_2CH_2CH_2COC_3H_7 \qquad (\mathbf{167})$$

$$HOCH_2CH\!=\!CH[C\!\equiv\!C]_2[CH\!=\!CH]_2CH_2CH_2\underset{\underset{OH}{|}}{CH}C_3H_7 \qquad (\mathbf{168})$$

$$HOCH_2CH\!=\!CH[C\!\equiv\!C]_2[CH\!=\!CH]_2CH_2CH_2CH_2C_3H_7 \qquad (\mathbf{169})$$

Das gemeinsame Vorkommen von (159) und (156) läßt vermuten, daß Verbindungen vom Typ (148) zunächst wiederum in β,γ-Stellung dehydriert werden und daß dann Allyloxydation erfolgt, wobei eventuell auch Allylumlagerung möglich ist und sich eine Hydrierung der Doppelbindung anschließen kann:

$$H_3C\!-\!CH_2CH_2[C\!\equiv\!C]_2\!- \xrightarrow{-[H]} H_2C\!=\!CHCH_2[C\!\equiv\!C]_2\!- \xrightarrow{[O]}$$
$$(\mathbf{170}) \qquad\qquad\qquad (\mathbf{171})$$

$$H_2C\!=\!CHCH[C\!\equiv\!C]_2\!- \rightarrow HOCH_2CH\!=\!CH[C\!\equiv\!C]_2\!- \xrightarrow{[H]} HO(CH_2)_3[C\!\equiv\!C]_2\!-$$
$$(\mathbf{172}) \quad | \qquad\qquad\qquad (\mathbf{173}) \qquad\qquad\qquad\qquad (\mathbf{174})$$
$$\phantom{(\mathbf{172})}\;OH$$

2*

Dieses Schema könnte auch bei den Verbindungen (165—169) von Bedeutung sein. Als Alternative kämen noch die Isomerisierung von (171) oder Allyloxydation von (170) und Wasserabspaltung in Betracht. Die Bildung von C_{17}-Verbindungen mit Vinylendgruppe scheint bei Compositen ganz allgemein bedeutungsvoll zu sein. Die Biogenese von Centaur X_4 (178) (73) sowie seiner Folgeprodukte könnte nach *Schema 14* ablaufen:

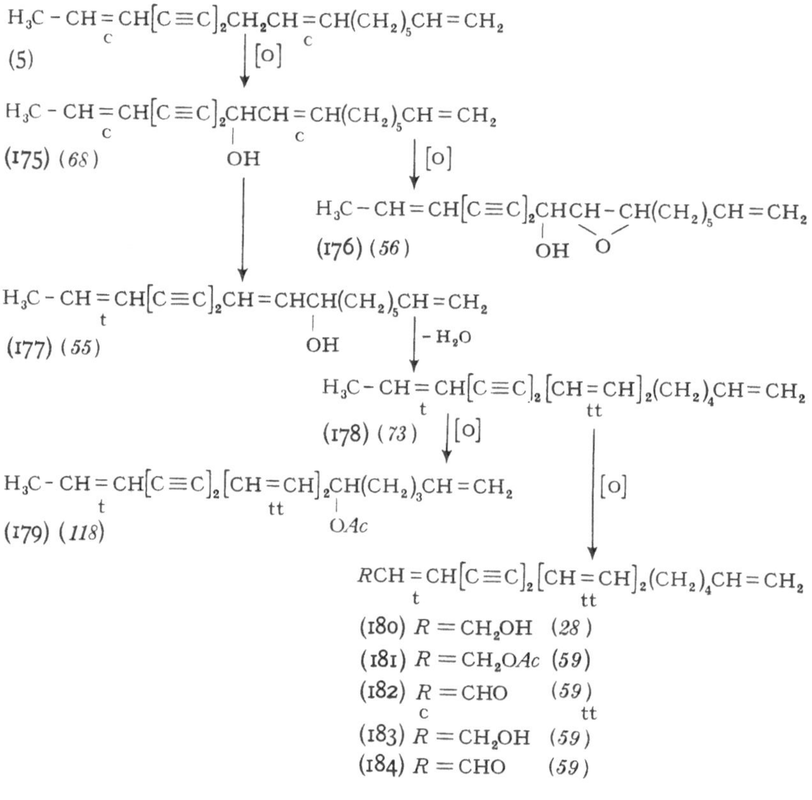

$$H_3C - CH = CH[C \equiv C]_2CH_2CH = CH(CH_2)_5CH = CH_2$$

(5) $\Big\downarrow [o]$

$$H_3C - CH = CH[C \equiv C]_2CHCH = CH(CH_2)_5CH = CH_2$$

(175) (68)   OH $\Big\downarrow [o]$

$$H_3C - CH = CH[C \equiv C]_2CHCH - CH(CH_2)_5CH = CH_2$$

(176) (56)   OH O

$$H_3C - CH = CH[C \equiv C]_2CH = CHCH(CH_2)_5CH = CH_2$$

(177) (55)   OH $\Big\downarrow -H_2O$

$$H_3C - CH = CH[C \equiv C]_2[CH = CH]_2(CH_2)_4CH = CH_2$$

(178) (73) $\Big\downarrow [o]$

$$H_3C - CH = CH[C \equiv C]_2[CH = CH]_2CH(CH_2)_3CH = CH_2$$

(179) (118)   OAc   [o]

$$RCH = CH[C \equiv C]_2[CH = CH]_2(CH_2)_4CH = CH_2$$

(180) $R = CH_2OH$ (28)
(181) $R = CH_2OAc$ (59)
(182) $R = CHO$ (59)
(183) $R = CH_2OH$ (59)
(184) $R = CHO$ (59)

Schema 14.

Wiederum wäre also die Allyloxydation der sehr reaktionsfähigen CH_2-Gruppe als wesentlicher Schritt anzunehmen. Diese Stufe ist durch die Isolierung von (175) und (176) belegt. Die Wasserabspaltung nach Allylumlagerung zu (177)* würde dann zu (178) führen, während es sich bei den Folgeprodukten von (178) wiederum um eine Allyloxydation der

---

* Die analoge Triinen-Verbindung läßt sich aus *Artemisia*-Arten isolieren (84).
Literaturverzeichnis: SS. 54—62.

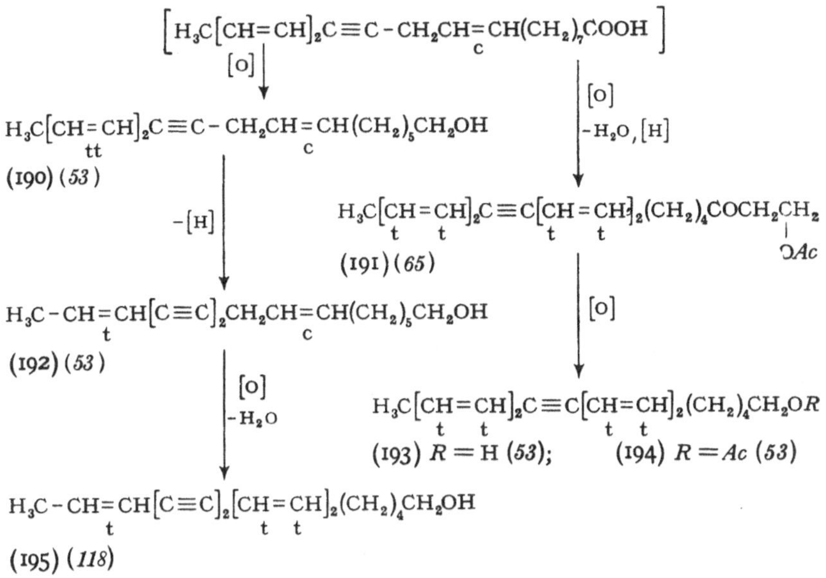

$$\left[H_3C[CH=CH]_2C\equiv C-CH_2\underset{c}{CH}=CH(CH_2)_7COOH \right]$$

$H_3C[CH=CH]_2C\equiv C-CH_2\underset{c}{CH}=CH(CH_2)_5CH_2OH$
tt
(190) (53)

$[O] \downarrow$ $\underset{-H_2O,\,[H]}{\overset{[O]}{\downarrow}}$

$H_3C[\underset{t}{CH}=\underset{t}{CH}]_2C\equiv C[\underset{t}{CH}=\underset{t}{CH}]_2(CH_2)_4COCH_2CH_2$
(191) (65) $\overset{|}{OAc}$

$-[H] \downarrow$

$H_3C-\underset{t}{CH}=CH[C\equiv C]_2CH_2\underset{c}{CH}=CH(CH_2)_5CH_2OH$
(192) (53)

$[O]$

$H_3C[\underset{t}{CH}=\underset{t}{CH}]_2C\equiv C[\underset{t}{CH}=\underset{t}{CH}]_2(CH_2)_4CH_2OR$
(193) $R = H$ (53);   (194) $R = Ac$ (53)

$\underset{-H_2O}{\overset{[O]}{\downarrow}}$

$H_3C-\underset{t}{CH}=CH[C\equiv C]_2[\underset{t}{CH}=\underset{t}{CH}]_2(CH_2)_4CH_2OH$
(195) (118)

Schema 15. Beziehungen zwischen C_{18}- und C_{16}-Verbindungen.

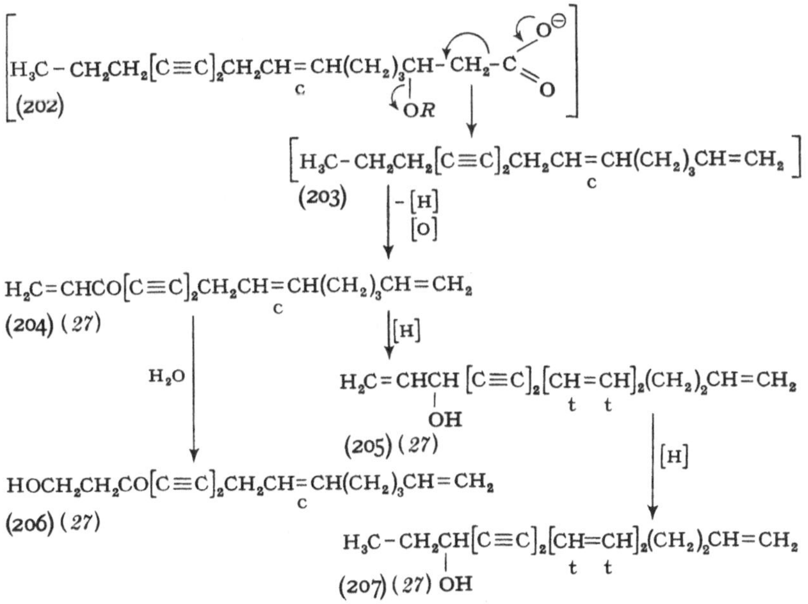

$$\left[H_3C-CH_2CH_2[C\equiv C]_2CH_2\underset{c}{CH}=CH(CH_2)_3\overset{|}{\underset{OR}{CH}}-CH_2-C\underset{O}{\overset{O^\ominus}{\diagdown}} \right]$$
(202)

$\left[H_3C-CH_2CH_2[C\equiv C]_2CH_2\underset{c}{CH}=CH(CH_2)_3CH=CH_2 \right]$
(203) $\underset{[O]}{\overset{-[H]}{\downarrow}}$

$H_2C=CHCO[C\equiv C]_2CH_2\underset{c}{CH}=CH(CH_2)_3CH=CH_2$
(204) (27)

$[H]$

$H_2C=CHCH[C\equiv C]_2[\underset{t}{CH}=\underset{t}{CH}]_2(CH_2)_2CH=CH_2$
$\overset{|}{OH}$
(205) (27)

H_2O

$[H]$

$HOCH_2CH_2CO[C\equiv C]_2CH_2\underset{c}{CH}=CH(CH_2)_3CH=CH_2$
(206) (27)

$H_3C-CH_2CH[C\equiv C]_2[\underset{t}{CH}=\underset{t}{CH}]_2(CH_2)_2CH=CH_2$
(207) (27) $\overset{|}{OH}$

Schema 16. Biogenese einiger C_{15}-Verbindungen.

endständigen Methylgruppe handelt. Dieser Schritt ist außerordentlich häufig zu beobachten. Da jedoch auch (185) natürlich vorkommt, ist auch die primäre Methylgruppenoxydation möglich:

$$(5, S. 20) \rightarrow HOCH_2CH=CH[C\equiv C]_2CH_2CH=CH(CH_2)_5CH=CH_2 \quad (185) \quad (56)$$
$$c c$$

Das Centaur X_3 (186) (70) dürfte analog wie Centaur X_4 (178) entstehen und entsprechend auch die sauerstoffhaltigen Derivate.

$$H_3C[C\equiv C]_3CH=CHCH=CH(CH_2)_4CH=CH_2$$
$$t t |$$
(186) (70)
$$t c [O]$$
(187) (70)
$$\downarrow$$
$$H_3C[C\equiv C]_3[CH=CH]_2CH(CH_2)_3CH=CH_2$$
$$tt |$$
$$OR$$

(188) $R = H$ (118, 84); (189) $R = Ac$ (118).

Neben der Fragmentierung von C_{18}-Fettsäuren zu C_{17}-Vinylverbindungen ist offenbar auch der Abbau durch β-Oxydation ein häufiger Biogeneseweg. Aus *Cosmos*-Arten isoliert man in sehr geringen Konzentrationen das Ketoacetat (191), das diesen Weg erkennen läßt. Nach Einführung der ungesättigten Bindungen erfolgt offenbar β-Oxydation, wobei in diesem Falle durch Reduktion der Carboxylgruppe der weitere Abbau unterbrochen wird. Aus Arten der Gattung *Dahlia*, die mit der Gattung *Cosmos* eng verwandt ist, isoliert man entsprechende C_{16}-Verbindungen, die zweifellos durch β-Oxydation aus C_{18}-Verbindungen entstanden sind (*Schema 15*, S. 21).

Die Reihenfolge der einzelnen Schritte ist naturgemäß nicht sicher anzugeben. So kann z. B. (195) aus (192) durch Allyloxydation und Wasserabspaltung entstehen oder durch Dehydrierung von (193). Die hier skizzierte β-Oxydation dürfte auch für die Bildung der zahlreichen C_{14}-Verbindungen (vgl. S. 24) wesentlich sein.

Aus *Centaurea*-Arten lassen sich verschiedene Aldehyde isolieren, die vermuten lassen, daß auch die α-Oxydation und der Abbau um ein C-Atom einen möglichen Biogeneseweg darstellen:

$$-CH_2-COOH \xrightarrow{[O]} -CH-COOH \rightarrow -CHO \xrightarrow{[H]} CH_2OH$$
$$\phantom{-CH_2-COOH \xrightarrow{[O]} -C}|$$
$$\phantom{-CH_2-COOH \xrightarrow{[O]} }O-H$$

Folgende Aldehyde mit 15 bis 17 Kohlenstoffatomen kommen in *Centaurea*-Arten vor (75):

$$H_3C-CH_2[CH=CH]_4(CH_2)_nCHO$$

$$n = 4, 5 \text{ oder } 6$$

$$H_3C-CH=CH[C\equiv C]_2[CH=CH]_2(CH_2)_nCHO$$
$$t tt$$

Literaturverzeichnis: SS. 54—62.

(196) $n = 3$; (197) $n = 4$; (198) $n = 5$

$$H_3C[C{\equiv}C]_3[CH{=}CH]_2(CH_2)_nCHO$$
$$\text{tt}$$

(199) $n = 3$; (200) $n = 4$.

Die Isolierung von (201) aus *Chrysanthemum*-Arten (*32*) läßt vermuten, daß dieser Kohlenwasserstoff, der eine CH_2-Gruppe weniger enthält als (186), aus einer C_{17}-Säure entstanden ist, die wahrscheinlich wiederum durch α-Oxydation aus einer C_{18}-Säure gebildet wird:

$$H_3C[C{\equiv}C]_3[CH{=}CH]_2(CH_2)_3CH{=}CH_2 \qquad \text{(201)} \quad (32)$$
$$\text{tt}$$

Die Bildung einiger C_{15}-Verbindungen, die aus *Cotula coronopifolia* isoliert wurden, läßt sich folgendermaßen deuten, wenn man als Vorstufe eine eventuell durch β-Oxydation gebildete C_{16}-Säure annimmt (*Schema 16*, S. 21).

Das Schema entspricht weitgehend dem der Bildung von (152) und (153) (S. 17) und ihren Folgeprodukten. Daneben erfolgt jedoch auch wieder Abbau durch β-Oxydation (siehe S. 24). Einige C_{15}-Verbindungen aus *Centaurea macrocephala* (*73*) lassen vermuten, daß hier wiederum Abbau durch α-Oxydation zu einer C_{15}-Säure erfolgt, die dann durch β-Oxydation und Reduktion der Carboxylgruppe die isolierten Substanzen liefern würde:

$$\left[H_3C{-}CH{=}CH[C{\equiv}C]_2[CH{=}CH]_2(CH_2)_3{-}COOH \right]$$

(208) $\downarrow$ [O]

$$\left[H_3C{-}CH{=}CH[C{\equiv}C]_2[CH{=}CH]_2CH_2CHCH_2COOH \right]$$

(209) $\Big\downarrow$ [H]   OH

$$H_3C{-}CH{=}CH[C{\equiv}C]_2[CH{=}CH]_2CH_2CHCH_2CH_2$$
$$\phantom{H_3C{-}CH{=}CH[C{\equiv}C]_2[}\text{t}\quad\text{t}|\quad\ \ |$$

(210) (*73*) OAc OAc

$$H_3C[C{\equiv}C]_3[CH{=}CH]_2CH_2CHCH_2CH_2$$
$$\phantom{H_3C[C{\equiv}C]_3[}\text{t}\quad\text{t}|\quad\ \ |$$

(211) (*73*) OAc OAc

3. C_{14}- und C_{13}-Verbindungen.

Die Gruppe der C_{14}-Verbindungen ist relativ groß. Das häufig zu beobachtende gemeinsame Vorkommen mit länger-kettigen Substanzen läßt vermuten, daß auch diese Gruppe durch Abbau über β-Oxydation von längeren Ketten gebildet wird. So findet man z. B. neben den C_{15}-Verbindungen (204—207) in *Cotula coronopifolia* die C_{14}-Polyine (214—218) (*27*). Die Bildung dieser Substanzen dürfte wahrscheinlich nach *Schema 17* verlaufen (S. 24).

(202, S. 21) $\overset{[O]}{\rightarrow}$ $\left[H_3C-CH_2CH_2[C\equiv C]_2CH_2CH=CH(CH_2)_3CH_2OH\right]$

(212) $\downarrow [\overset{c}{O}]$

$\left[H_3C-CH_2CH[C\equiv C]_2CHCH=CH(CH_2)_3CH_2OH\right]$

(213)   OH   OH $\downarrow -H_2O$

$H_3C-CH_2CH[C\equiv C]_2[CH=CH]_2CH_2CH_2CH_2OR'$

        |

        OR

(214) $R = R' = H$; (215) $R = R' = Ac$; (216) $R = H, R' = Ac$

         $\downarrow -H_2O$

$H_3C-CH=CH[C\equiv C]_2[CH=CH]_2CH_2CH_2CH_2OR$

(217) $R = H$; (218) $R = Ac$.

Schema 17. Bildung einiger C_{14}-Verbindungen.

Demnach würden eventuell sowohl (204—207) (S. 21) als auch (214 bis 218) aus einer gemeinsamen Vorstufe gebildet werden. (218) ist der Hauptinhaltsstoff, alle anderen Verbindungen kommen nur in sehr kleiner Menge vor, ebenso wie das Acetat (219) (27), das eine Dreifachbindung weniger enthält und wie (217) und (218) auch aus vielen anderen Compositen zu isolieren ist.

$$H_3C[CH=CH]_2C\equiv C[CH=CH]_2CH_2CH_2CH_2OAc \qquad (219)$$

Von (217) leiten sich einige weitere, sauerstoffreichere Derivate ab, die sowohl im Tribus Anthemideae als auch im Tribus Cynareae und Heliantheae vorkommen. Ihre Bildung erfolgt offensichtlich durch Allyl- bzw. β-Oxydation von (217):

(217) $\overset{[O]}{\longrightarrow}$ $CH_3CH=CH[C\equiv C]_2[CH=CH]_2CHCH_2CH_2$

                                OR   OR'

(220) $R = R' = H$ (67, 118), (221) $R = R' = Ac$ (67, 118),
(222) $R = Ac, R' = H$ (67), (223) $R = H, R' = Ac$ (67).

Eine weitere Gruppe von C_{14}-Polyinen leitet sich von (225) nach *Schema 18* ab.

(226) bzw. (227) sind sehr weit verbreitet, vor allem im Tribus Anthemideae und Heliantheae. Die sauerstoffreicheren Derivate (228—231) sind wiederum Allyloxydationsprodukte, die auch im Tribus Anthemideae sowie Cynareae zu finden sind. Die Epoxyde (232) und (233) sind biogenetisch interessante Vorstufen von Tetrahydropyranderivaten (siehe S. 37). Beim Übergang von (225) nach (226) sollte das Diol (234) als Zwischenprodukt durchlaufen werden. Dieses Diol ist

Literaturverzeichnis: SS. 54—62.

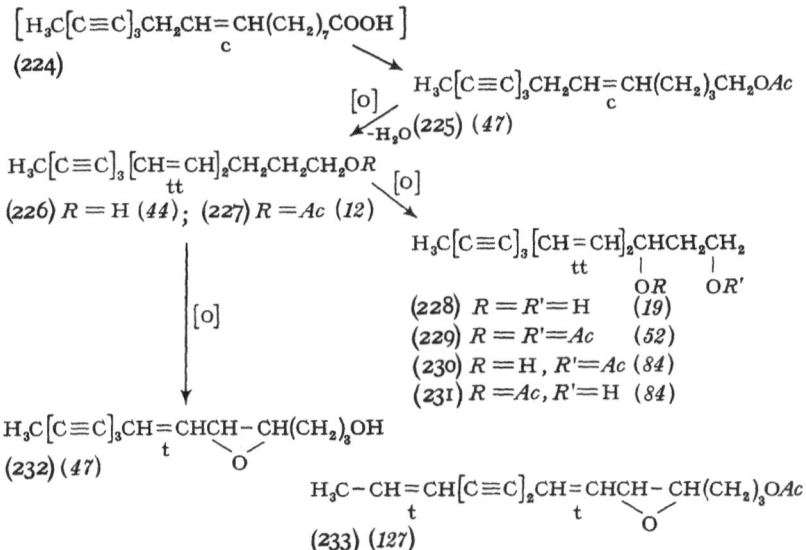

$$\left[H_3C[C\equiv C]_3CH_2CH=CH(CH_2)_7COOH \right]$$
$$(224) \quad\quad\quad c$$

$$[O] \qquad H_3C[C\equiv C]_3CH_2CH=CH(CH_2)_3CH_2OAc$$
$$\overset{}{\underset{-H_2O}{\searrow}}(225)\ (47) \qquad\qquad c$$

$$H_3C[C\equiv C]_3[CH=CH]_2CH_2CH_2CH_2OR$$
$$\underset{tt}{}$$
$$(226)\ R=H\ (44);\ (227)\ R=Ac\ (12) \quad \searrow [O]$$

$$H_3C[C\equiv C]_3[CH=CH]_2CHCH_2CH_2$$
$$\underset{tt}{} \qquad\qquad | \qquad |$$
$$OR \quad\ OR'$$
$$(228)\ R=R'=H\quad (19)$$
$$(229)\ R=R'=Ac\quad (52)$$
$$(230)\ R=H,\ R'=Ac\ (84)$$
$$(231)\ R=Ac,\ R'=H\ (84)$$

$$[O]$$

$$H_3C[C\equiv C]_3CH=CHCH-CH(CH_2)_3OH$$
$$t \qquad\qquad \searrow O \diagup$$
$$(232)\ (47)$$

$$H_3C-CH=CH[C\equiv C]_2CH=CHCH-CH(CH_2)_3OAc$$
$$t \qquad\qquad\qquad t \qquad \searrow O \diagup$$
$$(233)\ (127)$$

Schema 18. Biogenetische Beziehungen weiterer C_{14}-Verbindungen.

als Naturstoff isoliert worden (84) und ist eine wichtige Vorstufe für andere Verbindungen, wie durch Verfütterung von markiertem (234) gezeigt werden kann (siehe S. 36).

$$\left[H_3C[C\equiv C]_3CH=CHCHCH_2CH_2CH_2CH_2 \right] \qquad (234)$$
$$| \qquad\qquad\qquad |$$
$$OH \qquad\qquad\quad OH$$

Wichtiger als (234) ist das entsprechende C_{13}-Diol (236) (siehe S. 38), das als Zwischenstufe bei der Bildung des relativ verbreiteten Acetats (238) anzunehmen ist, aber noch nicht isoliert wurde. Bei diesen C_{13}-Verbindungen ist wahrscheinlich wiederum ein Abbau durch α-Oxydation beteiligt. Der Übergang von (236) in (238) konnte mit markiertem (236) sichergestellt werden (84).

$$(225) \xrightarrow{[O]} \left[H_3C[C\equiv C]_3CH_2CH=CHCH_2CH_2CH_2OH \right]$$
$$(235) \qquad\qquad\qquad\quad c \qquad \downarrow [O]$$

$$\left[H_3C[C\equiv C]_3CH=CHCHCH_2CH_2CH_2 \right]$$
$$| \qquad |$$
$$(236) \qquad\qquad\qquad OH \underset{-H_2O}{\downarrow} OH$$

$$H_3C[C\equiv C]_3[CH=CH]_2CH_2CH_2OR$$
$$\underset{tt}{}$$
$$(237)\ R=H\ (32);\ (238)\ R=Ac\ (36).$$

Die isomeren Acetate (239) und (240) sind bisher nur aus *Matricaria matricarioides* DC. isoliert worden, sie enthalten eine Dreifachbindung weniger als (238).

$$H_3C[C\equiv C]_2CH=CH-CH=CH-CH=CHCH_2CH_2OAc$$

                                  t   t   t

(239) (70)   t   c   t

(240) (*101*)

Der bisher einzige C_{14}-Polyinsäureester ist (241) (*82*). Diese Verbindung ist eventuell aus einem Nebenweg bei der Biogenese von (99) entstanden:

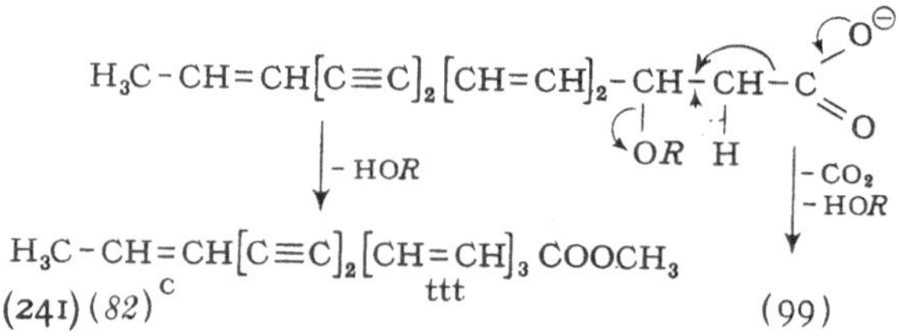

Zwei weitere Gruppen von C_{14}-Verbindungen leiten sich offenbar von (242) bzw. (247) ab, wobei jeweils β-Oxydation und Reduktion der Carboxylgruppe wesentliche Biogeneseschritte wären *(Schema 19)*.

(250) ist bisher nicht isoliert worden, jedoch mehrere Verbindungen, bei denen die endständige Methylgruppe fehlt. Derartige Fälle sind relativ häufig bei Compositen zu beobachten. Offenbar wird die Methylgruppe oxydativ zur Carboxylgruppe umgewandelt, so daß nach Decarboxylierung (251) entsteht, von dem sich wiederum mehrere Thiophenverbindungen ableiten *(Schema 20)*.

$$\left[CH_3CH=CH[C\equiv C]_2CH=CHCH_2CH_2CH_2CH_2COOH\right]$$

(242)   ↓ [O]

$$\left[CH_3-CH=CH[C\equiv C]_2CH=CHCH_2CH_2COCH_2COOH\right]$$

                                  ↓ [H]   (243)

$$CH_3-CH=CH[C\equiv C]_2CH=CHCH_2CH_2COCH_2CH_3$$

           t   t

(244) (*25*)   ↓ [H]

           ↓ [O] $CH_3-CH=CH[C\equiv C]_2CH=CHCH_2CH_2CHC_2H_5$

                                t   t |

           ↓   OH (245) (*61*)

$$CH_3-CH=CH[C\equiv C]_2CH=CHCHCH_2COC_2H_5$$

           t   t |

(246) (*84*)   OCOR

Literaturverzeichnis: SS. 54—62.

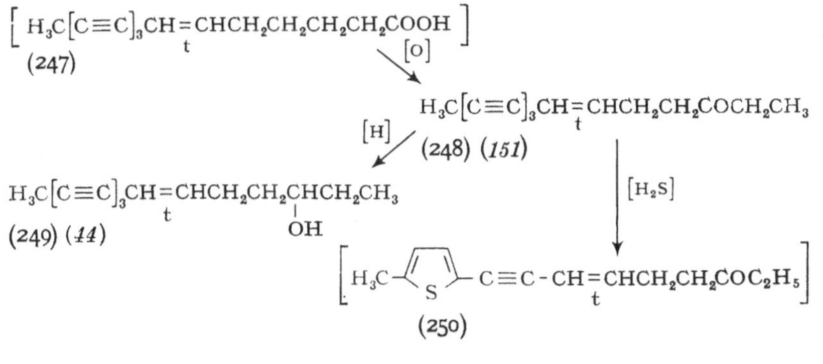

$$\left[\text{H}_3\text{C}[\text{C}\equiv\text{C}]_3\underset{t}{\text{CH}}=\text{CHCH}_2\text{CH}_2\text{CH}_2\text{CH}_2\text{COOH} \right]$$
(247)

$\searrow$ [o]

$$\text{H}_3\text{C}[\text{C}\equiv\text{C}]_3\underset{t}{\text{CH}}=\text{CHCH}_2\text{CH}_2\text{COCH}_2\text{CH}_3$$

[H] $\swarrow$ (248) (151)

$$\text{H}_3\text{C}[\text{C}\equiv\text{C}]_3\underset{t}{\text{CH}}=\text{CHCH}_2\text{CH}_2\underset{\text{OH}}{\text{CHCH}_2\text{CH}_3}$$
(249) (44)

[H$_2$S] $\downarrow$

$$\left[\text{H}_3\text{C}-\!\!\bigcirc_{\text{S}}\!\!- \text{C}\equiv\text{C}-\underset{t}{\text{CH}}=\text{CHCH}_2\text{CH}_2\text{COC}_2\text{H}_5 \right]$$
(250)

Schema 19.

$$\bigcirc_{\text{S}}\!\!- \text{C}\equiv\text{C}-\underset{t}{\text{CH}}=\text{CHCH}_2\text{CH}_2\text{COC}_2\text{H}_5$$

(251) (147) $\swarrow$ [H] | [o]

$$\bigcirc_{\text{S}}\!\!- \text{C}\equiv\text{C}-\underset{t}{\text{CH}}=\text{CHCH}_2\text{CH}_2\underset{\text{OR}}{\text{CHC}_2\text{H}_5}$$
(252) R = H (61)
(253) R = Ac (84)

$$\bigcirc_{\text{S}}\!\!- \text{C}\equiv\text{C}-\underset{t}{\text{CH}}=\text{CHCHCH}_2\text{COC}_2\text{H}_5$$
(254) (61) $\underset{\text{OCOCH}_2\text{CH(CH}_3)_2}{}$

$-$[H] | [o]

$$\bigcirc_{\text{S}}\!\!- \text{C}\equiv\text{C}\underset{tt}{[\text{CH}=\text{CH}]_2}\text{COC}_2\text{H}_5$$
(255) (147) tc
(256) (61)

$$\bigcirc_{\text{S}}\!\!- \text{C}\equiv\text{C}-\underset{t}{\text{CH}}=\text{CHCH}_2\text{CHCOC}_2\text{H}_5$$
(257) (61) $\underset{\text{(CH}_3)_2\text{CHCH}_2\text{OCO}}{}$

[H] $\downarrow$ [H$_2$O]

$$\bigcirc_{\text{S}}\!\!- \text{C}\equiv\text{C}\underset{tt}{[\text{CH}=\text{CH}]_2}\underset{\text{OH}}{\text{CHC}_2\text{H}_5}$$
(258) (61)

$$\bigcirc_{\text{S}}\!\!- \text{C}\equiv\text{C}-\text{CH}=\text{CHCHO}$$
(259) (61)

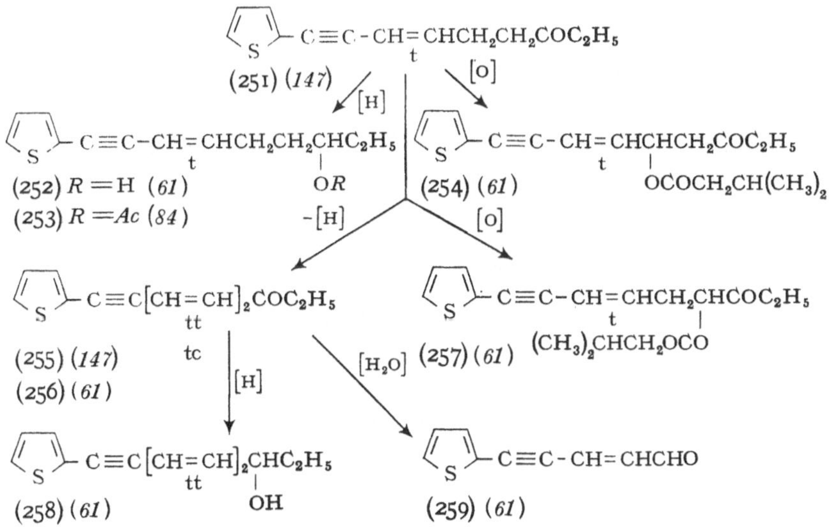

Schema 20. Biogenetische Beziehungen einiger Thiophenderivate.

Der Isovaleriansäureester (254) ist eventuell ein Nebenprodukt beim Übergang von (251) in (255), das wahrscheinlich durch Wasserabspaltung aus dem entsprechenden Alkohol gebildet wird.

Relativ häufig findet man bei den Compositen auch Isobutylamide. Aus *Anacyclus pyrethrum* DC. isoliert man die Amide (260) und (261) mit C_{14}-Kette (103, 104), während man in den Wurzeln von *Chrysanthemum frutescens* L. das C_{14}-Amid (262) und das C_{11}-Amid (263) neben dem Thiophenamid (266) findet (81). Das *trans,cis*-Isomere von (263) kommt zusammen mit dem C_{12}-Amid (265) in *Echinacea*-Arten vor (34).

$$CH_3CH_2CH_2[C\equiv C]_2CH_2CH_2\underset{tt}{[CH=CH]_2}CONHCH_2CH(CH_3)_2$$
(260)

$$\underset{t}{CH_3CH=CH}[C\equiv C]_2CH_2CH_2\underset{tt}{[CH=CH]_2}CONHCH_2CH(CH_3)_2$$
(261)

$$CH_3(CH_2)_8\underset{tt}{[CH=CH]_2}CONHCH_2CH(CH_3)_2$$
(262)

$$HC\equiv C-C\equiv C-CH_2CH_2\underset{tt}{[CH=CH]_2}CONHCH_2CH(CH_3)_2$$
(263)
(264)   tc

$$CH_3[C\equiv C]_2CH_2CH_2\underset{tc}{[CH=CH]_2}CONHCH_2CH(CH_3)_2$$
(265)

$$\underset{S}{\overset{\diagup\diagdown}{\Big\langle\ \ \Big\rangle}}-CH_2\underset{tt}{[CH=CH]_2}CONHCH_2CH(CH_3)_2$$
(266)

Über die biogenetischen Beziehungen dieser Amide zu anderen Acetylenverbindungen ist nichts bekannt.

Bei (266) konnte durch Verfütterung von I-^{14}C-Acetat gezeigt werden, daß die Carbonamidgruppe nicht aus dem Acetatcarboxyl stammt (42). Möglicherweise entsteht (266) aus dem durch Isomerisierung von (263) gebildeten (267):

$$(263)\ \longrightarrow\ \Big[H_3C[C\equiv C]_2CH_2[CH=CH]_2CONHR \Big]\qquad (267)$$

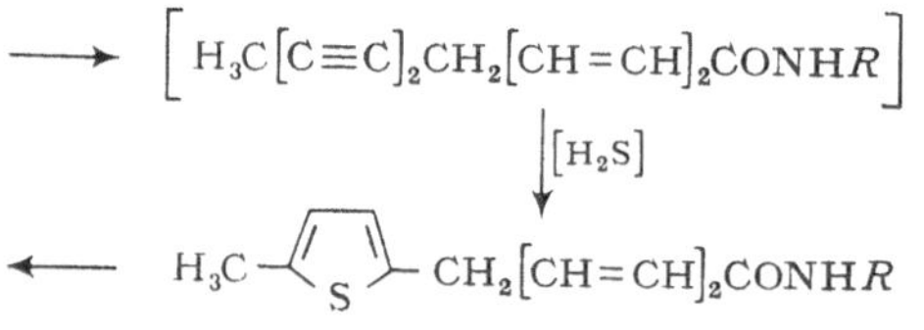

$$\Big\downarrow [H_2S]$$

$$(266)\ \longleftarrow\ H_3C-\underset{S}{\overset{\diagup\diagdown}{\Big\langle\ \ \Big\rangle}}-CH_2[CH=CH]_2CONHR\qquad (268)$$

Das primär zu erwartende Thiophenderivat (268) müßte dann durch Eliminierung der Methylgruppe in (266) übergehen, was im Einklang mit der Acetat-Fütterung stehen würde.

Ein C_{14}-Keton, das nicht in biogenetischer Beziehung zu anderen Acetylenverbindungen steht, kommt in *Centaurea*-Arten vor (75). Vielleicht ist die Ketosäure (269) eine Vorstufe, die durch Decarboxylierung und Eliminierung der endständigen Methylgruppe in (270) übergehen könnte:

$$\Big[CH_3[C\equiv C]_2CH_2CH=CH(CH_2)_5COCH_2COOH\Big]\qquad (269)$$
c

$$\downarrow [O]-CO_2$$

$$HC\equiv C-C\equiv C-CH_2CH=CH(CH_2)_5COCH_3\qquad (270)$$
c

Literaturverzeichnis: SS. 54—62.

4. C_{10}-Verbindungen.

Die Gruppe der C_{10}-Verbindungen ist nach den Substanzen mit 13 C-Atomen bei den Compositen die größte Gruppe. Besonders häufig kommen im Tribus Anthemideae die beiden isomeren Dehydromatricariaester (271, 272) vor. Hiervon leiten sich zahlreiche andere Substanzen ab, vor allem Schwefelverbindungen wie die Thiophene (273—276) und (280) *(Schema 21)*.

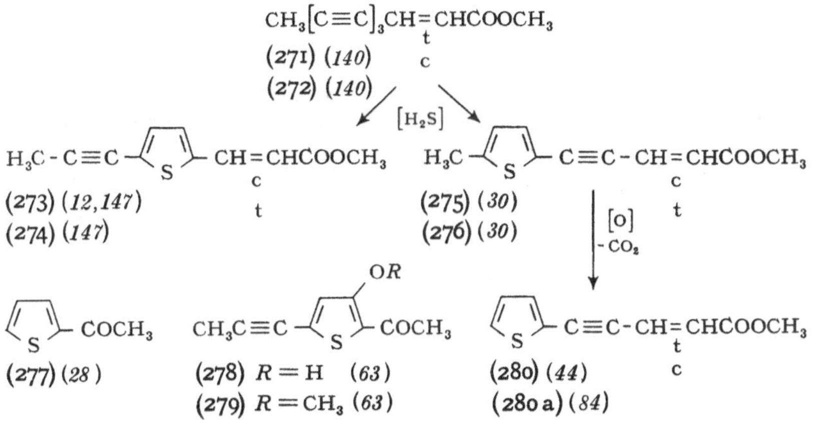

Schema 21. Thiophenderivate, die sich von Dehydromatricariaester ableiten.

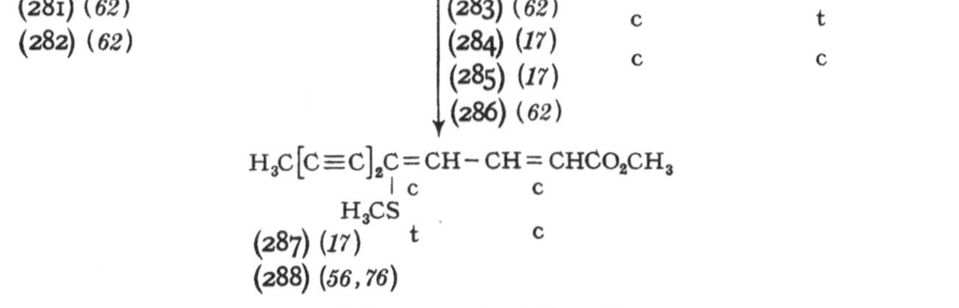

Schema 22. C_{10}-Thioenoläther.

Die Bildung von (276) aus (271) konnte durch Verfütterung von $1\text{-}^{14}C\text{-}(271)$ sichergestellt werden (45). (280) muß wiederum durch Eliminierung der Methylgruppe von (276) entstehen, während die biogenetischen

Zusammenhänge mit den Thiophenen (277—279) unklar sind. (277) sollte eventuell in Analogie zu den übrigen Thiophenverbindungen aus einem noch nicht bekannten C_7-Diinketon entstehen, während bei (278) eine β-Dicarbonylverbindung als Vorstufe in Betracht zu kommen scheint. Eine große Gruppe von Abkömmlingen des Dehydromatricariaesters stellen die Thioenoläther dar. Durch formale Methylmercaptan-Anlagerung könnten theoretisch 12 Isomere gebildet werden, von denen 8 natürlich vorkommen (*Schema 22*, S. 29).

Mit 1-^{14}C-(271) konnte gezeigt werden, daß (282) aus (271) gebildet wird (46).

Während die Bildung z. B. von (287) der formalen Methylmercaptan-Addition der Polarisierung der Dreifachbindungen entspricht, ist das bei den isomeren Thioäthern (289—291) (56) nicht der Fall. Eventuell werden diese Substanzen aus längeren Ketten durch nachträglichen Abbau zu C_{10}-Verbindungen gebildet:

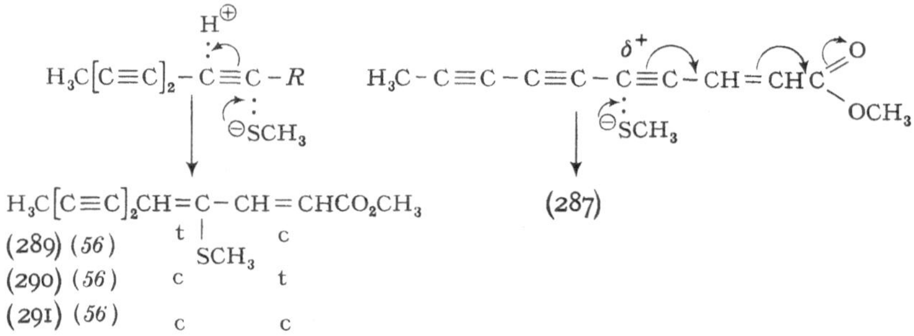

$$H_3C[C\equiv C]_2CH=C-CH=CHCO_2CH_3$$

(289) (56)   (290) (56)   (291) (56)

(287)

Versuche mit markierten Verbindungen müssen diese Frage klären. Neben (281—291) kommt auch das Dihydroderivat (292) und der Thioäther (294) in *Anthemis tinctoria* L. vor (23). Interessanterweise wird (294) aus (271) gebildet, wie durch Doppelmarkierung eindeutig gezeigt werden konnte (23, 69).

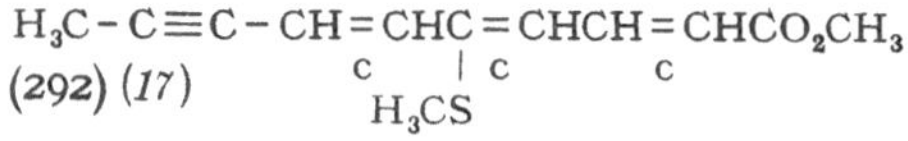

$$H_3C-C\equiv C-CH=CHC=CHCH=CHCO_2CH_3$$

(292) (17)

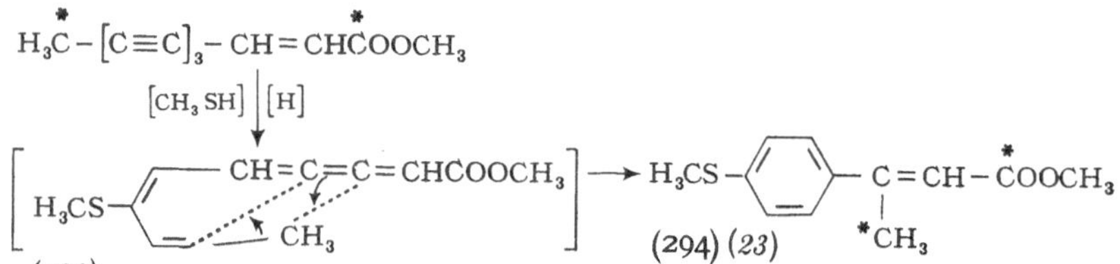

(293)

(294) (23)

Literaturverzeichnis: SS. 54—62.

Beim Übergang von (271) in (294) muß also neben der Bildung eines Phenylringes eine Methylgruppe wandern. Eventuell ist (293) ein mögliches Zwischenprodukt, das durch Isomerisierung und partielle Hydrierung aus (285) gebildet werden könnte.

Eine weitere biogenetische Umwandlung von (272) ist der Übergang in Lactone, die zweifellos wie folgt verläuft:

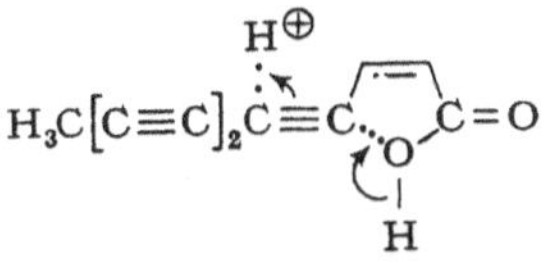

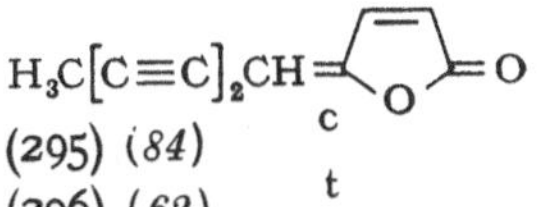

(295) (*84*)
(296) (*62*)

Außer den Butenoliden (295) und (296) kommen in *Anthemis*-Arten auch die Thioäther (297—303) vor.

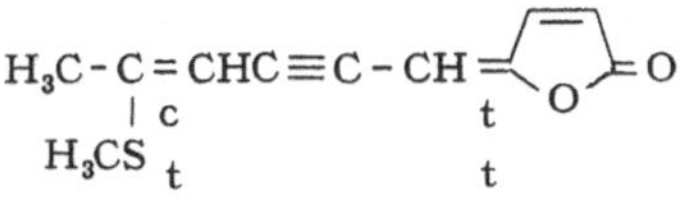

(297) (*62*)
(298) (*84*)

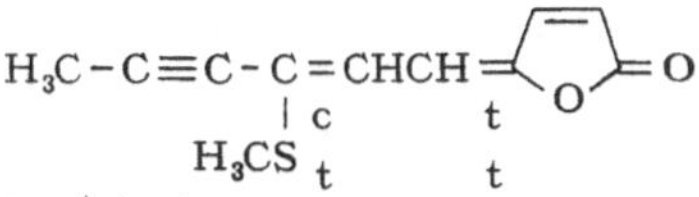

(299) (*62*)
(300) (*62*)

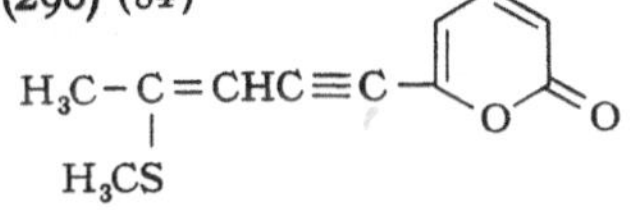

(301) (*60*)

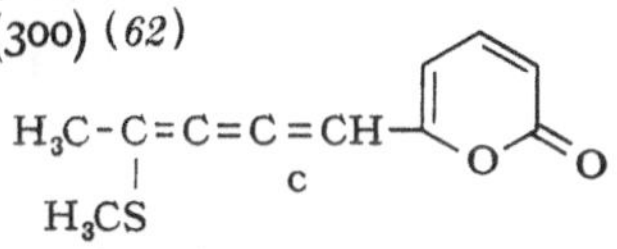

(302) (*60*)
(303) (*60*)

Während (297—300) durch nachträgliche formale Methylmercaptan-Addition an (295) bzw. (296) entstehen dürften, werden (301) bzw. (302 und 303) wahrscheinlich nach *Schema 23* gebildet.

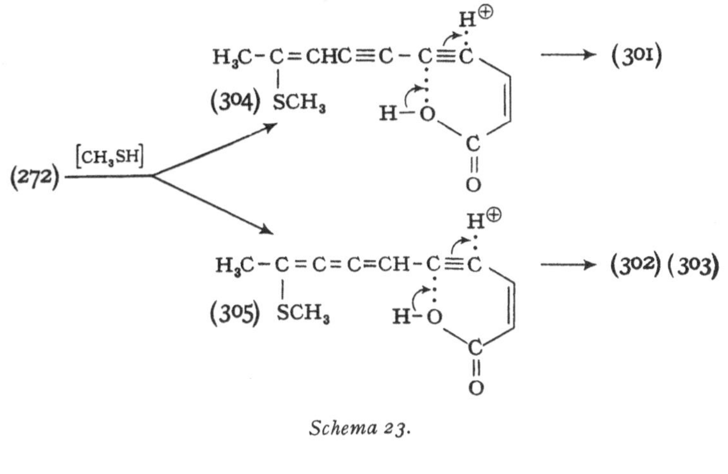

Schema 23.

Die formale Methylmercaptan-Anlagerung muß also zur Bildung des Zwischenproduktes (305) als 1.4-Addition ablaufen.

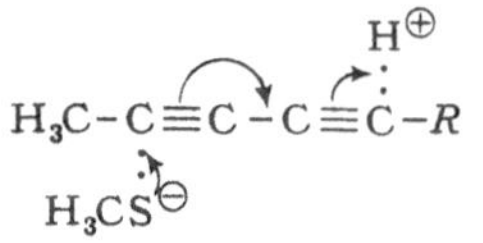

$$H_3C-C\equiv C-C\equiv C-R \xrightarrow{\quad} (305)$$
$$H_3CS^{\ominus}$$

Das Lacton (306) dürfte aus (275) entstehen.

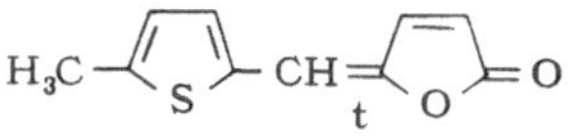

$$H_3C-\langle\!\langle{}_S\rangle\!\rangle-CH=\langle\!\langle{}_O\rangle\!\rangle=O \quad (306)\,(79)$$

Im Tribus Anthemideae findet man hin und wieder auch ein Isobutylamid (307), das sich vom Dehydromatricariaester ableitet. In diesem Falle konnte gezeigt werden, daß der Isobutylamin-Rest aus Valin gebildet wird (84). Offen ist jedoch die Frage, ob die Decarboxylierung erst nach Bildung der Acylaminosäure erfolgt oder schon vorher. Aus Mikroorganismen ist ein derartiges Aminosäurederivat (308) isoliert worden (96).

$$H_3C[C\equiv C]_3CH=CHCONHCH_2CH(CH_3)_2 \qquad (307)\ (41)$$

$$H[C\equiv C]_2CH=CHCONHCHCH(CH_3)_2 \qquad (308)$$
$$|$$
$$COOH$$

Der zweite C_{10}-Ester, der bei Compositen recht häufig vorkommt, ist der Matricariaester (309), der in drei verschiedenen Isomeren vorkommt. Auch von diesem Ester leiten sich eine Reihe von Folgeprodukten ab (*Schema 24*).

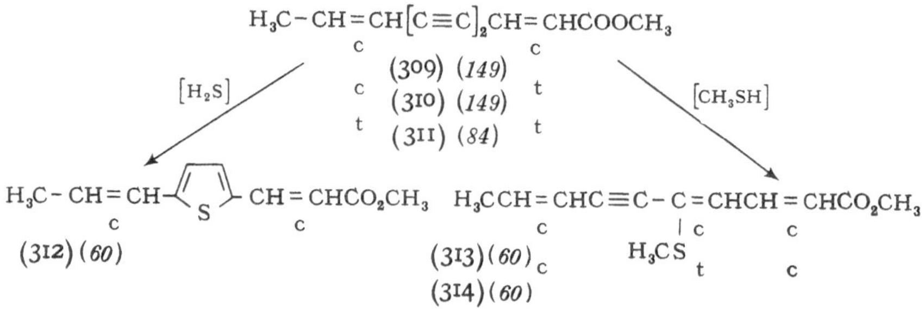

$$H_3C-CH=CH[C\equiv C]_2CH=CHCOOCH_3$$

(309) (149)
(310) (149)
(311) (84)

[H₂S]   [CH₃SH]

$$H_3C-CH=CH-\langle\!\langle{}_S\rangle\!\rangle-CH=CHCO_2CH_3 \qquad H_3CCH=CHC\equiv C-C=CHCH=CHCO_2CH_3$$

(312) (60)

(313) (60)
(314) (60)

$$H_3CS$$

Schema 24.

Die Mannigfaltigkeit der Schwefelverbindungen ist hier jedoch nicht so ausgeprägt wie bei (271). Auch die freie Säure (315) kommt vor, die die Vorstufe einer Reihe von Butenoliden darstellt:

Literaturverzeichnis: SS. 54—62.

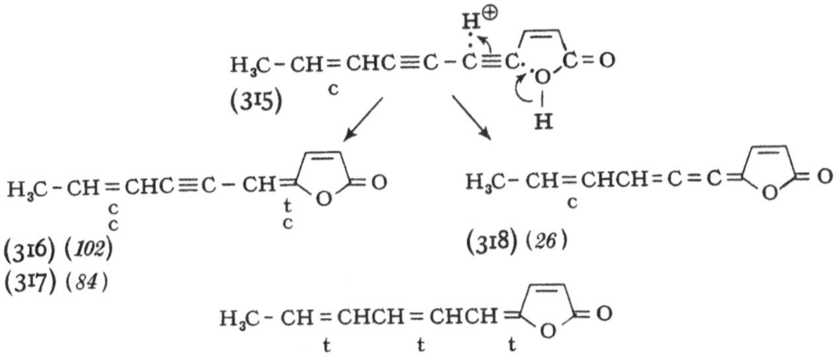

$$H_3C - CH = CHC \equiv C - \overset{\oplus}{\underset{(315)}{C}} \equiv C \diagdown_{O} \diagup C = O$$

$$H_3C - CH = CHC \equiv C - CH \diagup_{O} = O \qquad H_3C - CH = CHCH = C = C \diagup_{O} = O$$

(316) (*102*)
(317) (*84*)

(318) (*26*)

$$H_3C - CH = CHCH = CHCH \diagup_{O} = O$$

(319) (*84*)

Das Kumulen-Derivat (318) wird offenbar analog aus (315) gebildet, wobei jedoch der Ringschluß als 1.4-Addition abläuft:

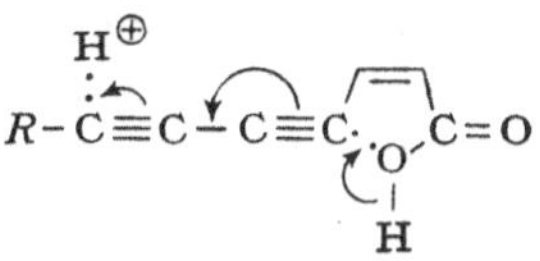

$$R - C \equiv C - C \equiv C \diagdown_{O} C = O \longrightarrow (318)$$

Dem Lacton (319) müßte ein noch nicht bekannter Dihydromatricariaester zugrunde liegen. Derartige Chromophore kommen als Alkohol, Acetat und Aldehyd in *Grindelia*-Arten vor:

$H_3C—CH=CHC\equiv C[CH=CH]_2CH_2OAc$

(320) (*148*)

$H_3C[CH=CH]_2C\equiv C—CH=CH—R$

(321) $R = CH_2OH$ (*64*)
(322) $R = CH_2OAc$ (*64*)
(323) $R = CHO$ (*64*)

Als weitere formale Hydrierungs- bzw. Reduktionsprodukte von (309) sind folgende Substanzen isoliert worden:

$H_3C—CH=CH[C\equiv C]_2CH=CH—R$

(324) $R = CH_2OAc$ (*154*)
(325) $R = CHO$ (*66*)

$H_3C—CH=CH[C\equiv C]_2CH_2CH_2CO_2CH_3$

(328) (*5*)

$H_3C—CH_2CH_2[C\equiv C]_2CH=CHCO_2CH_3$

(326) (*156*)
(327) (*113*)

$H_3C—CH=CH[C\equiv C]_2CH_2CH_2CH_2OR$

(329) $R = H$ (*71*)
(330) $R = Ac$ (*71*)

Noch nicht entschieden ist, ob z. B. (324) bzw. (325) eine Vorstufe oder ein Reduktionsprodukt von (309) ist. Das gleiche gilt naturgemäß auch für die anderen Verbindungen dieser Gruppe, wie überhaupt grundsätzlich die Frage zu klären ist, ob Dehydrierungen bzw. Hydrierungen relativ früh, eventuell noch auf dem Stadium der C_{18}-Säure oder erst nachträglich, wenn bereits durch oxydativen Abbau die Kette weitgehend verkürzt ist, erfolgen.

Ein Folgeprodukt von (326) ist das Butenolid (331) und der Angelica-
ester (332), der stets zusammen mit (309) vorkommt. Das deutet eventuell darauf hin, daß (309) aus (326) gebildet werden kann, wobei wiederum
durch Allyloxydation zunächst eine Sauerstoffunktion eingeführt werden
müßte, die durch Eliminierung zur Doppelbindung führen würde:

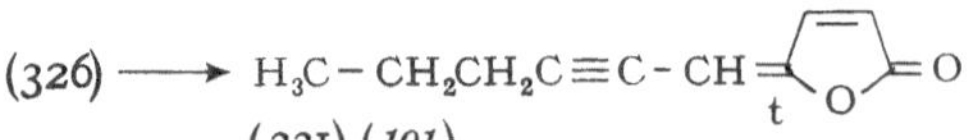

$$(326) \longrightarrow H_3C-CH_2CH_2C\equiv C-CH \underset{t}{=}\overset{O}{\diagdown}=O$$
$$(331)\ (101)$$

$$H_3C-CH_2CH[C\equiv C]_2\underset{|}{CH}=\underset{c}{CHCO_2CH_3}$$
$$(332)\ (58)\ OCOC=CHCH_3$$
$$\underset{|}{CH_3}$$

Die Tatsache, daß C_{10}-Säureester und ihre Derivate so häufig vorkommen, was insbesondere auch für die Polyine aus Mikroorganismen
gilt (siehe S. 46), läßt die Vermutung aufkommen, daß möglicherweise
ein direkterer Übergang der C_{18}-Säuren in C_{10}-Säuren erfolgt. Diese
Vermutung wird gestützt durch die Tatsache, daß nur sehr wenige
C_{12}-Derivate aufgefunden worden sind, die bei Annahme eines Abbaues
von C_{18}-Säuren durch normale β-Oxydation zwangsläufig Zwischenprodukte wären. Interessant ist in diesem Zusammenhang das Vorkommen des Triglycerids (333) im Samenfett von *Sapium sebiferum* (150),
dessen Struktur vermuten läßt, daß die Diestergruppierung durch eine
Art Baeyer-Villiger-Oxydation aus einem Umwandlungsprodukt der
Linolsäure entstanden ist (119):

$$CH_2-OCO(CH_2)_7CH=CHCH_2CH=CHCH_2CH=CHCH_2CH_3$$
$$CH-OCO(CH_2)_7CH=CHCH_2CH=CH(CH_2)_4CH_3$$
$$CH_2-OCO(CH_2)_3CH=C=CHCH_2OCOCH=CHCH=CH(CH_2)_4CH_3$$
$$(333)$$
$$HO_2C(CH_2)_7CH=CHCH_2CH=CH(CH_2)_4CH_3$$
$$(7)\qquad \downarrow[O]$$
$$[HO_2C(CH_2)_7CHCH=CHCH=CH(CH_2)_4CH_3]$$
$$(334)\quad OH\quad \downarrow-[H]$$
$$[HO_2C(CH_2)_3CH=C=CHCH_2COCH=CHCH=CH(CH_2)_4CH_3]$$
$$(335)\qquad \downarrow[O]$$
$$HO_2C(CH_2)_3CH=C=CHCH_2OCOCH=CHCH=CH(CH_2)_4CH_3$$
$$(336)$$

Literaturverzeichnis: SS. 54—62.

Der Übergang von Linolensäure in (334) und die Stereochemie von (334) entsprechen vollkommen dem Biogeneseschema, das kürzlich an verschiedenen ungesättigten Fettsäuren nachgewiesen werden konnte (*110*). Die Bildung der Allenbindung ist dagegen unklar, obwohl eine einfache Allen-Fettsäure (146, S. 17) bekannt ist (*6*). Für die Biogenese z. B. von Dehydromatricariaester wäre folgende analoge Reaktionsfolge denkbar:

$$\left[H_3C[C\equiv C]_3CH_2CH=CH(CH_2)_7CO_2H\right]$$

$$(337) \qquad \overset{c}{\underset{\downarrow}{}} [O]$$

$$\left[H_3C[C\equiv C]_3CH=CHC(CH_2)_7CO_2H\right]$$

$$(338) \qquad \overset{\|}{O}$$

$$\downarrow [O]$$

$$\left[H_3C[C\equiv C]_3CH=CHCOO(CH_2)_7CO_2H\right] \xrightarrow{H_2O} (271)$$

$$(339)$$

Bemerkenswert ist in diesem Zusammenhang, daß die Umwandlung der Triinsäure (340) in (271, S. 21) bei Verfütterung an *Artemisia vulgaris* L. wesentlich besser erfolgt als die der Triinensäure (342) (*84*), was mit der obigen Annahme vereinbar wäre, da (340) durch Oxydation in β-Stellung zum Triin und anschließende Wasserspaltung oder durch direkte Dehydrierung von (340) in 9.10-Stellung (337) liefern könnte, während bei (342) eine vielleicht weniger günstige Hydratisierung der Doppelbindung notwendig wäre. Wenn (271) durch Abbau nach dem Schema der üblichen β-Oxydation entstehen würde, sollte (342) die bessere Vorstufe sein. Das steht im Einklang mit dem Ergebnis der Fütterung der Diinsäure mit isolierter *cis*-Doppelbindung, die relativ glatt in (271) umgewandelt wird (*84*).

$$H_3\overset{*}{C}[C\equiv C]_3(CH_2)_{10}CO_2H$$

$$(340) \qquad \downarrow [O]$$

$$\left[H_3C[C\equiv C]_3CH_2CH(CH_2)_8CO_2H\right] \xrightarrow{-H_2O} (337) \rightarrow (271) \ (10\%)$$

$$(341) \qquad \overset{|}{OH}$$

$$H_3\overset{*}{C}[C\equiv C]_3CH=CH(CH_2)_8CO_2H \rightarrow (271) \ (1\%)$$

$$(342)$$

Weitere Versuche mit markierten Verbindungen müssen zeigen, wieweit diese Vermutungen zu Recht bestehen.

5. Sauerstoff-Heterocyclen.

In der Reihe der natürlichen Acetylenverbindungen nehmen Sauerstoffheterocyclen einen breiten Raum ein. Wie schon mehrfach erwähnt (siehe S. 8, 11—15, 20, 25), kommen sehr häufig Epoxyde und ihre Folgeprodukte vor. Zu diesen gehören auch die 5-Ring- und 6-Ring-Sauerstoffheterocyclen (70, 71, 74—77, S. 12) (105, S. 15), und die Lactone (295, 296, 297—303, 306, 316—319 und 331) sind weitere bereits erwähnte Heterocyclen. Für die Bildung anderer Heterocyclen sind die Diole (234 und 236) wichtige Vorstufen. Wie durch Verfütterung von 14-³H-(343), das durch Braunstein-Oxydation aus (234) erhalten wird, gezeigt werden kann, bilden sich die Spiroketalenoläther auf folgendem Wege (33):

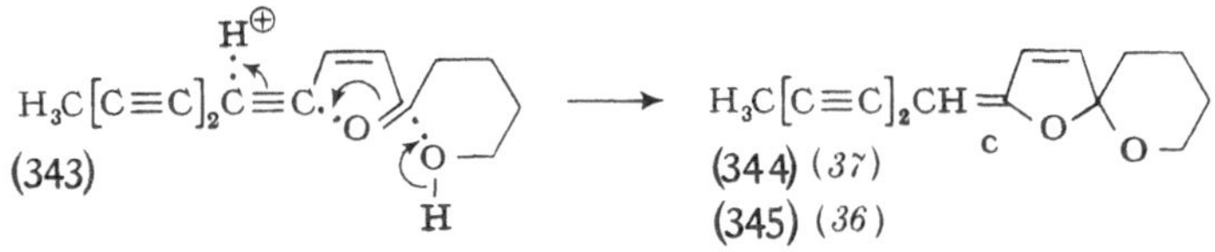

$$H_3C[C\equiv C]_2\overset{..}{C}\equiv C. \qquad \longrightarrow \qquad H_3C[C\equiv C]_2CH$$

(343)   (344) (37)
            (345) (36)

Diese Reaktion läßt sich auch in vitro mit Butylat und UV-Licht durchführen (33). Von (344) bzw. (345) leiten sich einige sauerstoffreichere Derivate ab:

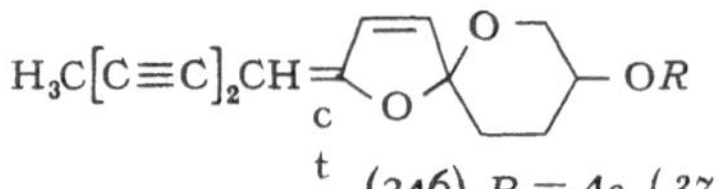

$$H_3C[C\equiv C]_2CH \qquad\qquad OR$$

(346) R = Ac (37); (347) R = COCH$_2$CH(CH$_3$)$_2$ (74)
(348) R = Ac (37); (349) R = COCH$_2$CH(CH$_3$)$_2$ (74)

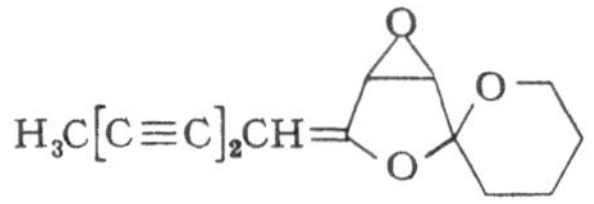

$$H_3C[C\equiv C]_2CH$$

(350) (18)

(346—349) werden offenbar durch x-Oxydation und anschließende Veresterung gebildet, während (350) ein weiteres Beispiel für das häufige Vorkommen von Epoxyden in dieser Reihe ist, was vermuten läßt, daß diese Pflanzen entsprechende Enzyme besitzen. Wiederum leiten sich auch Schwefelverbindungen von diesen Typen ab:

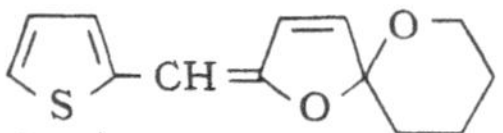

$$CH \qquad\qquad H_3CSCH=CHC\equiv C-CH$$

(351) (30) 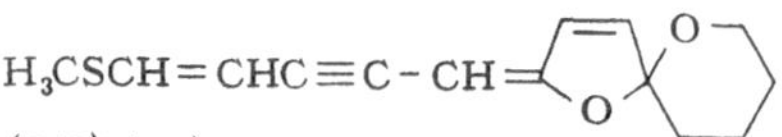   (352) (84)

In beiden Fällen fehlt also wiederum die endständige Methylgruppe.

Literaturverzeichnis: SS. 54—62.

Einige anders gebaute Ringverbindungen werden offensichtlich auf folgendem Wege aus den schon erwähnten Epoxyden (232) und (233) gebildet (siehe S. 25):

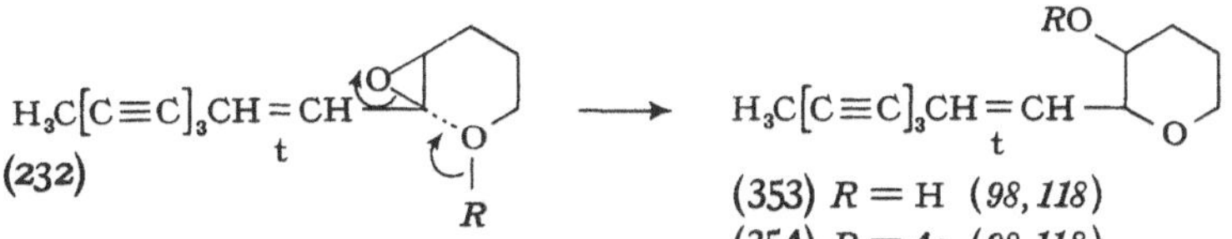

$$H_3C[C \equiv C]_3CH = CH \underset{t}{} \longrightarrow H_3C[C \equiv C]_3CH = CH \underset{t}{}$$

(232)

(353) $R = H$ (98, 118)
(354) $R = Ac$ (98, 118)

(232) kommt zusammen mit (353) und (354) in *Chrysanthemum serotinum* vor (47). (354) liegt jedoch auch in *Dahlia*-Arten sowie einer *Ichthyothere*-Art vor (98), aus der es als wirksame Komponente für die Fischgiftwirkung dieser Pflanze isoliert wurde. Die analogen Verbindungen mit Endiinen-Chromophor kommen ebenfalls in *Dahlia*-Arten vor (*118*):

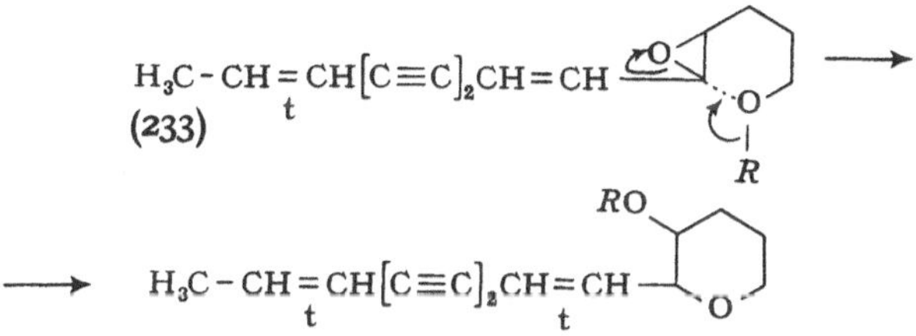

$$H_3C - CH = CH[C \equiv C]_2CH = CH \underset{t}{} \longrightarrow$$

(233)

$$\longrightarrow H_3C - CH = CH[C \equiv C]_2CH = CH \underset{t}{} \underset{t}{}$$

(355) $R = H$ (*118*) ; (356) $R = Ac$ (*118*)

Noch sehr viel zahlreicher sind die Verbindungen, die sich von (236) ableiten, das aus der noch nicht natürlich gewonnenen Verbindung (235) mit isolierter *cis*-Doppelbindung gebildet wird, wie Fütterungsversuche zeigen (*84*) (siehe auch S. 13). Wie ebenfalls durch Verfütterung von markiertem (236) bzw. (235) (*84*) oder (357) (*33*) gezeigt werden kann, bilden sich (358, 359, 361, 363, 364 und 366) analog den Verbindungen mit einem Sechsring alle aus (235) bzw. (236) als Vorstufe (*Schema 25*, S. 38).

Der offensichtlich sehr rasch erfolgende Übergang von (236) zu (357) ist mit einer *cis,trans*-Isomerisierung verbunden. Da auch die stärker ungesättigten Spiroketale aus (236) gebildet werden (*84*), muß ebenfalls die Weiteroxydation von (357) zu (360) rasch erfolgen. Die Bildung von (361) und (362) ist am besten über die Enolform von (360) zu verstehen. Die sauerstoffreicheren Derivate werden dann offenbar durch Allyl-

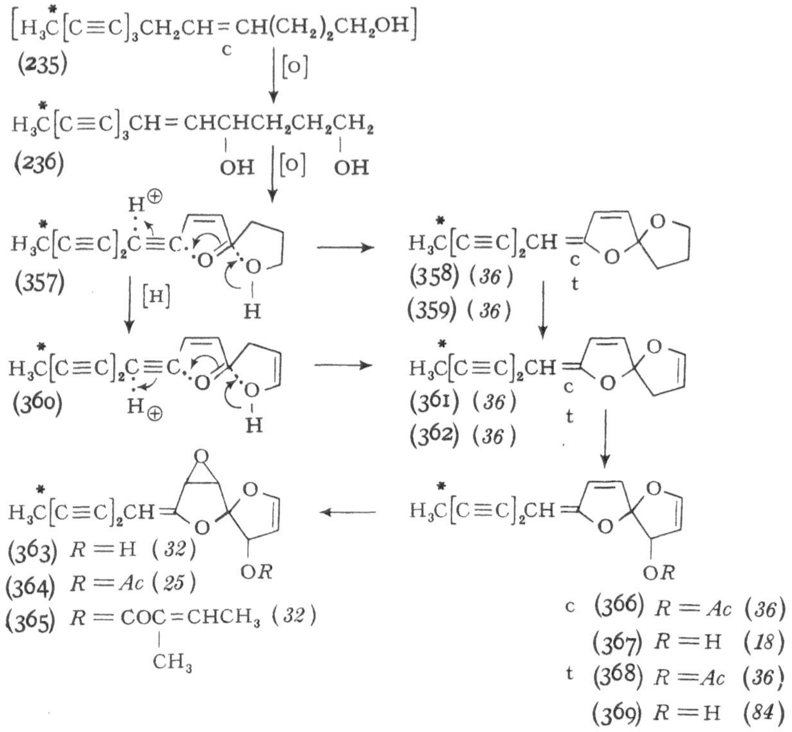

$[H_3\overset{*}{C}[C\equiv C]_3CH_2CH=CH(CH_2)_2CH_2OH]$
(235) c $\downarrow [O]$

$H_3\overset{*}{C}[C\equiv C]_3CH=CHCHCH_2CH_2CH_2$
(236) $\underset{OH}{|}$ $\downarrow [O]$ $\underset{OH}{|}$

$H_3\overset{*}{C}[C\equiv C]_2\overset{..}{C}\equiv C$ $\overset{H^{\oplus}}{}$ $\longrightarrow$ $H_3\overset{*}{C}[C\equiv C]_2CH$
(357) $\downarrow [H]$ (358) (36) c $\downarrow t$
                                                      (359) (36)

$H_3\overset{*}{C}[C\equiv C]_2\overset{..}{C}\equiv C$ $\overset{}{}$ $\longrightarrow$ $H_3\overset{*}{C}[C\equiv C]_2CH$
(360) $H_{\oplus}$ (361) (36) c
                                                      (362) (36) t $\downarrow$

$H_3\overset{*}{C}[C\equiv C]_2CH$ $\longleftarrow$ $H_3\overset{*}{C}[C\equiv C]_2CH$
(363) $R=H$ (32)
(364) $R=Ac$ (25) OR OR
(365) $R=COC=CHCH_3$ (32) c (366) $R=Ac$ (36)
         $|$ (367) $R=H$ (18)
         CH_3 t (368) $R=Ac$ (36)
                                                          (369) $R=H$ (84)

Schema 25. Biogenese der Spiroketalenoläther.

oxydation und eventuell anschließende Veresterung gebildet. Als weitere Abkömmlinge dieser Verbindungen sind die folgenden zu nennen:

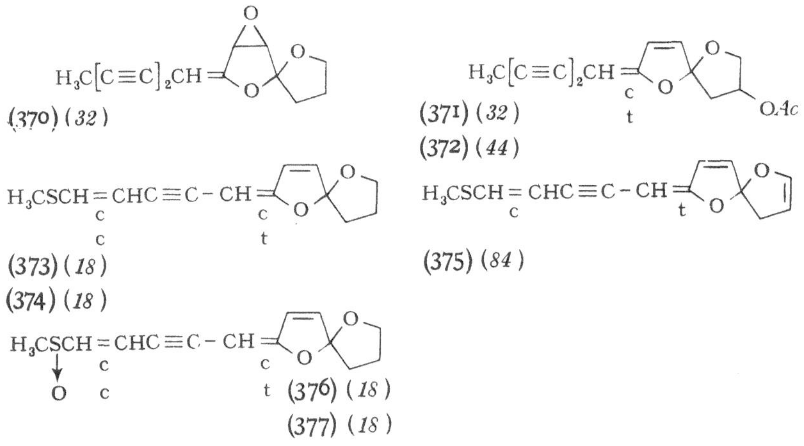

$H_3C[C\equiv C]_2CH$
(370) (32)

$H_3C[C\equiv C]_2CH$
(371) (32) c OAc
(372) (44) t

$H_3CSCH=CHC\equiv C-CH$
c c t
(373) (18)
(374) (18)

$H_3CSCH=CHC\equiv C-CH$
c t
(375) (84)

$H_3CSCH=CHC\equiv C-CH$
$\downarrow$ c c
O c t (376) (18)
                              (377) (18)

Literaturverzeichnis: SS. 54—62.

Für ihre Bildung dürfte das gleiche gelten wie für die entsprechenden 6-Ring-Verbindungen.

(236) ist auch Vorstufe für die Bildung von Furanverbindungen, wie durch Verfütterung von markiertem (236) an *Chrysanthemum silvaticum*

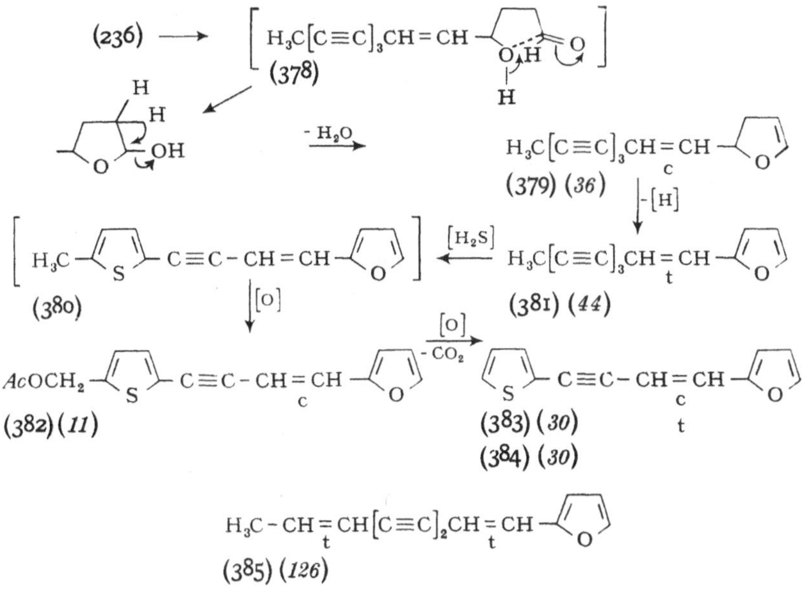

Schema 26. Biogenese von Furanpolyinen.

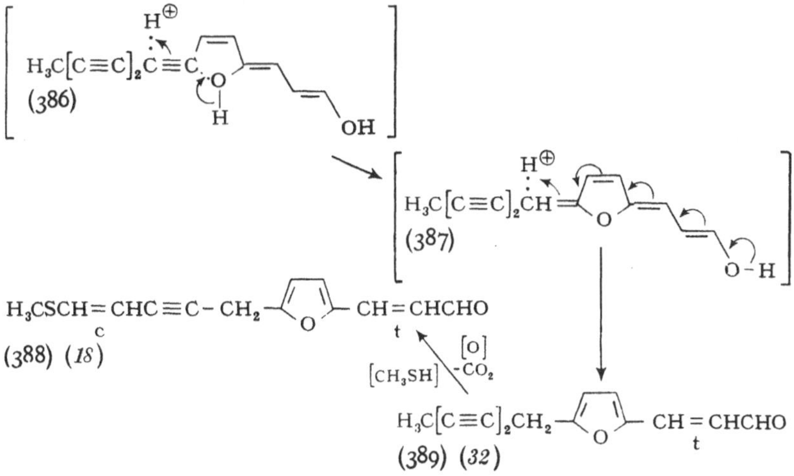

Schema 27. Biogenese disubstituierter Furane.

gezeigt werden kann. Offenbar wird das aus (235) gebildete Diol (236) nach Oxydation der endständigen OH-Gruppe nach *Schema 26* (S. 39) cyclisiert. Als weitere Folgeprodukte sind die Thiophenderivate (382—384) zu erwähnen, die nebeneinander in *Santolina*-Arten vorkommen. Offenbar erfolgt ihre Bildung über die noch nicht isolierte Zwischenstufe (380). Durch oxydative Eliminierung über (382) werden dann die beiden Isomeren (383) und (384) gebildet. (385), das bisher nur aus *Atractylis*-Arten isoliert werden konnte, dürfte analog gebildet werden.

Die beiden disubstituierten Furanderivate (388) und (389) könnten nach *Schema 27* (S. 39) ebenfalls aus (236) entstehen, wenn man (360) in der zweifach enolisierten Form (386) als Zwischenstufe annimmt. Das gemeinsame Vorkommen mit den Spiroketalenoläthern spricht für diese Annahme.

Eine weitere disubstituierte Furanverbindung kommt zusammen mit (248) vor. Die Bildung dieses Ketons könnte eventuell wie folgt ablaufen:

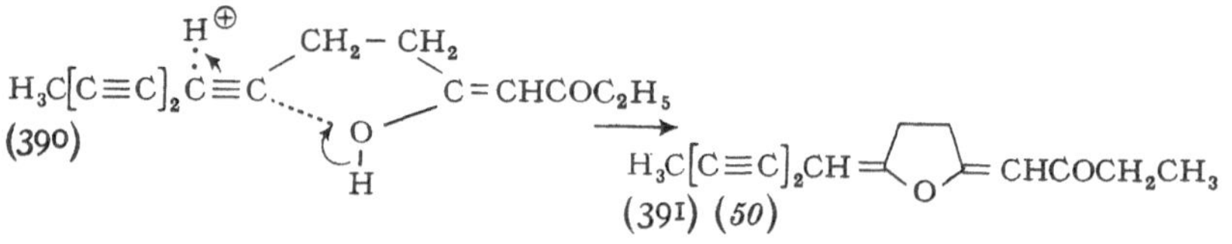

(390) wäre das β-Oxydationsprodukt von Dihydro-(248). Welche biogenetischen Zusammenhänge hier vorliegen, muß noch geklärt werden.

6. Phenylpolyine.

Relativ häufig sind bei den Compositen auch Phenylverbindungen zu finden. Die Biogenese dieser Verbindungen ist weitgehend geklärt. Wie

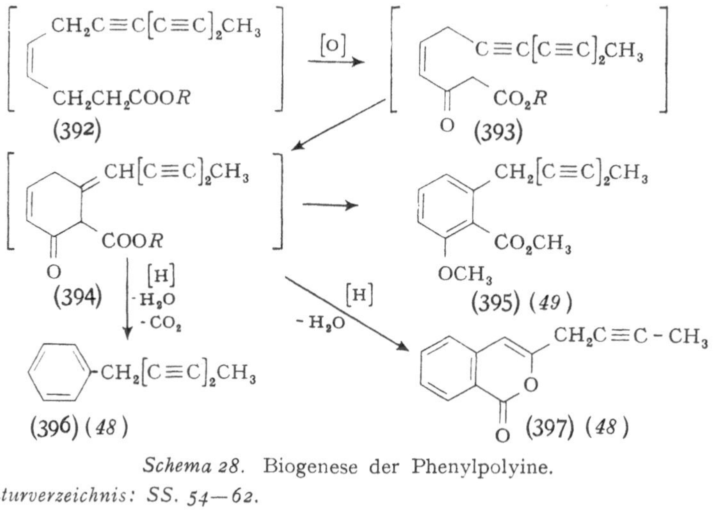

Schema 28. Biogenese der Phenylpolyine.

zunächst durch Acetat-Fütterungen gezeigt wurde, war die Bildung aus offenkettigen Verbindungen sehr wahrscheinlich (*42, 43*). Sowohl bei *Chrysanthemum frutescens* L. als auch bei *Coreopsis lanceolata* L. wird das Acetat gleichmäßig eingebaut, während vorgebildete Phenylverbindungen nicht inkorporiert werden. Durch Fütterungen mit markiertem (**392**) bzw. dem Acetat von (**235**) kann gezeigt werden, daß diese die biologischen Vorstufen sind und daß *Schema 28* wahrscheinlich ist (*84*).

Damit wäre (**235**) gemeinsame Vorstufe für die Phenylverbindungen und die Enolätherspiroketale, die in verschiedenen *Chrysanthemum*-Arten gemeinsam vorkommen. Die Cyclisierung des Allyloxydationsproduktes (**393**) wäre eine quasi-Michael-Addition. Nach Aromatisierung von (**394**) würde sich nach Methylierung (**395**) bilden. Erfolgt jedoch vorher eine biologische Reduktion von (**394**), so erhält man nach Wasserabspaltung und Lactonbildung das Isocumarin-Derivat (**397**). Analog könnte (**396**) durch Decarboxylierung der nach Hydrierung und Wasserabspaltung gebildeten Zwischenstufe entstehen.

Von (**395**) und (**396**) sind zahlreiche Umwandlungsprodukte bekannt, deren Bildung nach bekannten Mechanismen erfolgen sollte:

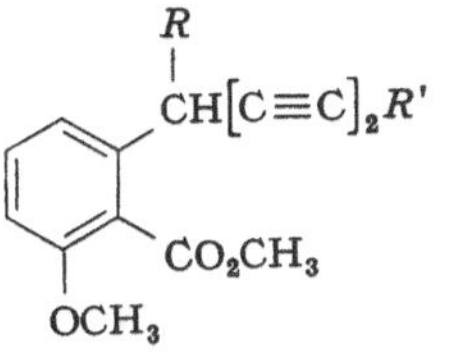

$$\underset{OCH_3}{\overset{\overset{\displaystyle R}{\underset{\displaystyle |}{}}}{\text{CH}[C\equiv C]_2 R'}}, CO_2CH_3$$

(**398**) $R = H, R' = CH_3$ (*49*)
(**399**) $R = R' = H$ (*49*)
(**400**) $R = OAc, R' = CH_3$ (*49*)
(**401**) $R = OAc, R' = H$ (*49*)

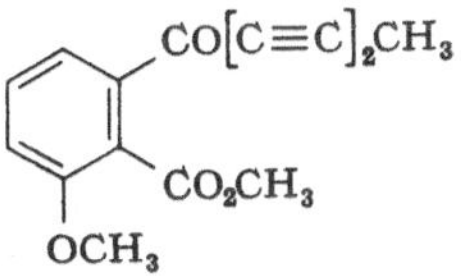

$$\underset{OCH_3}{}{\text{CO}[C\equiv C]_2 CH_3}, CO_2CH_3$$

(**402**) (*49*)

$PhCH_2C\equiv C-C\equiv CH$
(**403**) (*48*)

$PhCH_2C\equiv C-CH=CHSCH_3$ c
(**405**) (*37*)

$PhCOC\equiv C-CH=CHSCH_3$ c
(**407**) (*37*)
(**408**) (*37*)   t

$PhCOCH_2CH_2C\equiv C-CH_3$
(**410**) (*111*)

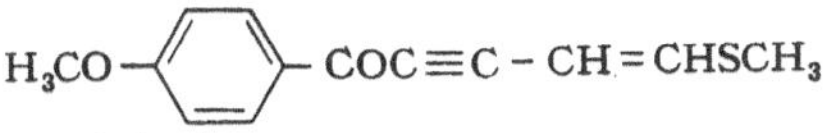

$PhCO[C\equiv C]_2CH_2OR$
(**412**) $R = H$ (*80*)
(**413**) $R = COCH=C(CH_3)_2$ (*80*)

$$\underset{}{\overset{\overset{\displaystyle OAc}{\underset{\displaystyle |}{}}}{Ph\text{CH}[C\equiv C]_2 CH_3}}$$
(**404**) (*49*)

$PhCO[C\equiv C]_2CH_3$
(**406**) (*49*)

$PhCO[C\equiv C]_2H$
(**409**) (*37*)

$H_3CO-\!\!\!\langle\ \rangle\!\!\!-COC\equiv C-CH=CHSCH_3$

(**411**) (*84*)

Bemerkenswert ist das häufige Vorkommen von Desmethylverbindungen. Wie durch Fütterungsversuche gezeigt werden konnte, wird z. B. (406) in (409) übergeführt (46), wahrscheinlich durch oxydative Eliminierung der Methylgruppe, wie sie bei Mikroorganismen mit zellfreien Extrakten sichergestellt werden konnte (107). Aus (409) werden dann (407) und (408) gebildet, wie ebenfalls durch Fütterungsexperimente gezeigt werden kann (40).

$$(406) \xrightarrow{[O]} PhCO[C\equiv C]_2CO_2H \xrightarrow{-CO_2} (409)$$
$$(414)$$

Die sauerstoffreicheren Derivate von (395) und (396) werden zweifellos durch enzymatische Oxydation der reaktionsfähigen Methylengruppe gebildet. Bei der Biogenese von (412) erfolgt zusätzlich noch eine Oxydation der Methylendgruppe, was als erster Schritt zur Bildung von (409) über (414) angesehen werden kann.

Die übrigen, bisher bekannten Phenylverbindungen besitzen alle ein C_{13}-Skelett. Als Biogeneseweg wäre hier z. B. für die Bildung des Phenylheptadiinens *Schema 29* möglich.

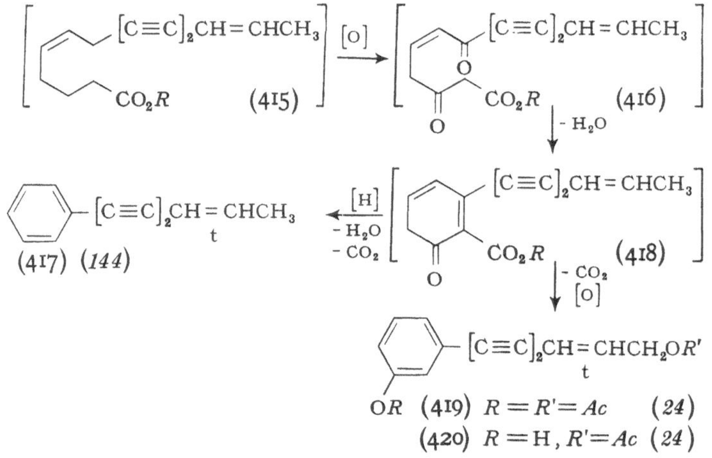

Schema 29.

Dieses Schema wird durch das Vorkommen von (419) gestützt. Die durch Oxydation von Dihydro-(225) gebildete Vorstufe (415) könnte durch Allyloxydation und β-Oxydation (416) liefern. Durch Aldolkondensation, biologische Reduktion der Ketogruppe, Wasserabspaltung und Decarboxylierung würde schließlich (417) entstehen. Das obige Schema wird weitgehend wahrscheinlich gemacht durch das Ergebnis der Verfütterung des

Literaturverzeichnis: SS. 54—62.

entsprechenden Triinesters von (415) an *Coreopsis lanceolata.* Dieser Ester wird glatt in (424) umgewandelt *(84).*

Weitere derartige Phenylverbindungen sind die folgenden, deren Biogenese analog bzw. nach bereits diskutiertem Weg ablaufen sollte:

$$Ph[C \equiv C]_2\underset{t}{CH} = CHR$$

(421) $R = CH_2OH$ *(68)*;   (422) $R = CH_2OAc$ *(142)*;   (423) $R = CHO$ *(68)*

$$Ph[C \equiv C]_3R$$

(424) $R = CH_3$ *(145)*;   (425) $R = CH_2OH$ *(28)*,   (426) $R = CH_2OAc$ *(28)*

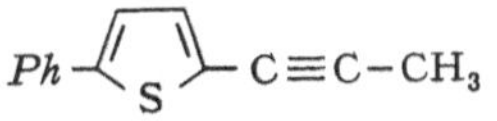

(427) *(146)*

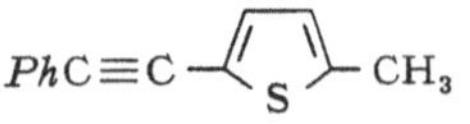

(428)

Die Thiophenverbindung (427) entsteht nicht aus (424) und Natriumsulfid, sondern das entsprechende Isomere (428) *(135)*.

Eine Sonderstellung nimmt die am längsten bekannte Acetylenverbindung, das Carlinaoxyd (431), ein. Dieses Phenylfuran-Derivat aus *Carlina acaulis* L. entsteht aus dem *cis*-Isomeren von (421), wie durch Verfütterung von markiertem Material gezeigt werden konnte *(43)*. Als wahrscheinlich ist folgender Weg anzunehmen *(Schema 30)*:

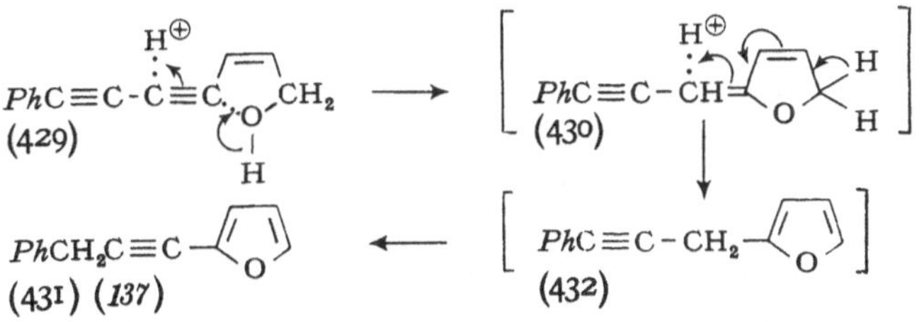

Schema 30.

Die Bildung von (430) und (432) als Zwischenstufen wird weiter wahrscheinlich durch die Isolierung analoger Verbindungen aus *Carlina*-Arten (siehe auch 91, S. 13):

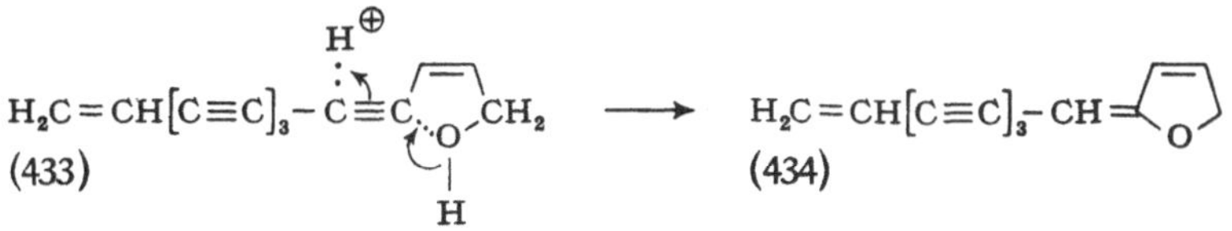

7. Verbindungen aus Mikroorganismen.

Die Acetylenverbindungen aus Mikroorganismen stellen eine relativ große Gruppe von Substanzen dar, die sich von den Verbindungen aus höheren Pflanzen im allgemeinen durch ihre größere Polarität unterscheiden. Im wesentlichen handelt es sich hier um Acetylenverbindungen, die von bestimmten Pilzen in die Kulturflüssigkeit ausgeschieden werden und daher oft wasserlöslich sind. Vorherrschend sind C_{10}- bzw. C_9-Verbindungen. Bemerkenswert sind die zahlreichen Allene sowie viele Substanzen mit einer Acetylen-H-Endgruppe (*117*). Wie schon erwähnt (vgl. S. 5), werden die Polyine bei den Mikroorganismen aus Ölsäure gebildet. Es gibt jedoch, im Gegensatz zu den Polyinen aus höheren Pflanzen nur ein Beispiel, das die *cis*-Doppelbindung der Ölsäure erkennen läßt. Die Dicarbonsäure (436) aus *Poria sinusa* (96) könnte wie folgt aus Ölsäure entstehen, wenn man eine zweifache β-Oxydation und eine Oxydation der Methylgruppe von (435) annimmt:

$$H_3C(CH_2)_7CH=CH(CH_2)_7COOH$$

$$\overset{c}{}\downarrow [O]$$

$$\left[H_3C[C\equiv C]_3CH_2CH=CH(CH_2)_7COOH\right] \qquad (435)$$

$$\overset{c}{}\downarrow [O]$$

$$HOOC[C\equiv C]_3CH_2CH=CH(CH_2)_3COOH \qquad (436)$$

$$\downarrow\ c$$

$$HC\equiv C-C\equiv C-CH=C=CHCH=CHCH=CHCH_2COOH \qquad (437)$$
$${c}{t}$$

$$HC\equiv C-C\equiv C-CH=C=CHCH_2CHCH_2CH_2CH_2COOCH_3 \qquad (438)$$
$$|$$
$$OH$$

Das Mycomycin (437) (99, *100*) könnte eventuell aus (436) durch Dehydrierung, Isomerisierung und Eliminierung der endständigen Methylgruppe gebildet werden. (438) (*119*) könnte ebenfalls aus (435) entstehen, wenn man Isomerisierung, Hydratisierung der Doppelbindung, zweifache β-Oxydation und Eliminierung der Methylgruppe annimmt.

Aus *Fistulina hepatica* lassen sich einige C_{13}-Verbindungen isolieren (*122*), die in enger Beziehung zu Verbindungen aus Compositen stehen:

$$H_3C[C\equiv C]_4CH_2CH_2CH_2CH_3 \qquad (439)$$

$$H_3C[C\equiv C]_4CH-CH-CH-CH_2 \qquad (440)$$
$$|\quad\ \ |\quad\ \ |\quad\ \ |$$
$$OH\ \ OH\ \ OH\ \ OH$$

$$HOCH_2CH=CH[C\equiv C]_3CH_2CH_2CH_2CH_3 \qquad (441)$$
$$t$$

(440) könnte durch Hydroxylierung von (67, S. 12) entstehen.

Literaturverzeichnis: SS. 54—62.

Aus Mikroorganismen sind auch einige Verbindungen mit C_{12}- bzw. C_{11}-Kette isoliert worden, die in enger biogenetischer Beziehung zueinander stehen:

$$H_3C[C\equiv C]_2CH=C=CHCHCH_2CH_2COOH$$
$$|$$
$$OR$$

(442) $R = H$ (90);   (443) $R = D$ - Xylose (88)

$$H[C\equiv C]_2CH=C=CHCHCH_2CH_2COOR'$$
$$|$$
$$OR$$

(444) $R = R' = H$ (89, 91);   (445) $R = D$ - Xylose, $R' = H$ (88)
(446) $R = H, R' = CH_3$ (8)

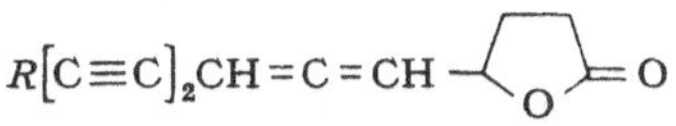

$$R[C\equiv C]_2CH=C=CH \overset{}{\underset{O}{\diagdown}}=O$$

(447) $R = CH_3$ (90);   (448) $R = H$ (89, 91)

$$H[C\equiv C]_2CH=C=CHCH_2CH_2CH_2CH_2OH \qquad (449)\ (9)$$

$$H[C\equiv C]_2CH=C=CHCHCH_2CH_2CH_2OH \qquad (450)\ (9)$$
$$|$$
$$OH$$

$$H[C\equiv C]_2CH=C=CHCH_2CHCH_2CH_2OH \qquad (451)\ (9)$$
$$|$$
$$OH$$

$$H[C\equiv C]_2CH=C=CHCHCH_2CH_2CH_2CH_2OH \qquad (452)\ (9)$$
$$|$$
$$OH$$

Wiederum ist die Eliminierung der endständigen Methylgruppe eine typische Umwandlung. Die Lactone (447) und (448) sind direkte Folgeprodukte von (442) bzw. (444). Die Verbindungen (449) und (450) sind eventuell Vorstufen von (444).

Die möglichen Isomerisierungsprodukte in dieser Reihe sind sehr gut am Beispiel der C_{11}-Säuren aus *Drosophila subatrata* (120) zu erkennen:

$$HC\equiv C-CH_2[C\equiv C]_2CH=CHCH_2COOH \qquad (453)$$
$$c$$

$$H_2C=C=CH[C\equiv C]_2CH=CHCH_2COOH \qquad (454)$$
$$c$$

$$H_3C[C\equiv C]_3CH=CHCH_2COOH \qquad (455)$$
$$c$$

$$HO_2C[CH_2]_3[C\equiv C]_2CH=CHCO_2H \qquad (456)$$
$$t$$

(456) ist als C_{11}-Verbindung eventuell durch oxydativen Abbau einer länger-kettigen Verbindung entstanden.

Die meisten Verbindungen aus Mikroorganismen haben eine C_{10}-bzw. C_9-Kohlenstoffkette. Ausgehend von C_{10}-Diinenen, -Triinenen, -Endiinenen und -Triinen kennt man zahlreiche mehr oder weniger sauerstoffreiche Derivate, wie *Tabelle 2* zeigen möge:

Tabelle 2. C_9- und C_{10}-Verbindungen aus Mikroorganismen.

$H_3C—CH_2CH_2[C{\equiv}C]_2CH{=}CH—R$ .... (457) $R = CH_2OH$ (*119*)
          t
                                                (458) $R = CO_2CH_3$ (*119*)

$HOCH_2CH{=}CH[C{\equiv}C]_2CH_2CO_2H$ .... (459) (*8*)
         t

$HOCH_2CH_2CH_2[C{\equiv}C]_2CH{=}CHCH_2OH$ .... (460) (*97*)
                        t

$HOCH_2CH_2[C{\equiv}C]_2CH{=}CHCH_2OH$ .... (461) (*96*)

$HOCH_2CH_2CH_2[C{\equiv}C]_2CH{=}CHCO_2R$ .... (462) $R = H$ (*93*)
                        t
                                         (463) $R = CH_3$ (*93*)

$RO_2C—CH_2CH_2[C{\equiv}C]_2CH{=}CHCO_2R$ .... (464) $R = H$ (*93*)
                       t
                                         (465) $R = CH_3$ (*93*)

$H_3C—CH{=}CH[C{\equiv}C]_2CH{=}CHCH_2OH$ .... (466) (*93*)
      t   t
      t   c   (467) (*106*)

$HOCH_2CH{=}CH[C{\equiv}C]_2CH{=}CHCH_2OH$ .... (468) (*119*)
         t   t

$HOCH_2CH{=}CH[C{\equiv}C]_2CH{=}CHCO_2CH_3$ .... (469) (*93*)
         t   t

$H_3C—CH{=}CH[C{\equiv}C]_2CH{=}CHCO_2R$ .... (470) $R = H$ (*93*)
      t   t
                                         (471) $R = CH_3$ (*93*)

$H_3C—CH{=}CH[C{\equiv}C]_2CH{=}CH—CO$ (472) (*93*)
        t   t |
$H_3CCOCH{=}CH[C{\equiv}C]_2CH_2CH_2CH_2O$
        t

$H_3C—CH{=}CH[C{\equiv}C]_2CH{=}CHCO$ (473) (*93*)
        t   t |
$H_3C—CH{=}CH[C{\equiv}C]_2CH{=}CHCH_2O$
        t   t

$RO_2C—CH{=}CH[C{\equiv}C]_2CH{=}CHCO_2R$ .... (474) $R = H$ (*120*)
        t   t
                                         (475) $R = CH_3$ (*93*)

$HO_2C—CH_2CH_2[C{\equiv}C]_3CO_2H$ .... (476) (*106*)

$HO_2C—CHCH_2[C{\equiv}C]_3CH_2OH$ .... (477) (*96*)
       |
       OH

$HO_2C—CHCH_2CH{=}CH[C{\equiv}C]_2CH_2OH$ .... (478) (*96*)
       | t
       OH

$H_3C[C{\equiv}C]_3CH{=}CH—CH_2OH$ .... (479) (*106*)
           t
           c   (480) (*122*)

Literaturverzeichnis: SS. 54—62.

$H_3C[C \equiv C]_3CH = CH - R$ . . . . (481) $R = CHO$ (*106*)
       t   (482) $R = CO_2H$ (*106*)

$HOCH_2[C \equiv C]_3CH = CH - R$ . . . . (483) $R = CH_2OH$ (*124*)
         t   (484) $R = CO_2H$ (*106*)

$OCH[C \equiv C]_3CH = CHCO_2H$ . . . . (485) (*96, 125*)
    t

$HO_2C[C \equiv C]_3CH = CHCO_2H$ . . . . (486) (*96*)
        t

$H_3C[C \equiv C]_2CH = C = CHCH_2CH_2OH$ . . . . (487) (*9*)

$H[C \equiv C]_3CH = CH - R$ . . . . (488) $R = CH_2OH$ (*124*)
       t   (489) $R = CHO$ (*124*)
                                                   (490) $R = CO_2H$ (*96*)
                                                   (491) $R = CO_2CH_3$ (*119*)

$H[C \equiv C]_3CH \underset{\diagdown \; \diagup}{O} CHCH_2OH$ . . . . (492) (*125*)

$H[C \equiv C]_3\underset{\underset{OH}{|}}{C}H - \underset{\underset{OH}{|}}{C}H - \underset{\underset{OH}{|}}{C}H_2$ . . . . (493) (*124*)

$H[C \equiv C]_2CH = CH\underset{\underset{OH}{|}}{C}H - \underset{\underset{OH}{|}}{C}H - \underset{\underset{OH}{|}}{C}H_2$ . . . . (494) (*97*)
    t

$H[C \equiv C]_3CH = CHCONHCH\underset{\underset{COOH}{|}}{C}H(CH_3)_2$ . . . . (495) (*96*)
    t

$H[C \equiv C]_2CH = CH(CH_2)_3CH_2OH$ . . . . (496) (*97*)
       t

$H[C \equiv C]_2CH = CHCH_2CH_2CO_2H$ . . . . (497) (*120*)
       c

$CH_3[C \equiv C]_2CH = CHCH_2CO_2CH_3$ . . . . (498) (*9*)
         t

$HC \equiv C - C \equiv C - CH = C = CHCH_2CO_2H$ . . . . (499) (*97*)

$HC \equiv C - C \equiv C - CH = C = CHCH_2CO_2CH_3$ . . . . (500) (*97*)

$H[C \equiv C]_2CH = C = CHCH_2OH$ . . . . (501) (*9*)

$H[C \equiv C]_2CH = C = CHCH_2CH_2OH$ . . . . (502) (*7, 97*)

$H[C \equiv C]_2CH = C = CHCH_2CH_2CH_2OH$ . . . . (503) (*9*)

$H[C \equiv C]_2CH = C = CHCH\underset{\underset{OH}{|}}{}CH_2CH_2OH$ . . . . (504) (*9*)

$\underset{\underset{OH}{|}}{C}H_2 - \underset{\underset{OH}{|}}{C}H[C \equiv C]_2COCH_2CH_3$ . . . . (505) (*121*)

$\underset{\underset{OH}{|}}{C}H_2 - \underset{\underset{OH}{|}}{C}H[C \equiv C]_2\underset{\underset{OH}{|}}{C}HCH_2CH_3$ . . . . (506) (*121*)

Man erkennt, daß sich die verschiedenen Verbindungen im wesentlichen durch Variationen in der Oxydationsstufe unterscheiden. Neben Allyloxydation ist auch α-Oxydation zu beobachten (477, 478). Wahrscheinlich entsteht auf diesem Wege durch Abbau z. B. (498). Wie schon bei (440) vermutet, entsteht offenbar das Triol (493) über (492) aus einer entsprechenden Verbindung mit einer Doppelbindung (488). Durch Verfütterung von 1-^{14}C-(271) kann gezeigt werden (*112*), daß folgender Biogeneseweg beschritten wird:

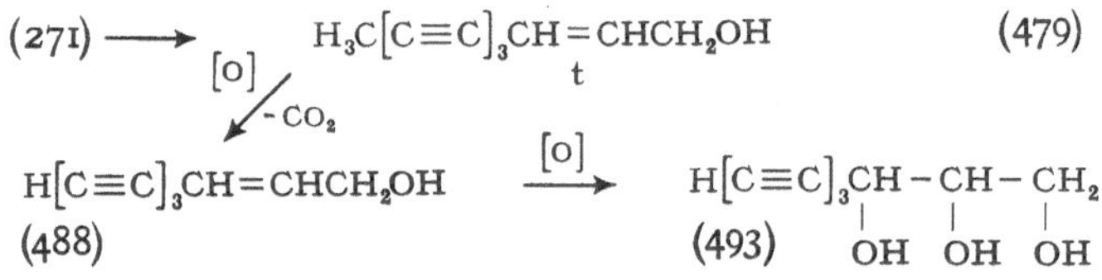

$$(271) \longrightarrow_{[O]} \quad \underset{t}{H_3C[C\equiv C]_3CH=CHCH_2OH} \qquad (479)$$

$$\swarrow_{-CO_2}$$

$$\underset{(488)}{H[C\equiv C]_3CH=CHCH_2OH} \xrightarrow{[O]} \quad \underset{(493)}{H[C\equiv C]_3CH-CH-CH_2} \atop \underset{OH\ OH\ OH}{\quad}$$

Die Eliminierung von Methylgruppen ist am Beispiel der Oxysäure (507) mit zellfreien Extrakten aus *Coprinus quadrifidus* gezeigt worden (*107*):

$$\underset{(507)}{\underset{t}{HO_2C[C\equiv C]_2CH=CHCH_2OH}} \xrightarrow{-CO_2} \underset{(508)}{\underset{t}{H[C\equiv C]_2CH=CHCH_2OH}}$$

Mit 1-^{14}C-(271) konnte weiter bewiesen werden, daß die Methylgruppe definiert zur Carboxylgruppe oxydiert wird, wobei in dem speziellen Fall interessanterweise auch eine Hydrierung der Doppelbindung erfolgt (*112*):

$$1\text{-}^{14}C\text{-}(271) \xrightarrow[\text{[H]}]{\text{[O]}} HO_2C[C\equiv C]_3CH_2CH_2CO_2H \qquad (476)$$

Damit dürfte das Schema der Methylgruppen-Entfernung eindeutig gesichert sein. Das Aminosäurederivat (495) ist eng verwandt mit dem Isobutylamid (307, S. 32). Die Biogenese der Allene ist bisher noch nicht

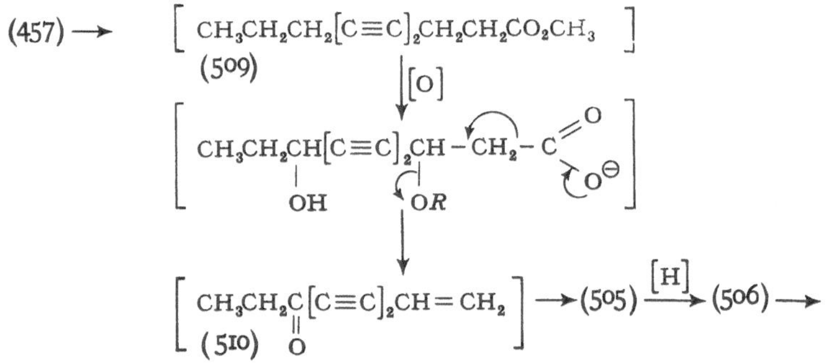

$$(457) \rightarrow \left[\underset{(509)}{CH_3CH_2CH_2[C\equiv C]_2CH_2CH_2CO_2CH_3} \right]$$

$$\downarrow [O]$$

$$\left[\underset{OH}{CH_3CH_2CH[C\equiv C]_2CH-CH_2-C \overset{O}{\underset{O^\ominus}{\diagdown}}} \right]$$

$$\downarrow_{OR}$$

$$\left[\underset{(510)}{CH_3CH_2C[C\equiv C]_2CH=CH_2} \atop \underset{O}{\|} \right] \rightarrow (505) \xrightarrow{[H]} (506) \rightarrow$$

Schema 31.

Literaturverzeichnis: SS. 54—62.

geklärt worden. Das Beispiel (453/454) läßt jedoch vermuten, daß sie eventuell allgemein durch analoge Isomerisierung einer isolierten Dreifachbindung entstehen. Das Marasmin (502) kommt in beiden Antipoden vor. (505) und (506) sind wahrscheinlich ebenfalls aus einer C_{10}-Säure entstanden. Vielleicht ist (509) eine Vorstufe, die über (505) das Triol (506) liefern könnte *(Schema 31)*.

Einige Verbindungen mit weniger als 9 C-Atomen kommen ebenfalls in den Kulturflüssigkeiten von Mikroorganismen vor:

$H[C\equiv C]_3H$ (511) *(108)*

$HOCH_2CH=CH[C\equiv C]_2CO_2R$ (512) $R=H$ *(8)* (513) $R=CH_3$ *(8)*

$HO_2C-CH=CH[C\equiv C]_2CONH_2$ (514) *(3,92)*

$HO_2C-CH=CH[C\equiv C]_2C\equiv N$ (515) *(3,92)*

$HOCH_2[C\equiv C]_3CONH_2$ (516) *(3, 92)*

$HOCH_2[C\equiv C]_3CO_2H$ (517) *(1)*

$CH_3O_2C-CH=CHC\equiv C-CH=CHCO_2CH_3$ (518) *(93)*

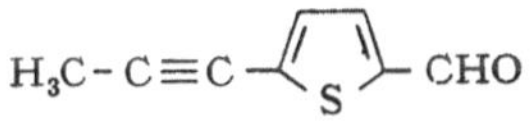

$H_3C-C\equiv C-\langle\underset{S}{}\rangle-CHO$ (519) *(139)*

$H_2NCOC\equiv C-CONH_2$ (520) *(152)*

Bemerkenswert sind in dieser Reihe die Stickstoffverbindungen. Neben Amiden ist auch ein Nitril aufgefunden worden. (519) ist bisher die einzige in ihrer Struktur aufgeklärte Schwefelverbindung aus Mikroorganismen, während aus Compositen zirka 100 schwefelhaltige Acetylenverbindungen isoliert werden konnten. Das extrem instabile Hexatriin (511) ist ein interessantes Stoffwechselprodukt der Basidiomycete *Fomes annosus* *(108)*. Möglicherweise wird es aus (517) nach Oxydation zur Dicarbonsäure durch doppelte Decarboxylierung gebildet.

8. Spezielle Verbindungen.

Neben den bereits besprochenen Substanzen gibt es einige Einzelvorkommen von Acetylenverbindungen aus den verschiedensten Pflanzen. Aus *Panax Ginseng* C. A. Meyer (Fam. Araliaceae) isoliert man z. B. das Carbinol (521), das eine gewisse Ähnlichkeit mit dem Falcarinon (152, S. 17) besitzt, das ebenfalls in der Familie Araliaceae vorkommt.

$H_3C(CH_2)_6[CH=CH]_2CH_2C\equiv C-CHCH=CH_2$

(521) *(153)*   OH

Aus einer Lauraceae-Art lassen sich eine Reihe von homologen Methyläthern isolieren, von denen (522) aufgeklärt wurde. Irgendwelche Zusammenhänge sind nicht erkennbar, ebenso wie bei der Bromverbindung (524), die aus einer Alge isoliert wurde. Offensichtlich ist hier lediglich die Bildung aus dem Epoxyd (523) als Vorstufe:

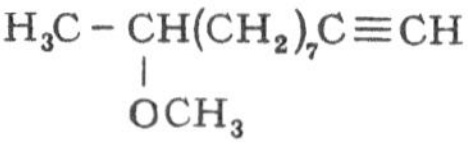

$$H_3C - CH(CH_2)_7C\equiv CH \qquad (522)\ (129)$$
$$\qquad\qquad |$$
$$\qquad\qquad OCH_3$$

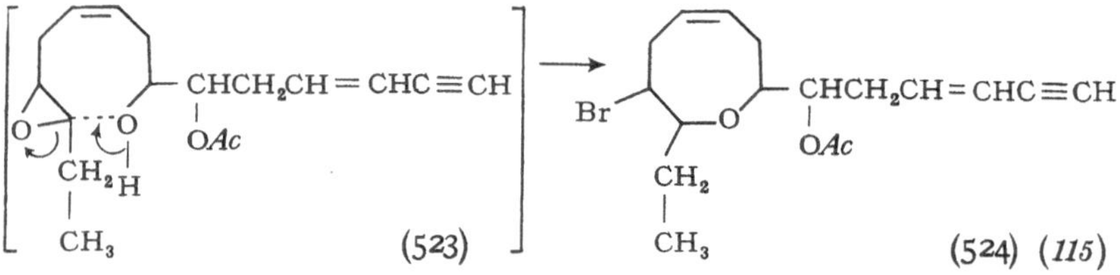

(523)   →   (524) (115)

Aus der Bohnenart *Vicia faba* läßt sich das Furan-Derivat (525) isolieren:

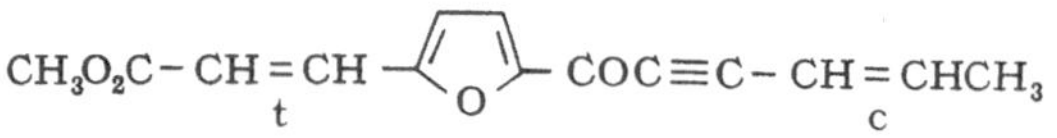

$$CH_3O_2C - CH = CH \underset{t}{}\!\!-\!\!\!\left\langle\!\!\!\begin{array}{c}\\O\end{array}\!\!\!\right\rangle\!\!- COC\equiv C - \underset{c}{CH} = CHCH_3 \qquad (525)\ (119)$$

Eine weitere Furanverbindung ist aus *Eremophila freelingii* (Fam. Myocarporaceae) isoliert worden (*128*). Es ist die bisher einzige natürliche Acetylenverbindung mit Kettenverzweigung. Offensichtlich ist (526) ein Sesquiterpen-Derivat:

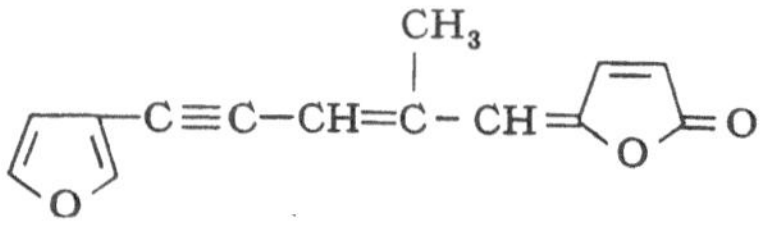

(526)

IV. Überblick über die erforderlichen Reaktionsschritte zum Aufbau der natürlichen Acetylenverbindungen.

Obwohl noch zahlreiche Fragen zur Biogenese der natürlichen Acetylenverbindungen offen sind, kann man doch schon jetzt etwa zusammenfassend angeben, welche wesentlichen Biogeneseschritte erforderlich sind, um die große Variationsbreite dieser Naturstoffklasse zu deuten. Der entscheidende Schritt für die Bildung der Dreifachbindung scheint in der Dehydrierung von Äthylenbindungen zu bestehen, eine Reaktion, die eventuell reversibel ist:

$$-CH=CH- \rightleftharpoons -C\equiv C-$$

Literaturverzeichnis: SS. 54—62.

Ungeklärt ist jedoch, in welchem Stadium der Biogenese dieser Schritt erfolgt. Vieles deutet darauf hin, daß bereits auf der Stufe der Fettsäure die entsprechenden Dehydrierungen erfolgen, wenngleich auch die Möglichkeit zu bestehen scheint, daß Endprodukte durch Weiterdehydrierung in andere Inhaltsstoffe umgewandelt werden können (siehe z. B. 16 → 19, S. 7). Die Tatsache, daß die C_{18}-Triinsäure in Dehydromatricariaester umgewandelt wird (vgl. S. 5), deutet darauf hin, daß nicht notwendigerweise stets Ölsäure als Vorstufe vorliegen muß.

Ein charakteristischer Biogeneseschritt ist offenbar der Abbau von ungesättigten Fettsäuren zu Vinylverbindungen, ebenso wie die β-Oxydation und die α-Oxydation *(Schema 32)*.

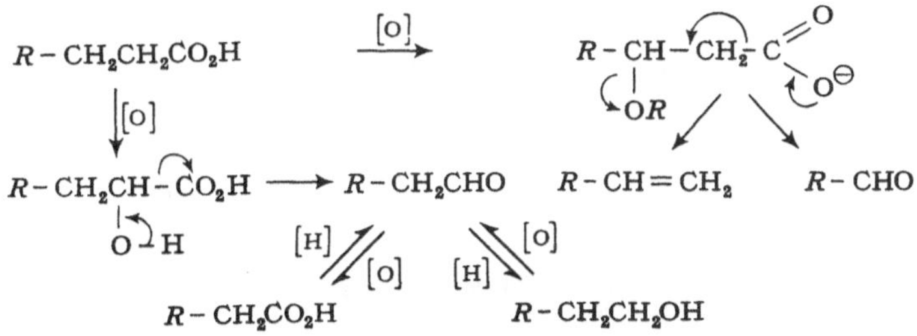

Schema 32.

Sehr häufig ist auch die weitere Umwandlung der Vinyl-Endgruppe *(Schema 33)*.

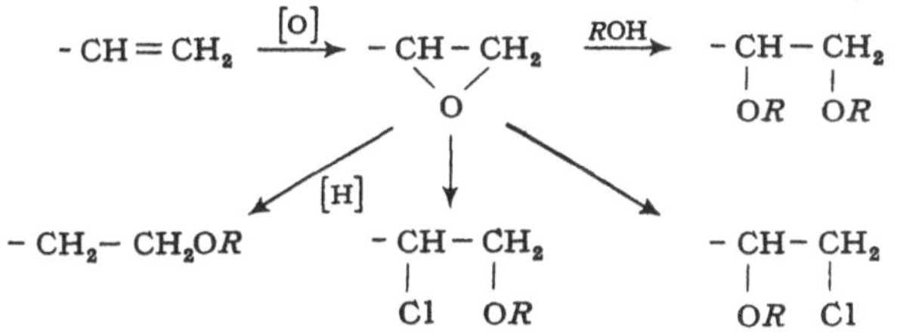

Schema 33.

Eine weitere wichtige Reaktion ist die Allyl-Oxydation, die im Falle einer Methylgruppe bis zur Eliminierung derselben führen kann:

$$\left.\begin{array}{r} -C\equiv C- \\ -CH=CH- \end{array}\right\} -CH_3 \rightarrow -CH_2OR \rightleftharpoons -CHO \rightleftharpoons -CO_2H \rightarrow -H$$

Wesentlich scheint auch die folgende Umwandlung zu sein:

$$-C\equiv C-CH_2-CH\underset{c}{=}CH- \xrightarrow[{[O]}]{} -C\equiv C-CH\underset{t}{=}CH-\underset{\displaystyle OH}{\underset{|}{CH}}-$$

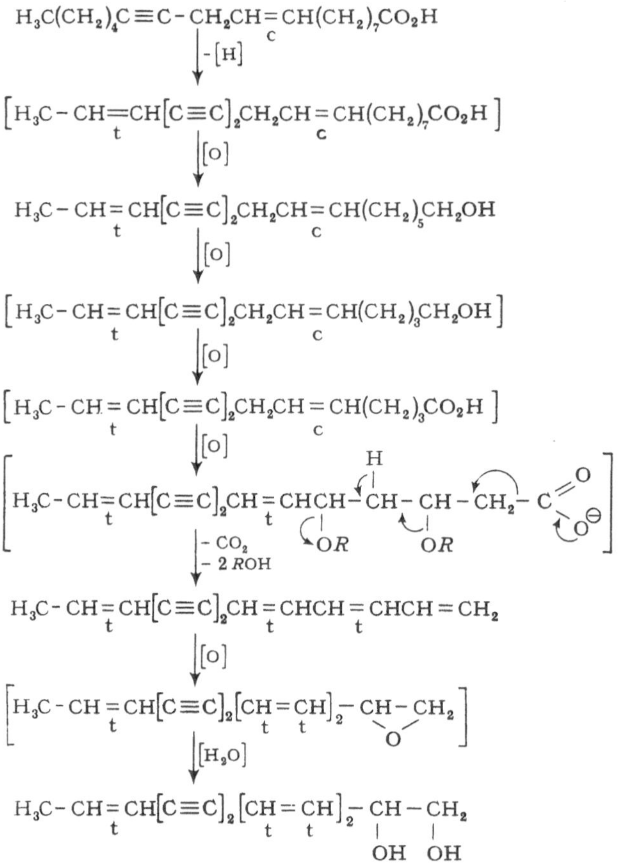

$$H_3C(CH_2)_4C\equiv C-\underset{c}{CH_2CH=CH}(CH_2)_7CO_2H \qquad (135)$$

$$\downarrow -[H]$$

$$\left[H_3C-\underset{t}{CH=CH}[C\equiv C]_2CH_2\underset{c}{CH=CH}(CH_2)_7CO_2H\right]$$

$$\downarrow [O]$$

$$H_3C-\underset{t}{CH=CH}[C\equiv C]_2CH_2\underset{c}{CH=CH}(CH_2)_5CH_2OH \qquad (192)$$

$$\downarrow [O]$$

$$\left[H_3C-\underset{t}{CH=CH}[C\equiv C]_2CH_2\underset{c}{CH=CH}(CH_2)_3CH_2OH\right]$$

$$\downarrow [O]$$

$$\left[H_3C-\underset{t}{CH=CH}[C\equiv C]_2CH_2\underset{c}{CH=CH}(CH_2)_3CO_2H\right]$$

$$\downarrow [O]$$

$$\left[H_3C-\underset{t}{CH=CH}[C\equiv C]_2\underset{t}{CH=CHCH}-\overset{H}{\underset{OR}{C}}H-CH-CH_2-C\overset{O}{\underset{O^{\ominus}}{}}\right]$$

$$\downarrow \begin{array}{l}-CO_2\\ -2\,ROH\end{array}$$

$$H_3C-\underset{t}{CH=CH}[C\equiv C]_2\underset{t}{CH=CHCH}=\underset{t}{CHCH}=CH_2 \qquad (99)$$

$$\downarrow [O]$$

$$\left[H_3C-\underset{t}{CH=CH}[C\equiv C]_2[\underset{t}{CH=CH}]_2-CH-CH_2\atop O\right]$$

$$\downarrow [H_2O]$$

$$H_3C-\underset{t}{CH=CH}[C\equiv C]_2[\underset{t}{CH=CH}]_2-\underset{OH}{\underset{|}{CH}}-\underset{OH}{\underset{|}{CH_2}} \qquad (102)$$

Schema 34. Umwandlung von Crepissäure in C_{13}-Polyine.

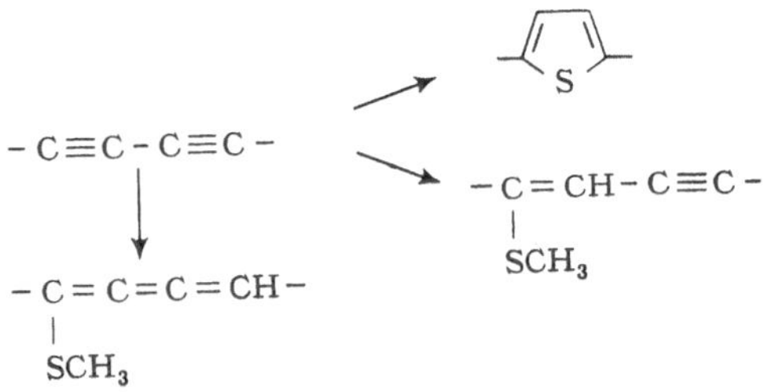

Schema 35. Bildung von Schwefelverbindungen.

Literaturverzeichnis: SS. 54—62.

Alle diese Reaktionen geben ein Gesamtschema für die Biogenese der offenkettigen Polyine, wie z. B. die von (102), wobei viele Zwischenstufen wiederum Vorstufen für zahlreiche andere natürliche Acetylene darstellen *(Schema 34)*.

Die große Zahl der Schwefelverbindungen zeigt, daß auch die Umwandlungen der Diingruppierung nach *Schema 35* wichtige Biogeneseschritte sind.

Der genaue Einblick in diese Reaktionen wird jedoch erst nach Isolierung der entsprechenden Enzyme möglich sein.

Der Übergang von *cis*- in *trans*-konfigurierte Doppelbindung (bzw. umgekehrt) läuft offenbar in der Pflanze sehr rasch ab, so daß eventuell Gleichgewichte vorliegen:

$$-CH=CH- \rightleftharpoons -CH=CH-$$
$$t c$$

Charakteristisch für die natürlichen Acetylenverbindungen ist auch ihre Umwandlung in zahlreiche Sauerstoff-Heterocyclen und Aromaten. Die Mechanismen dieser Reaktionen sind weitgehend geklärt. Wie das Beispiel in *Schema 36* zeigt, wird die Natur des Endproduktes durch relativ geringfügige, meist oxydative Veränderungen der gemeinsamen Vorstufe bestimmt.

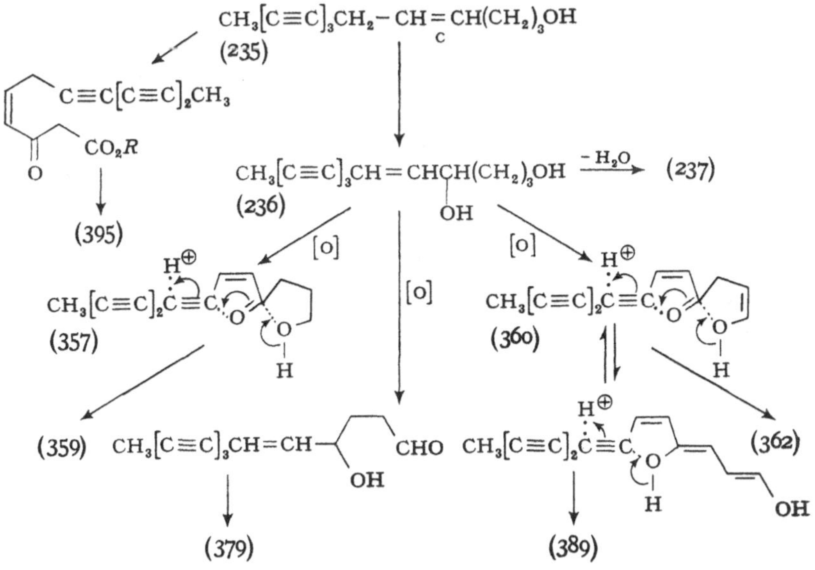

Schema 36.

Während die grundsätzlichen Reaktionswege geklärt sind, ist der genaue Verlauf noch durch weitere Versuche zu klären. Weiterhin ist die Frage interessant, wie die im sehr raschen Stoffwechsel stehenden Acetylenverbindungen weiter umgewandelt werden, um eventuell auch eine Antwort auf die naheliegende Frage zu erhalten: Welche Bedeutung haben diese reaktionsfähigen Substanzen im Rahmen der gesamten Pflanzenphysiologie?

Literaturverzeichnis.

1. ANCHEL, M.: Metabolic Products of *Clitocybe diatreta*. III. Characterization of Diatretyne 3 as *trans*-10-Hydroxy-dec-2-en-4,6,8-trynoic Acid. Arch. Biochem. Biophys. **85**, 569 (1959).
2. ANET, E. F. L. J., B. LYTHGOE, M. H. SILK and S. TRIPPETT: Oenanthotoxin and Cicutoxin. Isolation and Structures. J. Chem. Soc. (London) **1953**, 309.
3. ASHWORTH, P. J., E. R. H. JONES, G. H. MANSFIELD, K. SCHLÖGL, J. M. THOMPSON and M. C. WHITING: Researches on Acetylenic Compounds. LIX. The Synthesis of Three Polyacetylenic Antibiotics. J. Chem. Soc. (London) **1958**, 950.
4. ATKINSON, R. E., R. F. CURTIS and G. T. PHILLIPS: Naturally Occurring Thiophens. II. 5-(4-Chloro-3-hydroxybut-1-ynyl)-2,2'-bithienyl from *Tagetes minuta* L. J. Chem. Soc. (London) **C 1966**, 1101.
5. BAALSRUD, K. S., D. HOLME, M. NESTVOLD, J. PLÍVA, J. S. SÖRENSEN and N. A. SÖRENSEN: Studies Related to Naturally Occurring Acetylene Compounds. IX. The Occurrence of Methyl dec-8-*cis*-en-4 : 6-diynoate (= α,β-Dihydro-Matricaria Ester) and 2-*cis* : 8-*trans*-Matricaria Ester in Nature. Acta Chem. Scand. **6**, 883 (1952).
6. BAGBY, M. O., C. R. SMITH, Jr. and I. A. WOLFF: A Naturally Occurring Allenic Acid from *Leonotis nepetaefolia* Seed Oil. Chem. and Ind. **1964**, 1861.
7. BENDZ, G.: A Study of the Chemistry of Some *Marasmius* Species. Ark. Kemi **15**, 131 (1960).
8. BEW, R. E., R. C. CAMBIE, SIR E. R. H. JONES and G. LOWE: Natural Acetylenes. XIX. Metabolites from Some *Poria* Species. J. Chem. Soc. (London) **C 1966**, 135.
9. BEW, R. E., J. R. CHAPMAN, SIR E. R. H. JONES, B. E. LOWE and G. LOWE: Natural Acetylenes. XVIII. Some Allenic Polyacetylenes from Basidiomycetes. J. Chem. Soc. (London) **C 1966**, 129.
10. BOHLMANN, F. und C. ARNDT: Polyacetylenverbindungen. LXXVII. Über einen neuen Polyintyp aus *Anaphalis margaritacea* B. et H. Chem. Ber. **98**, 1416 (1965).
11. — — Polyacetylenverbindungen. LXXXIX. Über ein neues Thiophen-furan-Derivat. Chem. Ber. **99**, 135 (1966).
12. BOHLMANN, F., C. ARNDT und H. BORNOWSKI: Polyacetylenverbindungen. XXVIII. Über weitere Polyine aus dem Tribus Anthemideae L. Chem. Ber. **93**, 1937 (1960).
13. BOHLMANN, F., C. ARNDT, H. BORNOWSKI und P. HERBST: Polyacetylenverbindungen. XXVI. Die Polyine aus *Aethusa cynapium* L. Chem. Ber. **93**, 981 (1960).
14. BOHLMANN, F., C. ARNDT, H. BORNOWSKI, H. JASTROW und K.-M. KLEINE: Polyacetylenverbindungen. XXXVIII. Neue Polyine aus dem Tribus Anthemideae. Chem. Ber. **95**, 1320 (1962).

15. BOHLMANN, F., C. ARNDT, H. BORNOWSKI und K.-M. KLEINE: Polyacetylen-verbindungen. XXXI. Über Polyine aus der Familie der Umbelliferen. Chem. Ber. **94**, 958 (1961).

16. — — — — Polyacetylenverbindungen. XXXVII. Über die Polyine der Gattung *Bidens* L. Chem. Ber. **95**, 1315 (1962).

17. — — — — Polyacetylenverbindungen. XLVIII. Die Polyine der Gattung *Anthemis* L. Chem. Ber. **96**, 1485 (1963).

18. BOHLMANN, F., C. ARNDT, H. BORNOWSKI, K.-M. KLEINE und P. HERBST: Polyacetylenverbindungen. LVI. Neue Acetylenverbindungen aus *Chrysan-themum*-Arten. Chem. Ber. **97**, 1179 (1964).

19. BOHLMANN, F., C. ARNDT, K.-M. KLEINE und H. BORNOWSKI: Polyacetylen-verbindungen. LXIX. Die Acetylenverbindungen der Gattung *Echinops* L. Chem. Ber. **98**, 155 (1965).

20. BOHLMANN, F., C. ARNDT, K.-M. KLEINE und M. WOTSCHOKOWSKY: Poly-acetylenverbindungen. LXXV. Neue Inhaltsstoffe aus *Bidens*-Arten. Chem. Ber. **98**, 1228 (1965).

21. BOHLMANN, F., C. ARNDT und C. ZDERO: Polyacetylenverbindungen. CII. Über neue Enolätherpolyine aus *Anaphalis-* und *Gnaphalium*-Arten. Chem. Ber. **99**, 1648 (1966).

22. BOHLMANN, F. und E. BERGER: Polyacetylenverbindungen. LXXIII. Die Polyine der Gattung *Buphthalmum* L. Chem. Ber. **98**, 883 (1965).

23. BOHLMANN, F., D. BOHM und C. RYBAK: Polyacetylenverbindungen. LXXXVII. Über die Struktur und Biogenese eines aus *Anthemis*-Arten isolierten Thio-äthers. Chem. Ber. **98**, 3087 (1965).

24. BOHLMANN, F. und H. BORNOWSKI: Polyacetylenverbindungen. XCVIII. Über phenolisch substituierte natürliche Acetylenverbindungen. Chem. Ber. **99**, 1223 (1966).

25. BOHLMANN, F., H. BORNOWSKI und C. ARNDT: Polyacetylenverbindungen. XLIX. Weitere Polyine aus dem Tribus Anthemideae L. Liebigs Ann. Chem. **668**, 51 (1963).

26. — — — Polyacetylenverbindungen. LXXX. Über das erste natürlich vor-kommende Kumulen. Chem. Ber. **98**, 2236 (1965).

27. — — — Polyacetylenverbindungen. CX. Über die Polyine aus *Cotula coro-nopifolia* L. Chem. Ber. **99**, 2828 (1966).

28. BOHLMANN, F., H. BORNOWSKI und K.-M. KLEINE: Polyacetylenverbindungen. LXIII. Über neue Polyine aus dem Tribus Heliantheae. Chem. Ber. **97**, 2135 (1964).

29. BOHLMANN, F., H. BORNOWSKI und S. KÖHN: Polyacetylenverbindungen. LXIV. Die Polyine der Gattung *Cosmos*. Chem. Ber. **97**, 2583 (1964).

30. BOHLMANN, F., H. BORNOWSKI und H. SCHÖNOWSKY: Polyacetylenverbin-dungen. XXXIX. Über heterocyclisch substituierte Acetylenverbindungen aus dem Tribus Anthemideae L. Chem. Ber. **95**, 1733 (1962).

31. BOHLMANN, F. und E. BRESINSKY: Polyacetylenverbindungen. CXX. Um-setzungen reaktionsfähiger Acetylenverbindungen mit Schwefelverbindungen. Chem. Ber. **100**, 107 (1967).

32. BOHLMANN, F., L. FANGHÄNEL, K.-M. KLEINE, H.-D. KRAMER, H. MÖNCH und J. SCHUBER: Polyacetylenverbindungen. LXXXI. Über neue Polyine der Gattung *Chrysanthemum* L. Chem. Ber. **98**, 2596 (1965).

33. BOHLMANN, F. und G. FLORENTZ: Polyacetylenverbindungen. XCVI. Über die Biogenese der Spiroketalenolätherpolyine. Chem. Ber. **99**, 990 (1966).

34. BOHLMANN, F. und M. GRENZ: Polyacetylenverbindungen. CXII. Über die Inhaltsstoffe aus *Echinacea*-Arten. Chem. Ber. **99**, 3197 (1966).

35. Bohlmann, F. und P. Herbst: Polyacetylenverbindungen. XLIII. Über die Inhaltsstoffe von *Tagetes*-Arten. Chem. Ber. **95**, 2945 (1962).

36. Bohlmann, F., P. Herbst, C. Arndt, H. Schönowsky und H. Gleinig: Polyacetylenverbindungen. XXXIV. Über einen neuen Typ von Polyacetylenverbindungen aus verschiedenen Vertretern des Tribus Anthemideae L. Chem. Ber. **94**, 3193 (1961).

37. Bohlmann, F., P. Herbst und I. Dohrmann: Polyacetylenverbindungen. XLIV. Über neue Acetylenverbindungen aus der Gattung *Chrysanthemum* L. Chem. Ber. **96**, 226 (1963).

38. Bohlmann, F. und U. Hinz: Polyacetylenverbindungen. LI. Biogenetische Beziehungen zwischen natürlich vorkommenden Polyinen. Chem. Ber. **97**, 520 (1964).

39. — — Polyacetylenverbindungen. LXXII. Über biogenetische Umwandlungen des Tridecen-pentains. Chem. Ber. **98**, 876 (1965).

40. Bohlmann, F., U. Hinz, A. Seyberlich und J. Repplinger: Polyacetylenverbindungen. LIV. Über biogenetische Umwandlungen natürlicher Polyine. Chem. Ber. **97**, 809 (1964).

41. Bohlmann, F. und H. Jastrow: Polyacetylenverbindungen. XL. Die Polyine der Gattung *Achillea* L. Chem. Ber. **95**, 1742 (1962).

42. Bohlmann, F. und R. Jente: Polyacetylenverbindungen. XCVII. Zur Biogenese der Phenylpolyine. Chem. Ber. **99**, 995 (1966).

43. Bohlmann, F. und W. v. Kap-herr: Polyacetylenverbindungen. XCII. Zur Biogenese des Carlinaoxyds. Chem. Ber. **99**, 148 (1966).

44. Bohlmann, F., W. v. Kap-herr, L. Fanghänel und C. Arndt: Polyacetylenverbindungen. LXXVI. Über einige neue Inhaltsstoffe aus dem Tribus Anthemideae. Chem. Ber. **98**, 1411 (1965).

45. Bohlmann, F., W. v. Kap-herr, R. Jente und G. Grau: Polyacetylenverbindungen. CV. Über die Biogenese natürlicher Acetylenverbindungen. Chem. Ber. **99**, 2091 (1966).

46. Bohlmann, F., W. v. Kap-herr, C. Rybak und J. Repplinger: Polyacetylenverbindungen. LXXIX. Synthese und Biogenese von *Anthemis*-thioäthern. Chem. Ber. **98**, 1736 (1965).

47. Bohlmann, F. und H.-G. Kapteyn: Polyacetylenverbindungen. CIII. Über die Polyine aus *Chrysanthemum serotinum* L. Chem. Ber. **99**, 1830 (1966).

48. Bohlmann, F. und K.-M. Kleine: Polyacetylenverbindungen. XXXV. Die Polyine aus *Chrysanthemum frutescens* L. und *Artemisia Dracunculus* L. Chem. Ber. **95**, 39 (1962).

49. — — Polyacetylenverbindungen. XXXVI. Über neue Polyintypen aus *Chrysanthemum frutescens* L. Chem. Ber. **95**, 602 (1962).

50. — — Polyacetylenverbindungen. XLVI. Die Polyine aus *Anacyclus radiatus* Lois. Chem. Ber. **96**, 588 (1963).

51. — — Polyacetylenverbindungen. XLVII. Die Polyine aus *Flaveria repanda* Lag. Chem. Ber. **96**, 1229 (1963).

52. — — Polyacetylenverbindungen. LVII. Über zwei neue Polyinacetate. Chem. Ber. **97**, 1193 (1964).

53. — — Polyacetylenverbindungen. LXXI. Über die Inhaltsstoffe von *Dahlia Merckii* Lehm. Chem. Ber. **98**, 872 (1965).

54. — — Polyacetylenverbindungen. LXXXVI. Über rote natürliche Schwefelacetylenverbindungen. Chem. Ber. **98**, 3081 (1965).

55. — — Polyacetylenverbindungen, XCIV. Die Polyine aus *Cousinia hystrix* C. A. May. Chem. Ber. **99**, 590 (1966).

56. BOHLMANN, F. und K.-M. KLEINE: Polyacetylenverbindungen. CVI. Über einige neue Acetylenverbindungen aus der Gattung *Anthemis* L. Chem. Ber. **99**, 2096 (1966).

57. BOHLMANN, F., K.-M. KLEINE und C. ARNDT: Polyacetylenverbindungen. LXII. Über natürlich vorkommende Thiophenacetylenverbindungen. Chem. Ber. **97**, 2125 (1964).

58. — — — Polyacetylenverbindungen. LXVIII. Über einen Polyinester aus *Aster Novi Belgii* L. Chem. Ber. **97**, 3469 (1964).

59. — — — Polyacetylenverbindungen. LXXIV. Die Polyine aus *Tridax trilobata* Hemsl. Chem. Ber. **98**, 1225 (1965).

60. — — — Polyacetylenverbindungen. C. Über ein natürlich vorkommendes Kumulen sowie einige von Matricariaester abgeleitete Schwefelverbindungen. Liebigs Ann. Chem. **694**, 149 (1966).

61. — — — Polyacetylenverbindungen. CI. Die Inhaltsstoffe aus *Anthemis saguramica* Sosn. Chem. Ber. **99**, 1642 (1966).

62. BOHLMANN, F., K.-M. KLEINE, C. ARNDT und S. KÖHN: Polyacetylenverbindungen. LXXVIII. Neue Inhaltsstoffe der Gattung *Anthemis* L. Chem. Ber. **98**, 1616 (1965).

63. BOHLMANN, F., K.-M. KLEINE und H. BORNOWSKI: Polyacetylenverbindungen. XLI. Über zwei Thiophenketone aus *Artemisia arborescens* L. Chem. Ber. **95**, 2934 (1962).

64. — — — Polyacetylenverbindungen. LXX. Die Acetylenverbindungen der Gattung *Grindelia* W. Chem. Ber. **98**, 369 (1965).

65. — — — Polyacetylenverbindungen. XCI. Struktur und Synthese eines C_{18}-Ketoacetats aus *Cosmos sulphureus* Cav. Chem. Ber. **99**, 142 (1966).

66. BOHLMANN, F., H.-J. KOCH, S. KÖHN und W. HERFURT: Polyacetylenverbindungen. LXVI. Die Polyine aus *Aethusa cynapium* L. Chem. Ber. **97**, 2598 (1964).

67. BOHLMANN, F., S. KÖHN und C. ARNDT: Polyacetylenverbindungen. CXIV. Die Polyine der Gattung *Carthamus* L. Chem. Ber. **99**, 3433 (1966).

68. BOHLMANN, F., S. KÖHN und E. WALDAU: Polyacetylenverbindungen. CXIII. Die Polyine des Subtribus Carduinae. Chem. Ber. **99**, 3201 (1966).

69. BOHLMANN, F. und J. LASER: Polyacetylenverbindungen. CIV. Zur Biogenese eines Thioäthers aus *Anthemis tinctoria* L. Chem. Ber. **99**, 1834 (1966).

70. BOHLMANN, F. und H. J. MANNHARDT: Acetylenverbindungen im Pflanzenreich. Fortschr. Chem. organ. Naturstoffe **14**, 1 (1957).

71. BOHLMANN, F., H. MÖNCH und U. NIEDBALLA: Polyacetylenverbindungen. XCIII. Struktur und Synthese einiger Acetylenverbindungen aus *Chrysanthemum maximum* Ramond. Chem. Ber. **99**, 586 (1966).

72. BOHLMANN, F., U. NIEDBALLA und K.-M. RODE: Polyacetylenverbindungen. CXVIII. Über neue Polyine mit C_{17}-Kette. Chem. Ber. **99**, 3552 (1966).

73. BOHLMANN, F., S. POSTULKA und J. RUHNKE: Polyacetylenverbindungen. XXIV. Die Polyine der Gattung *Centaurea* L. Chem. Ber. **91**, 1642 (1958).

74. BOHLMANN, F. und K.-M. RODE: Polyacetylenverbindungen. CVIII. Über die Inhaltsstoffe von *Artemisia pedemontana* Balb. Chem. Ber. **99**, 2416 (1966).

75. BOHLMANN, F., K.-M. RODE und C. ZDERO: Polyacetylenverbindungen. CXVII. Neue Polyine der Gattung *Centaurea* L. Chem. Ber. **99**, 3544 (1966).

76. BOHLMANN, F. und A. SEYBERLICH: Polyacetylenverbindungen. XC. Synthese des Thioäthers aus *Anthemis carpatica* Willd. Chem. Ber. **99**, 138 (1966).

77. BOHLMANN, F., W. SUCROW, H. JASTROW und H.-J. KOCH: Polyacetylenverbindungen. XXXII. Über weitere Polyine aus *Centaurea ruthenica* Lam. Chem. Ber. **94**, 3179 (1961).

78. Bohlmann, F., M. Wotschokowsky, U. Hinz und W. Lucas: Polyacetylen-verbindungen. XCV. Über die Biogenese einiger Thiophenverbindungen. Chem. Ber. **99**, 984 (1966).

79. Bohlmann, F. und C. Zdero: Polyacetylenverbindungen. IC. Über ein Thiophenlacton aus *Chamaemelum nobile* L. Chem. Ber. **99**, 1226 (1966).

80. — — Polyacetylenverbindungen. CVII. Die Inhaltsstoffe von *Lonas annua*. Chem. Ber. **99**, 2413 (1966).

81. — — Polyacetylenverbindungen. CXIX. Über zwei neue Isobutylamide aus *Chrysanthemum frutescens* L. Chem. Ber. **100**, 104 (1967).

82. Bohlmann, F., C. Zdero und P.-H. Bonnet: Polyacetylenverbindungen. CXI. Über einen neuen Polyinester aus *Sanvitalia procumbens* Lam. Chem. Ber. **99**, 3194 (1966).

83. Bohlmann, F., H. Ziegenhirt, R. Jente und M. Wotschokowsky: Polyacetylenverbindungen. LXVII. Untersuchungen über die Biogenese des Phenylheptatriins. In: Beiträge zur Biochemie und Physiologie von Naturstoffen, S. 93. Jena: Veb Gustav Fischer Verl. 1965.

84. Bohlmann, F. und Mitarb.: unveröffentlicht.

85. Bu'Lock, J. D.: Polyacetylenes and Related Compounds in Nature. Progr. Organ. Chem. **6**, 86 (1964).

86. Bu'Lock, J. D., D. C. Allport and W. B. Turner: The Biosynthesis of Polyacetylenes. III. Polyacetylenes and Triterpenes in *Polyporus anthracophilus*. J. Chem. Soc. (London) **1961**, 1654.

87. Bu'Lock, J. D., and H. Gregory: Biosynthesis of Polyacetylenic Antibiotics. Biochem. J. **72**, 322 (1959).

88. — — The Structure and Reactions of a Polyacetylenic Glycoside. J. Chem. Soc. (London) **1960**, 2280.

89. Bu'Lock, J. D., E. R. H. Jones and P. R. Leeming: Chemistry of the Higher Fungi. V. The Structures of Nemotinic Acid and Nemotin. J. Chem. Soc. (London) **1955**, 4270.

90. — — — Chemistry of the Higher Fungi. VII. Odyssic Acid and Odyssin. J. Chem. Soc. (London) **1957**, 1097.

91. Bu'Lock, J. D., E. R. H. Jones, P. R. Leeming and J. M. Thompson: Chemistry of the Higher Fungi. VI. Isomerisation Reactions of Naturally Occurring Allenes. J. Chem. Soc. (London) **1956**, 3767.

92. Bu'Lock, J. D., E. R. H. Jones, G. H. Mansfield, J. W. Thompson and M. C. Whiting: The Structures of Two Polyacetylenic Antibiotics. Chem. and Ind. **1954**, 990.

93. Bu'Lock, J. D., E. R. H. Jones and W. B. Turner: Chemistry of the Higher Fungi. VIII. A Series of Acetylenic Compounds from *Polyporus anthracophilus*. J. Chem. Soc. (London) **1957**, 1607.

94. Bu'Lock, J. D. and H. M. Smalley: The Biosynthesis of Polyacetylenes. V. The Role of Malonate Derivatives, and the Common Origin of Fatty Acids, Polyacetylenes, and "Acetate-derived" Phenols. J. Chem. Soc. (London) **1962**, 4662.

95. Bu'Lock, J. D. and G. N. Smith: The Origin of Naturally Occuring Acetylenes. J. Chem. Soc. (London) **C 1967**, 332.

96. Cambie, R. C., J. N. Gardner, E. R. H. Jones, G. Lowe and G. Read: Chemistry of the Higher Fungi. XIV. Polyacetylenic Metabolites of *Poria sinuosa* Fr. J. Chem. Soc. (London) **1963**, 2056.

97. Cambie, R. C., A. Hirschberg, E. R. H. Jones and G. Lowe: Chemistry of the Higher Fungi. XVI. Polyacetylenic Metabolites from *Aleurodiscus roseus*. J. Chem. Soc. (London) **1963**, 4120.

98. CASCON, S. C., W. B. MORS, B. M. TURSCH, R. T. APLIN and L. J. DURHAM: Ichthyothereol and Its Acetate, the Active Polyacetylene Constituents of *Ichthyothere terminalis* (Spreng.) Malme, a Fish Poison from the Lower Amazon. J. Amer. Chem. Soc. 87, 5237 (1965).

99. CELMER, W. D. and I. A. SOLOMONS: The Structure of the Antibiotic Mycomycin. J. Amer. Chem. Soc. 74, 1870 (1952).

100. — — Mycomycin. IV. Stereoisomeric 3,5-Diene Fatty Acid Esters. J. Amer. Chem. Soc. 75, 3430 (1953).

101. CHRISTENSEN, P. K.: Structural and Spectroscopical Studies on Naturally Occurring Acetylenes of Composites. Norges Tekn. Vitenskapsakad. [2] No. 7, (1959).

102. CHRISTENSEN, P. K., N. A. SÖRENSEN, I. BELL, E. R. H. JONES and M. C. WHITING: The Constitution of the So-Called "Composit-Cumulene I" from Scentless Mayweed (*Matricaria inodora* L.). Festschrift Arthur Stoll, p. 545. Basel: Birkhäuser. 1957.

103. CROMBIE, L.: Amides of Vegetable Origin. IV. The Nature of Pellitorine and Anacyclin. J. Chem. Soc. (London) 1955, 999.

104. CROMBIE, L. and M. MANZOOR-I-KHUDA: Amides of Vegetable Origin. IX. Total Synthesis of Anacyclin and Related Trienediynamides. J. Chem. Soc. (London) 1957, 2767.

105. FLEMING, I. and J. HARLEY-MASON: A New Synthesis of Acetylenic Bonds and its Biosynthetic Implications. Proc. Chem. Soc. (London) 1961, 245.

106. GARDNER, J. N., E. R. H. JONES, P. R. LEEMING and J. S. STEPHENSON: Chemistry of the Higher Fungi. X. Further Polyacetylenic Derivatives of Decane from Various Basidiomycetes. J. Chem. Soc. (London) 1960, 691.

107. GARDNER, J. N., G. LOWE and G. READ: Chemistry of the Higher Fungi. XII. The Enzymic Decarboxylation of an α,β-Acetylenic Acid. J. Chem. Soc. (London) 1961, 1532.

108. GLEN, A. T., S. A. HUTCHINSON and N. J. McCORKINDALE: Hexa-1,3,5-triyne — a Metabolite of *Fomes annosus*. Tetrahedron Letters 1966, 4223.

109. GUNSTONE, F. D. and A. J. SEALY: Fatty Acids. XII. The Acetylenic Acids of Isano (Boleko) Oil. J. Chem. Soc. (London) 1963, 5772.

110. HAMBERG, M. and B. SAMUELSSON: Novel Biological Transformations of 8,11,14-Eicosatrienoic Acid. J. Amer. Chem. Soc. 88, 2349 (1966).

111. HARADA, R.: Essential Oil of *Artemisia capillaris*. II. The Structure of Capillon. Nippon Kagaku Zasshi 77, 990 (1956) [Chem. Abstr. 51, 18489 (1957)].

112. HODGE, P., Sir E. R. H. JONES and G. LOWE: Natural Acetylenes. XXII. *trans*-Dehydromatricaria Ester as a Biosynthetic Precursor of Some Fungal Polyacetylenes. J. Chem. Soc. (London) C 1966, 1216.

113. HOLME, D. and N. A. SÖRENSEN: Studies Related to Naturally Occurring Acetylene Compounds. XV. The Isolation of *trans*-Lachnophyllum Ester from *Bellis perennis* L. Acta Chem. Scand. 8, 280 (1954).

114. HOPKINS, C. Y. and M. J. CHISHOLM: Occurrence of Stearolic Acid in a Seed Oil. Tetrahedron Letters 1964, 3011.

115. IRIE, T., M. SUZUKI and T. MASAMUNE: Laurencin, a Constituent from *Laurencia* Species. Tetrahedron Letters 1965, 1091.

116. JENSEN, S. L. and N. A. SÖRENSEN: Studies Related to Naturally Occurring Acetylene Compounds. XXIX. Preliminary Investigations in the Genus *Bidens*. I. *Bidens radiata* Thuill and *Bidens ferulaefolia* (Jacq.) DC. Acta Chem. Scand. 15, 1885 (1961).

117. JONES, Sir E. H. R.: Polyacetylenes. Proc. Chem. Soc. (London) 1960, 199.

118. — Natural Polyacetylenes and their Precursors. Chem. in Britain 1966, 6.

119. Jones, Sir E. H. R.: Privatmitteilung.
120. Jones, E. R. H., P. R. Leeming and W. A. Remers: Chemistry of the Higher Fungi. XI. Polyacetylenic Metabolites of *Drosophila subatrata.* J. Chem. Soc. (London) **1960**, 2257.
121. Jones, Sir E. R. H., B. E. Lowe and G. Lowe: Chemistry of the Higher Fungi. XVII. Polyacetylenic Metabolites from *Clitocybe rhizophora* Velen. J. Chem. Soc. (London) **1964**, 1476.
122. Jones, Sir E. R. H., G. Lowe and P. V. R. Shannon: Natural Acetylenes. XX. Tetra-acetylenic and Other Metabolites from *Fistulina hepatica* (Huds) Fr. J. Chem. Soc. (London) **C 1966**, 139.
123. Jones, Sir E. R. H., S. Safe and V. Thaller: Natural Acetylenes. XXIII. A C_{18} Polyacetylenic Keto-aldehyde Related to Falcarinone from an Umbellifer (*Pastinaca sativa* L.). J. Chem. Soc. (London) **C 1966**, 1220.
124. Jones, E. R. H. and J. S. Stephenson: Chemistry of the Higher Fungi. IX. Polyacetylenic Metabolites from *Coprinus quadrifidus.* J. Chem. Soc. (London) **1959**, 2197.
125. Jones, E. R. H., J. S. Stephenson, W. B. Turner and M. C. Whiting: Chemistry of the Higher Fungi. XIII. Synthesis of (*a*) a C_9 Triacetylenic Epoxyalcohol, a *Coprinus quadrifidus* Metabolite and (*b*) a C_9 Triacetylenic 1,2-Diol. The Structure of Biformyne 1. J. Chem. Soc. (London) **1963**, 2048.
126. Josioko, J., H. Hikono und Y. Sasaki: Über die Inhaltsstoffe von Atractylodes. 8. Die Struktur von Actractylodin. 2. Struktur. Chem. pharm. Bull. (Japan) **8**, 949 (1960).
126a. Krishnaswamy, N. R., T. R. Seshadri and B. R. Sharma: The Structure of a New Polythienyl from *Eclipta alba.* Tetrahedron Letters **1966**, 4227.
127. Lam, J.: Privatmitteilung.
128. Massy-Westropp, R. A., G. D. Reynolds and T. M. Spotswood: Freelingyne, an Acetylenic Sesquiterpenoid. Tetrahedron Letters **1966**, 1939.
129. Mathews, W. S., G. B. Pickering and A. T. Umoh: Acetylenic Ethers in the Essential Oil from *Litsea odorifera* Val. Chem. and Ind. **1963**, 122.
130. Mikolajczak, K. L., C. R. Smith, Jr., M. O. Bagby and I. A. Wolff: A New Type of Naturally Occurring Polyunsaturated Fatty Acid. J. Organ. Chem. (USA) **29**, 318 (1964).
131. Morris, L. J.: The Oxygenated Acids of Isano (Boleko) Oil. J. Chem. Soc. (London) **1963**, 5779.
132. Powell, R. G., C. R. Smith, Jr., C. A. Glass and I. A. Wolff: *Helichrysum* Seed Oil. II. Structure and Chemistry of a New Enynolic Acid. J. Organ. Chem. (USA) **30**, 610 (1965).
133. — — — — New Enynolic Acids from *Acanthosyris.* Structures and Chemistry. J. Organ. Chem. (USA) **31**, 528 (1966).
134. Schulte, K. E. und S. Foerster: Ist Bithienylbutinen die biogenetische Vorstufe des α-Terthienyls? Tetrahedron Letters **1966**, 773.
135. Schulte, K. E., J. Reisch und L. Hörner: Thiophene aus Alkinen. I. Chem. Ber. **95**, 1943 (1962).
136. Schulte, K. E., G. Rücker and W. Meinders: The Formation of Methyl Propynylthienylacrylate from Dehydromatricaria Ester of *Chrysanthemum vulgare.* Tetrahedron Letters **1965**, 659.
137. Semmler, F. W.: Zusammensetzung des ätherischen Öls der Eberwurzel (*Carlina acaulis* L.). Ber. dtsch. chem. Ges. **39**, 726 (1906).
138. Shorland, F. B.: The Distribution of Fatty Acids in Plant Lipids. In: T. Swain (Edit.), Chemical Plant Taxonomy, p. 253. London and New York: Academic Press. 1963.

139. SKATTEBÖL, L.: Studies Related to Naturally Occurring Acetylene Compounds. XXVI. The Synthesis of 5-(1-Propynyl)-2-formyl-thiophene, Junipal, and *trans*-Methyl-5-(1-propynyl)-2-thienylacrylate. Acta Chem. Scand. **13**, 1460 (1959).

140. SÖRENSEN, J. S., T. BRUUN, D. HOLME and N. A. SÖRENSEN: Studies Related to Naturally Occurring Acetylene Compounds. XIII. The Occurrence of *trans*-Methyl-*n*-dec-2-en-4 : 6 : 8-triynoate in the Genus *Tripleurospermum* Schultz Bipontinus. Acta Chem. Scand. **8**, 26 (1954).

141. SÖRENSEN, J. S., D. HOLME, E. T. BORLAUG and N. A. SÖRENSEN: Studies Related to Naturally Occurring Acetylene Compounds. XX. A Preliminary Communication on Some Polyacetylenic Pigments from Compositae Plants. Acta Chem. Scand. **8**, 1769 (1954).

142. SÖRENSEN, J. S. and N. A. SÖRENSEN: Studies Related to Naturally Occurring Acetylene Compounds. XVII. Four New Polyacetylenes from Garden Varieties of *Coreopsis*. Acta Chem. Scand. **8**, 1741 (1954).

143. — — Studies Related to Naturally Occurring Acetylene Compounds. XIX. The Isolation of 1-Acetoxy-*n*-trideca-2 : 10 : 12-triene-4 : 6 : 8-triyne from *Carlina vulgaris* L. Acta Chem. Scand. **8**, 1763 (1954).

144. — — Studies Related to Naturally Occurring Acetylene Compounds. XXII. Correctional Studies on the Constitution of the Polyacetylenes of some Annual *Coreopsis* Species. Acta Chem. Scand. **12**, 756 (1958).

145. — — Studies Related to Naturally Occurring Acetylene Compounds. XXIII. 1-Phenylhepta-1 : 3 : 5-triyne from *Coreopsis grandiflora*, Hogg ex Sweet. Acta Chem. Scand. **12**, 765 (1958).

146. — — Studies Related to Naturally Occurring Acetylene Compounds. XXIV. 2-Phenyl-5-(α-propynyl)-thiophene from the Essential Oils of *Coreopsis grandiflora*, Hogg ex Sweet. Acta Chem. Scand. **12**, 771 (1958).

147. SÖRENSEN, N. A.: Some Naturally Occurring Acetylenic Compounds. Proc. Chem. Soc. (London) **1961**, 98.

148. — Chemical Taxonomy of Acetylenic Compounds. In: T. SWAIN (Edit.), Chemical Plant Taxonomy, p. 219, esp. p. 238. London and New York: Academic Press. 1963.

149. SÖRENSEN, N. A. und J. STENE: Über einen stark ungesättigten Ester aus *Matricaria inodora* L. Liebigs Ann. Chem. **549**, 80 (1941).

150. SPRECHER, H. W., R. MAIER, M. BARBER and R. T. HOLMAN: Structure of Optically Active Allene-containing Tetraester Triglyceride Isolated from the Seed Oil of *Sapium sebiferum*. Biochemistry **4**, 1856 (1965).

151. STAVHOLT, K. and N. A. SÖRENSEN: Studies Related to Naturally Occurring Acetylene Compounds. V. Dehydro Matricaria Ester (Methyl-*n*-decene-tri-ynoate) from the Essential Oil of *Artemisia vulgaris* L. Acta Chem. Scand. **4**, 1567 (1950).

152. SUZUKI, S., G. NAKAMURA, K. OKUMA and Y. TOMIYAMA: Cellocidin, a New Antibiotic. J. Antibiotics (Japan) **A 11**, 81 (1958) [Chem. Abstr. **54**, 1736 (1960)].

153. TAKAHASHI, M. and M. YOSHIKURA: Studies on the Components of *Panax Ginseng* C. A. Meyer. III. On the Ethereal Extract of Ginseng Radix Alta (3). On the Structure of a New Acetylene Derivative "Panaxynol". J. Pharmac. Soc. Japan **84**, 752, 757 (1964).

154. TRONVOLD, G. M., M. NESTVOLD, D. HOLME, J. S. SÖRENSEN and N. A. SÖRENSEN: Studies Related to Naturally Occurring Acetylene Compounds. XI. Further Investigations on the Composition of Essential Oils from the Genus *Erigeron*. Acta Chem. Scand. **7**, 1375 (1953).

155. Uhlenbroek, J. H. and J. D. Bijloo: Investigations on Nematicides. II. Structure of a Second Nematicidal Principle Isolated from *Tagetes* Roots. Rec. trav. chim. Pays-Bas **78**, 382 (1959).

156. Wiljams, W. W., W. S. Smirnow und W. P. Golmow: Über die Natur des kristallinischen Produkts aus dem ätherischen Öl von *Lachnophyllum gossypinum* Bge. Zhurn. Obschei Khimii (URSS) **5**, 1195 (1935) [Chem. Zbl. **1936** I, 3347].

157. Zechmeister, L. and J. W. Sease: A Blue-fluorescing Compound, Terthienyl, Isolated from Marigolds. J. Amer. Chem. Soc. **69**, 273 (1947).

(Eingelaufen am 21. Oktober 1966.)

The Chemistry of the Hop Resins.

By **P. R. Ashurst**, Nutfield, Surrey.

Acknowledgement. The author wishes to thank Dr. A. H. Cook, F. R. S., for his encouragement.

I. Introduction.

The hop plant (*Humulus lupulus, H. americanus* etc.) has for centuries been employed to give flavour to fermenting liquors. It is best known as the principal source of the aroma and bitter flavour of beer, and the hops which today are used for this purpose are cultivated plants which have been developed by continuous selection over at least the last century.

The useful constituents of hops are secreted by the lupulin glands which are found in the anthers of the male flower and at the base of the bracteoles of the female inflorescence or cone. These glands are only poorly developed on the male plants and in consequence it is the female plant which is harvested. Hops are grown in the temperate zones of both the Northern and Southern hemispheres and the particular growing areas are the South West Midlands and South East of England, Southern Germany and Czechoslovakia, the Pacific Coast of the U. S. A. and

Tasmania. English hops differ from the bulk of hops grown in other areas since both male and female plants are grown together resulting in the crop containing a substantial quantity (up to 30% of the dry weight) of seeds which have no economic value. Other hops are grown seedless (normally less than 2% dry weight of seeds).

On harvesting the crop, the cones are stripped from the bines, or stems, and dried immediately since any delay results in a rapid decomposition. Drying is carried out by passing air heated to 60–70° C. through the hops until the moisture content reaches about 10%. The hops are then compressed into sacks for storage.

Some of the chemical properties of the constituents of hops were recognised nearly a century and a half ago and since that time the main groups of compounds have been characterized as the hop resins, essential oils, tannins, pectins, nitrogenous constituents and carbohydrates. The chief functions of hops in brewing can be summarised as: providing a bitter flavour, providing an agreeable aroma, acting as a bacteriostatic preservative in beer and stabilising the "head" of beer. Of these functions all but providing an agreeable aroma which is ascribed to the essential oils, are due to the hop resins, and it is solely with the chemistry of this group of compounds that the present review is concerned.

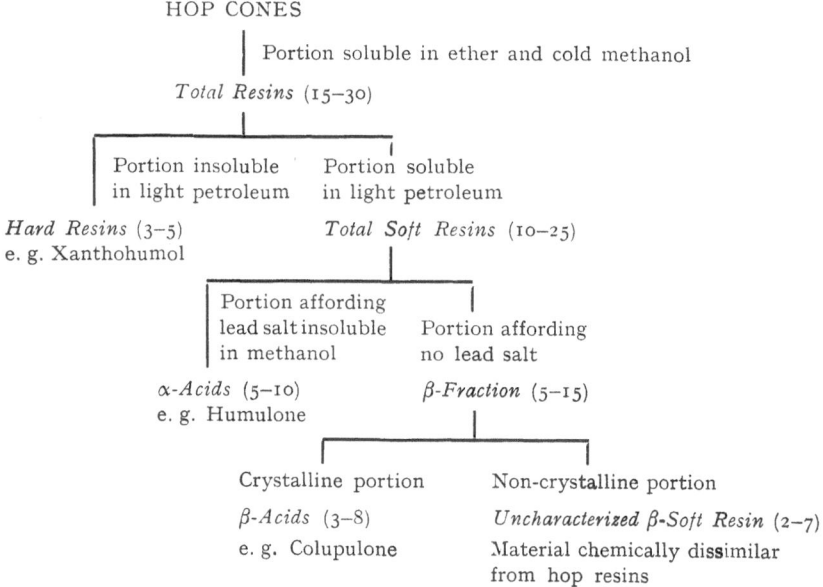

Scheme 1. Separation of Hop Resins by Extraction of Plant Material.

(Figures in parentheses refer to approximate percentages of the various fractions in freshly dried hop cones.)

References, pp. 85—89.

II. The Hop Resins.

Before embarking on a description of the chemistry of the hop resin constituents it is as well to refer briefly to the nomenclature of the various fractions obtained by the normal extraction of the hop resins, since this provides a suitable basis for the more detailed study of the constituents. The separation is detailed in *Scheme 1*.

All the simple hop resins are acyl phloroglucinols substituted with one, two, or three dimethylallyl side-chains.

The α-acids or humulones are oxidized derivatives of desoxyhumulones, di-isopentenyl phloracylphenones. There are three main α-acids: humulone itself with an isovaleryl side-chain, cohumulone with an isobutyryl side-chain and adhumulone with a 2-methylbutyryl side-chain. These three compounds together account for about 95% of the α-acids in all hops. Minor components include prehumulone (with an isocaproyl side-chain) and posthumulone, having a propionyl side-chain.

An early investigation was made with a material isolated as a copper salt from an ethereal extract of the hops by LERMER (56) in 1863. This was shown subsequently (20) to be identical with a crystalline compound obtained from an extract of hops made with light petroleum.

HAYDUCK (40) showed that the hop resins could be fractionated into the α- and β-resins, which were both soluble in light petroleum, and the γ-resin which was insoluble in it. Further, in contrast to the β-resin, the α-resin gave a yellow lead salt which was insoluble in most solvents. From the α- and β-resins crystalline compounds were isolated and termed α-acids and β-acids, respectively, and the material previously isolated was found to be β-acid.

The resin components are susceptible to oxidation and, due largely to their instability, partial characterization was not achieved until the work of WINDISCH (89). The material called α-acid was then designated as humulone and the β-acid as lupulone. The work of WINDISCH in conjunction with that of WIELAND and co-workers (88), enabled structures to be proposed for both humulone (1) and lupulone (2). The original structures of humulone and lupulone were modified (26, 81) in 1951 to those indicated in (3) and (4) on the basis of evidence obtained by ozonization. Ultimate confirmation of the structure of humulone was provided from a study of its proton magnetic resonance spectrum (72).

The synthesis of both humulone and lupulone was reported by RIEDL (63) in 1952. Reaction between phlorisovalerophenone, prepared by the Friedel-Crafts reaction, and 2,2-dimethylallylbromide gave desoxyhumulone (5) with two equivalents of the latter substance and lupulone with three equivalents. Oxidation of desoxyhumulone to humulone was achieved in poor yield using oxygen in the presence of lead acetate, the α-acid being isolated as a lead salt.

Desoxyhumulones, 3,5-di-(3-methylbut-2-enyl)phloracylphenones (5) have been reported (64) to occur as minor constituents of the soft resin of fresh hops. They are considered (91) to be the biogenetic precursors of both the α-acids, which are formed by a controlled oxidation, and of the β-acids, which are formed by a further addition of a 2,2-dimethylallyl side-chain.

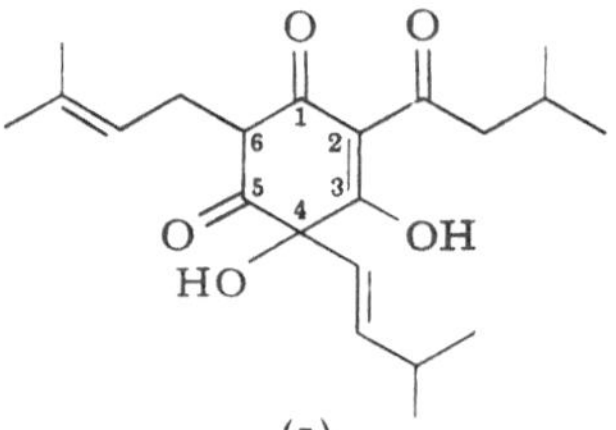

(1)

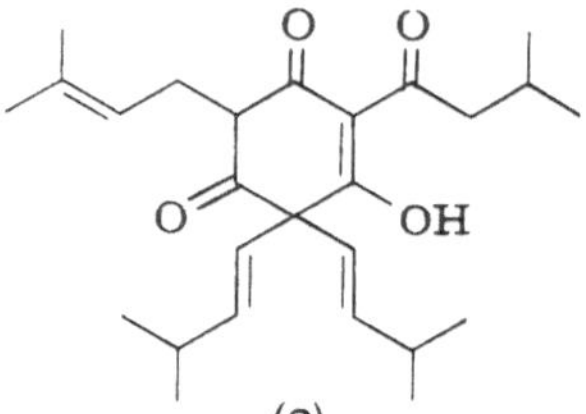

Humulone.

(2)

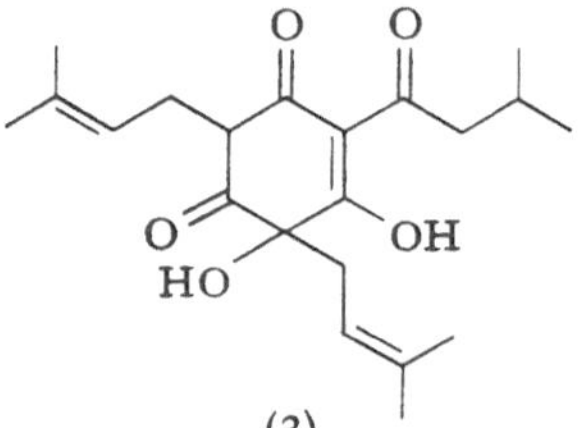

(3)

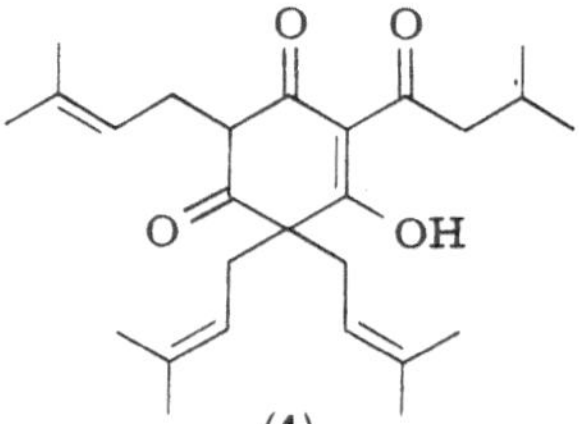

(4)

Lupulone.

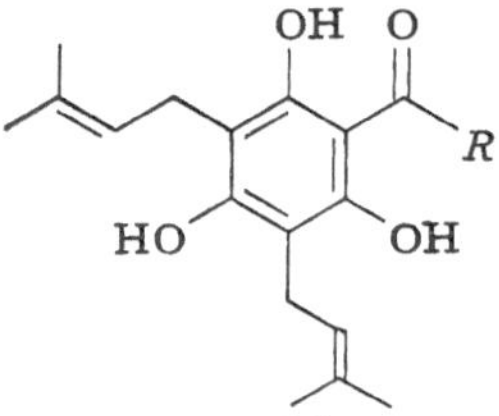

(5)

Desoxyhumulone $(R = CH_2 . CH (CH_3)_2)$.

During 1952, the discovery of another α-acid, cohumulone, was reported by RIGBY and BETHUNE (65). Cohumulone was very similar to humulone and studies on its structure (47) revealed an *iso*butyryl side-chain in comparison with the *iso*valeryl side-chain of humulone. Two years after the discovery of cohumulone a study (43) of the β-acid found in English hops revealed that it was largely colupulone, possessing the isobutyryl side-chain, whilst the β-acid of continental hops was largely lupulone.

A systematic examination of the α- and β-acids of many varieties of hops revealed the presence of compounds with other acyl side-chains at $C_{(2)}$. Compounds with a 2-methylbutyryl side-chain have been

References, pp. 85—89.

designated ad-analogues (*66*); those with a 4-methylpentanoyl side-chain pre-analogues (*68*) and those with an n-propionyl side-chain post-analogues (*67*). For example, if an α-acid possesses the 2-methylbutyryl side-chain at $C_{(2)}$ it is called adhumulone. Other unnamed α- and β-acids possessing the n-valeryl and n-caproyl side-chains have also been isolated (*67*).

In a typical English hop, humulone would normally account for about 70% of the α-acids, cohumulone 20%, adhumulone 5% and the remaining α-acids together 5%. Some hop varieties, e. g. Bullion, Brewers Gold, are known to contain up to 40% of cohumulone (*48*).

III. The Chemistry of the α-Acids.

During investigations on the structure of humulone (*88, 89*), alkaline hydrolysis to a crystalline compound called humulinic acid, which had been first noted by LINTNER and SCHNELL (*58*), was studied. Humulinic acid, formulated as (6), was considered to have been formed from humulone by two or possibly three distinct steps. The first step was considered to be a ring contraction of humulone to a compound called resin A, formulated as (7), which then proceeded to humulinic acid via a second intermediate, resin B (8). These resinous intermediates were of some importance since it was recognised that when hops were boiled with brewers' wort the α-acids became transformed into a bitter resinous material.

Hydrogenation studies on the "humulone boiling product", i. e. the resin A, by GOVAERT and VERZELE (*37*) afforded two products, both of which were crystalline. It was assumed that these compounds were the perhydro derivatives of resin A, for which the name isohumulone* was suggested, and of resin B. Later studies showed (*18*) that they were both derivatives of isohumulone and, accordingly, there is a considerable doubt as to the reality of resin B as a separate entity.

Humulinic Acids.

A close examination of the humulinic acids by ANTEUNIS and co-workers (*5*) was made using steady-state distribution. Four isomeric compounds were isolated and called humulinic acids A (the known isomer), B, C and D, but no structural formulations were made for the latter compounds.

* *Note on Nomenclature.* The term "isohumulones" which has unfortunately been used to denote the products of a ring contraction of humulone rather than, for instance, a simple chain branching has also been used in the literature to denote collectively the products of isomerization of the three principal α-acids. Within this review the term "iso-α-acids" will be used to denote the isomerization products of mixed α-acids whilst the term "isohumulones" will be reserved for the products of isomerization of humulone itself. The use of the more specific terms isohumulone A and B is discussed on p. 71.

Burton, Elvidge and Stevens (21) oxidized humulinic acid to dehydrohumulinic acid (9) and found that on reduction of the latter with sodium borohydride they obtained pure humulinic acid B. Examination of humulinic acids A and B by proton magnetic resonance spectroscopy

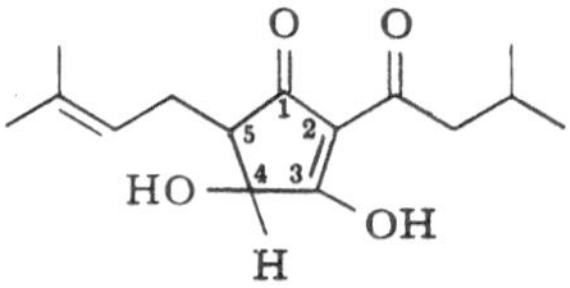

(6)
Humulinic acid.

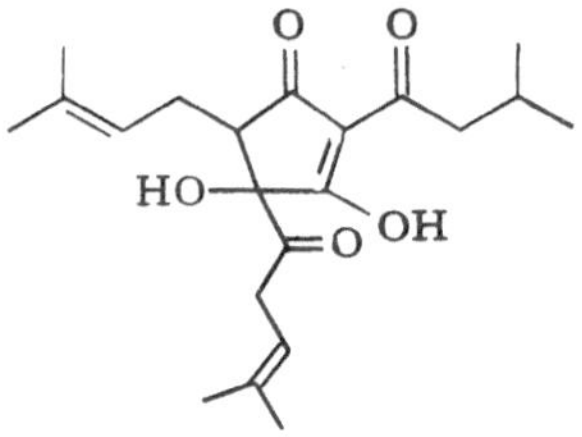

(7)
Resin A.

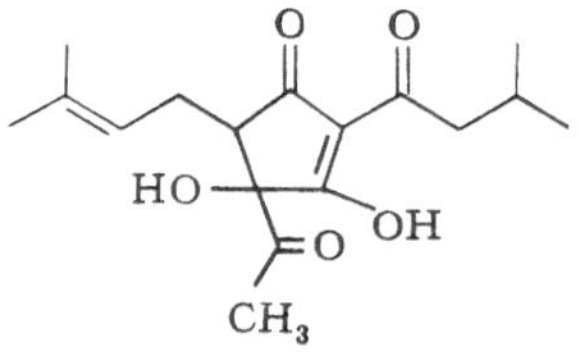

(8)
Resin B.

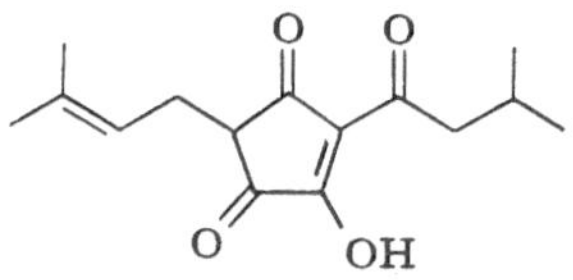

(9)

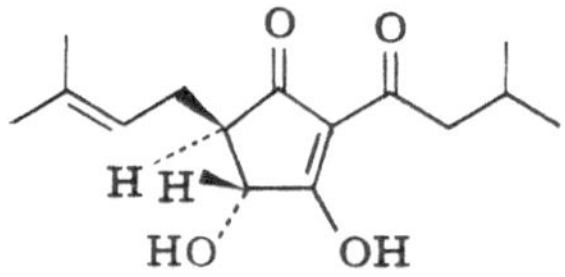

(10)
Humulinic acid A.

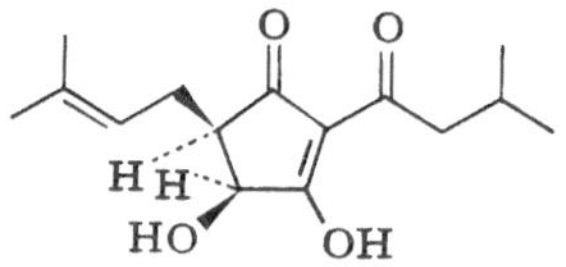

(11)
Humulinic acid B.

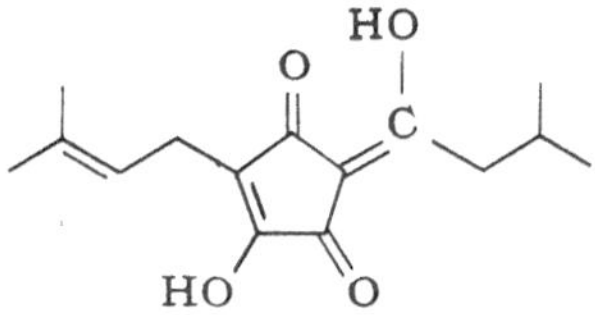

(12)
Dehydrohumulinic acid.

showed that they only differed by the stereochemical relationship between the hydroxyl group at $C_{(4)}$ and the isopentenyl group at $C_{(5)}$. These groups were found to possess a *trans* conformation in humulinic acid A (10) and a *cis* conformation in humulinic acid B (11). Similar conclusions regarding the structures of these acids were reached independently by Alderweireldt and Anteunis (1, 55).

Proton magnetic resonance studies (36) on dehydrohumulinic acid indicated that it did not possess the same enolisation system as did the other hop resin derivatives. Its formulation (9) was thus incorrect and it was shown to be an exocyclically enolised form (12) since the alternative cyclopentadienenone configuration is unstable (*38*).

In an attempt to further elucidate the chemistry of the humulinic acids, the effect of strong acids on humulinic acid A was studied by BURTON, ELVIDGE and STEVENS (*22*). The isomeric product, found to be humulinic acid C which possessed the same chromophore as humulinic acid A, lacked the characteristics of a free hydroxyl group and also failed to give reactions characteristic of an unsaturated compound. The structure of the product was accordingly formulated as (13) but proton magnetic resonance measurements suggested the alternative structure (14) for which chemical confirmation was subsequently obtained.

The product obtained by treating dehydrohumulinic acid (12) with strong acids was also investigated by these authors and found to possess the pyran structure (15) in contrast to the spirofuran structure of humulinic acid C. ANTEUNIS, BRACKE, ALDERWEIRELDT and VERZELE (*3*) also reached these conclusions regarding the structure of humulinic acid C, and observed that humulinic acid D was a rather unstable compound (16) which in the presence of acid or alkali was converted quantitatively to humulinic acid C. Humulinic acid D could, however, be stabilised as its methyl ether.

Methylation of the products obtained by boiling humulinic acid A or B in acidic methanol was found (*4*) to yield, in addition to methylhumulinic acid D, a small quantity of another methylhumulinic acid. This compound, methyl*allo*humulinic acid D, isolated from the mixture by counter-current distribution, was shown by proton magnetic resonance spectroscopy to possess structure (17). Methyl*allo*humulinic acid D was not isolated as the free acid.

Elucidation of the structural relationships of the humulinic acid series has also provided a pointer to stereochemical relationships in the iso-α-acid series. It should, however, be noted that, as yet, no humulinic acids have been isolated as hop constituents or as occurring in beer. In the latter their presence would be unwelcome since they are non-bitter, yet would interfere in the spectrophotometric estimation of bitter substances (*15*).

Iso-α-acids.

The iso-α-acids, isohumulones, isocohumulones, etc., are the most important substances derived from the hop resins since it has been shown (*30*) that they are responsible for at least 70% of the bitterness of a beer brewed with hops which have been stored for less than 6 months.

Whilst the gross structure for isohumulone can be represented as (7, p. 68), further studies on the nature of the material (*33*) gave rise to suspicions that isohumulone was a mixture. These views were supported by SPETSIG et al. (75), who examined both the material prepared by alkaline isomerization of humulone and the iso-α-acids obtainable from beer, by reversed-phase partition chromatography operating under

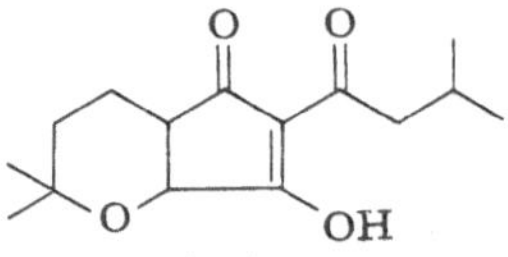

(13)

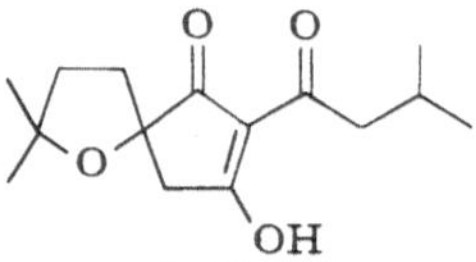

(14)
Humulinic acid C.

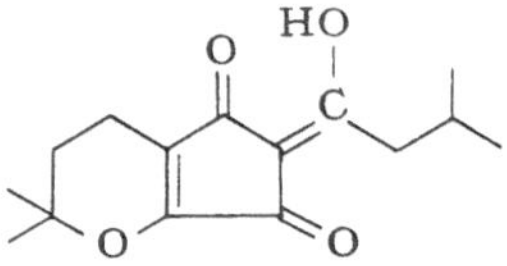

(15)

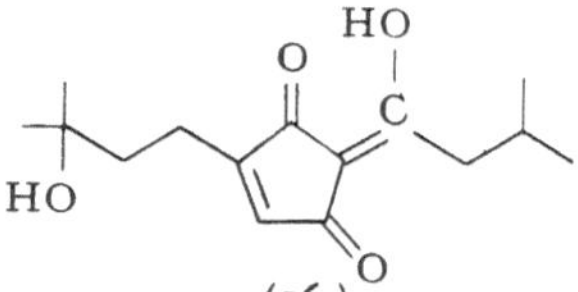

(16)
Humulinic acid D.

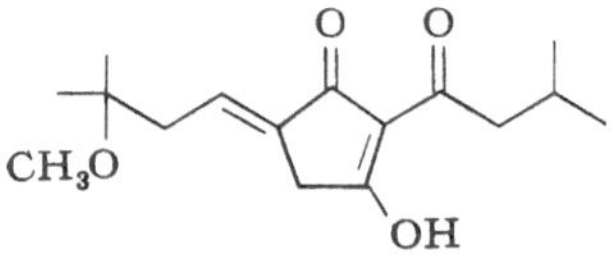

(17)
Methylallohumulinic acid D.

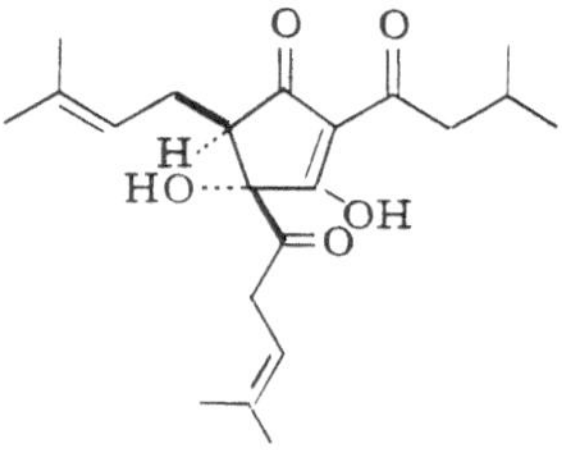

(18)
Isohumulone A (cis at $C_{(4)} - C_{(5)}$).

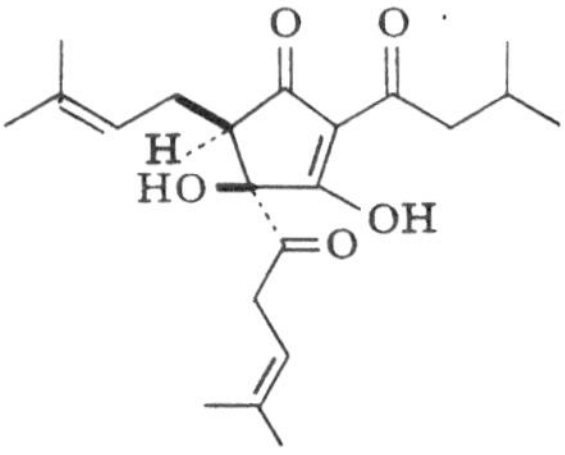

(19)
Isohumulone B (trans at $C_{(4)} - C_{(5)}$).

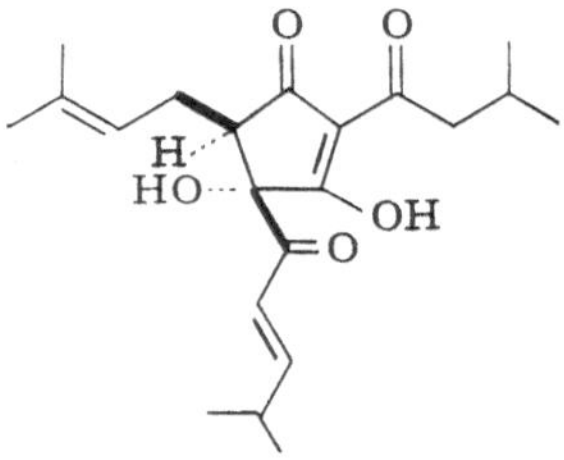

(20)
alloisoHumulone A.

controlled conditions. From the alkaline isomerization product of humulone they obtained evidence for the presence of two compounds, and from the iso-α-acids of beer, evidence for the presence of six compounds, presumably two each from humulone, cohumulone and adhumulone. WHITEAR and HUDSON (87) examined the distribution pattern obtained by counter-current distribution of the alkaline isomerization product of humulone and found that it would not fit the curve predicted for the

References, pp. 85—89.

distribution of a pure compound but did correspond with a theoretical distribution pattern for two similar compounds present in almost equal quantities.

Similar conclusions on the nature of the alkaline isomerization product of humulone were reached by BURTON, ELVIDGE and STEVENS (23), who examined the proton magnetic resonance spectrum of isohumulone and assigned structures to the component compounds on the basis of the previous work with humulinic acids. The two isohumulones present in the mixture were found to possess alternative conformations at $C_{(4)}$ and $C_{(5)}$. Thus the structures were those shown in (18) where the isopentenyl group at $C_{(5)}$ and the isohexenoyl group at $C_{(4)}$ are in *cis* disposition (i. e., related to humulinic acid A), and (19) where these groups are in *trans* disposition (related to humulinic acid B).

Using reversed-phase partition chromatography, SPETSIG (74) achieved the first separation of the two isohumulones. One isomer was found to be a crystalline solid, m. p. 62°, $[\alpha]_D$ — 6°, the other a pale yellow oil, $[\alpha]_D$ + 44° (in ethanol).

Indications had been obtained (28, 41) that humulone was isomerized when irradiated under the appropriate conditions, and in a detailed publication, CLARKE and HILDEBRAND (29) showed that a crystalline, bitter isohumulone was obtained as a major product when humulone was irradiated with light from a tungsten filament source. Although the authors suggested that the compound was a constituent of the bitter substances in beer, they did not make any suggestions as to its structure. In the same publication a successful separation, by chromatography on silicic acid, of the alkaline isomerization product of humulone into the two fractions described by SPETSIG, was reported. The "photo-iso-humulone" was then found to possess similar properties to that of the crystalline compound obtained from the alkaline isomerization of humulone, and it was concluded that the two were identical.

A detailed examination of the humulone isomerization product obtained by boiling the α-acid in a buffer solution was made by ALDER-WEIRELDT and co-workers (2), using counter-current distribution. Evidence was obtained for the presence of four isomers of humulone. In addition to the two isohumulones described previously, the crystalline compound named isohumulone A *(cis)* and the oily isomer isohumulone B *(trans)*, two further isomers were found as minor constituents. These were called *alloiso*humulones A and B (the letters designating the same stereochemical relationship as the "parent" isohumulones) and differed in position of the double bond in the isohexenoyl chain at $C_{(4)}$ and possessed structures (20) and (21), respectively.

The possibility of using non-aqueous alkali to achieve the isomerization of humulone was first studied by CARSON (27) who found the product, obtained by treating humulone with alcoholic alkali, to contain three crystalline fractions which were not bitter and further, two oily fractions both of which were bitter. Attempts to obtain pure compounds from the crystalline fractions were unsuccessful, and although all products were isomeric with humulone, no structural formulation could be made. All the fractions, however, afforded humulinic acid A (10, p. 68) on hydrolysis; and on hydrogenation, dihydro derivatives were obtained in contrast to the normal tetrahydro derivatives of isohumulones. The dihydro derivatives afforded dihydrohumulinic acid A on hydrolysis showing that the isohexenoyl group at $C_{(4)}$ rather than the isopentenyl group at $C_{(5)}$ had been cyclised. A further study (46) of this reaction, using countercurrent distribution, showed that the fractions were complex mixtures, from which no pure compounds could be isolated.

Examination of the crystalline products from the isomerization of humulone with alcoholic alkali was also made by Ashurst (7) and whilst he was unable to separate the components of the major crystalline fraction, one of the minor crystalline fractions afforded an apparently pure compound which proton magnetic resonance studies (9) indicated to contain a bicyclic isomer of isohumulone (22).

Howard and co-workers (46) studied the effects of potassium-t-butoxide in t-butanol and found that its use at room temperature would give the normal alkaline isomerization product. The effect of alcoholic alkoxides on α-acids in general, however, has not been studied.

Synthesis of Isohumulone.

The importance of substituted cyclopentane triones and their derivatives as degradation and transformation products of hop resins has prompted many synthetic studies on such compounds with the direct synthesis of isohumulone being an ultimate aim.

Harris, Howard and Pollock (39) synthesized a degradation product of isohumulinic acid (23) which itself was a degradation product of humuloquinone (24). Humuloquinone was obtained by hydrogenolysis of humulone to the quinol, followed by oxidation. Two equivalents of ethyl oxalate were condensed with one of 6-methylheptan-2-one in sodium ethoxide to give 3-ethoxalyl-5-(3-methylbutyl)cyclopentane-1,2,4-trione which was hydrolysed with aqueous methanolic hydrogen chloride to the required 3-(3-methylbutyl)-cyclopentane-1,2,4-trione (25) identical with the substance isolated by hydrolysis of isohumulinic acid (cf. 34).

The acylation of substituted cyclopentane triones and in particular the synthesis of isohumulinic acid from (25) and isovaleric anhydride with boron trifluoride was reported by Leucht and Riedl (57).

VANDEWALLE and co-workers (79), in an attempt to prepare acylcyclopentane triones, only prepared new compounds so that evidence for the structures rested entirely on spectroscopic data. These authors (80) later indicated, however, that they had misinterpreted the original data and had prepared the isomeric pyrone carboxylic acids rather than the required acylcyclopentane triones. ELVIDGE and STEVENS (35) also arrived at these conclusions and showed that acylcyclopentane triones could be prepared by the method of VANDEWALLE (79).

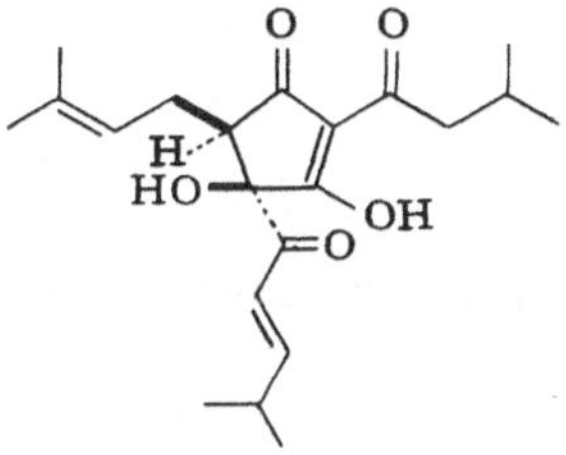

(21)
*alloiso*Humulone B.

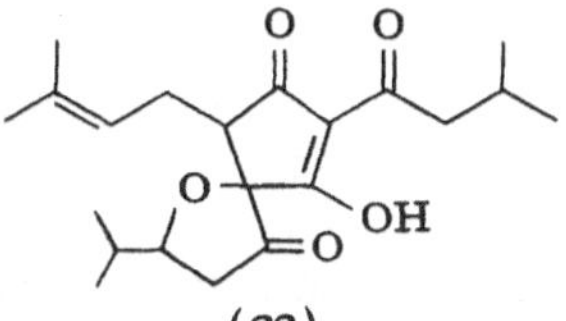

(22)

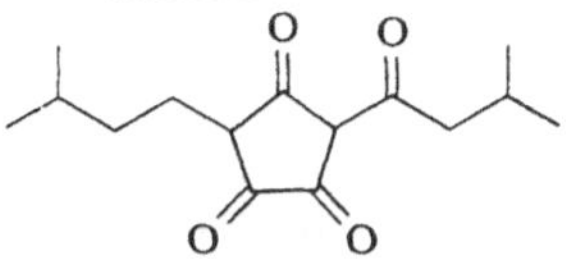

(23)
Isohumulinic acid.

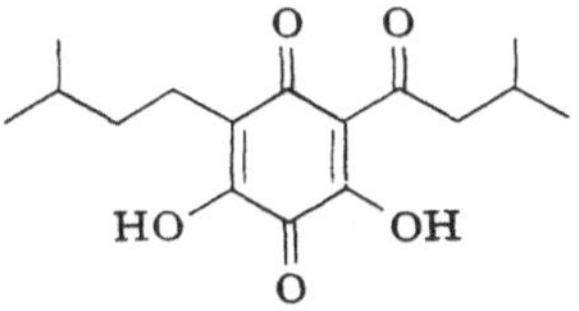

(24)
Humuloquinone.

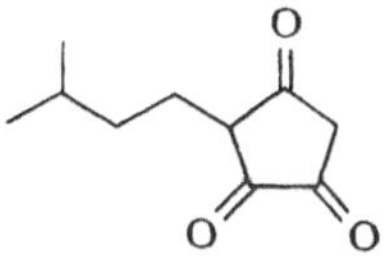

(25)
3-(3-Methylbutyl)cyclopentane - 1,2,4 -trione.

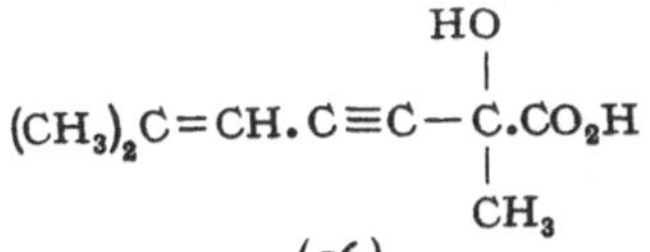

$(CH_3)_2C=CH.C\equiv C-\underset{\underset{CH_3}{|}}{\overset{\overset{HO}{|}}{C}}.CO_2H$

(26)
2,6 - Dimethyl - 2 -hydroxyhept - 3 - yn - 5 -
enoic acid.

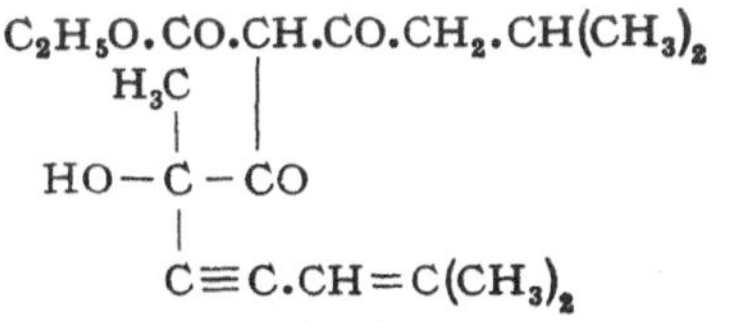

$C_2H_5O.CO.CH.CO.CH_2.CH(CH_3)_2$

$HO-\underset{\underset{C\equiv C.CH=C(CH_3)_2}{|}}{\overset{\overset{H_3C}{|}}{C}}-CO$

(27)

Syntheses of acylcyclopentane-1,3-diones have also been reported. During an unsuccessful attempt to prepare cyclopentane-1,3-dione from succinoyl chloride, vinyl acetate, and aluminium chloride, SIEGLITZ and HORN (73) isolated a small amount of 2-acetylcyclopentane-1,3-dione, whilst MERÉNYI and NILSSON (59) prepared the latter compound from succinic anhydride, aluminium chloride and isopropenyl acetate. A further synthesis of acylcyclopentane diones was reported (78) and 2-acylcyclopentane-1,3-diones together with 2-acyl-4-carbethoxycyclopentane-1,3-diones were obtained by a Dieckmann condensation of 1,4-dicarbo-ethoxyhexane-3,5-diones.

The direct synthesis of ($\pm$)-isohumulone was achieved by Ashurst and Laws (*10*) who employed a novel route for the synthesis of cyclo-pentane dione derivatives. A key intermediate in the synthesis, 2-methyl-pent-2-en-4-yne, was obtained with considerable difficulty. Attempts to dehydrate 2-methylpent-4-yn-2-ol or 2-methylpent-4-yn-3-ol either directly, via a hydrolysis of the *p*-tosyl derivatives, or via a dehydro-bromination of the corresponding bromo derivatives, were all unsuccessful. A novel 1,4-elimination reaction of 1-bromoallenes with cuprous cyanide reported by Laws (*52*) afforded the corresponding conjugated enynes; thus, 2-methylpent-2-en-4-yne was obtained in about 50% yield from 1-bromo-4-methylpenta-1,2-diene. Addition of the hydrocarbon to ethyl pyruvate followed by hydrolysis afforded 2,6-dimethyl-2-hydroxy-hept-3-yn-5-enoic acid (**26**), the acid chloride of which was added to the β-ketoester ethyl 3-oxo-5-methylhexanoate in the presence of magnesium ethoxide. The product of this reaction (**27**) underwent cyclisation under strongly basic conditions (potassium-t-butoxide in t-butanol) to the compound (**28**) due presumably to the electron withdrawing effect of adjacent groups. Then alkenylation with 1-bromo-3-methylbut-2-ene in the presence of sodium ethoxide to give the isohumulone analogue (**29**) was followed by hydration of the triple bond to give ($\pm$)-isohumulone (**30**) in low overall yield.

The Acid Degradation of Humulone.

Although as will be shown, the acid degradation of the β-acids has been well documented, little is known of the acid degradation of humulone.

Verzele and Govaert (*82*) noted that a solution of humulone in benzene was decomposed slowly by silica gel, and Howard and co-workers (*46*) found that mineral acids decomposed humulone rapidly. Neither group was able to characterize any products from the mixture. A re-examination (*7*) of the effect of acids on humulone showed that two major products were obtained, only one of which was characterized. The effect of acids at higher temperatures gave a predominance of a compound possessing a gross structure represented in (**31**) together with the release of isoprene. The other compound obtained defied attempts at characterization. A study of the proton magnetic resonance spectrum of the characterized acid degradation product showed (*9*) that it contained a substantial proportion of the alternative tautomeric form (**32**).

The Oxidation of α-Acids.

Humulinones. It has long been recognised (*16*) that during the storage of hops the resins, and in particular the α-acids, undergo oxidation, although more recent studies (*86*) have shown that the actual "brewing value" of the hop decreases little even on prolonged storage. There has, nevertheless, been considerable interest in the oxidation products of the α-acids, and the first definite compound isolated as such, humulinone, was obtained in 1950 by COOK and HARRIS (31). Conflicting evidence as to whether the structure was as shown in (33) or (34) was only finally resolved when the proton magnetic resonance spectrum (72) showed that humulinone possessed the five-membered ring structure (34).

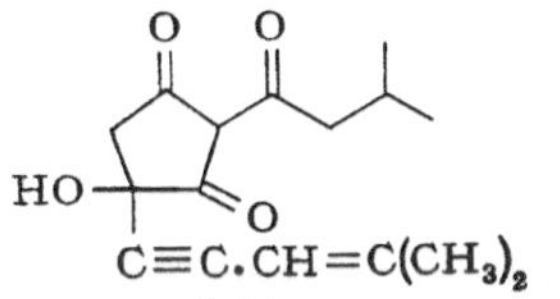

$C \equiv C.CH = C(CH_3)_2$
(28)

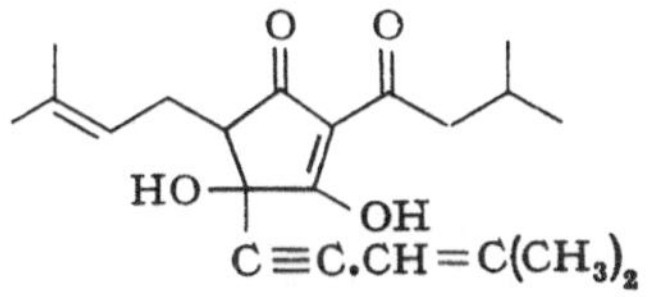

$C \equiv C.CH = C(CH_3)_2$
(29)

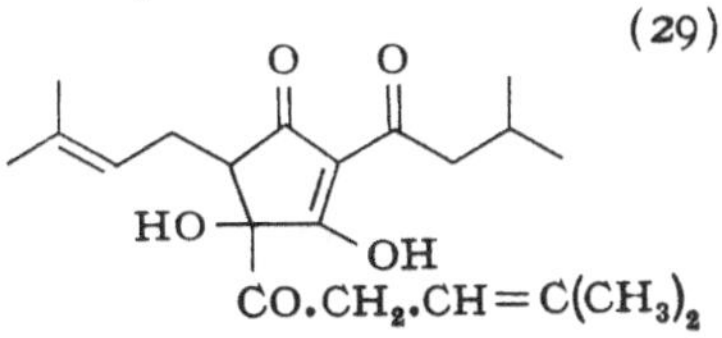

$CO.CH_2.CH = C(CH_3)_2$
(30)
(±)-Isohumulone.

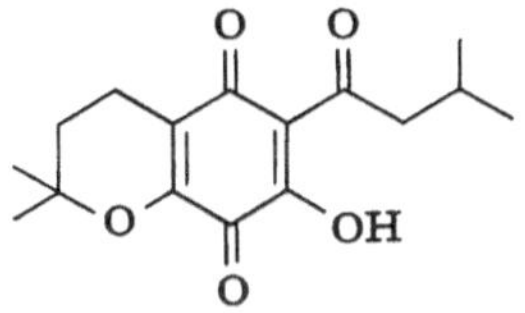

(31)

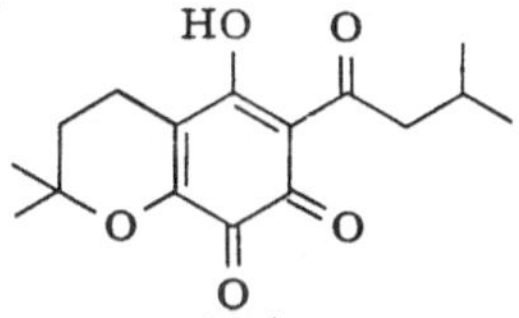

(32)

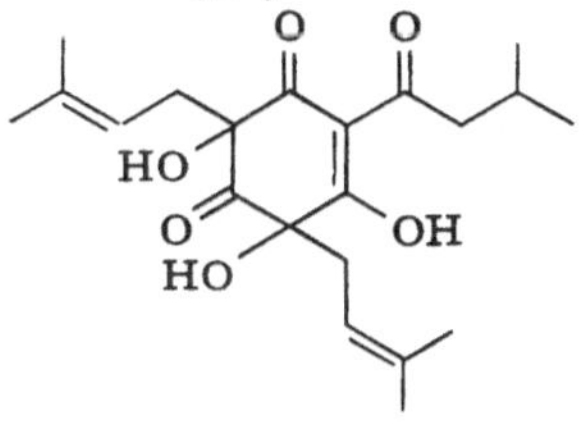

(33)

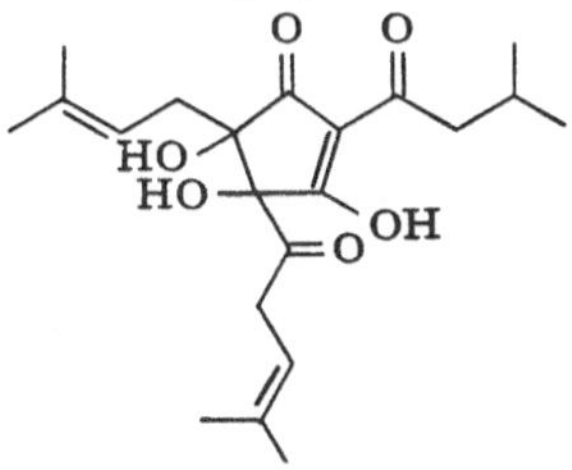

(34)
Humulinone.

Humulinone was very bitter and was thought for some time to be an important autoxidation product of humulone. It has later been found, however, that for humulinone to be produced, the oxidation must be carried out at pH 8 or above and it seems very unlikely that it is produced in hops. Further, the autoxidation products of the α-acids are largely insoluble in light petroleum (i. e., hard resins), in contrast to humulinone.

Humulinone is a useful model compound for possible reactions of isohumulone, since it is crystalline and structurally similar. Humulinone is optically inactive and is said to be produced by a *cis* oxidation of humulone and a stereospecific ring contraction mechanism (2) to give only a single compound and not a mixtures of conformers which occurs with isohumulones. Thus, using the nomenclature for the iso-α-acids and humulinic acid, the compound could be called humulinone A.

Humulinone is readily isomerized under acidic conditions. The attempted separation of cohumulinone and humulinone by chromatography on silica gel led to the discovery (32) of isohumulinone, and a re-examination of the chemistry of the latter compound led ASHURST and WHITEAR (13) to postulate its structure as (35). Attempts to hydrolyse isohumulinone, which was called isohumulinone A by the latter authors, produced a second isomeric compound which was thought to possess the tricyclic structure (36) and was named isohumulinone B.

The amount of isohumulinone B formed when isohumulinone A was heated in aqueous sodium hydroxide may represent the equilibrium mixture between the form possessing hydroxyls at $C_{(4)}$ and $C_{(5)}$ in a *cis* conformation and the form possessing the groups in a *trans* conformation, although there are no data available to confirm this feature.

The use of alkaline hydrogen peroxide under more drastic conditions than those used to prepare humulinone has been employed (49) as a means of estimating the proportions of various α- or β-acids in a mixture. Under the conditions used, the acyl side-chains are removed and the proportions of the various acids, isovaleric, isobutyric, etc., are estimated by gas-liquid chromatography.

Autoxidation*. A study of the effect of passing oxygen through solutions of humulone and colupulone in light petroleum was made by ASHURST and WHITEAR (14) who found that humulone was rapidly oxidized in about 95% yield to a crystalline product insoluble in light petroleum. Colupulone, on the other hand, afforded only a small amount of product which was insoluble in light petroleum. A subsequent examination of the petroleum insoluble oxidation product of humulone (8) showed that the relatively mild conditions used had brought about, not only the loss of a five-carbon side-chain and ring contraction, but also the oxidation of the remaining olefinic bond to a glycol, the compound possessing structure (37). Such an oxidation in high yield under mild conditions is very unusual. Extraction of old hops with the appropriate solvents afforded the compound (37), as a normal oxidation product of humulone. It would, however, be wrong to assume that this compound is the sole autoxidation product, since a study (11) of the total oxidation

---

* The definition of autoxidation as used here is that given by W. A. WATERS, in The Chemistry of Free Radicals, Oxford University Press, as "Oxidations which can be brought about by oxygen gas at normal temperatures without the intervention of a visible flame or an electric spark".

References, pp. 85—89.

products, using ion-exchange chromatography with spherical bead resin under controlled conditions, showed the mixture to be complex, with compound (37) as one product.

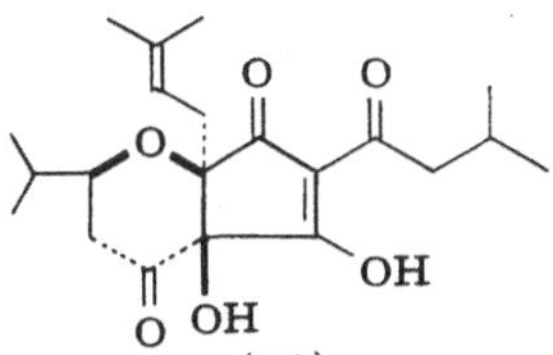

(35)
Isohumulinone A.

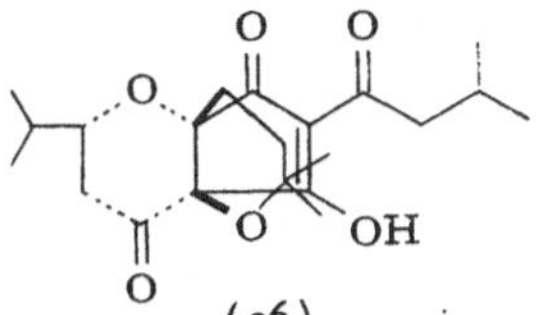

(36)
Isohumulinone B.

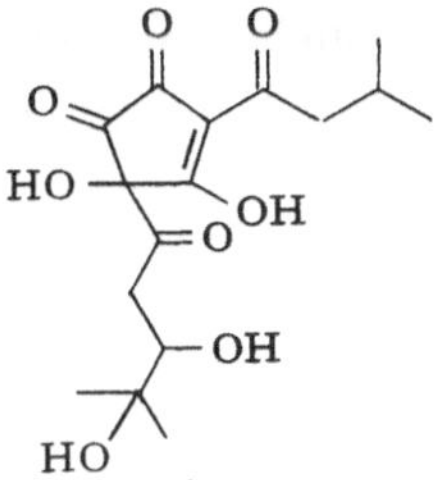

(37)

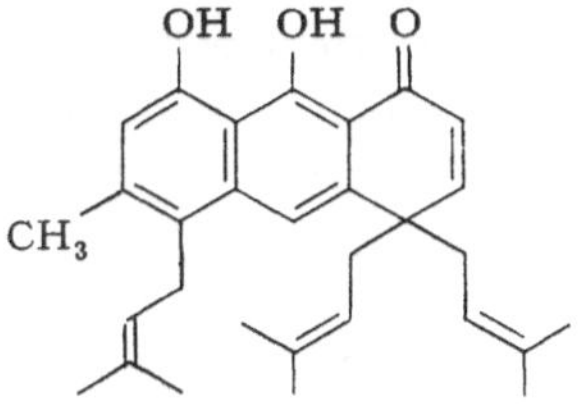

(38)
Harunganin.

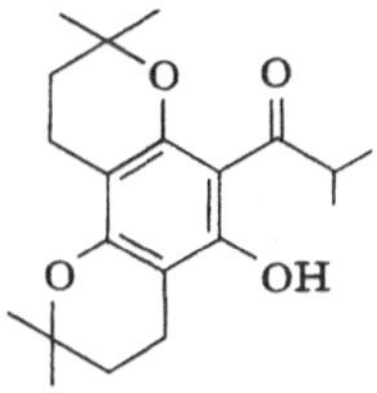

(39)

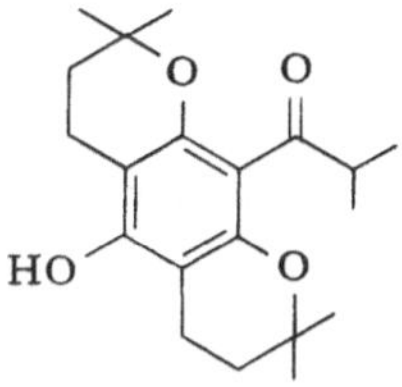

(40)

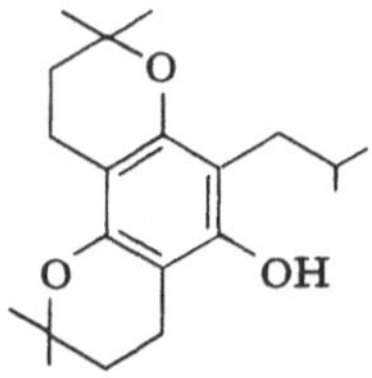

(41)

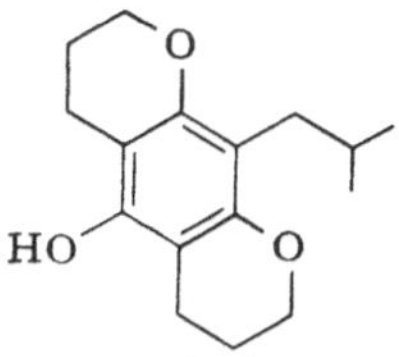

(42)

IV. The Chemistry of the β-Acids.

Early workers with hop resins who had isolated both lupulone and humulone observed that whereas the α-acids were changed by the action of alkali, the β-acids could be recovered unchanged. The β-acids, which of themselves possessed little bitterness, were not thought to contribute much to the bitter compounds of beer.

· It is noteworthy that since elucidation of the structure of the lupulones (4) which showed that the compounds possessed, among other features, gem-substituted dimethyl allyl side-chains, only one compound, harunganin (38) showing this particular feature has been reported (69).

The Acid Degradation of Colupulone.

The effect of mineral acids on colupulone was first studied by Walker (85) who noted the loss of a hydrocarbon, which he believed to be a methylbutene, and recovered two isomeric crystalline compounds, one acidic, the other neutral. Howard, Pollock and Tatchell (44) were able to formulate the products as the tricyclic compounds (39) and (40), the linear dipyran (40) being acidic and the other compound neutral. Their structures were confirmed by synthesis.

The close relationship of the two isomers was emphasized by the ready conversion of (39) to (40) by the action of warm sulphuric acid, the reversal occurring with hydrochloric acid; and further confirmation of the structural relationship was obtained by a study of their ultraviolet spectra which had analogies in the rottlerin series (60). These isomerizations are strictly analogous to those described (42) for the isomeric α- and β-lapachones. The acidic compound (40) was further characterized (6) by reduction under the conditions of the Clemmensen reaction to the derivative (42), whereas the neutral compound (38) only underwent the Clemmensen reduction to the corresponding compound (41) with difficulty.

The Oxidation of β-Acids.

Although it has long been realised that β-acids are susceptible to autoxidation both in the hop and in an isolated state, few attempts to examine any of the products have been made until the past ten years.

Lupuloxinic Acid. During attempts to establish the exact nature of the acyl side-chains of various β-acids by oxidation with alkaline hydrogen peroxide, Howard and Pollock (43) isolated an oxidation product which they called lupuloxinic acid, formulated as (43). Lupuloxinic acid was found (44) to undergo decarboxylation to lupulenol (44); and hydrogenation of lupulenol to the hexahydro derivative, followed by oxidation with bismuth oxide gave 3,3,5-(3-methylbutyl)cyclopentane-1,2,4-trione (45).

Under different conditions, however, alkaline hydrogen peroxide produced different products from colupulone. Thus, if a solution of colupulone in warm reagent was allowed to cool very slowly, subsequent heating to 100° afforded a neutral oil formulated as 7-hydroxy-2,11-dimethyl-8-(3-methylbut-2-enyl)dodeca-2,10-dien-6-one (46), whereas the

same reagent used solely at a temperature of 100° gave rise to hydro-
lytic fission of colupulone affording isobutyric acid and 2,9-dimethyl-
dec-8-en-3,5-dione (47).

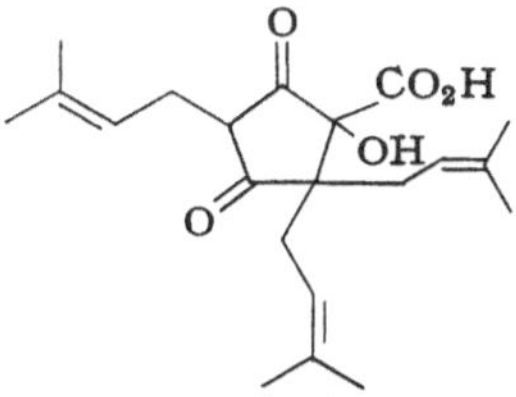

(43)

Lupuloxinic acid.

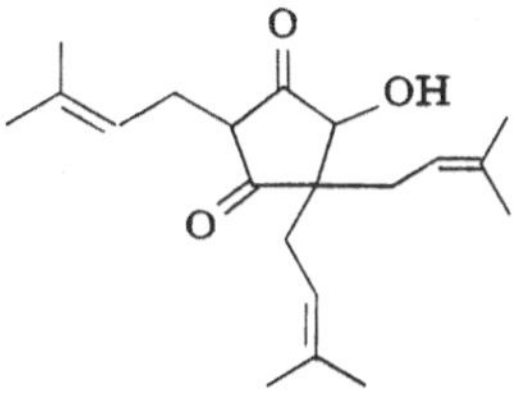

(44)

Lupulenol.

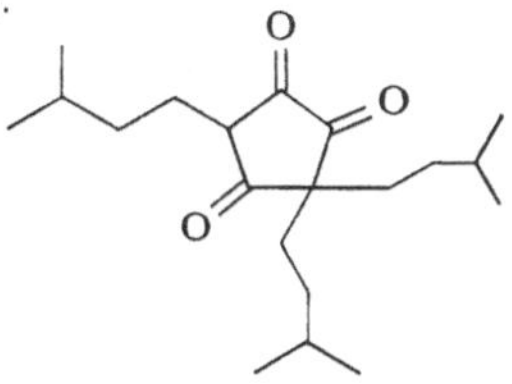

(45)

3,3,5 − (3 − Methylbutyl)cyclopentane −
1,2,4 − trione.

$(CH_3)_2C=CH.CH_2.CH_2$ ⌐CO⌐ CH.OH
|
CH.CH_2.CH:C(CH_3)_2
|
CH_2CH:C(CH_3)_2

(46)

7 − Hydroxy − 2,11−dimethyl− 8 −
(3 − methylbut−2−enyl) dodeca −
2,10 − dien − 6 − one.

$(CH_3)_2C:CH.(CH_2)_2.CO.CH_2.CO.CH(CH_3)_2$

(47)

2,9 −Dimethyl−dec−8−en−3,5−dione.

Autoxidation. The first examination of a naturally occurring autoxidation product of colupulone was made by SALEH (70) who studied a
fraction of the hard resin of old hops and found it to be a bacteriostatic
substance. From chemical analyses it was postulated to have structure
(48) but no confirmation of this structures was made.

Hulupones. The application of reversed-phase partition chromatography to the soft resin fraction of hops yielded a group of compounds
called hulupones (75), present in hops at about 0.1% dry weight, for
which the general structure (49) was proposed (17). These compounds
were considered to have arisen by autoxidation of the lupulones.

A study of the relationship between the lupulones and hulupones was made by STEVENS and WRIGHT (76) and in order to simplify the overall picture of this oxidation the initial experiments were made with

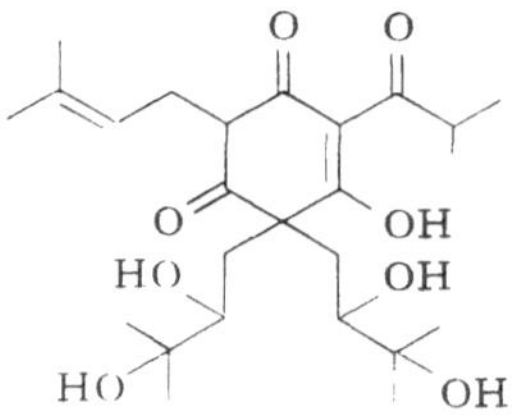

(48)

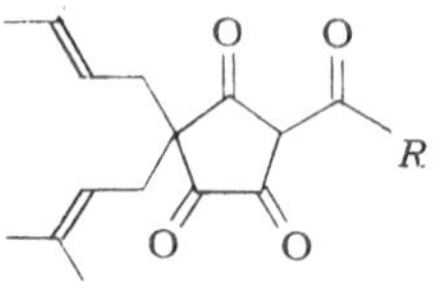

(49)
Hulupones.

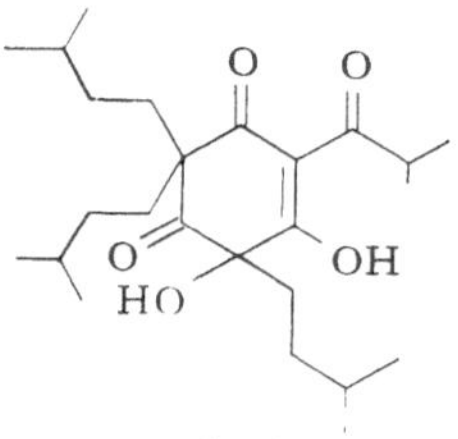

(50)
6-(3-Methylbutyl) tetrahydrocohumulone.

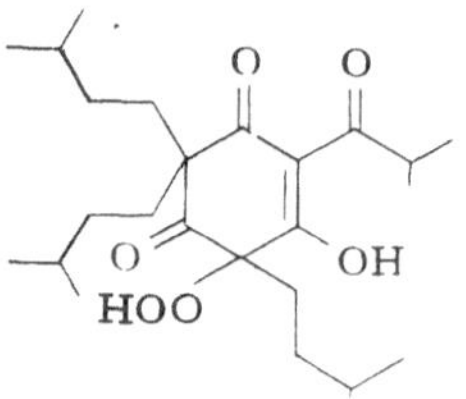

(51)

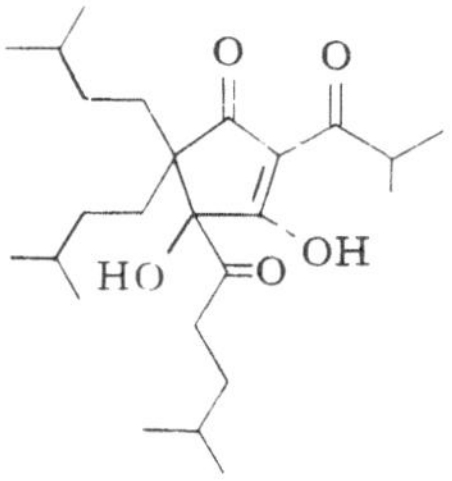

(52)

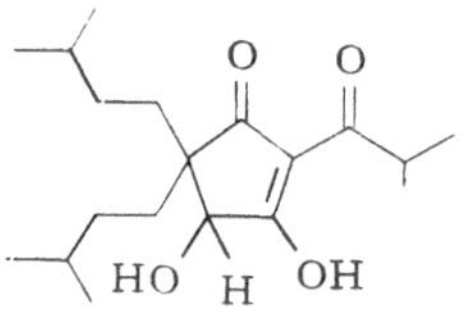

(53)
5-(3-Methylbutyl)dihydrocohumulinic acid.

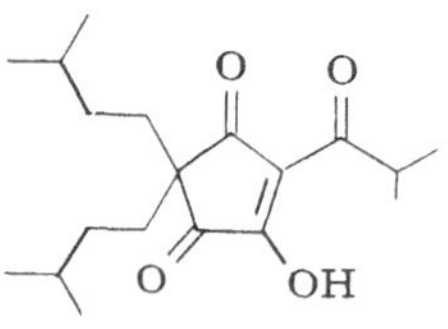

(54)
Tetrahydrocohulupone.

hexahydrocolupulone. The first product of this oxidation to be isolated was a hydroperoxide which on subsequent hydrogenation afforded 6-(3-methylbutyl)tetrahydrocohumulone (50) indicating that the hydroperoxide possessed structure (51). Although the cohumulone analogue (50) failed to give an insoluble lead salt, it was readily isomerized under alkaline conditions to the isocohumulone analogue (52) which was isolated via the sodium salt.

References, pp. 85—89.

On prolonged alkaline hydrolysis the isocohumulone analogue (52) afforded 5-(3-methylbutyl)dihydrocohumulinic acid (53) which in turn was oxidized to tetrahydrocohulupone (54). Autoxidation of (52) afforded tetrahydrocohulupone directly in about 20% yield but the major product of this oxidation remained uncharacterized.

Attention was then turned to the oxidation of the β-acids themselves by WRIGHT (90). An attempt to obtain a hydroperoxide of colupulone was unsuccessful, as were attempts to prepare cohulupone by treatment of colupulone with cumene hydroperoxide using the conditions employed for obtaining humulinone from humulone. During the estimation of the iodine value of colupulone, which involved the use of sodium sulphite in methanol, the ultraviolet spectrum of the resin was found to have altered and become similar to that of cohulupone. This prompted a study of the effect of oxygen on colupulone in the presence of sodium sulphite, when it was found that cohulupone was produced in excellent yield. It was also noted that, in common with other hop resins, cohulupone was oxidized in air to a resin insoluble in light petroleum. A study of the fate of β-acids during the brewing process (77) indicated that hulupones were formed in low yield from lupulones and gave rise to a small amount of the total bitterness in beer.

A study of the behaviour of the hulupones under various conditions of hydrogenation was made by BURTON and STEVENS (24). It was found that whereas hydrogen in the presence of palladium catalyst suspended on barium sulphate gave rise to a hexahydro derivative, 5-(3-methyl-butyl)dihydrocohumulinic acid (53), further hydrogenation with the same catalyst and solvent afforded an unexpected product (55) where the acyl side-chain had been reduced to alkyl. Compound (55) was obtained as the major product when cohulupone was reduced catalytically using Adams catalyst in methanol.

Similar abnormal reductions have been observed with callophyllolide and 5,7-dihydroxy-8-isopentenyl-6-isovaleryl-4-phenyl-coumarin (61).

There has been speculation (51) that in addition to the hulupones, 5-isopentenyl humulinic acids (56) are also obtained as normal oxidation products of lupulones, but there seems little evidence to substantiate this claim.

In an attempt to establish the chemical nature of the oxidation products of the hulupones, BURTON and co-workers (25) subjected cohulupone to autoxidation in refluxing ethanol and obtained a crystalline product which they called hulupinic acid and formulated as (57). Samples of old hops were examined for the presence of hulupinic acid and from many of them very small quantities were isolated. Hulupinic acid was thus considered to be a normal oxidation product of the hulupones. The only other natural product which has been isolated and which possesses a chromophore similar to that of hulupinic acid is linderone (50) (58).

Following the work of Ashurst and co-workers (8, 14) on the autoxidation of α-acids and on certain aspects of the β-acids, a re-examination of the pattern of the autoxidation of β-acids has been made recently (11). Using a system of chromatography on ion-exchange resin under controlled conditions based on the method of Howard and Slater (45), the reaction products of various fractions from an autoxidation of β-acids were examined. The conditions were considered to be essentially similar to those prevailing during the normal storage of hops. It was found that most of the oxidation products of the non-bitter colupulone were bitter materials which were soluble in light petroleum. Cohulupone was present in substantial quantity. No evidence for the presence of hulupinic acid as a normal autoxidation product of β-acids was found.

The overall pattern which is emerging regarding the oxidation of hop resins, a feature of very considerable importance when considering the economic aspects of hop usage, is that the bitter α-acids are autoxidized to less bitter products, whereas the bitter iso-α-acids become oxidized to non bitter products. The β-acids which themselves have little if any bitterness are, however, autoxidized mainly to bitter, light petroleum soluble substances of which hulupones account for about 20%. The other substances are as yet uncharacterized. The small proportion of light petroleum insoluble material which has been obtained from the autoxidation of colupulone is a well defined crystalline solid. Although the detailed structure of this substance has yet to be elucidated, it has been established that at least one of the dimethyl allyl side-chains has been oxidized to give a glycol system. The autoxidation of isohumulone leads to a non-bitter product and for obvious economic reasons is best avoided in the brewing process. There may, however, be economic advantages in specifically oxidizing the β-acid fraction from hops to increase the available total bitterness.

V. Tetra-(3-methylbut-2-enyl) Derivatives of Phloracylphenones.

As has been shown, the humulones are disubstituted and the lupulones trisubstituted derivatives of phloracylphenones. Tetraisoprenylated phloracylphenones were first reported by Riedl (63) who attempted to synthesize tetraisoprenyl-phlorisovalerophenone (59). Later investigations (19) on the steam-volatile fraction of hops, mainly essential oils, indicated that tetraisoprenylated phloracylphenones might be normal constituents of fresh hops. The synthesis of these compounds was accomplished by Laws (53) who termed them lupones and showed that the material obtained by Riedl was in fact lupulone. The lupones were characterized as crystalline solids with low melting points and also as their crystalline amino derivatives which were shown to possess structure (60) (54). A detailed examination of both fresh and old hops failed, however, to reveal their presence.

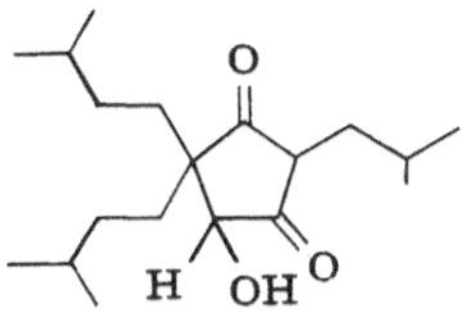

(55)

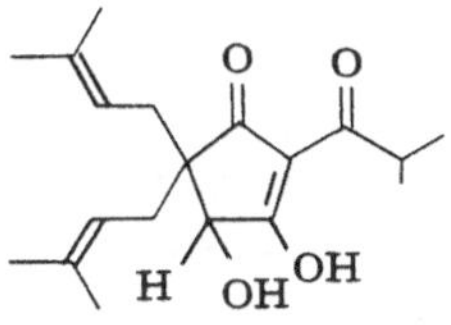

(56)

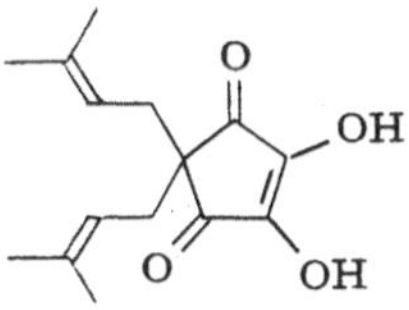

(57)
Hulupinic acid.

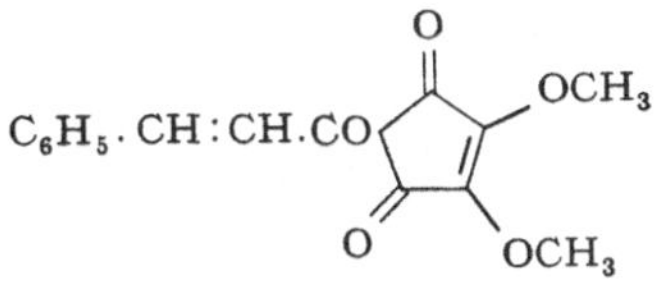

$C_6H_5 \cdot CH:CH.CO$

(58)
Linderone.

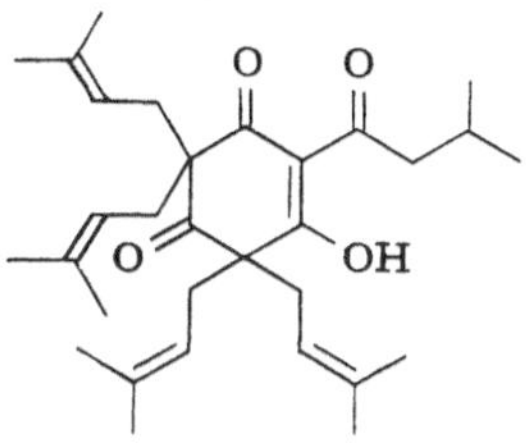

(59)
Tetraisoprenyl–phlorisovalerophenone.

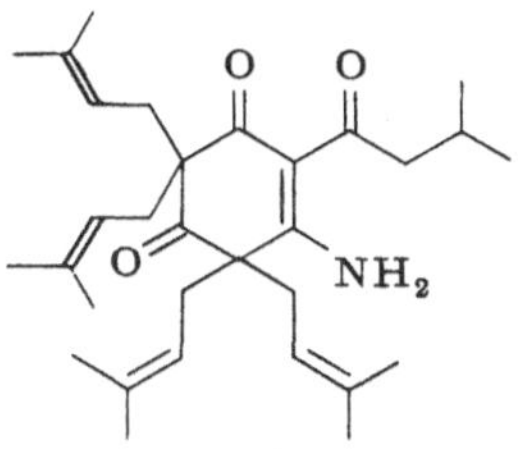

(60)
Lupones (aminoderivatives).

The most striking chemical difference between the lupones and the lupulones is their behaviour under conditions of hydrogenation. Whereas the lupulones undergo hydrogenolysis in the presence of palladium catalyst with the removal of one of the gem-substituted isopentenyl groups, lupones are simply hydrogenated to octahydrolupones.

VI. Naturally Occurring Hard Resin Constituents.

The hard resin, i. e., petroleum insoluble, fraction of hops may be arbitrarily divided into the native hard resin, i. e. that which is present in fresh hops, and the hard resin which occurs as a result of oxidation of soft resin components by storage of the hops. The nature of some of the hard resin produced by autoxidation of soft resins has been discussed under the oxidation of the various soft resin fractions (p. 64), although it should be emphasized that the complex nature of the oxidative hard resin is not fully understood.

6*

An examination of the naturally occurring hard resin components was made by ASHURST, LAWS and STEVENS (*12*) who freeze-dried hops before examining this fraction. The principal component proved to be the chalkone xanthohumol. The presence of an orange coloured material in hops which had properties similar to those of quercitin was first noted by WAGNER (*84*) although the material was not isolated until some years later by SEYFFERT (*71*). Larger quantities were isolated by POWER, TUTIN and ROGERSON (*62*) who called the compound xanthohumol.

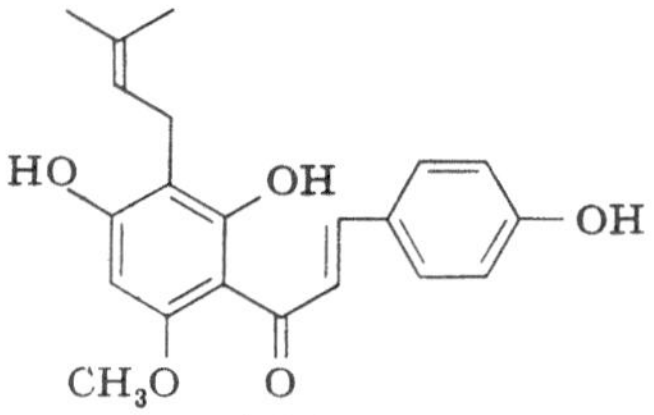

(61)
Xanthohumol.

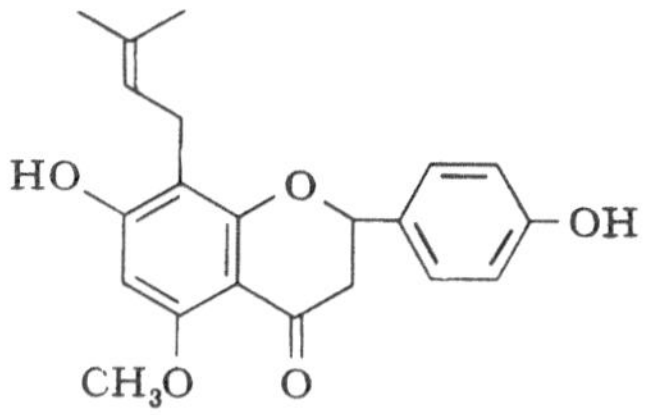

(62)
Isoxanthohumol.
(humulol)

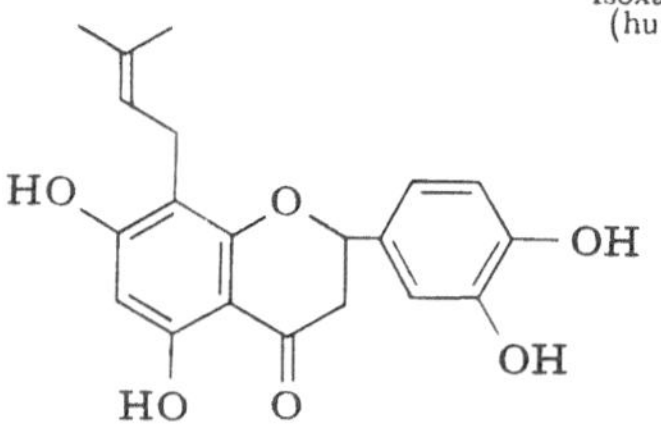

(63)

A related compound which they called humulol was also isolated. A detailed study of these compounds was made by VERZELE and co-workers (*83*) who found that xanthohumol was a chalkone with structure (61) and humulol, which they called isoxanthohumol, was the isomeric flavanone (62). On treatment of xanthohumol with dilute aqueous alkali an equilibrium mixture was obtained which consisted of 95% isoxanthohumol with 5% of xanthohumol. In the presence of acids the equilibrium was re-established in favour of the chalkone.

Since the earlier extraction techniques used by POWER and co-workers to obtain humulol (isoxanthohumol) had employed treatments with alkali, it was concluded by VERZELE and co-workers that this compound was probably an artefact. Xanthohumol itself was present in hops at approximately 0.25% of the dry weight but apparently had no brewing value.

In a re-examination of the xanthohumol content of hops, it was found in this laboratory that in green or unkilned hops which had been freeze-dried, xanthohumol was present to almost 1% of the dry weight. The compound was, however, largely lost on kilning and no corresponding degradation products could be found. The isolation of racemic iso-xanthohumol from freeze-dried and kiln-dried hops by a method which did not involve treatment with alkali was also described, suggesting that the compound may occur naturally in hops.

Considerable evidence for the presence of other chalkones in hops similar in structure to xanthohumol was obtained by ASHURST, LAWS and STEVENS (12), but no structural formulations were made. A tentative structure (63) for another flavanone isolated from hops and believed to be isomeric with one of these other chalkones was described by ASHURST (7).

References.

1. ALDERWEIRELDT, F. and M. ANTEUNIS: The Humulinic Acids. III. The Nuclear Magnetic Resonance Spectra and the Structure of Humulinic Acid A and Humulinic Acid B. Bull. soc. chim. Belges 73, 285 (1964).
2. ALDERWEIRELDT, F., M. VERZELE, M. ANTEUNIS and J. DIERCKENS: The Isomerization Products of Humulone. Bull. soc. chim. Belges 74, 29 (1965).
3. ANTEUNIS, M., M. BRACKE, F. ALDERWEIRELDT and M. VERZELE: The Humulinic Acids. IV. Humulinic Acids C, D. Bull. soc. chim. Belges 73, 910 (1964).
4. ANTEUNIS, M., M. BRACKE, A. LEPOIVRE, M. VERZELE and F. ALDERWEIRELDT: The Humulinic Acids. V. Specific Catalytic Reductions. Methyl alloHumulinic Acid D. Bull. soc. chim. Belges 74, 629 (1965).
5. ANTEUNIS, M., M. BRACKE, M. VERZELE and F. ALDERWEIRELDT: The Humulinic Acids. Part I. Isolation and Preparation of Humulinic Acid Isomers. Bull. soc. chim. Belges 71, 623 (1962).
6. ARNOLD, B. H., J. J. H. HASTINGS and T. K. WALKER: Degradation of Colupulone by Methanolic Hydrochloric Acid. Chem. and Ind. 1955, 323.
7. ASHURST, P. R.: New Aspects of the Chemistry of Hop Resins. Ph. D. Thesis, London, 1966.
8. ASHURST, P. R. and J. A. ELVIDGE: Chemistry of Hop Constituents. Part XXVIII. The Structure of an Autoxidation Product of Humulone. J. Chem. Soc. (London) C 1966, 675.
9. — — Unpublished.
10. ASHURST, P. R. and D. R. J. LAWS: The Synthesis of (±)Isohumulone. J. Chem. Soc. (London) C 1966, 1615.
11. ASHURST, P. R., D. R. J. LAWS and M. A. PINNEGAR: Further Studies on the Oxidation of Hop Resins. J. Inst. Brewing 1966, 561.
12. ASHURST, P. R., D. R. J. LAWS and R. STEVENS: Chemistry of Hop Constituents. Part XXV. Xanthohumol Content. J. Inst. Brewing 1965, 492.
13. ASHURST, P. R. and A. L. WHITEAR: Chemistry of Hop Constituents. Part XXIV. The Structure of the Isohumulinones. J. Chem. Soc. (London) 1965, 1283.
14. — — Hop Resins and Beer Flavour. IV. Observations Concerning Hard Resin. J. Inst. Brewing 1965, 46.
15. BRENNER, M. W., C. VIGILANTE and J. L. OWADES: A Study of Hop Bitters (Isohumulones) in Beer. Proc. Annu. Meeting Amer. Soc. Brewing Chem. 1956, 48.

16. Briant, L. and C. S. Meacham: Some Remarks on Hops and their Storage. J. Inst. Brewing 1897, 481.
17. Brohult, S., R. Ryhage, L.-O. Spetsig and E. Stenhagen: Mass Spectrometric Studies of Hop Bitter Substances. Proc. European Brewery Conv., Rome 1959, 121.
18. Brown, P. M., G. A. Howard and A. R. Tatchell: Chemistry of Hop Constituents. Part XIII. The Hydrogenation of Isohumulone. J. Chem. Soc. (London) 1959, 545.
19. Brown, P. M. and D. R. Maule: Private communication.
20. Bungener, M. H.: Des principes amers du Houblon. Bull. soc. chim. Paris 45, 487 (1886).
21. Burton, J. S., J. A. Elvidge and R. Stevens: Chemistry of Hop Constituents. Part XX. The Constitution of Humulinic Acid A and B as Indicated by Proton Magnetic Resonance Spectroscopy. J. Chem. Soc. (London) 1964, 3816.
22. — — — Chemistry of Hop Constituents. Part XXIII. The Structure of Humulinic Acid C and other Acid-transformation Products of Humulinic Acids. J. Chem. Soc. (London) 1965, 1276.
23. — — — Chemistry of Hop Constituents. Part XXII. Proton Magnetic Resonance Spectroscopy of Isohumulone A. J. Inst. Brewing 1964, 345.
24. Burton, J. S. and R. Stevens: Chemistry of Hop Constituents. Part XVII. Hydrogenation of the Hulupones. J. Chem. Soc. (London) 1963, 4382.
25. Burton, J. S., R. Stevens and J. A. Elvidge: Chemistry of Hop Constituents. Part XVIII. Hulupinic Acid. J. Chem. Soc. (London) 1964, 952.
26. Carson, J. F.: The Structure of Humulone and Lupulone as Revealed by Ozonization. J. Amer. Chem. Soc. 73, 4652 (1951).
27. — The Alkaline Isomerization of Humulone. J. Amer. Chem. Soc. 74, 4615 (1952).
28. Clarke, B. J., F. V. Harold, R. P. Hildebrand and P. J. Murray: Trace Volatile Constituents of Beer. II. Hop Wax. J. Inst. Brewing 1961, 529.
29. Clarke, B. J. and R. P. Hildebrand: The Isomerization of Humulone. I. Isolation of Photoisohumulone. J. Inst. Brewing 1965, 26.
30. Cook, A. H.: Changing Outlooks on Hops. Proc. 8th Conv. Inst. Brewing, Austral. Sect., April 1964. Adelaide 1965, 113.
31. Cook, A. H. and G. Harris: The Chemistry of Hop Constituents. Part I. Humulinone, a New Constituent of Hops. J. Chem. Soc. (London) 1950, 1873.
32. Cook, A. H., G. A. Howard and C. A. Slater: Chemistry of Hop Constituents. IX. Isomerization of Humulinone and Cohumulinone. J. Inst. Brewing 1956, 220.
33. David, S. et J. Duchemin: Recherches sur les constituants du Houblon et leur transformation au brassage. Proc. European Brewery Conv. Baden-Baden 1955, 128.
34. Diels, O., J. Sielisch und E. Müller: Über Methyl-1-cyclopentantrion-(2.4.5). Ber. dtsch. chem. Ges. 39, 1328 (1906).
35. Elvidge, J. A. and R. Stevens: Comments on a Recent Synthesis and Observations from Proton Resonance on the Enolisation and Isomerism in 5-Methoxy-hept-4-en-3-one. J. Chem. Soc. (London) 1965, 2251.
36. Forsén, S., M. Nilsson, J. A. Elvidge, J. S. Burton and R. Stevens: Spectroscopic Studies on Enols. Part 6. Enolisation and Hydrogen Bonding in 3-Acyl-cyclopentane-1,2,4-triones. Acta Chem. Scand. 18, 513 (1964).
37. Govaert, F. et M. Verzele: Sur la transformation de l'humulone. Congrès intern. ind. ferment., Gand 1947, 297.
38. Hafner, K. und K. Goliasch: Zur Kenntnis des Cyclopentadienons. Chem. Ber. 94, 2909 (1961).

39. HARRIS, G., G. A. HOWARD and J. R. A. POLLOCK: The Chemistry of Hop Constituents. Part III. The Structures of Humulinic and Isohumulinic Acids. J. Chem. Soc. (London) 1952, 1906.

40. HAYDUCK, M.: Über die bitteren und harzigen Bestandtheile des Hopfens. Wochschr. Brau. 1888, 937.

41. HILDEBRAND, R. P.: The Isomerization of Humulone. Proc. 7th Conv. Inst. Brewing, Austral. Sect., Sept. 1962. Adelaide 1963, 157.

42. HOOKER, S. C.: The Constitution of Lapachic Acid (Lapachol) and its Derivatives. J. Chem. Soc. (London) 82, 611 (1892).

43. HOWARD, G. A. and J. R. A. POLLOCK: The Natural Occurrence of Lupulone Analogues. Chem. and Ind. 1954, 991.

44. HOWARD, G. A., J. R. A. POLLOCK and A. R. TATCHELL: The Chemistry of Hop Constituents. Part VII. Colupulone. J. Chem. Soc. (London) 1955, 174.

45. HOWARD, G. A. and C. A. SLATER: Ion Exchange Chromatography of Hop Resins and its Application to the Preparation of Extracts for Bittering Beer. J. Inst. Brewing 1960, 305.

46. HOWARD, G. A., C. A. SLATER and A. R. TATCHELL: Chemistry of Hop Constituents. XI. Some Observations on the Isomerization of Humulone. J. Inst. Brewing 1957, 237.

47. HOWARD, G. A. and A. R. TATCHELL: The Chemistry of Cohumulone. Chem. and Ind. 1953, 436.

48. — — Further Observations on the Occurrence and Isolation of Cohumulone. J. Inst. Brewing 1953, 491.

49. — — Evaluation of Hops: New Approach to the Detailed Analysis of Hop Resins. J. Inst. Brewing 1956, 20.

50. KIANG, A. K., H. H. LEE and K. Y. SIM: The Structures of Linderone and Methyl-linderone. Proc. Chem. Soc. (London) 1961, 455.

51. KUROIWA, Y. and H. HASHIMOTO: Luputriones — New Bitter Substance Present Merely in Beer. Report Res. Lab. Kirin Brewery Co. Ltd. 1963, 27.

52. LAWS, D. R. J.: The Synthesis and Reactivity of Allenic and Acetylenic Intermediates. Ph. D. Thesis, London 1964.

53. — Chemistry of Hop Constituents. Part XXVI. 2-Acyl-4,4,6,6-tetra(3-methylbut-2-enyl)cyclohexane-1,3,5-triones (Lupones). J. Chem. Soc. (London) 1965, 6542.

54. LAWS, D. R. J. and J. A. ELVIDGE: unpublished.

55. LEPOIVRE, A., F. ALDERWEIRELDT, M. ANTEUNIS and M. VERZELE: The Humulinic Acids. II. Hydrogenation Products. Chemical Evidence for the Structure of Humulinic Acids A and B. Bull. soc. chim. Belges 73, 275 (1964).

56. LERMER, J. C.: Die Bitterstoffe des Hopfens, crystallinisch rein dargestellt. Dinglers Polyt. J. 169, 54 (1863).

57. LEUCHT, E. und W. RIEDL: C-Acylierung von 3-Alkylcyclopentantrion-(1. 2. 4). Liebigs Ann. Chem. 669, 55 (1963).

58. LINTNER, C. J. and J. SCHNELL: The Bitter Principles of Hops. Z. ges. Brauw. 27, 666 (1904).

59. MERÉNYI, F. and M. NILSSON: Facile Preparation of 2-Acetylcyclopentane-1,3-dione and 2-Acetylcyclohexane-1,3-dione. Acta Chem. Scand. 17, 1801 (1963).

60. MORTON, R. A. and Z. SAWIRES: Rottlerin. Part VI. A Spectrographic Study of Rottlerin and its Derivatives. J. Chem. Soc. (London) 1940, 1052.

61. POLONSKY, J. et J. RONDEST: Sur une oxydation d'un dérivé du callophyllolide accompagnée d'un réarrangement du squelette carboné. Bull. soc. chim. France 1962, 1560.

62. Power, F. B., F. Tutin and H. Rogerson: The Constituents of Hops. J. Chem. Soc. (London) 103, 1267 (1913).
63. Riedl, W.: Konstitution und Synthese der Hopfenbitterstoffe d,l-Humulon und Lupulon sowie einiger Analoga. Chem. Ber. 85, 692 (1952).
64. Riedl, W. und H. Hübner: Zur Kenntnis des 4-Desoxyhumulons. Chem. Ber. 90, 2870 (1957).
65. Rigby, F. L. and J. L. Bethune: Cohumulone, a New Hop Constituent. J. Amer. Chem. Soc. 74, 6118 (1952).
66. — — Components of the Lead-precipitable Fraction of Humulus lupulus. Adhumulone. J. Amer. Chem. Soc. 77, 2828 (1955).
67. Rigby, F. L., E. Sihto and A. Bars: Additional Constituents of Hops. J. Inst. Brewing 1962, 60.
68. Rillaers, G. and M. Verzele: Prehumulone, a New α-Acid. Bull. soc. chim. Belges 71, 438 (1962).
69. Ritchie, E. and W. C. Taylor: The Constituents of Harungana madagascariensis (Poir). Tetrahedron Letters 1964, 1431.
70. Saleh, M. S. E.: The Bacteriostatic Activity of a Fraction of Hops. Ph. D. Thesis, Manchester 1952.
71. Seyffert, H.: Beitrag zur Kenntnis der chemischen Bestandteile des Lupulins. Z. ges. Brauw. 15, 31 (1892).
72. Shoolery, J. N., M. Verzele and F. Alderweireldt: On the Structure of Humulinone. Tetrahedron 9, 271 (1960).
73. Sieglitz, A. und O. Horn: Eine neue Darstellungsweise für 1.3-Diketone aus Vinylacetat und Säurechloriden. Chem. Ber. 84, 607 (1951).
74. Spetsig, L.-O.: Isolation and Characterization of Two Isohumulone Isomers. J. Inst. Brewing 1964, 440.
75. Spetsig, L.-O., M. Steninger and S. Brohult: The Resolution of Hop Bitter Substances by Reversed-Phase Partition Chromatography. Proc. European Brewery Conv. Copenhagen 1957, 22.
76. Stevens, R. and D. Wright: The Chemistry of Hop Constituents. Part XV. Tetrahydrocohulupone. J. Chem. Soc. (London) 1963, 1763.
77. — — Evaluation of Hops. X. Hulupones and the Significance of ß-Acids in Brewing. J. Inst. Brewing 1961, 496.
78. Vandewalle, M.: Acylcyclopentanones. III. The Synthesis of 2-Acetylcyclopentane-1,3-diones. Bull. soc. chim. Belges 73, 628 (1964).
79. Vandewalle, M., S. Dewaele, F. Alderweireldt and M. Verzele: Acylcyclopentanones. Part I. The Synthesis of 3-Acylcyclopentane-1,2,4-triones. J. Chem. Soc. (London) 1964, 367.
80. — — — — Acylcyclopentanones. The Synthesis of 3-Acylcyclopentane-1,2,4-triones; a Correction. J. Chem. Soc. (London) 1965, 2258.
81. Verzele, M. and F. Govaert: On the Constitution of the Hops ß-Acid. Bull. soc. chim. Belges 58, 432 (1949).
82. — — Analysis of the "Humulone Complex" by a Chromatographic Partition Procedure. Wallerstein Lab. Commun. 18, 187 (1955).
83. Verzele, M., J. Stockx, F. Fontijn and M. Anteunis: Xanthohumol, A New Natural Chalkone. Bull. soc. chim. Belges 66, 452 (1957).
84. Wagner, R.: Über einige Bestandteile des Hopfens. Dinglers Polyt. J. 154, 65 (1859).
85. Walker, T. K.: Report on the Preservative Principles of Hops. Part V. The Chemical Constitution of Lupulon. J. Inst. Brewing 1924, 712.
86. Whitear, A. L.: Brewing Behaviour of Stored Hops. J. Inst. Brewing 1966, 177.

87. WHITEAR, A. L. and J. R. HUDSON: Hop Resins and Beer Flavour. III. Hop Resins in Beer. J. Inst. Brewing **1964**, 24.

88. WIELAND, H.: Über die chemische Natur der Hopfenhartz-Säuren (II). Ber. dtsch. chem. Ges. **58**, 2012 (1925).

89. WINDISCH, W., P. KOLBACH and R. SCHLEICHER: Transformation of the α-Bitter Acid of Hops in Boiling Aqueous Solutions of Various Reactions and the Nature of the Products Formed. Wschr. Brauerei **44**, 453 (1927).

90. WRIGHT, D.: The Chemistry of the Hop Constituents. Part XVI. Preparation of Hulupone and Cohulupone. J. Chem. Soc. (London) **1963**, 1769.

91. WRIGHT, D. and G. A. HOWARD: Biosynthesis of the Hop Resins. J. Inst. Brewing **1961**, 236.

(Received, July 22, 1966.)

The Pseudoguaianolides.

By J. Romo and A. Romo de Vivar, Mexico City.

With 4 Figures.

Contents.

Acknowledgement. We wish to express our sincere gratitude to Professor Werner
Herz of Florida State University for critically reading the manuscript and suggesting a number of corrections and changes. We are also much obliged to the
Director of our Institute, Dr. Alberto Sandoval, for his interest in our work
and for the colored photographs of plants.

I. Introduction.

The term "pseudoguaianolide" has been applied to a rapidly increasing
number of sesquiterpene lactones whose common feature is an abnormal
guaiane skeleton resulting from migration of the $C_{(4)}$ methyl group to
$C_{(5)}$ in the guaianolide system during biogenesis. The latter system can
be deduced from a cyclization process of an ionic intermediary of *trans*-
farnesyl pyrophosphate (27, 60, 73, 74). These transformations (1, 3)
have not yet been verified experimentally. The guaiane structure pre-

References, pp. 127—130.

viously assigned to the first members of the pseudoguaianolides was later revised by Herz and coworkers in a series of interesting publications (*42*, *47*, *38*, *41*). Establishment of their correct structure formed the basis for a systematic investigation in this field. The large number of substances which have been shown to possess this abnormal structure justifies their classification in a different group from that of the guaianolides. Pseudoguaianolides have so far been isolated from the following genera of the Compositae family: *Ambrosia, Iva* and *Parthenium* of the tribe Heliantheae and from *Helenium, Gaillardia* and *Balduina* of the tribe Heleniae.

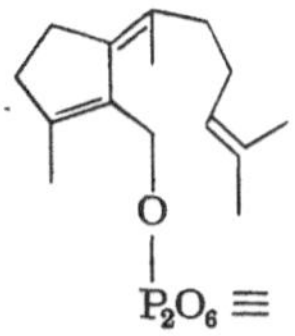

(I)

trans — Farnesyl pyrophosphate.

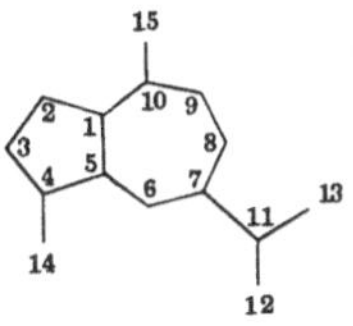

(2)

Guaiane.

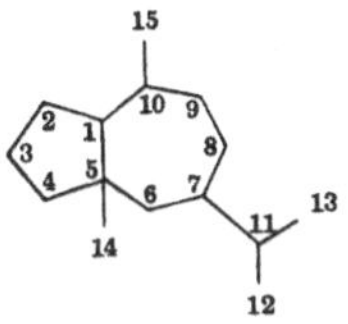

(3)

Pseudoguaiane.

The purpose of this review is to describe briefly the evidence which led to the establishment of the structures and stereochemistry of the pseudoguaianolides and to discuss the relationships between the various members of this new group of sesquiterpenes.

The pseudoguaianolides are described in two sections. Chapter II is concerned with the substances isolated from genera of the tribe Heliantheae which possess a lactone grouping oriented to $C_{(6)}$, and have been correlated to each other. In Chapter III are treated the constituents of genera of the tribe Heleniae. Correlation has been established in a large number of these lactones. Their stereochemistry differs from that of the compounds discussed in Chapter II and the lactone group is closed to $C_{(8)}$.

Previous reviews (*8*, *58*), had dealt with some members of this group of azulogenic lactones before their pseudoguaianolide structure was recognized. A recent publication (*76*) describes various compounds of the pseudoguaianolide series.

II. Pseudoguaianolides Isolated from the Tribe Heliantheae.

Ambrosin.

This sesquiterpene lactone, $C_{15}H_{18}O_3$, (*7*) was isolated by Abu-Shady and Soine (*1*) from *Ambrosia maritima* L. and later from *Parthenium incanum* H. B. K. (*29*), *A. hispida* Pursh (*44*) and *P. hysterophorus* L. (*68*). Preliminary studies were carried out by Abu-Shady and Soine (*2*).

Chemical and spectral properties suggested the presence in ambrosin of two isolated chromophores: a cyclopentenone and an exocyclic methylene group conjugated with the lactone (9, 77). Dehydrogenation experiments of ambrosin derivatives resulted in formation of chamazulene (4) (2, 9)

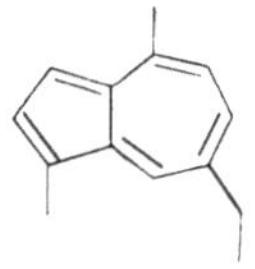

(4)
Chamazulene.

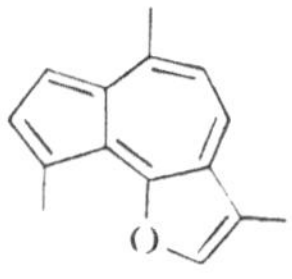

(5)
Artemazulene

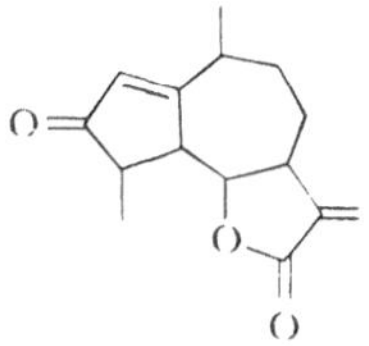

(6)

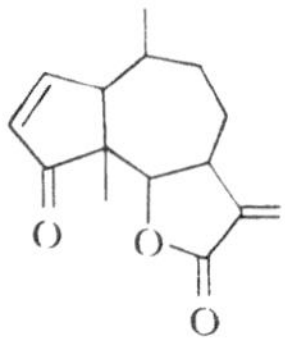

(7)
Ambrosin.

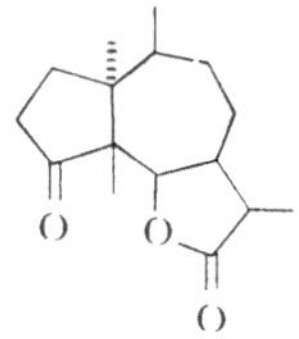

(8)
Tetrahydroambrosin.

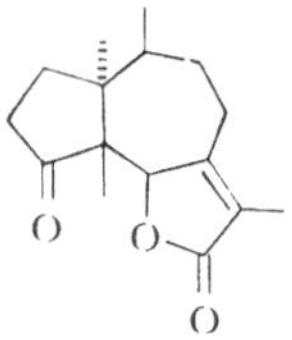

(9)
Dihydroisoambrosin.

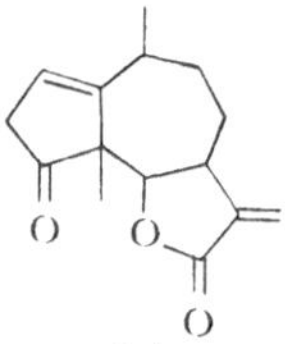

(10)

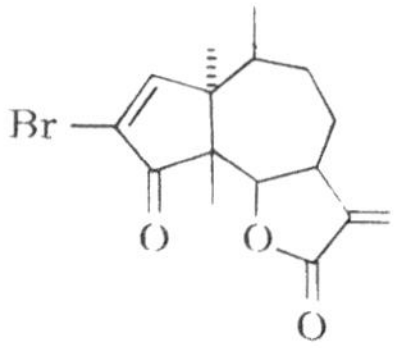

(11)
Bromoambrosin

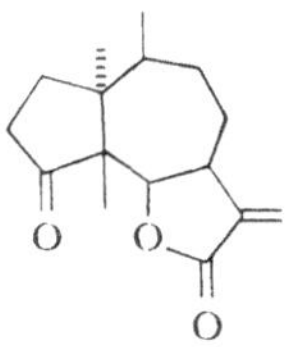

(12)
Damsin.

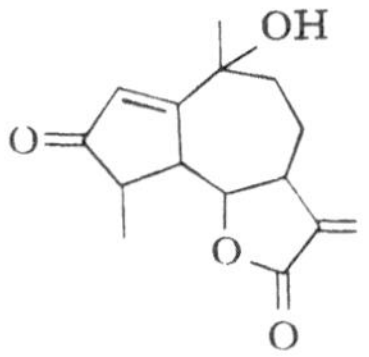

(13)

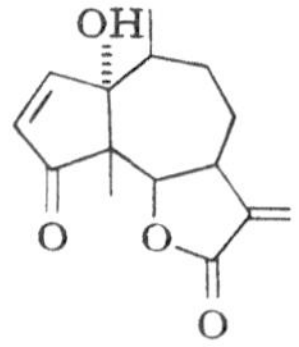

(14)
Parthenin.

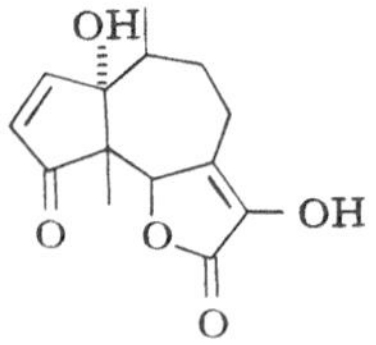

(15)
Norparthenone.

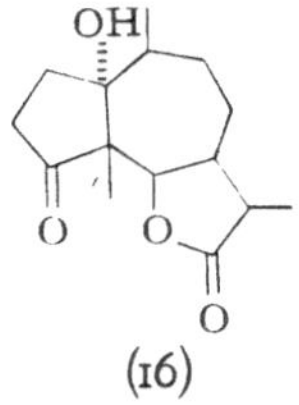

(16)
Tetrahydroparthenin.

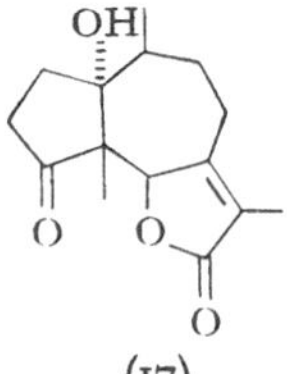

(17)
Dihydroisoparthenin.

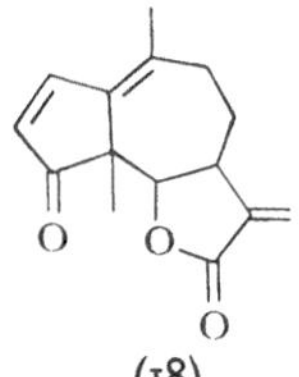

(18)

and artemazulene (5) (77), thus establishing the perhydroazulene skeleton and the lactone orientation at $C_{(6)}$ of ambrosin. Interpretation of the chemical and spectral evidence (2) led to the assignment of a guaianolide structure (6) by Bernardi and Büchi (9). Further work carried out by Šorm, Suchý and Herout (77) appeared to confirm the formula (6) previously postulated for ambrosin.

References, pp. 127—130.

Examination of the NMR spectrum of ambrosin by HERZ, MIYAZAKI and KISHIDA (*38*) disclosed the presence of four vinyl protons and a tertiary methyl group, disproving the guaianolide formula (6). This fact and the correlation of ambrosin with the parthenin series (*47*) led to the proposal by HERZ et al. (*47, 38*) of the "abnormal" guaianolide structure (7) for ambrosin. Aromatization of the pseudoguaianolides results in formation of 4-methyl-azulenes due to migration of the angular methyl group.

Hydrogenation of ambrosin (7) gave tetrahydroambrosin (8) and dihydroisoambrosin (9) (*2, 77*). The remaining double bond of the latter is in endocyclic conjugation with the lactone carbonyl and resisted hydrogenation.

Tetrahydroambrosin (8) and dihydroisoambrosin (9) were identified with a tetrahydro and an dihydro derivative resulting from catalytic hydrogenation of the dienone (18) of the parthenin series (*47*). This correlation confirmed the pseudoguaianolide structure (7) of ambrosin.

A *trans* ring junction was assigned to ambrosin since the optical rotatory dispersion (ORD) curves of dihydroisoambrosin and tetrahydroambrosin had positive Cotton effects of similar shape and amplitude as those shown by appropriate steroid models (*21, 75*). The β-orientation of the $C_{(6)}$ oxygen bond of ambrosin was postulated by ŠORM et al. (*77*) by applying the modified Hudson-Klyne rule. β-Configuration at $C_{(10)}$ in ambrosin was suggested by its correlation with parthenin (14) (*47*) and proved by dehydration of coronopilin (19) (*69*) of known stereochemistry at $C_{(10)}$ which afforded the ketone (10), obtained also by isomerization of ambrosin under acid conditions. This observation, combined with the establishment of the relative configuration of the asymmetric centers of ambrosin as a result of an X-ray analysis of bromoambrosin (11) carried out by EMERSON, HERZ, CAUGHLAN and WITTERS (*24*), led to the absolute stereochemistry of ambrosin as shown in (7).

Damsin.

This lactone, $C_{15}H_{20}O_3$, (12) was isolated by ABU-SHADY and SOINE (*1*) from *Ambrosia maritima* L. It is also a constituent of *A. hispida* Pursh (*44*) and *A. deltoidea* Torr (*50*). Damsin is a 2,3-dihydroambrosin (12) (*9, 77, 78*). Hydrogenation of damsin gave tetrahydroambrosin (8) and dihydroisoambrosin (9) (*78*). Its stereochemistry follows from that of ambrosin (7).

Parthenin.

This pseudoguaianolide, $C_{15}H_{18}O_4$, (14) is a bitter principle isolated by ARNY from *Parthenium hysterophorus* L. (*6*). Parthenin is also a constituent of *A. psilostachya* DC (*55*). HERZ and WATANABE (*46*) have established the functionalities and proposed structure (13) for parthenin as a result of the formation of artemazulene (5) by dehydrogenation of the crude product obtained by lithium aluminum hydride reduction of parthenin.

The NMR spectrum of parthenin (38) exhibited signals corresponding to four vinyl protons and a tertiary methyl group; this coupled with extensive investigations carried out by Herz, Watanabe, Miyazaki and Kishida (47) led to the revision of structure (13) and to the establishment of formula (14).

The exocyclic methylene group of parthenin (14) formed an adduct with diazomethane and is responsible for the liberation of formaldehyde on ozonolysis in acetic acid. When this reaction was carried out in methanol an enolic ketolactone (15) was obtained. Norparthenone (15) gave positive ferric chloride and Tollens tests. Catalytic hydrogenation of parthenin (47) furnished a mixture of two products. One of them was tetrahydroparthenin (16), the main product was, however, the dihydro derivative (17) which resisted further hydrogenation. Its spectral properties and formation of acetic acid on ozonolysis indicated that the cyclopentenone ring was saturated and the exocyclic double bond had migrated to endocyclic conjugation with the lactone carbonyl. Such isomerizations are frequently observed (42).

The dienone (18) resulted from dehydration of the tertiary hydroxyl group of parthenin (14) with formic acid (47).

Degradative oxidation of norparthenone (15) with potassium permanganate afforded S-(+)-α-methylglutaric acid (47). This result is in accordance with the postulated structure for parthenin (14) and establishes the absolute configuration at the asymmetric center $C_{(10)}$ as methyl β. Correlation of ambrosin (7) and coronopilin (19) with parthenin (29, 47, 69) led to the establishment of the stereochemistry of the other asymmetric centers of the latter as shown in formula (14).

Coronopilin.

Herz and Högenauer (29) isolated this sesquiterpene, $C_{15}H_{20}O_4$, (19) from *Ambrosia psilostachya* DC var. *coronopifolia* Farw. Coronopilin is also a constituent of *A. artemisifolia* L. var. *elatior* (29), *Parthenium incanun* H. B. K. (29), *A. psilostachya* DC (25) and *A. dumosa* (Gray) (25).

Coronopilin contains a five-membered ketone, an exocyclic methylene group conjugated with a γ-lactone and a tertiary hydroxyl group (29). The relationship of coronopilin (19) with parthenin (14) as 2,3-dihydro derivative of the latter was established when, under hydrogenation conditions, the exocyclic double bond of (19) was isomerized to endocyclic conjugation yielding dihydroisoparthenin (17) (29). The stereochemistry of coronopilin (19) resulted from its correlation with ambrosin (7) (69) and parthenin (14) (47).

References, pp. 127—130.

GEISSMAN and TURLEY (*25*) confirmed the pseudoguaianolide constitution of coronopilin (**19**) by way of an interesting transformation of the latter under acid conditions. Coronopilic acid (**22**) resulted by treatment of an acetic solution of (**19**) with sulfuric acid. It gave a deep red 2,4-

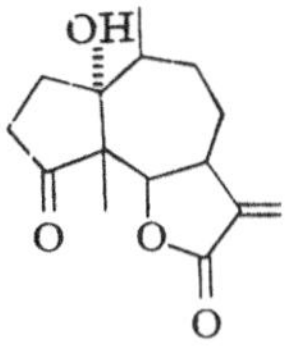

(19)
Coronopilin

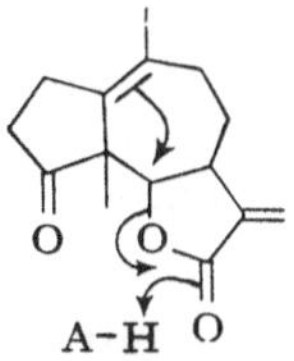

(20)

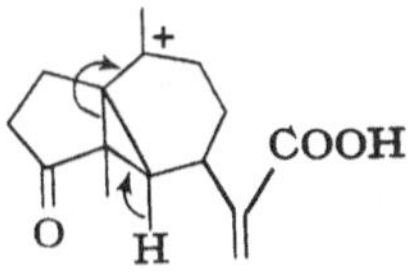

(21)

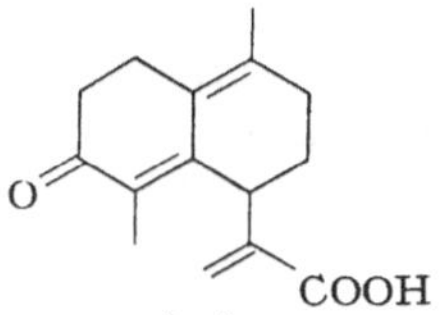

(22)
Coronopilic acid

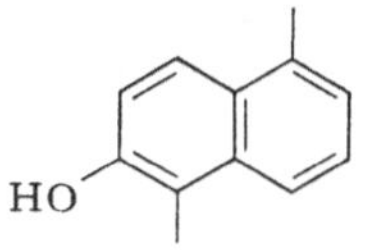

(23)

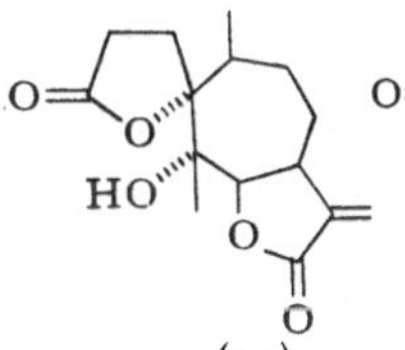

(24)
Psilostachyin.

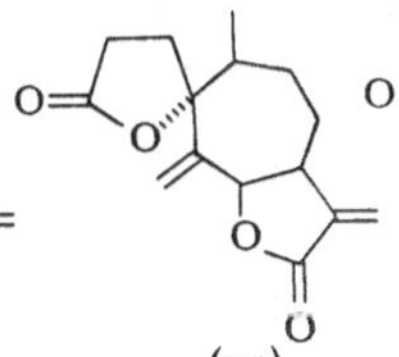

(25)
Anhydropsilostachyin

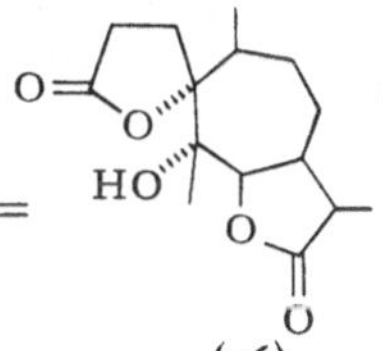

(26)
Dihydropsilostachyin

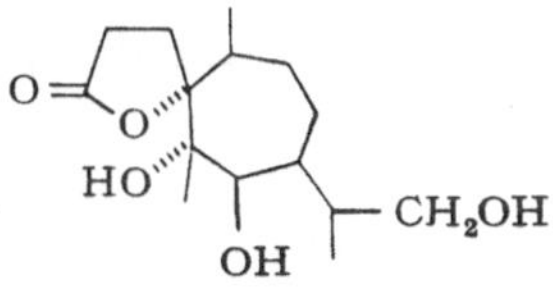

(27a, b)

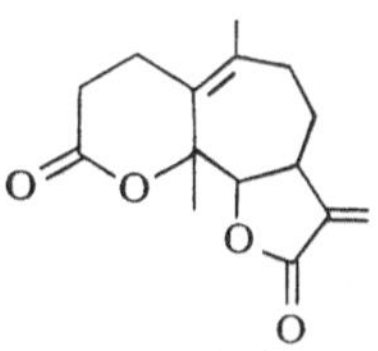

(28)
Psilostachyin B

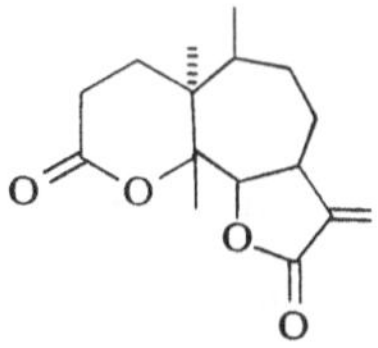

(29)
Psilostachyin C.

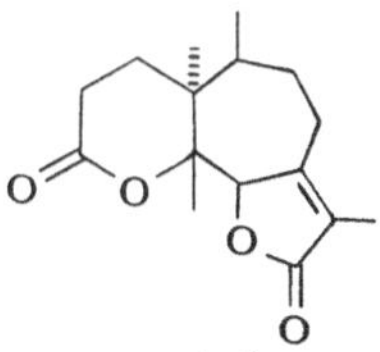

(30)
Isopsilostachyin C.

dinitrophenylhydrazone and liberated formaldehyde on ozonolysis. Treatment of (**22**) with diazomethane yielded the pyrazoline of its methyl ester. Confirmation of the structure (**22**) of coronopilic acid was obtained by dehydrogenation. The resulting phenol was identified as 1,5-dimethyl-2-naphthol (**23**). The sequence (**20** → **21** → **22**) was postulated by GEISSMAN and TURLEY (*25*) in order to rationalize the conversion of coronopilin (**19**) into coronopilic acid (**22**).

Psilostachyins.

The psilostachyins are constituents of a variety of *Ambrosia psilostachya* DC collected in Galveston island (Texas). Although they are not pseudoguaianolides, they deserve to be described here due to their close relationship to coronopilin (**19**) and damsin (**12**).

Psilostachyin. This sesquiterpene, $C_{15}H_{20}O_5$, (**24**) was isolated by MILLER, KAGAN, RENOLD and MABRY (*54, 57*) who established its structure as the dilactone (**24**). Psilostachyin (**24**) was converted to the anhydro derivative (**25**) by treatment with acetic and sulfuric acid. Reduction with sodium borohydride demonstrated the presence of two lactone groups in (**24**) and afforded the dihydro derivative (**26**). Further treatment with sodium borohydride resulted in reduction of one of the lactone groups, that closed to $C_{(6)}$, and yielded two isomeric hexahydropsilostachyins, (**27a** and **27b**), which appeared to differ in configuration at $C_{(11)}$. The synthesis of psilostachyin by Baeyer-Villiger oxidation of coronopilin (**19**) of established absolute configuration confirmed the structure and configuration (**24**).

Psilostachyin B. MABRY, KAGAN and MILLER (*53*) isolated this dilactone, $C_{15}H_{18}O_4$, (**28**). Its structure and stereochemistry (**28**) were established by correlation with psilostachyin C (**29**) (*50*). Hydrogenation of (**28**) saturated the quaternary double bond and caused migration of the exocyclic methylene group to endocyclic conjugation yielding isopsilostachyin C (**30**), obtained by isomerization of psilostachyin C under hydrogenation conditions.

Psilostachyin C. This sesquiterpene, $C_{15}H_{20}O_4$, (**29**) has been isolated by KAGAN et al. (*50*). Chemical and spectral investigation led to the establishment of psilostachyin C as a dilactone. Baeyer-Villiger oxidation of damsin (**12**, p. 92) yielded psilostachyin C (**29**) which afforded definitive proof of its structure and stereochemistry.

Ambrosiol.

The isolation and determination of the structure of ambrosiol, $C_{15}H_{22}O_4$, (**31a**) were carried out by MABRY et al. (*55*). This pseudoguaianolide is a constituent of a variety of *Ambrosia psilostachya* DC. collected in Austin (Texas).

The four oxygen atoms of ambrosiol (**31a**) are contained in an α,β-unsaturated five-membered lactone and two secondary hydroxyl groups. Ambrosiol forms a diacetate (**31b**) and a monotosylate (**31c**). The vicinal character and the *cis* orientation of the hydroxyl groups of (**31a**) was established as a result of a positive periodic acid test and the formation of the acetonide (**32**). The position of these functions in the five-membered ring of ambrosiol was deduced from the NMR spectrum. The structure

References, pp. 127—130.

and configuration of the $C_{(1)}$, $C_{(5)}$, $C_{(6)}$, $C_{(7)}$ and $C_{(10)}$ asymmetric centers of ambrosiol (31 a) was fully elucidated when treatment of the monotosylate (31 c) with formic acid by way of a pinacol rearrangement afforded damsin (12, p. 92). The α-configuration of the asymmetric centers at $C_{(3)}$ and $C_{(4)}$ was established by the method of asymmetric esterification described by HOREAU (48).

Hysterin.

Hysterin, $C_{17}H_{24}O_5$, (33 a) is a constituent of a variety of *Parthenium hysterophorus* L. from the Valley of Mexico and was discovered by ROMO DE VIVAR, BRATOEFF and RÍOS (68) who have elucidated the structure (33 a) of this pseudoguaianolide. The presence in hysterin of a $C_{(6)}$ oriented α,β-unsaturated lactone, of a primary hydroxyl group and of an acetyl was secured by spectral methods. Hysterin gave the acetate (33 b) and the tosylate (33 c). Hydrogenation of the anhydro derivative (34) obtained by dehydration of hysterin (33 a) with phosphorous oxychloride led to the dihydroisoderivative (35). Hydrolysis of the latter, followed by oxidation, furnished dihydroisoambrosin (9, p. 92) (2, 77). This correlation established the gross structure as well as the configuration at $C_{(1)}$, $C_{(5)}$, $C_{(6)}$ and $C_{(7)}$ of hysterin as those in ambrosin (7, p. 92).

The α-configuration at the $C_{(4)}$ asymmetric center of (33 a) was deduced as a result of the following transformations. Platinum oxide catalyzed hydrogenation of ambrosin (7) in acetic acid containing perchloric acid yielded tetrahydroisoambrosin (36 a) and its acetate (36 b). The latter was also obtained by reduction of dihydroisoambrosin (9) with sodium borohydride followed by acetylation. The acetate (36 b) was isomeric at $C_{(4)}$ with the dihydroisoderivative (35) which was assumed to possess α-configuration at the $C_{(4)}$ asymmetric center, since the derivatives (36 a) and (36 b) of ambrosin must be produced by attack from the side opposite to the angular methyl group.

Cumanin.

Collections of *Ambrosia cumanensis* H. B. K. and *A. peruviana* Willd. from the Valley of Mexico afforded the pseudoguaianolides cumanin, peruvin and peruvinin which in lactone orientation differ from the other sesquiterpenes isolated from Heliantheae. The lactone groupings in cumanin, peruvin and peruvinin are closed at $C_{(8)}$.

The pseudoguaianolide cumanin, $C_{15}H_{22}O_4$, (37 a) is a constituent of *Ambrosia cumanensis* H. B. K. Its isolation and constitution was described by ROMO, JOSEPH-NATHAN and SIADE (65). The nature of the oxygen atoms of cumanin (37 a) follows from its spectral properties; they are distributed as an unsaturated γ-lactone and two secondary hydroxyl groups. Cumanin yielded a dimesylate (37 c) and a diacetate

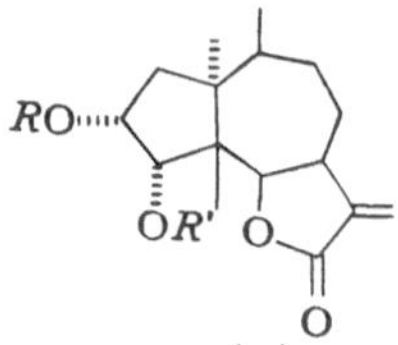

(31)

(a). R=R'=H, Ambrosiol.
(b). R=R'=Ac
(c). R=Ts, R'=H

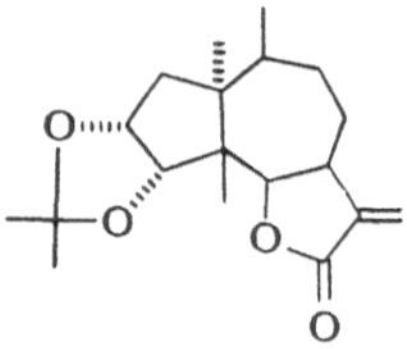

(32)

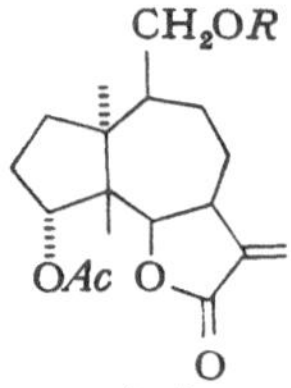

CH₂OR

(33)

(a), R=H. Hysterin.
(b), R=Ac
(c), R=Ts

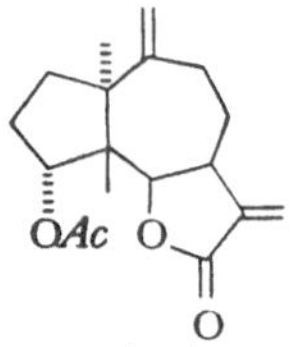

(34)

Anhydrohysterin.

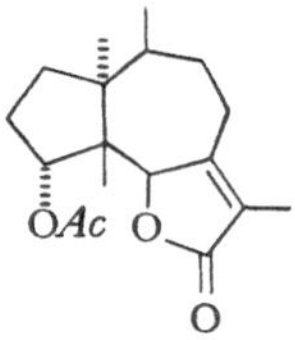

(35)

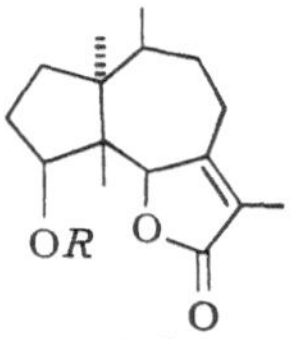

(36)

(a), R=H
(b), R=Ac

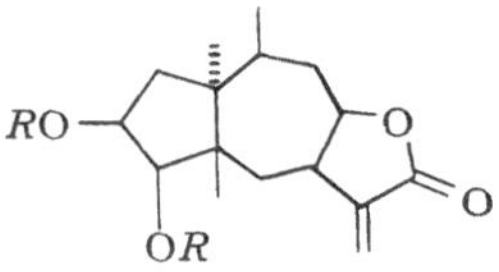

(37)

(a), R=H Cumanin
(b), R=Ac
(c), R=CH₃SO₂

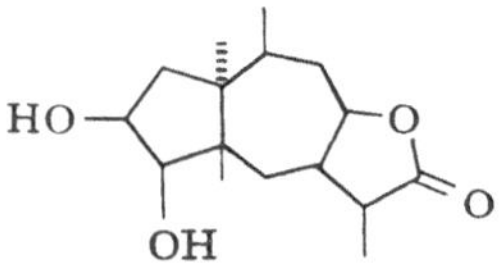

(38)

Dihydrocumanin.

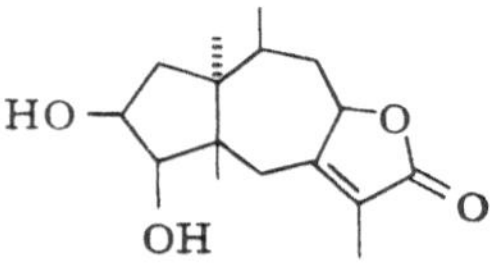

(39)

Isocumanin.

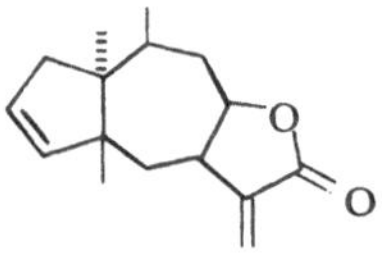

(40)

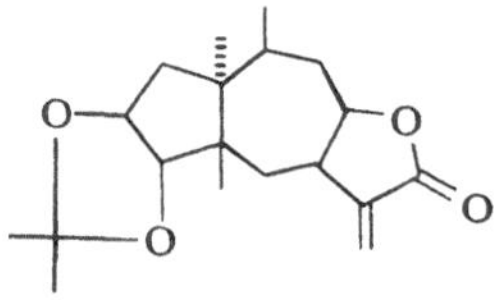

(41)

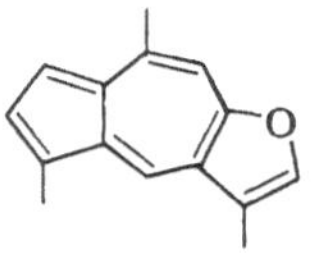

(42)

Linderazulene.

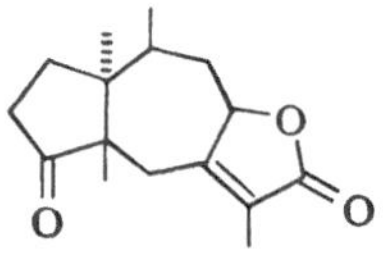

(43)

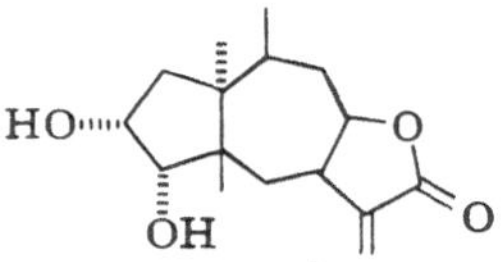

(44)

(37 b). The latter liberated formaldehyde on ozonolysis. Hydrogenation of cumanin (37 a) resulted in dihydrocumanin (38) and isocumanin (39). Cumanin consumed periodic acid. Confirmation of the vicinal position of the hydroxyl groups resulted from treatment of cumanin dimesylate (37 c) with sodium iodide which gave the lactone (40). Formation of the acetonide (41) indicated that the hydroxyl groups were *cis*. Dehydrogenation of the crude product obtained by reduction of (37 a) afforded linderazulene (42), a fact which established the azulogenic character of cumanin.

Dehydration of isocumanin (39) with potassium bisulfate yielded the ketone (43) which was identical with a compound of known structure and stereochemistry prepared from peruvin (45) (49). This established the gross structure and the configuration at $C_{(1)}$, $C_{(5)}$, $C_{(7)}$ and $C_{(10)}$.

Hydroxylation of the $C_{(3)}$ double bond of the lactone (40) with osmium tetroxide, which attacks the less hindered position opposite to the angular methyl group of (40), gave the diol (44), isomeric with cumanin. If it is assumed that the osmium tetroxide attacks from the α-side, the hydroxyl groups of the new diol (44) are α-oriented. Hence, cumanin must be epimeric at $C_{(3)}$ and $C_{(4)}$ as shown in formula (37 a).

Peruvin.

The pseudoguaianolide peruvin, $C_{15}H_{20}O_3$, (45) was isolated from *Ambrosia peruviana* Willd. by JOSEPH-NATHAN and ROMO (49) who determined its structure. Peruvin (45) contains a cyclopentanone ring, a tertiary hydroxyl group, an exocyclic methylene conjugated with a $C_{(8)}$ oriented γ-lactone and it forms a pyrazoline.

When peruvin (45) was submitted to hydrogenation only the iso derivative (46) was isolated. However, a small amount of a dihydro compound was also formed, since thionyl chloride dehydration of the crude product, obtained from the above hydrogenation, followed by treatment with selenium dioxide, resulted in the formation of the dihydrodienone (47) and the isodienone (48) whose structures were inferred from the spectral properties. Dehydration of isoperuvin (46) led to the anhydroderivatives (49) and (50). Platinum oxide catalyzed hydrogenation of the dienone (48) yielded a saturated ketone, identical with the cyclopentanone (43) obtained from cumanin (37 a).

Peruvin (45) was correlated with mexicanin A (102 a) through the following steps. Treatment of the crude mesylate of dihydromexicanin A (103, p. 109) with collidine, followed by reaction with selenium dioxide, yielded the dienone (48). Therefore, peruvin has the pseudoguaianolide structure (45) with the configuration of its asymmetric centers at $C_{(5)}$ and $C_{(8)}$ as in mexicanin A (102 a) (34). Peruvin (45) was assumed to have a *trans* ring junction, since the ORD curves of (45), isoperuvin (46) and the lactone (43) displayed positive Cotton effects similar to those of other

7*

trans-fused pseudoguaianolides (*42*). Since the ketone (43) differed from the lactone (97, p. 108) of the helenalin series, which has the same gross structure and the same stereochemistry at $C_{(1)}$, $C_{(5)}$ and $C_{(8)}$, and an *x*-oriented $C_{(10)}$ methyl group, the $C_{(10)}$ methyl group of (43) must be β-oriented. Furthermore, peruvin (45) must have the same configuration at $C_{(10)}$ as the ketone (43), because the latter was also formed by hydrogenation of the β,γ-unsaturated cyclopentenone (49).

Peruvinin.

This lactone was isolated from *Ambrosia peruviana* Willd. by Romo, Romo de Vivar, Joseph-Nathan and Alvarez (*64*). Peruvinin, $C_{15}H_{20}O_4$, (51 a) is closely related to cumanin (37 a). It contains a ketol grouping since (51 a) consumed periodic acid and yielded the ketone (52) on reduction with calcium in liquid ammonia. Treatment of peruvinin acetate (51 b) with sodium borohydride afforded a product identical with dihydrocumanin (38). From this the structure and stereochemistry of peruvinin were established.

III. Pseudoguaianolides Isolated from the Tribe Heleniae.
Tenulin.

(a) *Structure and Configuration.* This lactone, $C_{17}H_{22}O_5$, (55) is a bitter principle first isolated by Clark (*13*) from *Helenium amarum* (Raf.) H. Rock (previously known as *H. tenuifolium* Nutt), *H. elegans* DC. and *H. Badium* Greene. It has been found also in *H. montanum* Nutt (*14*), *H. Bloomquistii* Rock (*28*), *H. Bigelovii* Gray (*59*), and *H. thurberi* Gray (*33*).

Earlier investigations (*13–15*) afforded only a limited knowledge concerning the structure of this compound. On spectral grounds Ungnade et al. (*79, 80*) determined the presence in tenulin of an x,β-unsaturated ketone and a lactone group. The preparation of chamazulene (4, p. 92) and linderazulene (*42*, p. 98) (*7, 10*) by dehydrogenation of tenulin or of its reduction products, coupled with additional chemical and spectral data, led Barton and de Mayo (*7*) to postulate structure (53). Formula (54) was proposed by Braun, Herz and Rabindran (*10*), both structures being based on a guaiane skeleton. Further studies carried out by Herz et al. (*41*) permitted them to establish the pseudoguaianolide structure (55) for tenulin. The available experimental data will be interpreted in terms of formula (55).

The NMR spectrum (*41*) of tenulin and of isotenulin (56 b) (*13*), an isomeric lactone obtained by mild alkaline treatment of (55) and isolated from *H. arizonicum* Blake (*28*) and *H. Bigelovii* Gray (*59*), indicated that both products possess two vinyl protons associated in an

ABX system and a tertiary methyl group in the perhydroazulene skeleton. The multiplicity of the signal assigned to the hydrogen on the carbon bearing the ethereal oxygen of the lactone indicated ring closure toward $C_{(8)}$ (41). BARTON and DE MAYO (7) deduced that tenulin (55) had a masked acetate grouping, since it did not produce acetic acid on acid hydrolysis, whereas isotenulin (56 b) liberated acetic acid under the same conditions. Pyrolisis of tenulin or a treatment with acetic

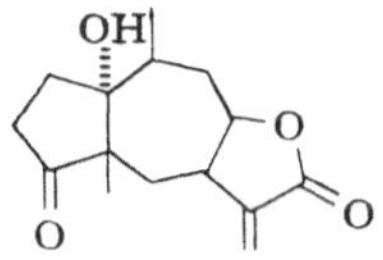

OH|

O

O

(45)

Peruvin.

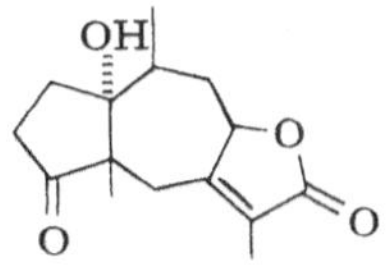

OH|

O

O

(46)

Isoperuvin.

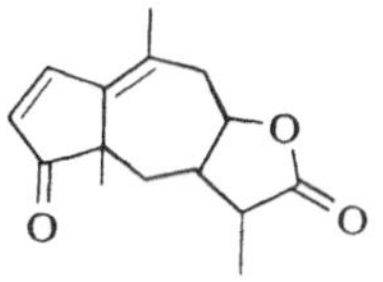

O

O

(47)

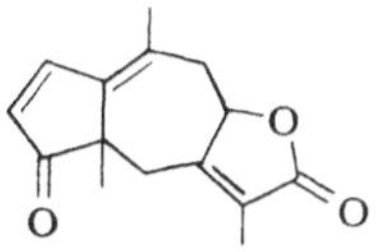

O

O

(48)

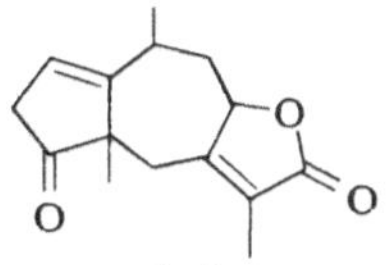

O

O

(49)

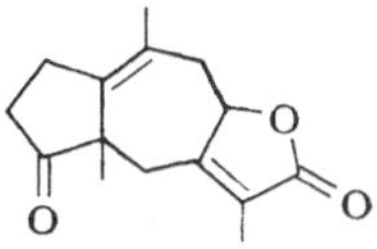

O

O

(50)

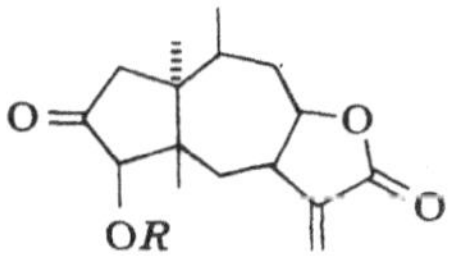

O=

O

O

OR

(51)

(a). R = H. Peruvinin.
(b). R = Ac

O=

O

O

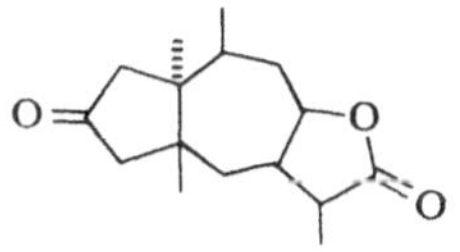

(52)

anhydride and sodium acetate (7, 13) caused dehydration of the tertiary hydroxyl group with simultaneous migration of the double bond to $C_{(1)}$ and yielded pyrotenulin now formulated as (57) (7, 41). Dehydration of dihydrotenulin (58) (13) obtained by catalytic hydrogenation of tenulin gave the anhydrodihydro derivative (59) which on ozonolysis afforded the dilactone (60) (7). Treatment of tenulin or isotenulin (56 b) with alkaline hydrogen peroxide caused oxidation of the cyclopentenone ring and yielded tenulinic acid (61 a) (13). The latter formed the acetate (61 b) which can be obtained by potassium permanganate oxidation of (55) or (56 b) (7). Ozonolysis of the double bond of tenulin resulted in formation of the lactol (62 a) (41) which could also be prepared by osmium tetroxide hydroxylation of tenulin, followed by periodic acid oxidation of the intermediary glycol (41). The lactol (62 a) gave a positive Tollens test, formed the methyl ether (62b) on treatment with diazomethane, and could be reduced by sodium borohydride to the bis-γ-lactone (63) (*41*).

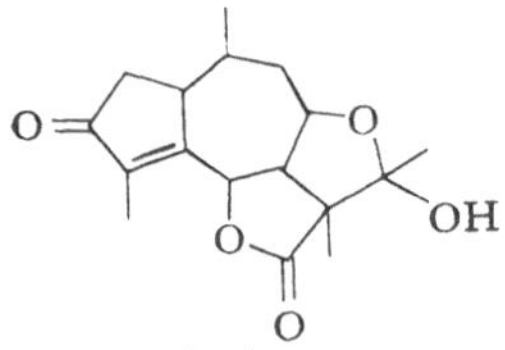

(53)

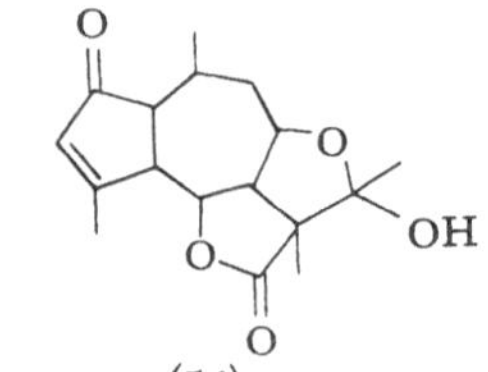

(54)

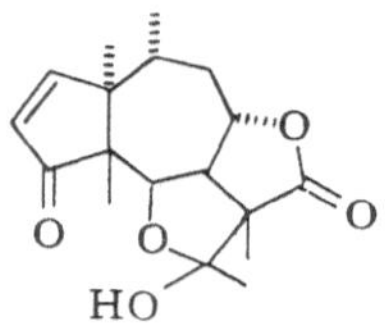

(55)

Tenulin.

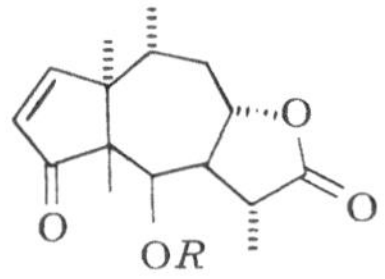

(56)

(a), R=H
(b), R=Ac, Isotenulin.

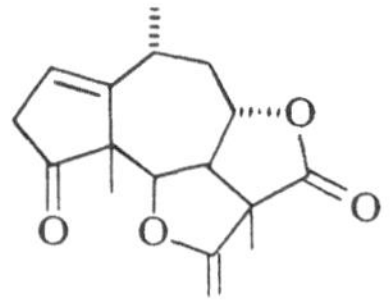

(57)

Pyrotenulin

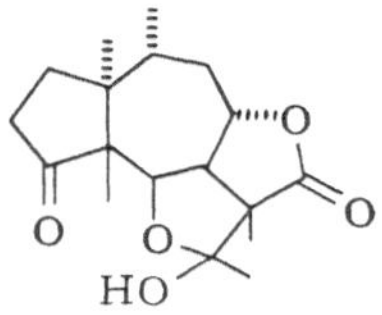

(58)

Dihydrotenulin.

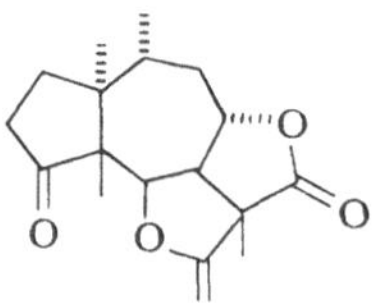

(59)

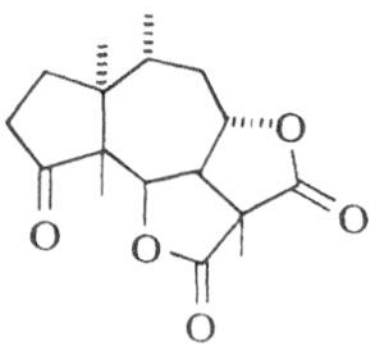

(60)

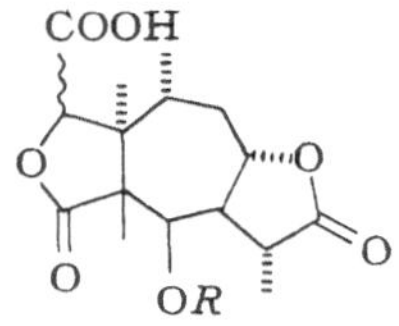

(61)

(a), R=H, Tenulinic acid
(b), R=Ac

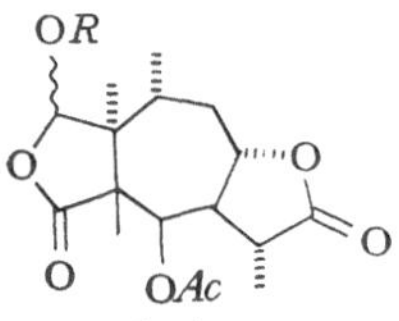

(62)

(a), R=H
(b), R=CH₃

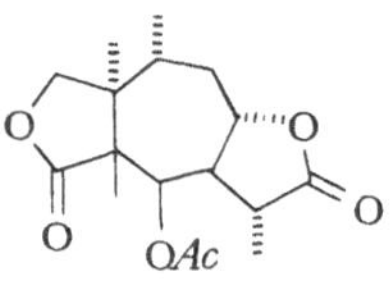

(63)

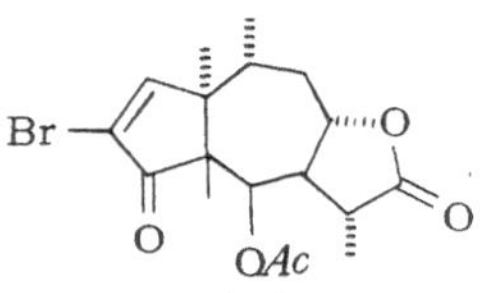

(64)

Bromoisotenulin.

The X-ray analysis of bromoisotenulin (64) by Rogers and Mazhar-ul-Haque (*61*) proved that the structure (55) of tenulin advanced by Herz et al. (*41*) was correct. Combination of this result with evidence afforded by optical rotatory dispersion measurements (*20, 42*) led to the absolute configuration of bromoisotenulin depicted in (64).

References, pp. 127—130.

(b) *Further Transformations.* The following conversions of tenulin derivatives, apart from the light shed on the chemistry of tenulin (55), have resulted in correlation with some other pseudoguaianolides. Dihydro-isotenulin (65 b), prepared by catalytic hydrogenation of isotenulin (56 b) (*13*), gave on alkaline or acid hydrolysis desacetyldihydroisotenulin (65 a)

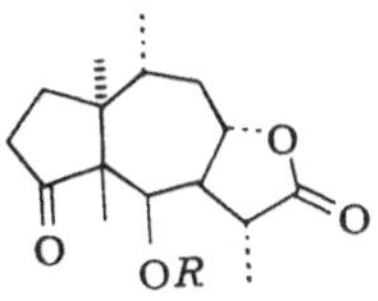

(65)

(a), R = H
(b), R = Ac, Dihydroisotenulin.
(c), R = CH₃SO₂

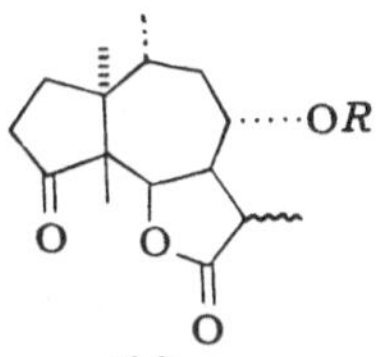

(66)

(a), R = H
(b), R = Ac

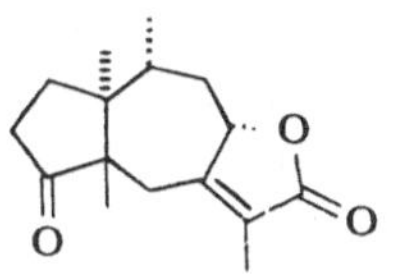

(67)

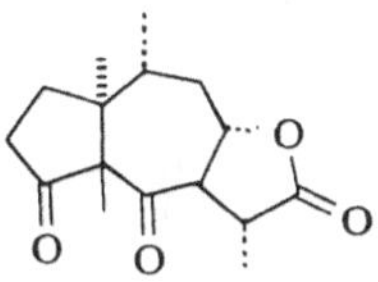

(68)

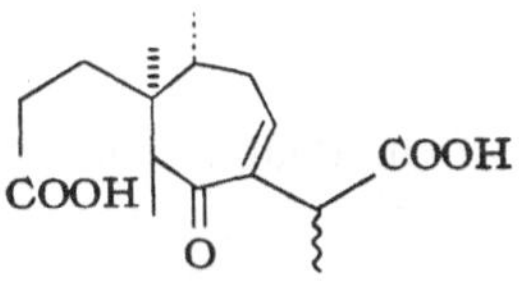

(69)

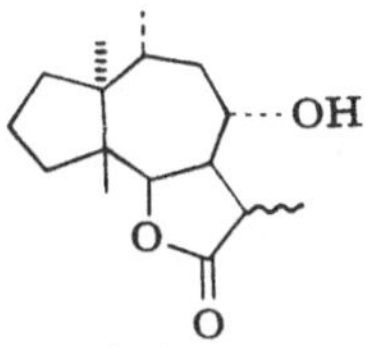

(70)

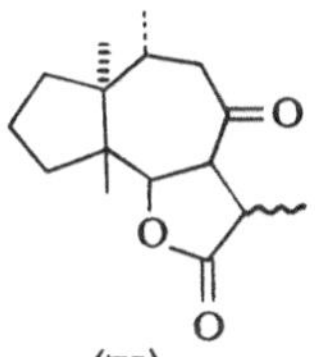

(71)

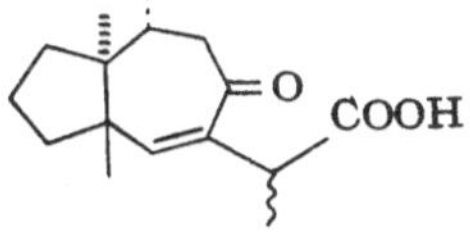

(72)

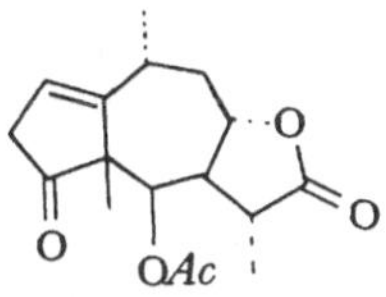

(73)

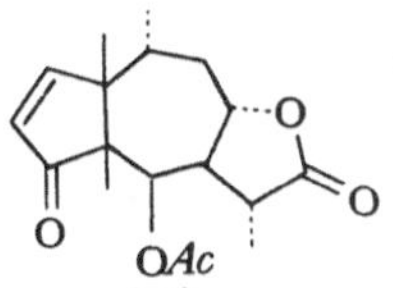

(74)

Epiisotenulin.

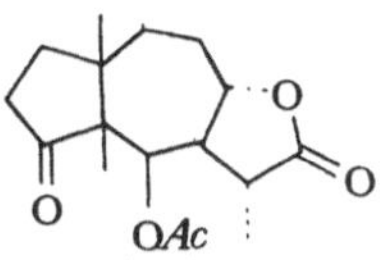

(75)

Dihydroepiisotenulin.

(*13*) which regenerated (65 b) on acetylation. In the alkaline hydrolysis of dihydroisotenulin (65 b), a by-product with the lactone oriented at $C_{(6)}$, was obtained by HERZ et al. (*10*) and named "dihydroalloisotenulin" (66 a), the term "allo" being applied to all products with the lactone group reoriented to $C_{(6)}$. Acetylation of (66 a) afforded the acetate (66 b).

Treatment of the mesylate of desacetyldihydroisotenulin (65 c) with lutidine yielded the unsaturated lactone (67) (*41*).

Chromium trioxide oxidation of desacetyldihydroisotenulin (65 a) gave the dehydro derivative (68) (*10*). Treatment with sodium carbonate of the latter caused a β-diketone cleavage resulting in the α,β-unsaturated ketodicarboxylic acid (69) (*10*) whose formation had been difficult to interpret on the basis of the former guaianolide structures (53) and (54) proposed for tenulin. The NMR spectrum of the acid (69) indicated that the double bond is in endocyclic conjugation with the ketone (*43*).

Wolff-Kishner reduction of dihydroisotenulin (65 b) was accompanied by lactone ring reorientation and yielded the hydroxylactone (70) (*10*) of the allo series. The same derivative (70) resulted from a similar reduction carried out with desacetyldihydroalloisotenulin (66 a) (*10*). Chromic acid oxidation of (70) gave the ketolactone (71) (*10*) and treatment with sodium carbonate of the latter afforded the acid (72).

Interaction of isotenulin (56 b) in chloroform solution and hydrogen chloride gave the unconjugated cyclopentenone (73) (*34*) as established by spectral examination. Further reaction of (73) with hydrogen chloride resulted in conjugation of the double bond yielding the compound (74). Although the spectral features of (74) are very similar to those of iso-tenulin (56 b), these compounds differ in the stereochemistry of the asymmetric center at $C_{(1)}$, with $H_{(1)}$ of the isomer (74) occupying the β-configuration. Platinum oxide catalyzed hydrogenation of the un-conjugated cyclopentenone (73) yielded 1-epidihydroisotenulin (75). The latter could also be obtained by hydrogenation of 1-epiisotenulin (74) catalyzed with palladium on charcoal. The ORD curves of the compounds of the $C_{(1)}$ epi-series showed Cotton effects for which appropriate cis-fused 17-ketosteroids do not serve as models since the cis-fused pseudoguaianolides do not possess the rigid conformation of these steroids (*34*).

(c) *Desacetylneotenulin*. Barton and de Mayo (*7*) had observed earlier that treatment of tenulin (55) with sodium bicarbonate furnished two products. One of these was desacetylisotenulin (56 a) previously prepared by hydrolysis of isotenulin (56 b) with sulfuric acid (*13*) and also obtained from *Helenium Bigelovii* Gray (*59*). The other product named desacetylneotenulin was formulated as (76) on the basis of earlier evidence. Further work carried out by Herz et al. (*41*) after the establishment of the pseudoguaianolide structure (55, p. 102) of tenulin (*41*) led to the revision of the desacetylneotenulin formula to (80 a).

The spectral properties of desacetylneotenulin (80 a) indicated the presence of an α-methyl-β,β-disubstituted-α,β-unsaturated cyclopentenone (*7, 41*). It formed acetic acid on ozonolysis (*7*). The secondary

Fig. 1. *Helenium mexicanum* H. B. K.

Fig. 2. *Helenium mexicanum* H. B. K.

Fig. 3. *Ambrosia cumanensis* H. B. K.

Fig. 4. *Parthenium hysterophorus* L.

hydroxyl group of (80a) could be acetylated and was oxidized by chromium trioxide to the dehydro derivative (81) (7). The latter was not reduced by zinc in acetic acid; therefore, an enedione grouping was excluded (7).

Chemical proof of the location of the hydroxyl group at $C_{(6)}$ in desacetylneotenulin (80a) was obtained by ozonolysis followed by periodic acid oxidation which yielded the ketodilactone (82) (41), whereas treatment of the mesylate (80b) with lutidine gave the dienone (83) (41).

The transformation of tenulin (55) to desacetylneotenulin has been rationalized by HERZ et al. (41) as follows: Initial isomerization of tenulin (55) to isotenulin (56b) and hydrolysis of the latter to desacetylisotenulin (56a) which undergoes a retroaldol ring opening to (77); the presence of the free hydroxyl at $C_{(6)}$ is an essential requisite for this reaction. This is followed by rearrangement of the cyclopentenone chromophore to (79) which may or may not involve (78) as an intermediate. Alternatively, migration of the double bond to form the desacetyl derivative of (73) could precede the retroaldol reaction since neohelenalin (110) (42) has been isolated by similar treatment of mexicanin A (102a, p. 109). Furthermore, it has been found that desacetylisotenulin (56a) is converted to (80a) in alkali or acid (41); and treatment of (73) with sodium bicarbonate also affords desacetylneotenulin (80a) (34). The postulated conversions involve the asymmetric centers at $C_{(5)}$ and $C_{(6)}$ and possibly $C_{(7)}$, $C_{(8)}$ and $C_{(10)}$ of (80a). The resulting uncertainty concerning the configuration of desacetylneotenulin at these centers has not yet been cleared up.

Bigelovin.

PARKER and GEISSMAN (59) discovered bigelovin, $C_{17}H_{20}O_5$, (84) in *Helenium Bigelovii* Gray. It is also a constituent of *Gaillardia pinnatifida* Torr (39). PARKER and GEISSMAN proposed for this pseudoguaianolide the gross structure (84). Hydrolysis of bigelovin liberated acetic acid. Spectral analysis showed the presence of two isolated chromophores: a cyclopentenone and an exocyclic methylene group conjugated with a five-membered lactone.

The NMR spectrum of bigelovin exhibited features similar to those of other pseudoguaianolides that contain the same chromophores. The multiplicity of an NMR signal corresponding to the hydrogen in the carbon, bearing the ethereal oxygen of the lactone, indicated a $C_{(8)}$ orientation of the latter (38, 41). Platinum oxide catalyzed hydrogenation of bigelovin, interrupted after the uptake of one mole of hydrogen, resulted in the formation of the dihydro derivative (85); this result differs from that in the helenalin series where the exocyclic methylene group is saturated first (5). On oxidation by the Lemieux process bigelovin and dihydrobigelovin (85) afforded formaldehyde. Further hydrogenation of bigelovin yielded the tetrahydro derivative (86). The stereochemistry of bigelovin shown in (84) was established by correlation with thurberilin (33) and isotenulin (69).

Tetrahydrobigelovin (86) is identical with a product obtained by acetylation of the lactone (156a, p. 119) of the thurberilin series of known stereochemistry at $C_{(1)}$, $C_{(5)}$, $C_{(8)}$ and $C_{(10)}$. The configuration at the asymmetric centers $C_{(6)}$, $C_{(7)}$ and $C_{(11)}$ was elucidated when (86) was obtained from the mesylate of desacetyldihydroisotenulin (65c) by treatment with sodium acetate (69).

Helenalin.

Helenalin, $C_{15}H_{18}O_4$, (89a) is a constituent of the following *Helenium* species: *H. autumnale* (12), *H. quadridentatum* Labill (14, 26), *H. microcephalum* M. A. Curt (3), *H. vernale* Walt (37) *H. campestre* Small (31), *H. mexicanum* H. B. K. (70—72), *H. laciniatum* Gray (28) and *H. aromaticum* (Hook) Bailey (62, 63). It has also been isolated from *Gaillardia megapotamica* (Spreng) Baker (30), *G. multiceps* Greene (30), *G. pinnatifida* Torr (39) and from *Balduina angustifolia* (Pursch) Robins (35).

Earlier work carried out by Adams and Herz (3–5) defined the functionalities of helenalin as a cyclopentenone ring system, a five-membered lactone, an exocyclic methylene group and a secondary hydroxyl. Büchi and Rosenthal (11) found that reduction of helenalin followed by dehydrogenation yielded guaiazulene (87). This result combined with chemical and spectroscopic evidence led to the proposal of the helenalin formula (88) (11). The latter structure was revised to (89a) by Herz, Romo de Vivar, Romo and Viswanathan (42) as a result of the observation that tenulin which had been previously correlated with helenalin (35) (for details see below) has the "abnormal" guaianolide structure (55) (41). The high resolution NMR spectrum of helenalin (42) indicated that four vinyl protons were present instead of the three required by structure (88). Signals for a secondary and a tertiary methyl group were also observed, which was likewise not in accord with structure (88). Lactone orientation at $C_{(8)}$ of helenalin was confirmed by the multiplicity of the NMR signal corresponding to the $C_{(8)}$ hydrogen.

Partial hydrogenation of helenalin saturated the exocyclic methylene group and yielded dihydrohelenalin (90a) (5). As a by-product isohelenalin (91) (42), a constituent of *H. microcephalum* (11) was obtained. Hydrogenation of the latter afforded the dihydroisoderivative (92) (11). Tetrahydrohelenalin (93a) resulted from hydrogenation of helenalin (89a) or of dihydrohelenalin (90a) (3, 5). Alkaline treatment of dehydrotetrahydrohelenalin (94) (10), obtained by chromium trioxide oxidation of tetrahydrohelenalin (5), gave the α,β-unsaturated ketodicarboxylic acid

(95a), characterized as its dimethylester (95b) (42). As in the tenulin series, the interpretation of this reaction, which on the basis of the guaiane structure of helenalin (88) had remained obscure, was rationalized as a β-diketone cleavage and afforded chemical proof of the relative position

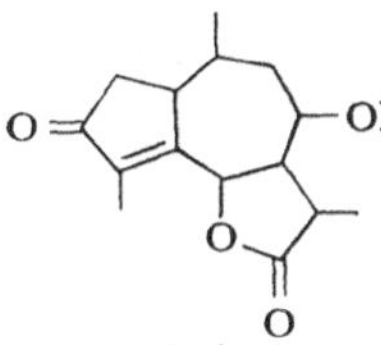

(76) 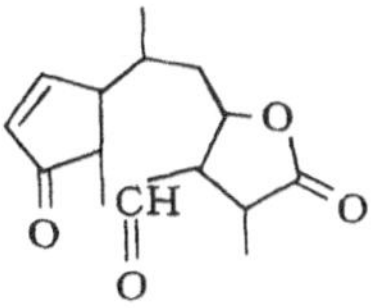 (77) 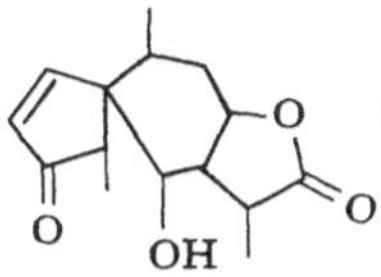 (78) 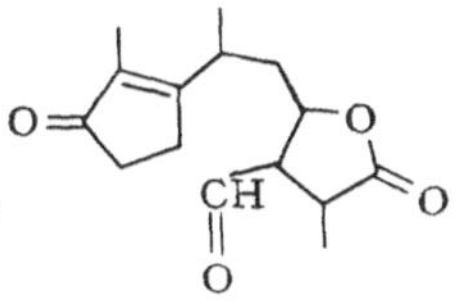 (79)

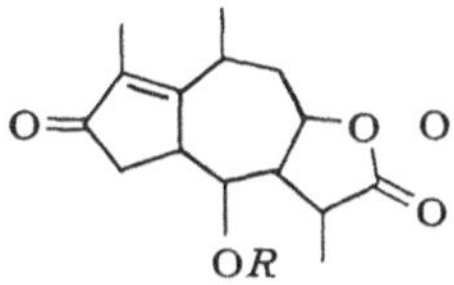

(80) (81) (82) (83)

(a), R = H . Desacetylneotenulin.
(b), R = CH₃SO₂

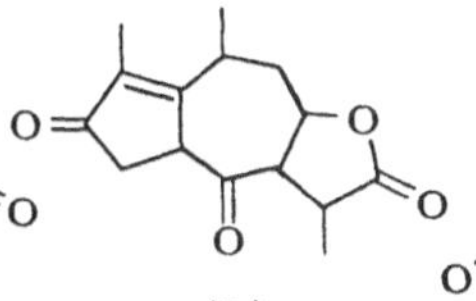

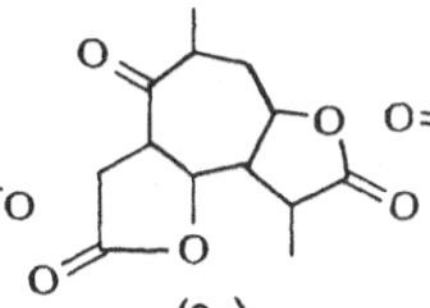

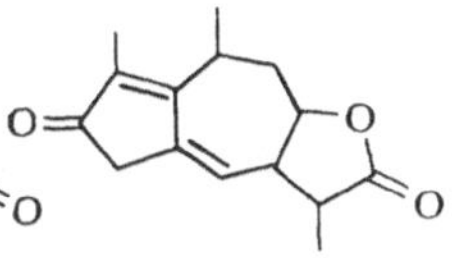

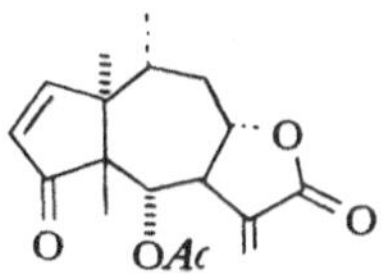

(84) (85) (86)

Bigelovin. Dihydrobigelovin. Tetrahydrobigelovin.

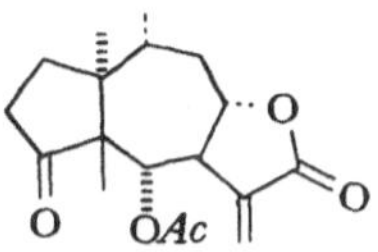

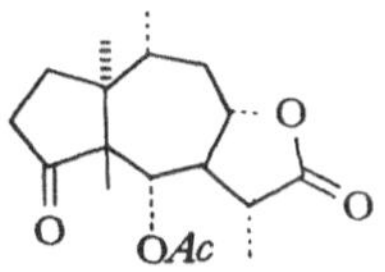

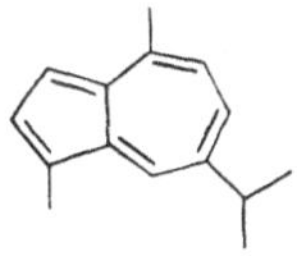

(87) (88) 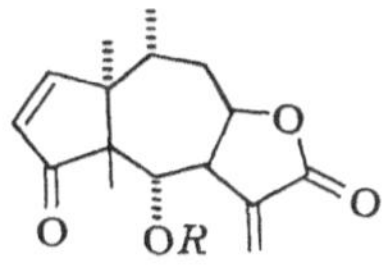 (89) (90)

Guaiazulene. (a), R = H . Helenalin. (a), R = H . Dihydrohelenalin
(b), R = Ac (b), R = Ac

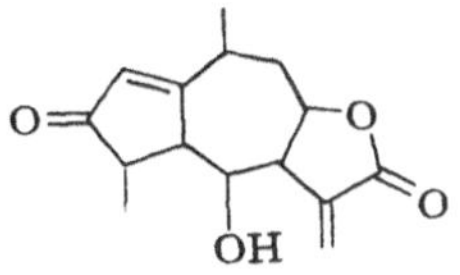

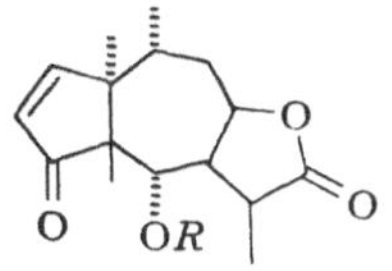

of the functional groups in helenalin (42). The dilactone (96) prepared by Baeyer-Villiger oxidation of tetrahydrohelenalin acetate (93b) (42) showed in its NMR spectrum a displacement to lower field of the singlet corresponding to the $C_{(5)}$ group, thus affording a further proof concerning the position of the keto group in helenalin (42). Treatment of tetrahydrohelenalin mesylate (93c) with lutidine furnished the α,β-unsaturated lactone (97) (31).

Tenulin (55) and helenalin (89a) which differ in the stereochemistry at the $C_{(6)}$ and $C_{(8)}$ asymmetric centers were correlated by Herz and Mitra (35) through the following steps. Desulfuration with Raney nickel of the ethylenethioketal of tetrahydrohelenalin (93a) afforded the desoxo-hydroxylactone (98). Mild alkaline treatment of (98) or passage through alumina reoriented its lactone group and yielded desoxoallotetrahydro-helenalin (99) (43, 35). Oxidation of (99) with chromium trioxide formed the dehydro derivative (100). Basic treatment of the latter yielded an α,β-unsaturated ketoacid identified as (72) of the allotenulin series. This

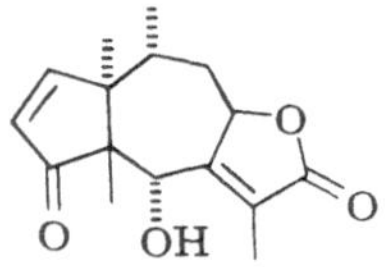

(91)
Isohelenalin.

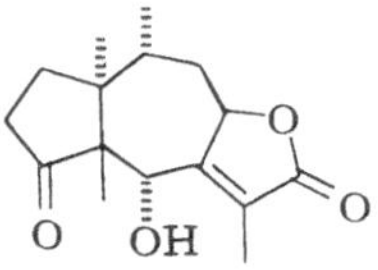

(92)
Dihydroisohelenalin.

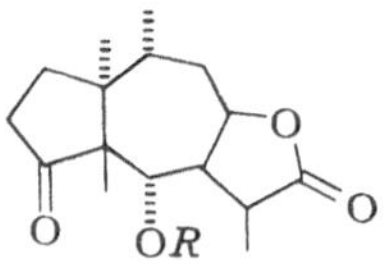

(93)
(a), $R = $ H. Tetrahydrohelenalin.
(b), $R = Ac$
(c), $R = CH_3SO_2$

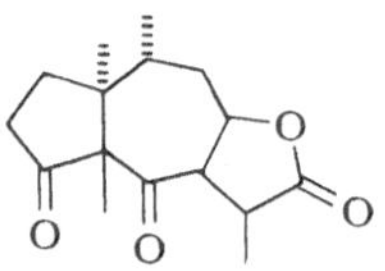

(94)

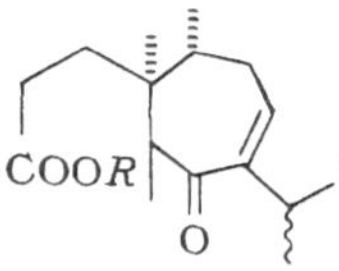

(95)
(a), $R = $ H
(b) $R = CH_3$

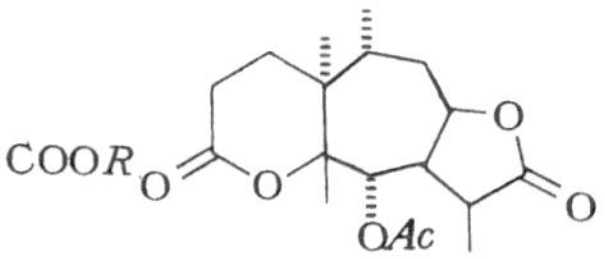

(96)

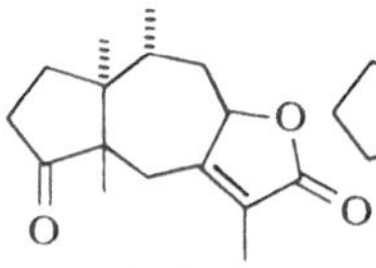

(97)

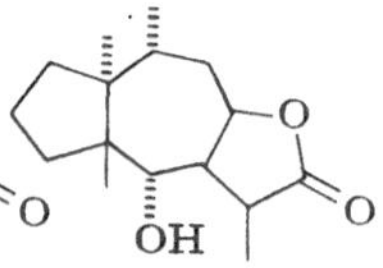

(98)

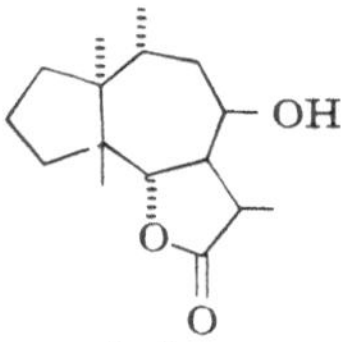

(99)

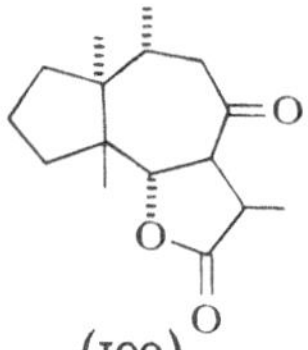

(100)

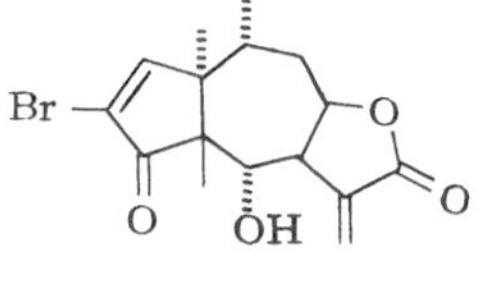

(101)
Bromohelenalin.

correlation gave further proof for the pseudoguaianolide structure of helenalin and established the identity of its asymmetric centers at $C_{(1)}$, $C_{(5)}$ and $C_{(10)}$ with those in tenulin (55).

X-ray analysis of bromohelenalin (101) carried out by Emerson, Caughlan and Herz (23) has confirmed structure (89a).

The relative configuration of (101) was in accord with that deduced earlier, and also represents the absolute configuration because of the correlation with tenulin (55).

References, pp. 127—130.

Mexicanin A.

Mexicanins A, C, D, E and H were isolated from *Helenium mexicanum* H. B. K. collected in the Valley of Mexico.

Mexicanin A, $C_{15}H_{18}O_4$, (102a) was isolated by Romo de Vivar and Romo (70, 71). The presence of an exocyclic methylene group conjugated

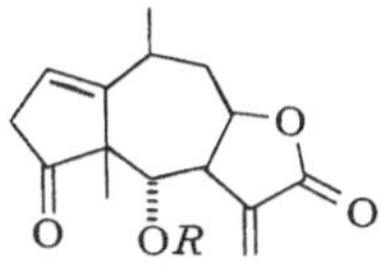

(102)
(a), R = H. Mexicanin A.
(b), R = Ac

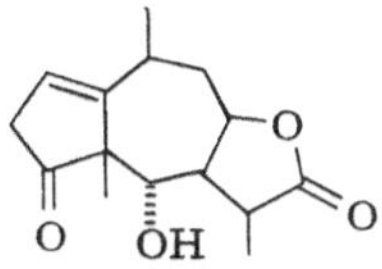

(103)
Dihydromexicanin A.

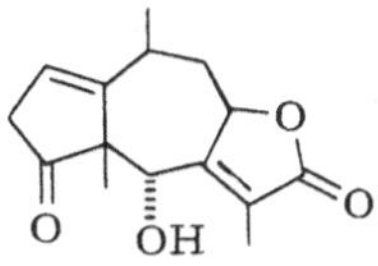

(104)
Isomexicanin A.

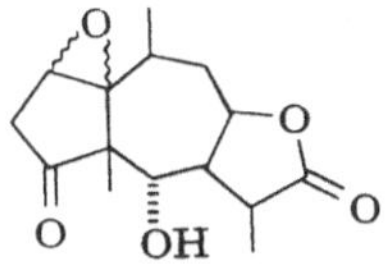

(105)

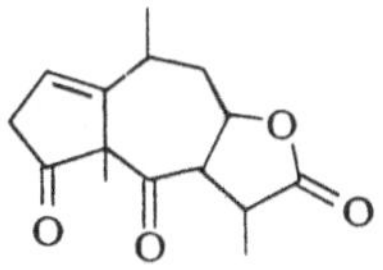

(106)

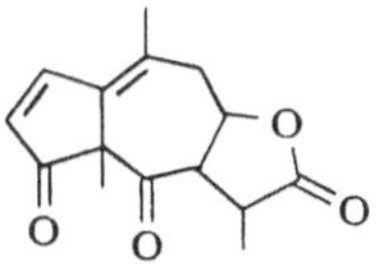

(107)

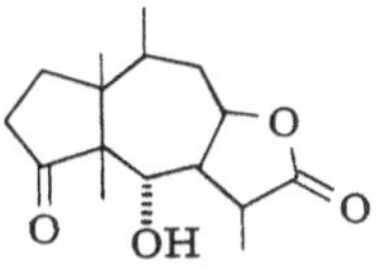

(108)
Tetrahydromexicanin A.

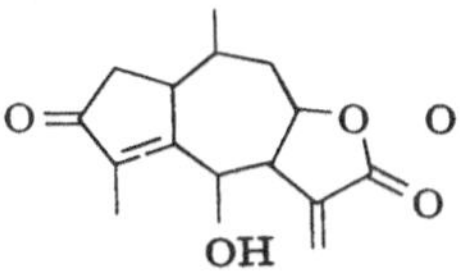

(109)

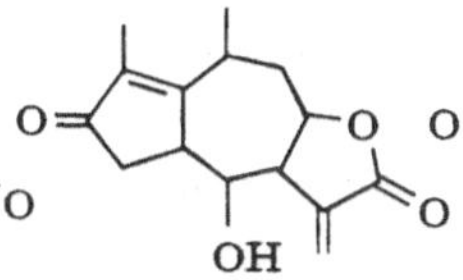

(110)
Neohelenalin.

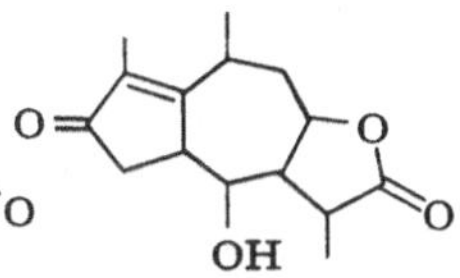

(111)

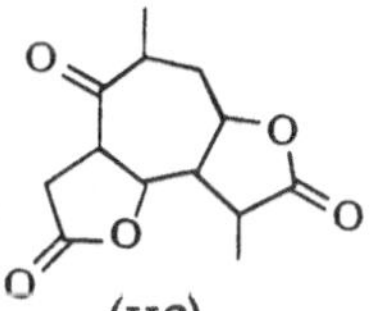

(112)

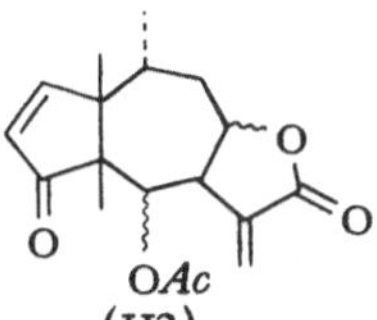

(113)
Linifolin A.

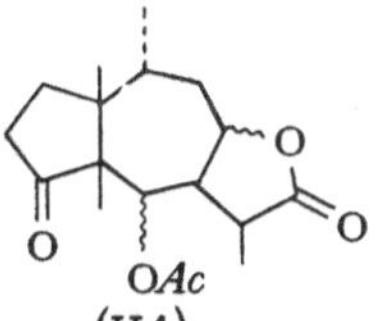

(114)
Tetrahydrolinifolin A.

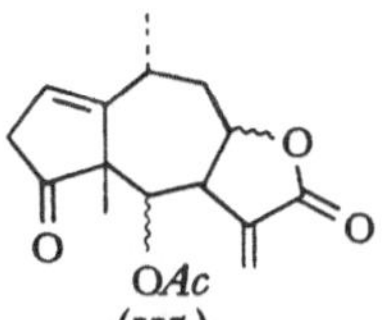

(115)
Linifolin B.

with a γ-lactone, of a hydroxyl group and of a ketone group was established by spectral methods (70). Further work carried out by Herz and coworkers (42) has demonstrated that mexicanin A is a pseudoguaianolide of structure (102a) and is closely related to helenalin. The NMR spectrum of (102a) exhibited only three vinyl protons and indicated that the lactone is oriented to $C_{(8)}$. Palladium-carbon catalyzed hydrogenation of mexicanin A saturated only the exocyclic methylene group, yielding dihydromexicanin A (103) and isomexicanin A (104). The dihydro derivative contained a double bond as shown by its spectrum and by the formation of the epoxide (105). Chromium trioxide oxidation of dihydromexicanin A gave the dehydro derivative (106). The spectroscopic properties of the dienone (107) prepared by selenium dioxide treatment of (106) were in accordance with the postulated position of the dienone chromophore.

The double bond of dihydromexicanin A (103) which resisted hydrogenation under neutral conditions was saturated in acidic medium in the presence of platinum oxide and yielded tetrahydromexicanin A (108). Mexicanin A (102a) was correlated with helenalin (89a) when a chloroform solution of the latter was treated with hydrogen chloride. This reaction afforded mexicanin A and neohelenalin (110) (see below). Treatment of helenalin (89a) with hydrochloric acid in acetic acid gave helenalin acetate (89b) and mexicanin A. From these observations it was concluded originally that the presence of the free hydroxyl group in helenalin was necessary for its transformation into mexicanin A and that the isomerization proceeded by way of a retroaldol cleavage followed by cyclization at the same position. On this basis the configuration of mexicanin A at the ring junction remained uncertain. However, the finding of Herz et al. (34) that treatment of helenalin acetate afforded mexicanin A acetate (102b) has indicated that the transformation involved only a shift of the double bond to $C_{(1)}$. Consequently, the stereochemistry of mexicanin A is as shown in (102a). Since tetrahydrohelenalin (93a) differs from tetrahydromexicanin A (108), the latter must possess a cis-β-configuration. As in the $C_{(1)}$ epitenulin series, the ORD curves of mexicanin A and of its derivatives showed Cotton effects which differed from those of model substances with rigid conformations (42, 34).

Neohelenalin.

Neohelenalin, $C_{15}H_{18}O_4$, (110) was isolated from an extract of Helenium flexuosum Raf. (37), by chromatography on basic alumina of an extract of Balduina angustifolia Pursch (31) and from H. ooclinium (28). Mexicanin D, a lactone isolated from H. mexicanum H. B. K. (70, 71) has been identified as neohelenalin (42). It is, however, not clear whether neohelenalin occurs as such or is an artifact formed from helenalin (89a) during the isolation.

Earlier work on neohelenalin (31) led to the proposal of structure (109) which was based on the erroneus guaianolide structure for helenalin (88). After the pseudoguaianolide formula of helenalin (89a) had been established and the correct constitution of desacetylneotenulin (80a) elucidated, Herz and coworkers (42) postulated structure (110) for neohelenalin in accordance with its chemical and spectral properties. On ozonolysis neohelenalin liberates acetic acid and formaldehyde.

The close relationship between neohelenalin, helenalin and mexicanin A was demonstrated by the facile isomerization of (89a) under acid conditions which resulted in mexicanin A (102a) and neohelenalin (110) (42). Furthermore, potassium bicarbonate treatment of dihydromexicanin A (103) gave dihydroneohelenalin (111) which could be prepared also by catalytic hydrogenation of (110) (42). Ozonolysis of neohelenalin followed by periodic acid oxidation afforded the ketodilactone (112) (42). Since the mechanism of the neohelenalin formation undoubtedly parallels that of desacetylneotenulin (80a), the stereochemistry of (110) at $C_{(5)}$ and $C_{(6)}$, perhaps also at $C_{(7)}$, $C_{(8)}$ and $C_{(10)}$, is uncertain.

Linifolin A.

From *Helenium linifolium* Rydb. HERZ (28) isolated a new sesquiterpene lactone, linifolin A, $C_{17}H_{20}O_5$, (113). Its spectral and chemical properties defined linifolin A as a new pseudoguaianolide with the same gross structure as acetylhelenalin (89b) and balduilin (117) (28) (see below). Ozonolysis of linifolin A gave formaldehyde. Catalytic hydrogenation yielded tetrahydrolinifolin A (114). HERZ et al. (39) suggested that linifolin A had a *cis* ring junction because the ORD curves of (113) and (114) were very similar to those of 1-epiisotenulin (74) and its tetrahydro derivative (75). Tetrahydrolinifolin A (114) differed from (75) and (108).

Linifolin B.

This lactone, $C_{17}H_{20}O_5$, (115) was also isolated by HERZ (28) from *H. linifolium* Rydb. Its spectrum indicated that it bears the same relationship to linifolin A (113) as the unconjugated cyclopentenone (73) does to 1-epiisotenulin (74) (34), and mexicanin A (102a) to helenalin (89a) (42).

Balduilin.

Balduilin, $C_{17}H_{20}O_5$, (117) was named by HERZ, MITRA and JAYARAMAN (36) who isolated it from *Balduina uniflora*. The nature of its oxygen functions was established by the same authors as a cyclopentenone, an exocyclic methylene group conjugated with a five-membered lactone and an acetate. Balduilin liberated formaldehyde on ozonolysis. Hydrolysis of several derivatives afforded acetic acid. HERZ et al. correlated balduilin with some compounds of the helenalin and tenulin series and proposed structure (116) before the pseudoguaianolide structures of these lactones had been established. On the basis of the formulation of tenulin (55) and helenalin (89a) as pseudoguaianolides, the structure of balduilin was revised to (117) (42, 43). Balduilin and helenalin acetate

(89b), which have the same gross structure, differ in their properties, so do tetrahydrobalduilin (118) obtained by catalytic hydrogenation of (117) and dihydroisotenulin (65b). Thus, the three series differ from each other at one or more asymmetric centers.

Saponification of tetrahydrobalduilin (118) with potassium hydroxide took place with simultaneous epimerization at $C_{(11)}$ and the resulting desacetyl derivative (119a) gave an acetate (119b) which differed from tetrahydrobalduilin (118). The $C_{(11)}$ asymmetric center of tetrahydro-helenalin (93a) was also epimerized by alkaline treatment yielding (120a) (36, 70). Chromic acid oxidation of the epimer (120a) afforded the dehydro derivative (121) (36) which was identified with the diketone prepared by oxidation of the hydroxyl group of desacetyltetrahydro-balduilin (119a). This finding demonstrates that balduilin (117) and helenalin (89a) differ in configuration at the $C_{(6)}$ hydroxyl group, that of balduilin being β-oriented (43).

Desoxotetrahydrobalduilin (122) was obtained by desulfuration with Raney nickel of the intermediary ethylenethioketal of tetrahydro-balduilin (118) (36). Saponification of (122) with sodium hydroxide was accompanied by lactone reorientation and gave the hydroxylactone (123). Chromium trioxide oxidation of the latter resulted in a dehydro derivative which was identified as the ketolactone (71, p. 103) of the alloisotenulin series. Therefore, balduilin (117) and tenulin (55) differ in the configuration at the $C_{(8)}$ asymmetric center, and the $C_{(8)}$ hydroxyl group of balduilin is β-oriented.

Mexicanin C.

This pseudoguaianolide, $C_{15}H_{20}O_4$, (124a) was found in *Helenium mexicanum* H. B. K. by Romo de Vivar and Romo (70, 71). Mexicanin C contains a cyclopentenone chromophore, a five-membered lactone and a secondary hydroxyl group. It formed the acetate (124b) and could be oxidized to the ketone (125) (43, 70). Mexicanin C (124a) was correlated with helenalin (89a) (36, 70). Dihydromexicanin C obtained by catalytic hydrogenation (36, 70) was identified with the $C_{(11)}$ epimer of tetrahydro-helenalin (120a). This correlation led to the structure and stereochemistry of mexicanin C (124a) (43).

Mexicanin I.

Mexicanin I, $C_{15}H_{18}O_4$, (126) was described by Domínguez and Romo (22) as a constituent of a variety of *Helenium mexicanum* H. B. K. collected in the neighborhood of the city of Oaxaca, isolated later from *H. aromaticum* (Hook) Bailey (62, 63) and *Gaillardia pinnatifida* Torr (39).

It is a pseudoguaianolide (**126 a**) with the gross structure of helenalin (**89 a**); however, its asymmetric centers are oriented as those of isotenulin (**56 b**, p. 102), as shown by the following transformations (*22*).

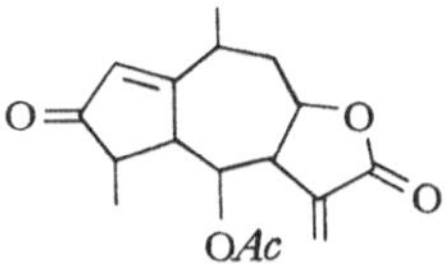

(**116**)

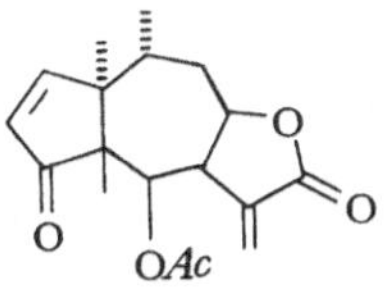

(**117**)
Balduilin.

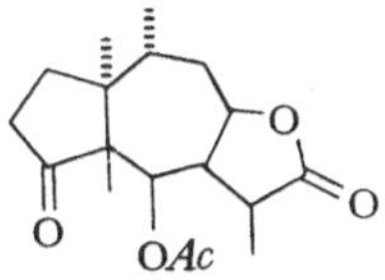

(**118**)
Tetrahydrobalduilin.

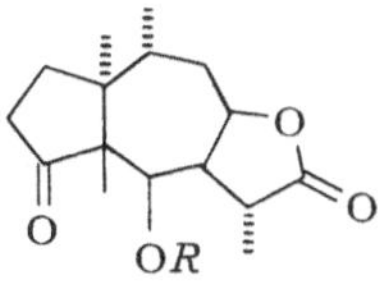

(**119**)
(a), R＝H
(b), R＝Ac

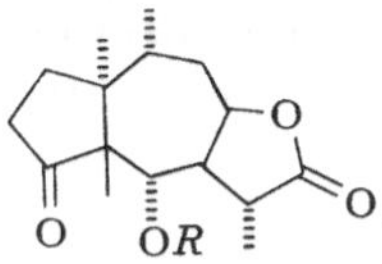

(**120**)
(a), R＝H. Epitetrahydrohelenalin.
(b), R＝Ac

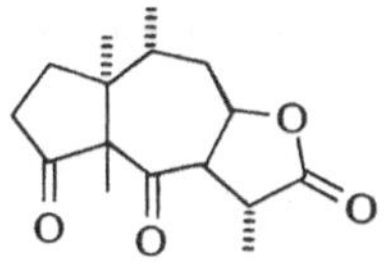

(**121**)

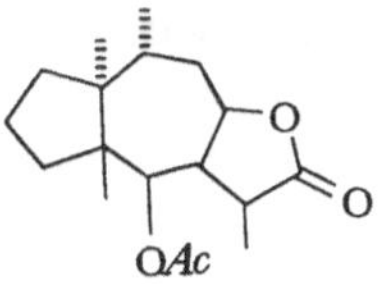

(**122**)

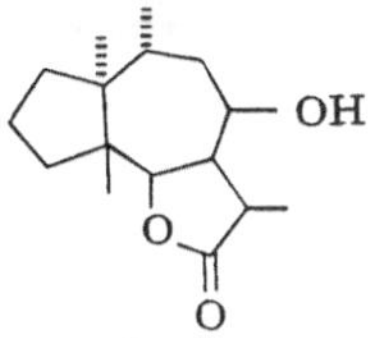

(**123**)

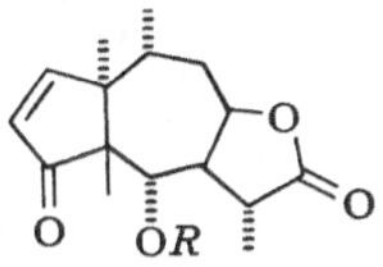

(**124**)
(a), R＝H. Mexicanin C.
(b), R＝Ac

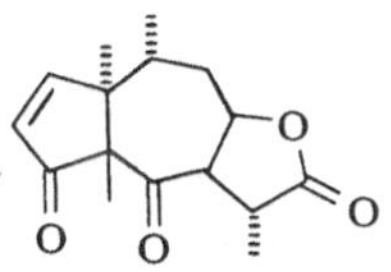

(**125**)
Dehydromexicanin C.

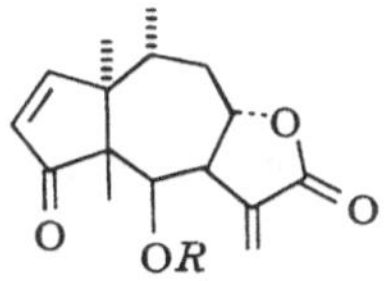

(**126**)
(a), R＝H. Mexicanin I.
(b), R＝Ac

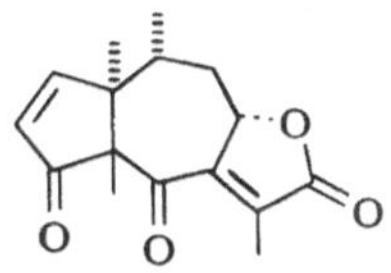

(**127**)
Dehydroisomexicanin I.

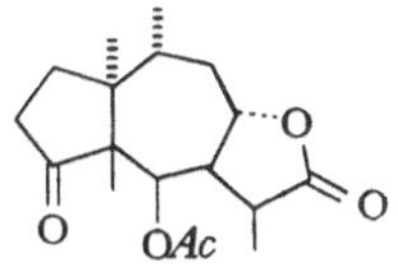

(**128**)
Tetrahydromexicanin I acetate.

The nature of the chromophores of mexicanin I (**126 a**) was deduced by spectral methods. Base catalyzed acetylation formed an acetate whose NMR spectrum was in accord with structure (**126 b**). Chromium trioxide oxidation of mexicanin I was accompanied by migration of the exocyclic methylene double bond to the endocyclic position, and resulted in dehydroisomexicanin I (**127**). Reduction of the latter with zinc in acetic acid saturated both double bonds and afforded a diketone which was identified as dehydrodesacetyldihydroisotenulin (**68**, p. 103). Desulfuration of the adduct of (**126 a**) with toluenethiol yielded a tetrahydro derivative which was identical with desacetyldihydroisotenulin (65 a, p. 103). Hydrogenation of acetylmexicanin I (126 b) formed a tetrahydro derivative (128) which was epimeric at $C_{(11)}$ with dihydroisotenulin (65 b).

Aromatin.

This lactone, $C_{15}H_{18}O_3$, (129) was isolated by Romo, Joseph-Nathan and Díaz (62, 63) from the Chilean species *Helenium aromaticum* (Hook) Bailey. The presence in aromatin (129) of two separate chromophores, viz. a cyclopentenone and an exocyclic methylene group conjugated with a γ-lactone was inferred from the spectral properties (62, 63). The relationship of aromatin to helenalin (89 a) has become evident when dihydro-isoaromatin, obtained by catalytic hydrogenation of (129), was identified as an anhydro derivative (97, p. 108) of the helenalin series. This correlation permitted the assignment of structure and stereochemistry to aromatin (129).

Aromaticin.

This sesquiterpene, $C_{15}H_{18}O_3$, (130) was discovered by Romo, Joseph-Nathan and Díaz in *Helenium aromaticum* (62, 63) and was isolated later also from *H. amarum* (Raf.) Rock (51). Aromaticin which differs in configuration at $C_{(8)}$ from aromatin (129) bears the same relationship to mexicanin I (126a) as (129) does to helenalin (89a) (62, 63). The structure of aromaticin was established by correlation with isotenulin (56 b) (51, 62, 63). Hydrogenation of aromaticin gave a dihydroisoderivative identical with the desoxylactone (67) derived from desacetyl-dihydroisotenulin (65a).

Mexicanin E.

This nor-sesquiterpene lactone, $C_{14}H_{16}O_3$, (131) was isolated from *Helenium mexicanum* H. B. K. by Romo de Vivar and Romo (71) and found later by Herz (28) in *H. ooclinium*. Although mexicanin E (131) is not a pseudoguaianolide (72, 66), its isolation from *Helenium* species and the relationship to some other lactones of this series (67, 66) merits its inclusion in the present review.

Earlier work (71, 72) had established the azulogenic character of this lactone and the nature of its chromophores; and determination of the molecular weight by mass spectrometry was in accord with the empirical formula. Structure (131) for mexicanin E was proposed by Romo, Romo de Vivar and Herz (66) as a result of further chemical and spectral evidence. The NMR spectrum demonstrated the presence of four vinyl protons and of only a single secondary methyl group. Mexicanin E afforded a red 2,4-dinitrophenylhydrazone (72) and liberated formaldehyde on ozonolysis. Partial hydrogenation saturated the exocyclic

References, pp. 127—130.

methylene group and gave dihydromexicanin E (**132**), a constituent of
H. autumnale L. isolated by LUCAS, SMITH and DORFMAN (*52*). Dihydromexicanin E furnished an orange-colored 2,4-dinitrophenylhydrazone
(*72*).

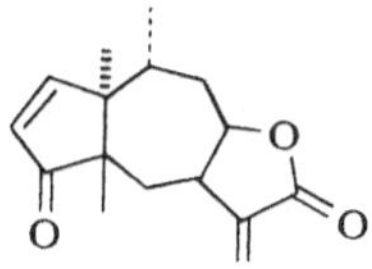

(**129**)
Aromatin.

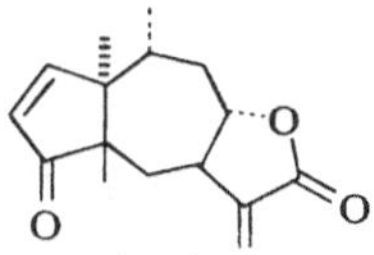

(**130**)
Aromaticin.

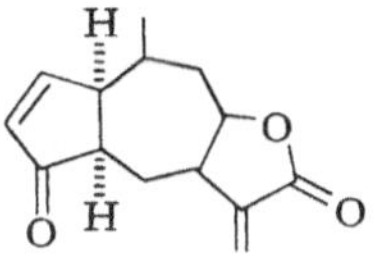

(**131**)
Mexicanin E.

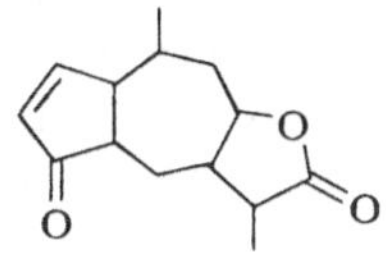

(**132**)
Dihydromexicanin E.

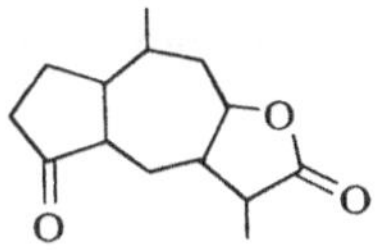

(**133**)
Tetrahydromexicanin E.

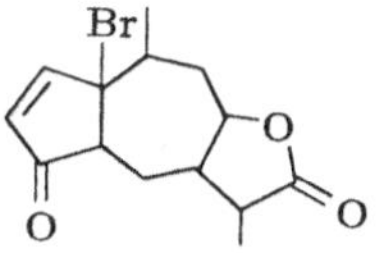

(**134**)

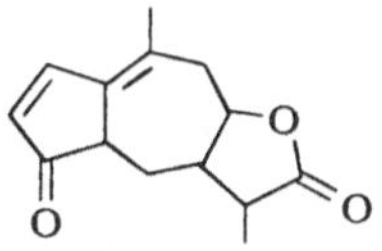

(**135**)

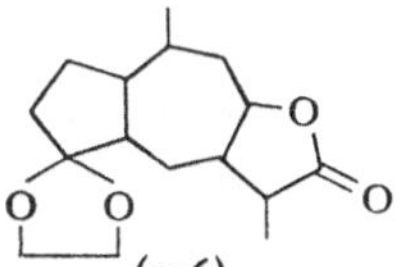

(**136**)

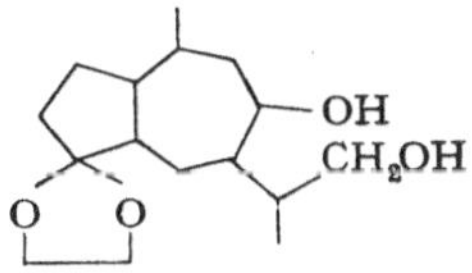

(**137**)

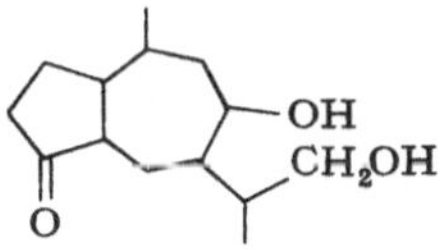

(**138**)

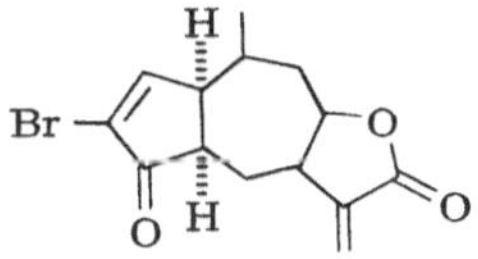

(**139**)
Bromomexicanin E.

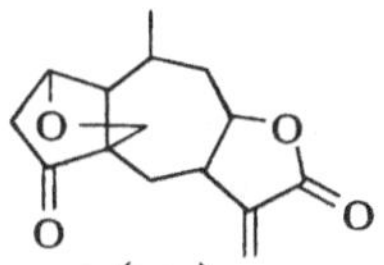

(**140**)
Mexicanin H.

Absorption of two equivalents of hydrogen by mexicanin E or further
hydrogenation of (**132**) led to tetrahydromexicanin E (**133**). Allylic
bromination of dihydromexicanin E (**132**) afforded the bromo derivative
(**134**) (*66*). Elimination of hydrobromic acid with potassium acetate from
(**134**) resulted in the dienone (**135**) whose spectral properties demonstrated
the relative position of the chromophore.

The following transformations (*66*) have indicated that the lactone
ring of mexicanin E was closed at $C_{(8)}$. Tetrahydromexicanin E was

converted to its cycloethyleneketal (136). Acid hydrolysis of the diol (137), obtained by lithium aluminum hydride reduction of (136), furnished the ketone (138). Treatment of the latter with methyl magnesium bromide, followed by dehydrogenation of the resulting adduct, yielded linderazulene (42, p. 98).

X-ray analysis of bromomexicanin E (139) by Mazhar-ul-Haque and Caughlan (56) indicates that the stereochemistry of mexicanin E as shown in (131) differs from that of the other pseudoguaianolides isolated in the genus *Helenium*.

Pseudoguaianolides with appropriate features, such as a $C_{(4)}$ keto group and an oxygen function at $C_{(14)}$, may serve as precursors in the biogenesis of mexicanin E, since they can undergo degradative elimination resulting in loss of the $C_{(5)}$ angular substituent. This assumption is supported by the structure assigned to mexicanin H (140) (67).

Mexicanin H.

Mexicanin H, $C_{15}H_{18}O_4$, (140) was isolated from *Helenium mexicanum* H. B. K. by Romo de Vivar and Romo (71). The spectral properties of this pseudoguaianolide (140) (67, 71) indicated that it contains an exocyclic methylene group conjugated with a γ-lactone closed at $C_{(8)}$, a cyclopentanone ring and a furane ring. The ethereal oxygen of the latter is attached to a carbon atom carrying a tertiary hydrogen and to a methylene group, the latter apparently resulting from partial oxidation of a tertiary methyl group. Romo, Romo de Vivar and Joseph-Nathan (67) postulated structure (140) for mexicanin H (65) on the basis of its mass spectrum. The fragmentation pattern showed a peak with mass 30 corresponding to formaldehyde, and otherwise possessed all the features of the mass spectrum of mexicanin E (131). The stereochemistry of mexicanin H has not yet been established.

Flexuosins.

Flexuosin A, $C_{17}H_{24}O_6$, (141a) and flexuosin B, $C_{20}H_{28}O_6 \cdot H_2O$, (146a) were isolated by Herz, Jayaraman and Watanabe (31) from *Helenium flexuosum* Raf.

Flexuosin A. Its structure (141a) was proposed by Herz, Kishida and Lakshmikantham (32). The six oxygen atoms of flexuosin were found to be distributed in an α,β-unsaturated γ-lactone closed toward $C_{(8)}$, a $C_{(6)}$ acetoxy grouping and two hydroxyl groups at $C_{(2)}$ and $C_{(4)}$ in the five-membered ring. Flexuosin A forms a diacetate (141b). Selective oxidation of flexuosin A gave the hydroxycyclopentanone (142) which could be dehydrated with methane-sulfonyl chloride to the α,β-unsaturated ketone (143). Desulfuration of the ethylene thioketal of (142), followed by oxidation of the resulting intermediary desoxo derivative, led to a ketone (144) isomeric with dihydrobigelovin (85, p. 107). Treatment of flexuosin A with sodium amalgam resulted in reduction of the exocyclic methylene group, hydrolysis of the acetate and lactone reorientation; thus, the hydroxylactone (145) of the alloflexuosin series was formed. Since flexuosin A has not yet been correlated with a sesquiterpene lactone of the helenalin, tenulin or balduilin series, its stereochemistry remains in doubt.

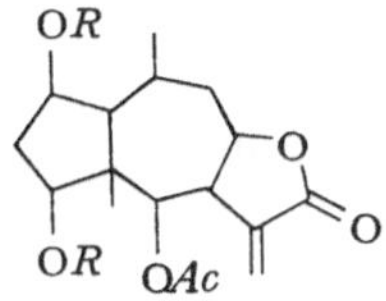

(141)
(a), R = H. Flexuosin A.
(b), R = Ac

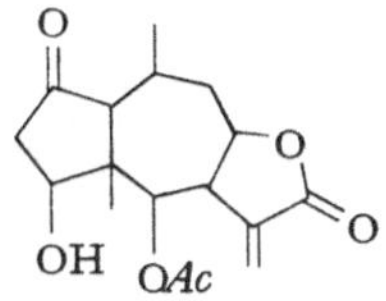

Dehydroflexuosin A.
(142)

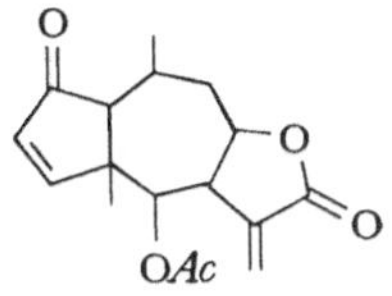

(143)

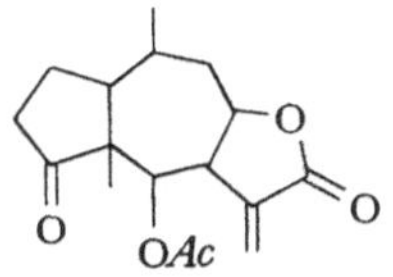

(144)

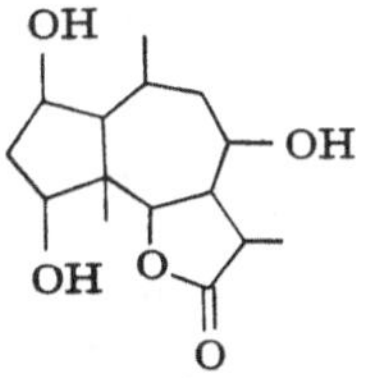

(145)

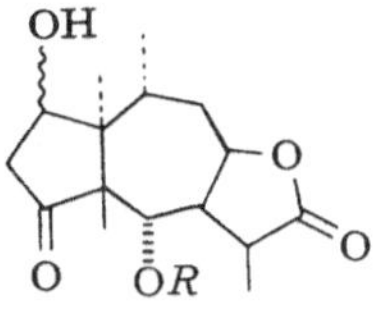

(146)
(a), R = (CH₃)₂C=CH−C=O
Flexuosin B.
(b), R = (CH₃)₂CH−CH₂−C=O

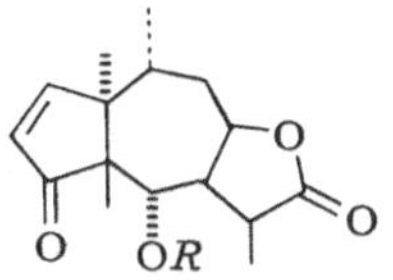

(147)
(a), R = (CH₃)₂CH=CH−C=O
(b), R = (CH₃)₂CH−CH₂−C=O

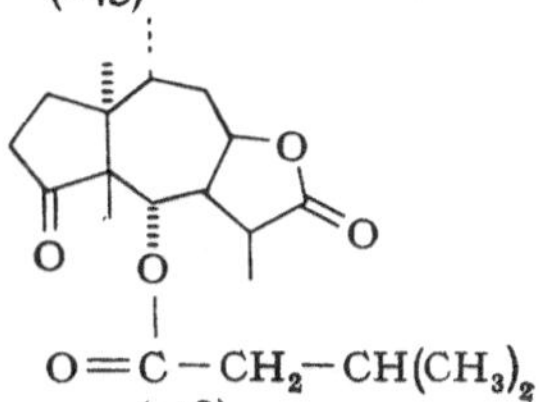

(148)
O=C−CH₂−CH(CH₃)₂

Flexuosin B was established by HERZ et al. (32) as a pseudoguaianolide of structure (146a) which is closely related to mexicanin C (124a). Spectral examination of flexuosin B indicated the presence of a five-membered lactone ring closed toward $C_{(8)}$, a cyclopentanone, a hydroxyl group and an ester grouping, the latter being due to esterification by a five-carbon unsaturated acid. NMR evidence suggested that this was senecioic acid, which was confirmed by ozonolysis producing acetone. Catalytic hydrogenation of (146a) yielded dihydroflexuosin B (146b). The cyclopentenones (147a) and (147b) resulted from the dehydration of flexuosin B or of its dihydro derivative (146b). Hydrogenation of the anhydro derivatives (**147a**) and (**147b**) yielded the same saturated ketone (**148**). Correlation of flexuosin B with mexicanin C (**124a**) was achieved when saponification of (**148**) gave isovaleric acid and dihydromexicanin C (**120a**, p. 113). The synthesis of (**148**) by esterification of tetrahydrohelenalin (**93a**, p. 108) with isovaleroyl chloride has demonstrated that in flexuosin B and its derivatives the center at $C_{(11)}$ possesses β-configuration. Evidence bearing on the configuration of the $C_{(2)}$ hydroxyl group is not yet available.

Amaralin.

Lucas et al. (*51*) isolated from *Helenium amarum* (Raf.) Rock a pseudoguaianolide named amaralin, $C_{15}H_{20}O_4$, and determined its structure (**149a**). Amaralin was the first epoxide described in the pseudoguaianolide series and is of special interest because of its analgesic activity. The presence of a hydroxyl and of an exocyclic methylene group conjugated with a γ-lactone was inferred from its spectral properties. The NMR spectrum of amaralin showed the $C_{(8)}$ orientation of the lactone.

Amaralin formed the acetate (**149b**). Catalytic hydrogenation or reduction with sodium borohydride afforded the dihydro derivative (**150**). Oxidation of (**150**) gave the ketone (**151**). The α-epoxyketone system of (**151**) was rearranged in alkali to the enolic α-diketone (**152a**). The ultraviolet spectra of the latter and of its acetate (**152b**) are in accord with the postulated structures.

The structure of amaralin (**149a**) was elucidated when chromous chloride reduction of the ketone (**153**), obtained by oxidation of (**149a**), resulted in aromaticin (**130**, p. 115). The above correlations have defined the structure and stereochemistry of amaralin at $C_{(1)}$, $C_{(5)}$, $C_{(7)}$, $C_{(8)}$ and $C_{(10)}$. The configurations at the asymmetric centers $C_{(2)}$, $C_{(3)}$ and $C_{(4)}$ as given in (**149a**) were deduced from the NMR spectrum of amaralin.

Thurberilin.

Herz and Lakshmikantham (*33*) discovered thurberilin, $C_{20}H_{26}O_5$, in *Helenium thurberi* Gray and proposed the structure (**154**) for this pseudoguaianolide. The empirical formula combined with the spectral properties indicated that this lactone is a pseudoguaianolide esterified with a C_5-carboxylic acid. The presence of a cyclopentenone and of a five-membered lactone was also secured by spectral methods. Since thurberilin produced acetaldehyde on ozonolysis, and hydrolysis of (**154**) led to the isolation of tiglic acid, it was concluded that thurberilin was a tiglate or angelate of a pseudoguaianolide. The NMR spectrum of (**154**) indicated that angelic acid was responsible for the ester function of thurberilin.

Catalytic hydrogenation of thurberilin with platinum oxide afforded tetrahydrothurberilin (155b), whereas hydrogenation in the presence of palladium on charcoal resulted in a mixture of dihydrothurberilin (155a) and (155b) which could not be resolved. However, from the ozonolysis of this mixture, tetrahydrothurberilin was recovered while (155a) was

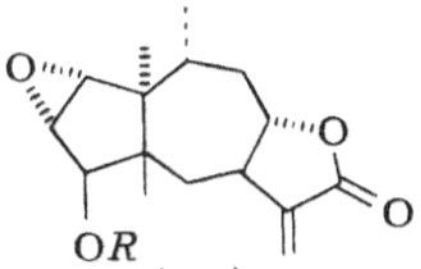

(149)
(a), R = H, Amaralin.
(b), R = Ac

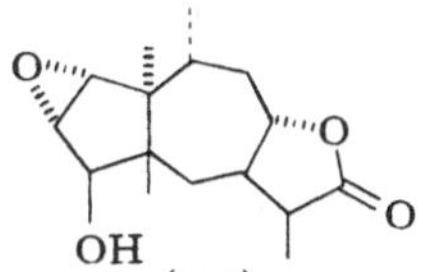

(150)
Dihydroamaralin.

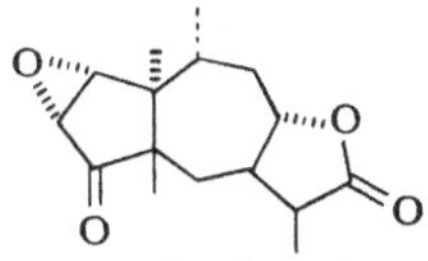

(151)
Dihydrodehydroamaralin.

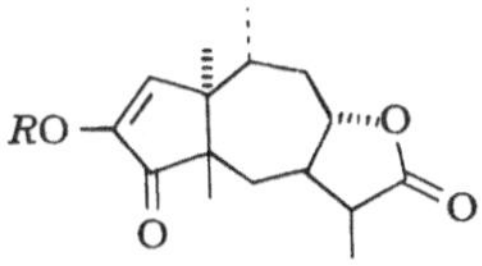

(152)
(a), R = H
(b), R = Ac

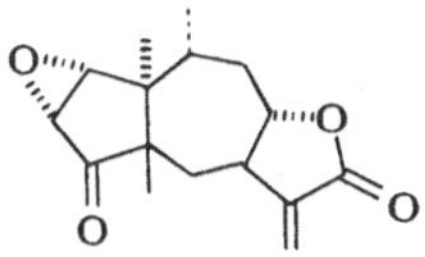

(153)
Dehydroamaralin.

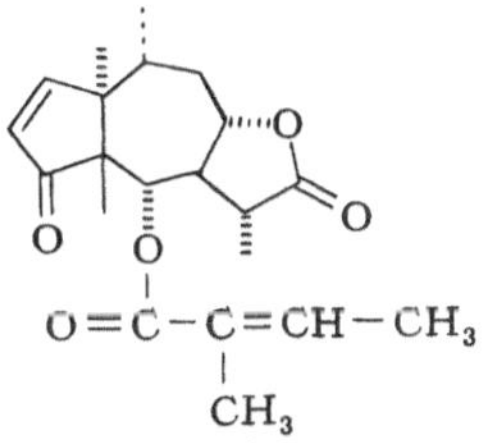

(154)
Thurberilin.

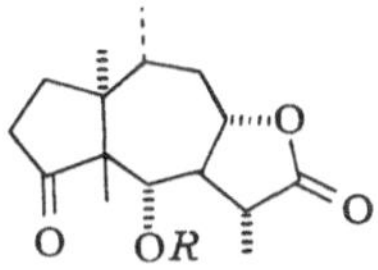

(155)
(a), $R = CH_3-CH=C-C=O$ Dihydrothurberilin.
            |
           CH_3
(b), $R = CH_3-CH_2-CH-C=O$ Tetrahydrothurberilin.
            |
           CH_3
(c), $R = CH_3-CO-C=O$

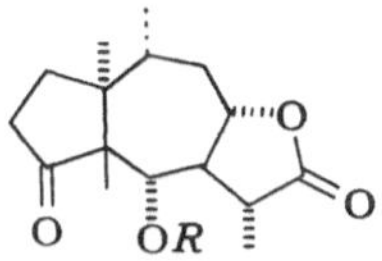

(156)
(a), R = H
(b), $R = CH_3SO_2$

degraded to the pyruvate (155c). The latter was smoothly hydrolyzed to the hydroxylactone (156a). Treatment with lutidine of the mesylate of the hydroxylactone (156b) furnished anhydrodesacetyldihydroisotenulin (67, p. 103). This correlation established the structure of thurberilin as (154) with the asymmetric centers at $C_{(1)}$, $C_{(5)}$, $C_{(8)}$ and $C_{(10)}$ having the same configuration as in tenulin (55, p. 102). The stereochemistry of the remaining asymmetric centers of thurberilin was elucidated when the acetate of the hydroxylactone (156a) was found to be identical with tetrahydrobigelovin (86, p. 107).

Pulchellins.

Pulchellin. Herz, Ueda and Inayama (45) isolated the pseudo-guaianolide pulchellin, $C_{15}H_{22}O_4$, (157 a) from the coastal variety of *Gaillardia pulchella* Foug. Pulchellin possesses an α,β-unsaturated lactone group closed toward $C_{(8)}$ as deduced by spectral methods. Two secondary hydroxyl groups of (157 a) were responsible for the formation of a di-acetate (157 b). Ozonolysis of pulchellin gave formaldehyde. The same

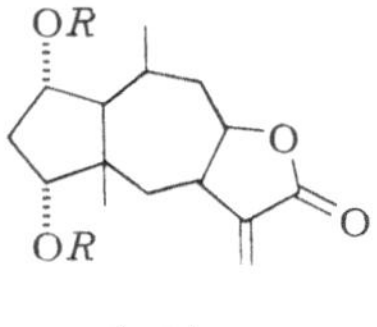

(157)
(a), R = H, Pulchellin.
(b), R = Ac

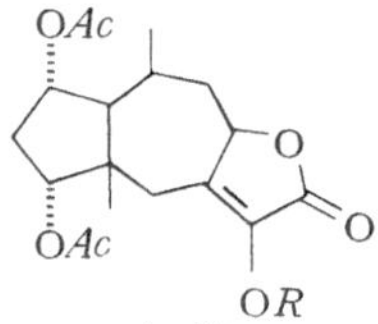

(158)
(a), R = H, Norpulchellonediacetate.
(b), R = Ac

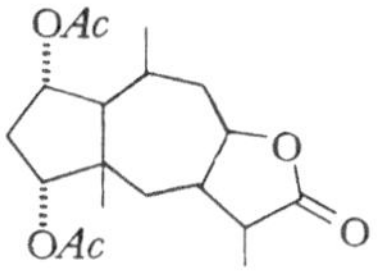

(159)
Dihydropulchellindiacetate.

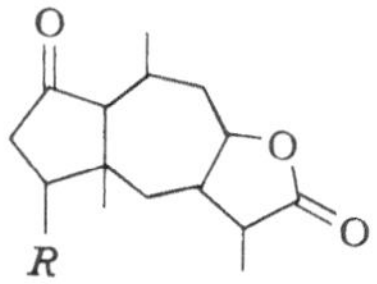

(160)
(a), R = OH
(b), R = AcO
(c), R = CH₃SO₂
(d), R = H

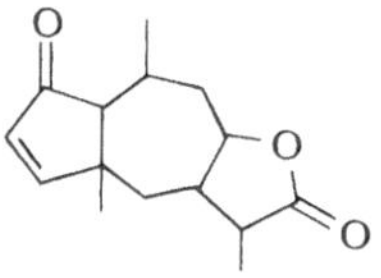

(161)

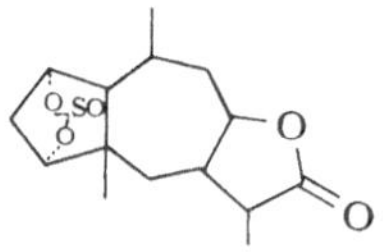

(162)

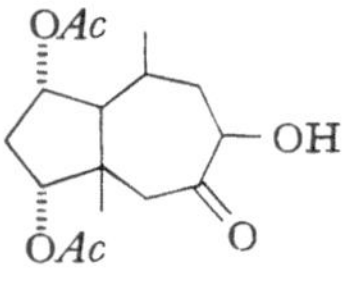

(163)

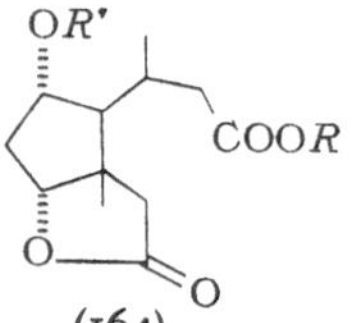

(164)
(a), R = R' = H
(b), R = CH₃, R' = H
(c), R = CH₃, R' = Ac

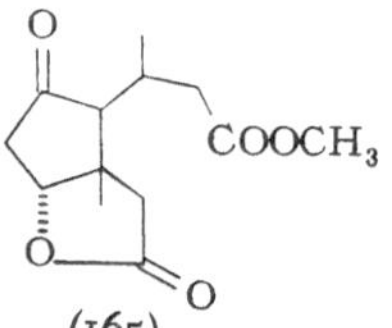

(165)

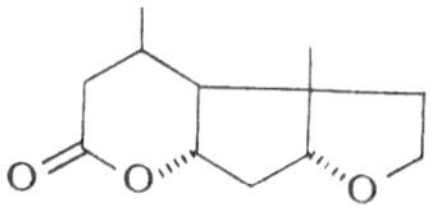

(166)

reaction carried out with the diacetate (157 b) led to the formation of the enolic ketolactone (158 a) which gave a positive ferric chloride test and yielded an acetate (158 b).

Hydrogenation of pulchellin afforded the dihydro derivative (159). Careful oxidation of dihydropulchellin led to the hydroxyketone (160 a). The latter afforded the acetate (160 b) and the mesylate (160 c). Elimination of the elements of methanesulfonic acid from (160 c) gave the cyclo-pentenone (161) whose hydrogenation yielded the ketone (160 d). The *cis* relationship of the hydroxyl groups of pulchellin was demonstrated by the formation of the cyclic sulfite (162) on treatment of dihydropulchellin with thionyl chloride.

Reduction of pulchellin with lithium aluminum hydride, followed by dehydrogenation, gave only traces of azulenes. However, the following degradations have afforded evidence of the skeletal structure and allowed configurational assignments to most of the asymmetric centers in (157a). Potassium permanganate oxydation of the enolic ketolactone (158a) resulted in formation of the α-ketol (163) which on treatment with base was further degraded to the lactone-acid (164a). The spontaneous lactonization involved in formation of (164a) indicated a cis relationship of the hydroxyl group and the carboxylic acid side-chain. The lactone acid (164a) gave a methyl ester (164b) and a methylester acetate (164c). Oxidation of (164b) led to the ketone (165) whereas treatment of (164a) with acetic anhydride and sodium acetate resulted in the δ-lactone (166). Application of the Hudson-Klyne rule to (164a) and (166) suggested the absolute configuration at $C_{(4)}$ which secures the configurations at $C_{(2)}$ and $C_{(5)}$ as well. The $C_{(7)}$ side-chain is presumably equatorial and β.

Pulchellins B and C. Pulchellin B, $C_{17}H_{22}O_5$, (167b) and C, $C_{15}H_{20}O_4$, (167a) were isolated by HERZ and INAYAMA (30) from *Gaillardia pulchella* Foug. These pseudoguaianolides are closely related (30). Pulchellin B was identified as the monoacetate (167b) of pulchellin C (167a) since both lactones afforded the same diacetate (167c). The positive periodate test given by pulchellin C demonstrated the vicinal position of its hydroxyl groups. Since both pulchellins formed pyrazolines, they must contain an unsaturated γ-lactone grouping. The presence of a second double bond in the pulchellins was inferred from their infrared spectra. That these sesquiterpenes were perhydroazulenes was proved by dehydrogenating the reduction product obtained from a mixture of the two pulchellins; this resulted in a low yield of a 1,7-dialkyl- and a 1,4,7-trialkylazulene as indicated by the spectrum.

Pd-CaCO₃ catalyzed hydrogenation of the diacetate (167c) saturated only the exocyclic methylene group conjugated with the lactone and yielded (168c). Hydrogenation of the pulchellins B and C and of the diacetate (167c) with platinum oxide saturated both double bonds and led to the tetrahydro derivatives (169b), (169a) and (169c), respectively.

The relative position of the vicinal glycol system of the pulchellins was established as follows. Oxidation of tetrahydropulchellin B (169b) with chromium trioxide in acetic acid yielded the ketone (170). Pyrolysis of the latter gave a crude product with the spectral properties of a cyclopentenone. Oxidation of tetrahydropulchellin C (169a) afforded an enolic α-diketone (171). Ozonolysis of dihydropulchellin C (168a) or of its diacetate (168c) liberated formaldehyde and furnished the apoketones

(172a) and (172b). Treatment of (172a) with acetic anhydride and pyridine resulted in dehydration, leading to the x,β-unsaturated ketone (173). Interaction of (172a) with formic acid produced by way of a pinacolic rearrangement the enolic β-diketone (174) which was also obtained by treatment of the dimesylate (172c) with sodium acetate.

The stereochemistry of the pulchellins is still unclarified.

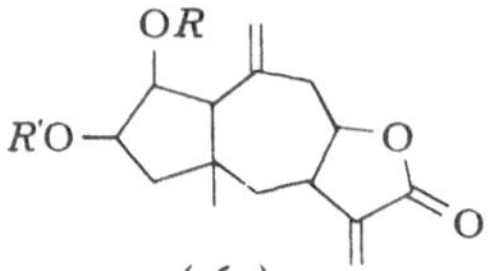

(167)

(a). $R=R'=H$ Pulchellin C.
(b). $R=H$, $R'=Ac$ Pulchellin B.
(c). $R=R'=Ac$

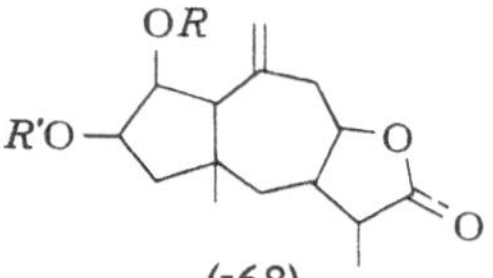

(168)

(a). $R=R'=H$ Dihydropulchellin C.
(b). $R=H$, $R'=Ac$ Dihydropulchellin B.
(c). $R=R'=Ac$

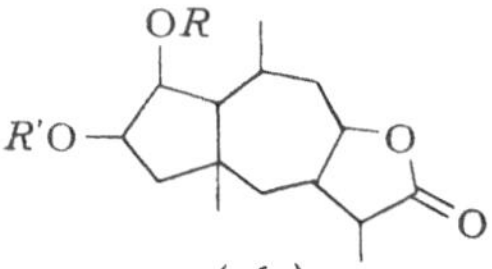

(169)

(a). $R=R'=H$ Tetrahydropulchellin C.
(b). $R=H$, $R'=Ac$ Tetrahydropulchellin B.
(c). $R=R'=Ac$

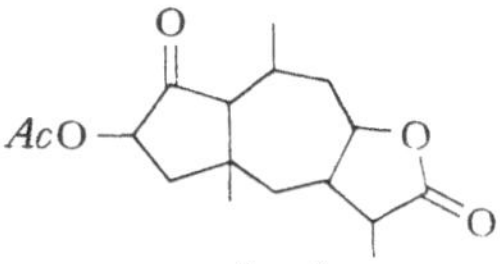

(170)

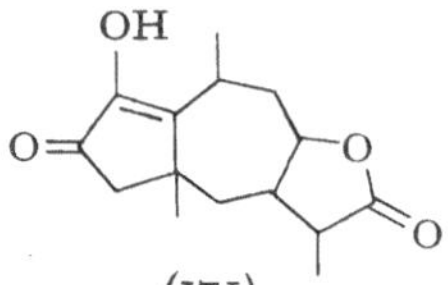

(171)

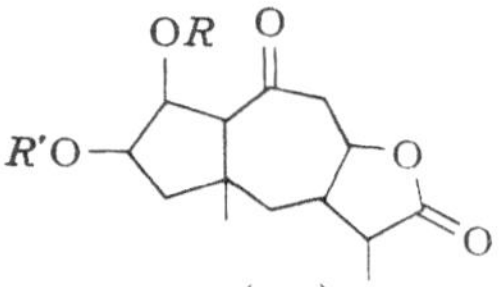

(172)

(a). $R=R'=H$
(b). $R=R'=Ac$
(c). $R=R'=CH_3SO_2$

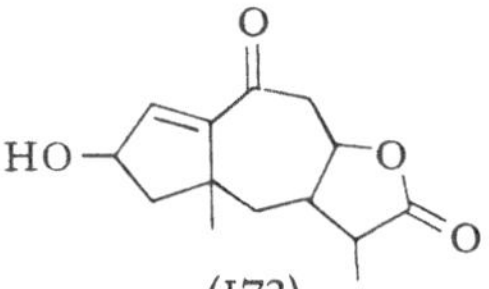

(173)

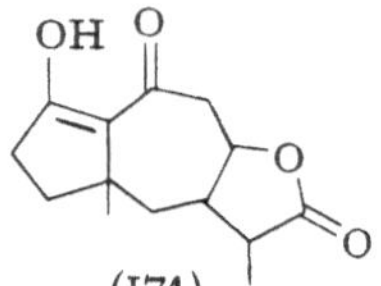

(174)

Gaillardilin.

Gaillardilin, $C_{17}H_{22}O_6$, (175) was isolated by HERZ et al. from *Gaillardia pinnatifida* Torr. and *G. arizonica* Gray (39). The same authors have established the constitution of gaillardilin (175). This sesquiterpene possesses an unsaturated lactone ring closed toward $C_{(8)}$, a hydroxyl group and an acetate as indicated by previous spectroscopic examination. On ozonolysis gaillardilin liberated formaldehyde. Catalytic hydrogenation yielded the dihydro derivative (176). Treatment of gaillardilin with boron trifluoride etherate resulted in the β,γ-unsaturated cyclopentenone (177). The same reaction afforded the ketone (178) from dihydrogaillardilin (176). Treatment of (178) with hydrochloric acid gave the conjugated cyclopentenone (179). Hydrogenation of (179) yielded the saturated compound (180).

Examination of the ORD curves of (179) and (180) which are stereoisomers of isotenulin (56 b) and of dihydroisotenulin (65 b, pp. 102, 103), respectively, indicated that these gaillardilin derivatives posses a β-cis ring junction. Coupling constants and chemical shifts suggested the stereochemistry as implied in (175).

Gaillardipinnatin.

This lactone, $C_{19}H_{22}O_9$, (181) was isolated from a collection of *Gaillardia pinnatifida* Torr. by HERZ et al. (39). The structure (181) of gaillardipinnatin was determined on spectral grounds (39, 40). A comparative study of the NMR spectra of (181) with those of the fastigilins by the irradiation technique, carried out by the same investigators, has confirmed the gaillardipinnatin structure and suggested the stereochemistry as shown in (181).

Fastigilins.

These closely related lactones were isolated by HERZ et al. (40) from *Gaillardia fastigiata* Greene. The structures of the fastigilins were determined by these authors mainly on spectral basis. The similarity of ORD curves and of the coupling constants in the NMR spectra of the fastigilins and derivatives and those exhibited by lactones of the helenalin series led to a tentative assignment of the stereochemistry depicted in the formulas (182a), (182b) and (183).

Fastigilin A. The constitution of fastigilin A, $C_{20}H_{26}O_6$, is (182a). The NMR spectrum permitted to deduce the presence of an angeloyl side-chain attached to $C_{(6)}$. The similarity of chemical shifts and coupling constants in the NMR spectra of (182a) and fastigilin B (182b) indicates that these lactones possess closely related structures and probably the same stereochemistry.

Fastigilin B. This lactone, $C_{20}H_{26}O_6$, (182b) differs from fastigilin A (182a) only in the nature of the side-chain attached to $C_{(6)}$; the $C_{(6)}$ hydroxyl is esterified by senecioic acid. Hydrogenation of (182b) gave the tetrahydro derivative (184).

Fastigilin C. This pseudoguaianolide, $C_{20}H_{24}O_6$, (183) is closely related to fastigilin B. Catalytic hydrogenation yielded a hexahydro derivative identified as tetrahydrofastigilin B (184).

Geigerinin.

The sesquiterpene geigerinin, $C_{15}H_{22}O_4$, (185) isolated by De Villiers (17) from the South African species *Geigeria aspera* Herv. and *G. africana*

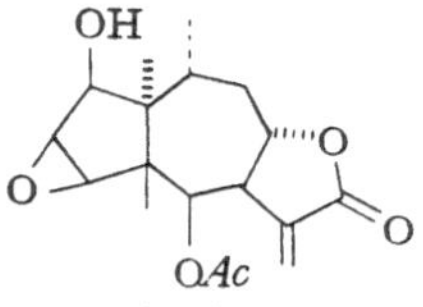

(175)
Gaillardilin.

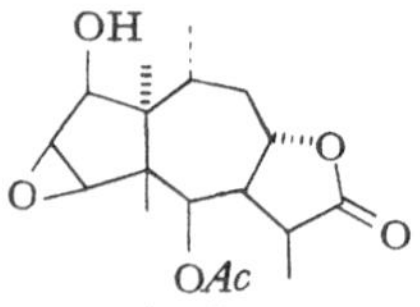

(176)
Dihydrogaillardilin.

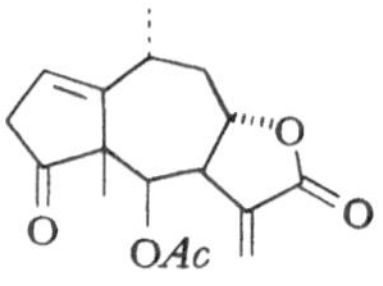

(177)

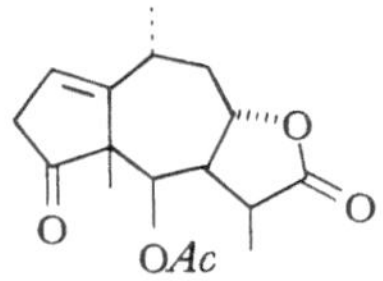

(178)

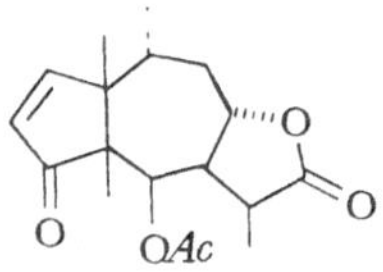

(179)

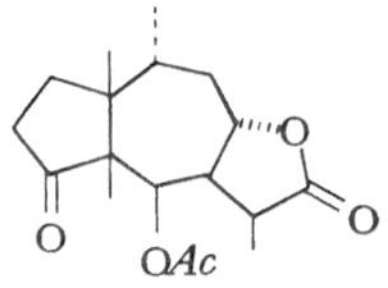

(180)

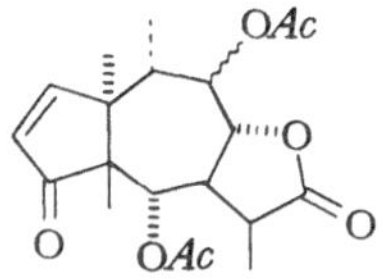

(181)
Gaillardipinnatin.

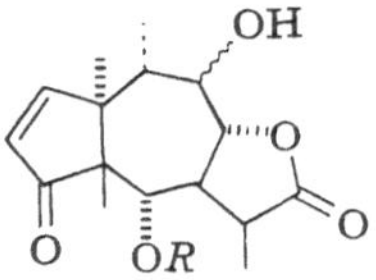

(182)
(a), $R = CH_3 - CH = C - C = O$ Fastigilin A.
            |
            CH₃
(b), $R = (CH_3)_2C = CH - C = O$ Fastigilin B.

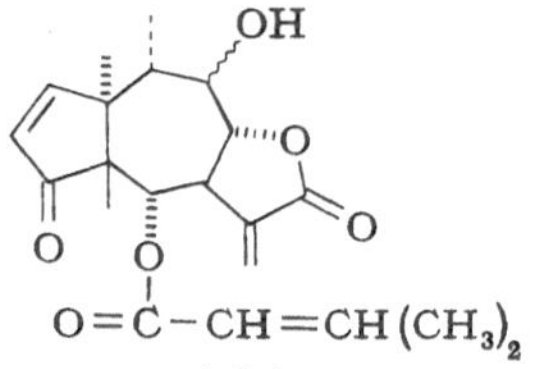

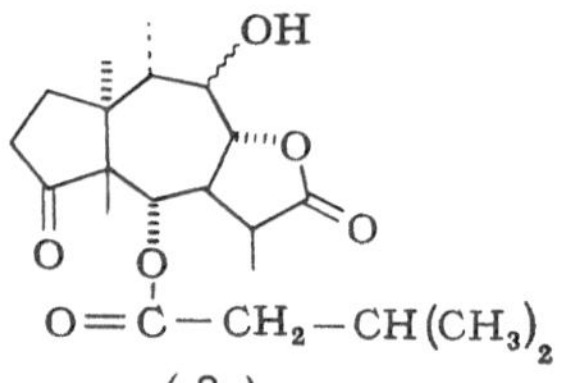

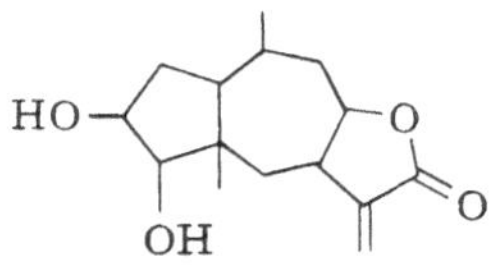

(183)
Fastigilin C.
$O = C - CH = CH(CH_3)_2$

(184)
$O = C - CH_2 - CH(CH_3)_2$

(185)
Geigerinin

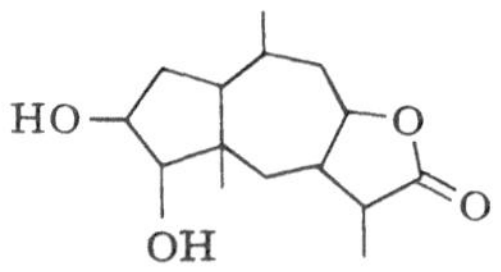

(186)
Dihydrogeigerinin.

Gries, is the first pseudoguaianolide found in a tribe other than Helian-theae or Heleniae. Earlier work was interpreted as indicating a guaia-nolide structure since on dehydrogenation geigerinin afforded chamazulene (4, p. 92) (*17*). After the presence of a pseudoguaianolide skeleton had been demonstrated in ambrosin and parthenin (*47, 38*), DE VILLIERS and PACHLER (*18*) postulated the pseudoguaianolide geigerinin formula (185) on the basis of the NMR spectrum. The chemical evidence for the functional groups of geigerinin can be summarized as follows. Geigerinin contains two secondary hydroxyl groups which could be acetylated under acid conditions and which must be vicinal because geigerinin reacts with potassium periodate. Platinum oxide catalyzed hydrogenation of geigerinin saturated the α,β-unsaturated lactone group and afforded the dihydro derivative (186) which on dehydration with alumina yielded a cyclopentanone derivative. Cumanin (37a, p. 98) differs from geigerinin. The stereochemistry of the latter has not yet been defined.

IV. Remarks on the Application of Spectral Methods.

Spectral methods offer a powerful tool in the elucidation of the structure of the pseudoguaianolides. Ultraviolet and infrared spectra afford a very important preliminary information in this field.

Studies of the NMR spectra of these substances (*19*) have led to the elucidation of the correct structure of the first members of the series (*47, 38*) which had been considered as guaianolides. Most of the pseudoguaianolides exhibit a characteristic singlet near 9 τ, corresponding to the $C_{(5)}$ angular methyl group. This technique has also been useful in the determination of the lactone orientation (*47, 42, 41*). When the lactone group is oriented to $C_{(6)}$, the hydrogen attached to $C_{(6)}$ is coupled only with the proton at $C_{(7)}$ and is responsible for a doublet in the NMR spectrum. In the pseudoguaianolides with a $C_{(8)}$ lactone closure, the signal due to the $C_{(8)}$ hydrogen frequently appears at lower field as a multiplet. The complexity of this signal results from further coupling of the $C_{(8)}$ hydrogen with the protons attached to $C_{(9)}$. A refinement of the NMR method is obtained by the irradiation technique which recently proved of great value in establishing the structure and stereochemistry of several pseudoguaianolides (*39, 40*).

Mass spectrometry has not been extensively applied to the field in structure determinations at the time of this writing.

Other optical methods have been very useful in the elucidation of the structure and stereochemistry of the pseudoguaianolides. Optical rotatory dispersion has been widely used in the assignment of configuration at the ring junction in the perhydroazulene skeleton (*16, 20, 42*). The Cotton effects shown by the ORD curves of *trans* fused pseudoguaianolides having a keto group at $C_{(3)}$ or $C_{(4)}$, correspond with those of appropriate steroid models, in contrast to the *cis* fused compounds which do not possess a rigid conformation.

The structure and relative stereochemistry of several pseudoguaianolides have been determined by X-ray analysis (*23, 24, 56, 61*) which, combined with previous findings, has led to the establishment of several absolute configurations (*23, 24, 61*).

V. List of Compounds.

Compound	M. p.	$[\alpha]_D$		References
Ambrosin	146°	— 154.5°	(CHCl$_3$)	(*1, 29*)
Ambrosiol	116–117°	— 111.3°	(CHCl$_3$)	(*55*)
Amaralin	195–198°	+ 5°	(CHCl$_3$)	(*51*)
Aromaticin	232–234°	+ 18°	(CHCl$_3$)	(*62*)
Aromatin	159–160°	— 6°	(CHCl$_3$)	(*62*)
Balduilin	231–232°	+ 57°	(CHCl$_3$)	(*36*)
Bigelovin	190–191°	+ 46.1°	(C$_2$H$_5$OH)	(*59*)
Coronopilin	177–178°	— 30.2°	(C$_2$H$_5$OH)	(*29*)
Cumanin	120°	+ 161°	(CHCl$_3$)	(*65*)
Damsin	111°	— 72°	*	(*78*)
Dihydromexicanin E	133–135°	— 188°	(CHCl$_3$)	(*52*)
Fastigilin A	175–177°	— 81.6°	(CHCl$_3$)	(*40*)
Fastigilin B	259–261°	*		(*40*)
Fastigilin C	197–199°	— 85.8°	(CHCl$_3$)	(*40*)
Flexuosin A	220–221.5°	+ 12.4°	(CHCl$_3$)	(*31*)
Flexuosin B	132–137°	+ 44.2°	(CHCl$_3$)	(*31*)
Gaillardilin	197–199°	— 2.03°	(CHCl$_3$)	(*39*)
Gaillardipinnatin	270°	*		(*39*)
Geigerinin	202–203°	— 10.7°	(C$_2$H$_5$OH)	(*17*)
Helenalin	169–172°	— 102.4°	(C$_2$H$_5$OH)	(*3*)
Hysterin	168°	— 80°	(CHCl$_3$)	(*68*)
Isohelenalin	260–262°	— 117°	(CHCl$_3$)	(*11, 42*)
Isotenulin	162°	+ 4°	(CHCl$_3$)	(*13, 59*)
Linifolin A	195–198°	+ 33°	(CHCl$_3$)	(*28*)
Linifolin B	149–151°	*		(*28*)
Mexicanin A	138–140°	— 27°	(CHCl$_3$)	(*70*)
Mexicanin C	251–252°	— 80°	(CHCl$_3$)	(*70*)
Mexicanin D	252–253°	+ 107°	(CHCl$_3$)	(*70*)
Mexicanin E	100–101°	— 47°	(CHCl$_3$)	(*72*)
Mexicanin H	150–151°	+ 44.6°	(CHCl$_3$)	(*71*)
Mexicanin I	257–260°	+ 42.5°	(CHCl$_3$)	(*22*)
Parthenin	163–166°	+ 7.02°	(CHCl$_3$)	(*46*)
Peruvin	191–193°	+ 155°	(CHCl$_3$)	(*49*)
Peruvinin	169–171°	+ 34°	(CHCl$_3$)	(*64*)
Pulchellin	165–168°	— 36.2°	(CHCl$_3$)	(*45*)
Pulchellin B	215–218°	+ 92.7°	(C$_2$H$_5$OH)	(*30*)
Pulchellin C	199–202°	+ 125°	(C$_2$H$_5$OH)	(*30*)
Psilostachyin	215°	— 125.2°	(CHCl$_3$)	(*57*)
Psilostachyin B	123°	— 5°	(CHCl$_3$)	(*53*)
Psilostachyin C	223–225°	— 82°	(CHCl$_3$)	(*50*)
Tenulin	variable 130°—215°	— 20°	(C$_2$H$_5$OH)	(*7*)
Thurberilin	162°	+ 20°	(C$_2$H$_5$OH)	(*33*)

* Not reported.

References.

1. Abu-Shady, H. A. and T. O. Soine: The Chemistry of *Ambrosia maritima*. I. The Isolation and Preliminary Characterization of Ambrosin and Damsin. J. Amer. Pharm. Assoc. **42**, 387 (1953).
2. — — The Chemistry of *Ambrosia maritima*. II. Hydrogenation, Oxidation and Dehydrogenation of Ambrosin and Damsin. J. Amer. Pharm. Assoc. **43**, 365 (1954).
3. Adams, R. and W. Herz: Helenalin. I. Isolation and Properties. J. Amer. Chem. Soc. **71**, 2546 (1949).
4. — — Helenalin. II. Helenalin Oxide. J. Amer. Chem. Soc. **71**, 2551 (1949).
5. — — Helenalin. III. Reduction and Dehydrogenation. J. Amer. Chem. Soc. **71**, 2554 (1949).
6. Arny, H. V.: J. Pharm. **1890**, 121; **1897**, 169.
7. Barton, D. H. R. and P. de Mayo: Sesquiterpenoids. VII. The Constitution of Tenulin, a Novel Sesquiterpenoid Lactone. J. Chem. Soc. (London) **1956**, 142.
8. — — Recent Advances in Sesquiterpenoid Chemistry. Quart. Rev. (Chem. Soc. London) **11**, 189 (1957).
9. Bernardi, L. and G. Büchi: The Structures of Ambrosin and Damsin. Experientia **13**, 466 (1957).
10. Braun, B. H., W. Herz and K. Rabindran: The Structure of Tenulin. J. Amer. Chem. Soc. **78**, 4423 (1956).
11. Büchi, G. and D. Rosenthal: The Structure of Helenalin and Isohelenalin. J. Amer. Chem. Soc. **78**, 3860 (1956).
12. Clark, E. P.: Helenalin. I. Helenalin, the Bitter Sternutative Substance Occurring in *Helenium autumnale*. J. Amer. Chem. Soc. **58**, 1982 (1936).
13. — The Constituents of Certain Species of *Helenium*. II. Tenulin. J. Amer. Chem. Soc. **61**, 1836 (1939).
14. — The Constituents of Certain Species of *Helenium*. III. The Ester Nature of Tenulin. J. Amer. Chem. Soc. **62**, 597 (1940).
15. — Constituents of Certain Species of *Helenium*. IV. Concerning the Compound Melting at 233–234° Obtained from *Helenium tenuifolium*. J. Amer. Chem. Soc. **62**, 2154 (1940).
16. Crabbé, P.: Optical Rotatory Dispersion and Circular Dichroism in Organic Chemistry. San Francisco, London, Amsterdam: Holden-Day. 1965.
17. De Villiers, J. P.: The Isolation and Structure of Geigerinin, a Guaianolide. J. Chem. Soc. (London) **1959**, 2412.
18. De Villiers, J. P. and K. Pachler: The Revised Structure of Geigerinin. J. Chem. Soc. (London) **1963**, 4989.
19. Díaz, E., P. Joseph-Nathan, A. Romo de Vivar y J. Romo: Análisis mediante resonancia magnética nuclear. I. Determinación de estructuras de lactonas sesquiterpénicas azulogénicas. Bol. inst. quím. univ. nac. auton. México **17**, 122 (1965).
20. Djerassi, C., J. Osiecki and W. Herz: Optical Rotatory Dispersion Studies. XIII. Assignment of Absolute Configuration to Certain Members of the Guaianolide Series of Sesquiterpenes. J. Organ. Chem. (USA) **22**, 1361 (1957).
21. Djerassi, C., R. Riniker and B. Riniker: Optical Rotatory Dispersion Studies. VII. Application to Problems of Absolute Configurations. J. Amer. Chem. Soc. **78**, 6362 (1956).
22. Domínguez, E. and J. Romo: Mexicanin I. A New Sesquiterpene Lactone Related to Tenulin. Tetrahedron **19**, 1415 (1963).
23. Emerson, M. T., C. N. Caughlan and W. Herz: The Crystal and Molecular Structure of Bromohelenalin. Tetrahedron Letters **1964**, 621.

24. Emerson, M. T., W. Herz, C. N. Caughlan and R. W. Witters: The Crystal and Molecular Structure of Bromoambrosin. Tetrahedron Letters 1966, 6151.

25. Geissman, T. A. and R. Turley: Sesquiterpene Lactones. Coronopilic Acid. J. Organ. Chem. (USA) 29, 2553 (1964).

26. Giral, F. y S. Ladabaum: Preparaciones fitoquímicas. IV. Helenalina. Ciencia (México) 21, 35 (1961).

27. Hendrickson, J. B.: Stereochemical Implications in Sesquiterpene Biogenesis. Tetrahedron 7, 82 (1959).

28. Herz, W.: Constituents of Helenium Species. XII. Sesquiterpene Lactones of Some Southwestern Species. J. Organ. Chem. (USA) 27, 4043 (1962).

29. Herz, W. and G. Högenauer: Isolation and Structure of Coronopilin, a New Sesquiterpene Lactone. J. Organ. Chem. (USA) 26, 5011 (1961).

30. Herz, W. and S. Inayama: Constituents of Gaillardia Species. II. Structures of Pulchellin B and Pulchellin C. Tetrahedron 20, 341 (1964).

31. Herz, W., P. Jayaraman and H. Watanabe: Constituents of Helenium Species. IX. The Sesquiterpene Lactones of H. flexuosum Raf. and H. campestre Small. J. Amer. Chem. Soc. 82, 2276 (1960).

32. Herz, W., Y. Kishida and M. V. Lakshmikantham: Constituents of Helenium Species. XVI. Structures of Flexuosin A and Flexuosin B. Tetrahedron 20, 979 (1964).

33. Herz, W. and M. V. Lakshmikantham: Constituents of Helenium Species. XVII. Sesquiterpene Lactones of Helenium thurberi Gray and the Stereochemistry of Bigelovin. Tetrahedron 21, 1711 (1965).

34. Herz, W., M. V. Lakshmikantham and R. N. Mirrington: Constituents of Helenium Species. XVIII. 1-Epiisotenulin and its Transformations. Tetrahedron 22, 1709 (1966).

35. Herz, W. and R. B. Mitra: Correlation of Helenalin and Alloisotenulin. J. Amer. Chem. Soc. 80, 4876 (1958).

36. Herz, W., R. B. Mitra and P. Jayaraman: Constituents of Helenium Species. VIII. Isolation and Structure of Balduilin. J. Amer. Chem. Soc. 81, 6061 (1959).

37. Herz, W., R. B. Mitra, K. Rabindran and W. A. Rohde: Constituents of Helenium Species. VII. Bitter Principles of H. pinnatifidum (Nutt) Rydb., H. vernale Walt., H. brevifolium (Nutt) A. Wood and H. flexuosum Raf. J. Amer. Chem. Soc. 81, 1481 (1959).

38. Herz, W., M. Miyazaki and Y. Kishida: Structures of Parthenin and Ambrosin. Tetrahedron Letters 1961, 82.

39. Herz, W., S. Rajappa, M. V. Lakshmikantham and J. J. Schmid: Constituents of Gaillardia Species. III. The Structure of Gaillardilin, a New Pseudo-guaianolide. Tetrahedron 22, 693 (1966).

40. Herz, W., S. Rajappa, S. K. Roy, J. J. Schmid and R. N. Mirrington: Constituents of Gaillardia Species. IV. The Sesquiterpene Lactones of Gaillardia fastigiata Greene. Tetrahedron 22, 1907 (1966).

41. Herz, W., W. A. Rohde, K. Rabindran, P. Jayaraman and N. Viswanathan: Revised Structure of Tenulin. J. Amer. Chem. Soc. 84, 3857 (1962).

42. Herz, W., A. Romo de Vivar, J. Romo and N. Viswanathan: Constituents of Helenium Species. XIII. The Structure of Helenalin and Mexicanin A. J. Amer. Chem. Soc. 85, 19 (1963).

43. — — — — Constituents of Helenium Species. XV. The Structure of Mexicanin C, Relative Stereochemistry of its Congeners. Tetrahedron 19, 1359 (1963).

44. Herz, W. and Y. Sumi: Constituents of Ambrosia hispida Pursh. J. Organ. Chem. (USA) 29, 3438 (1964).

45. HERZ, W., K. UEDA and S. INAYAMA: Constituents of *Gaillardia* Species. I. The Structure of Pulchellin. Tetrahedron **19**, 483 (1963).
46. HERZ, W. and H. WATANABE: Parthenin, a New Guaianolide. J. Amer. Chem. Soc. **81**, 6088 (1959).
47. HERZ, W., H. WATANABE, M. MIYAZAKI and Y. KISHIDA: The Structures of Parthenin and Ambrosin. J. Amer. Chem. Soc. **84**, 2601 (1962).
48. HOREAU, A.: Principe et applications d'une nouvelle méthode de détermination des configurations dites »par dédoublement partiel». Tetrahedron Letters **1961**, 506.
49. JOSEPH-NATHAN, P. and J. ROMO: Isolation and Structure of Peruvin. Tetrahedron **22**, 1723 (1966).
50. KAGAN, H. B., H. E. MILLER, W. RENOLD, M. V. LAKSHMIKANTHAM, L. R. TETHER, W. HERZ and T. J. MABRY: The Structure of Psilostachyin C, a New Sesquiterpene Dilactone from *Ambrosia psilostachya* DC. J. Organ. Chem. (USA) **31**, 1629 (1966).
51. LUCAS, R. A., S. ROVINSKI, R. J. KIESEL, L. DORFMAN and H. B. MACPHILLAMY: A New Sesquiterpene Lactone with Analgesic Activity from *Helenium amarum*. J. Organ. Chem. (USA) **29**, 1549 (1964).
52. LUCAS, R. A., R. G. SMITH and L. DORFMAN: The Isolation of Dihydromexicanin E from *Helenium autumnale* L. J. Organ. Chem. (USA) **29**, 2101 (1964).
53. MABRY, T. J., H. B. KAGAN and H. E. MILLER: Psilostachyin B, a New Sesquiterpene Dilactone from *Ambrosia psilostachya* DC. Tetrahedron **22**, 1943 (1966).
54. MABRY, T. J., H. E. MILLER, H. B. KAGAN and W. RENOLD: The Structure of Psilostachyin, a New Sesquiterpene Dilactone from *Ambrosia psilostachya*. Tetrahedron **22**, 1139 (1966).
55. MABRY, T. J., W. RENOLD, H. E. MILLER and H. B. KAGAN: The Structure of Ambrosiol. A New Sesquiterpene Lactone from *Ambrosia psilostachya*. J. Organ. Chem. (USA) **31**, 681 (1966).
56. MAZHAR-UL-HAQUE and C. N. CAUGHLAN: The Molecular and Crystal Structure of Bromomexicanin E ($C_{14}H_{15}O_3Br$). J. Chem. Soc. (London) **B 1967**, 355.
57. MILLER, H. E., H. B. KAGAN, W. RENOLD and T. J. MABRY: Psilostachyin, a New Type of Sesquiterpene Lactone. Tetrahedron Letters **1965**, 3397.
58. NOZOE, T. and S. ITÔ: Recent Advances in the Chemistry of Azulenes and Natural Hydroazulenes. Fortschr. Chem. organ. Naturstoffe **19**, 32 (1961).
59. PARKER, B. A. and T. A. GEISSMAN: The Sesquiterpenoid Lactones of *Helenium bigelovii* Gray. J. Organ. Chem. (USA) **27**, 4127 (1962).
60. RICHARDS, J. H. and J. B. HENDRICKSON: The Biosynthesis of Steroids, Terpenes and Acetogenins. New York: Benjamin. 1964.
61. ROGERS, D. and MAZHAR-UL-HAQUE: The Structure of Bromoisotenulin. Proc. Chem. Soc. (London) **1963**, 92.
62. ROMO, J., P. JOSEPH-NATHAN and F. DÍAZ A.: Aromatin and Aromaticin, New Sesquiterpene Lactones Isolated from *Helenium aromaticum*. Chem. and Ind. **1963**, 1839.
63. — — — The Constituents of *Helenium aromaticum* (Hook) Bailey. The Structures of Aromatin and Aromaticin. Tetrahedron **20**, 79 (1964).
64. ROMO, J., P. JOSEPH-NATHAN, A. ROMO DE VIVAR and C. ALVAREZ: The Structure of Peruvinin, a Pseudoguaianolide Isolated from *Ambrosia peruviana* Willd. Tetrahedron. **23**, 529 (1967).
65. ROMO, J., P. JOSEPH-NATHAN and G. SIADE: The Structure of Cumanin, a Constituent of *Ambrosia cumanensis*. Tetrahedron **22**, 1499 (1966).

66. Romo, J., A. Romo de Vivar and W. Herz: Constituents of *Helenium* Species. XIV. The Structure of Mexicanin E. Tetrahedron 19, 2317 (1963).
67. Romo, J., A. Romo de Vivar and P. Joseph-Nathan: The Structure of Mexicanin H. Tetrahedron Letters 1966, 1029.
68. Romo de Vivar, A., E. A. Bratoeff and T. Ríos: Structure of Hysterin, a New Sesquiterpene Lactone. J. Organ. Chem. (USA) 31, 673 (1966).
69. Romo de Vivar, A., L. Rodríguez-Hahn, J. Romo, M. V. Lakshmikantham, R. N. Mirrington, J. Kagan and W. Herz: Constituents of *Helenium* Species. XIX. Further Transformations of Helenalin and its Congeners. The 1-Epihelenalin and 1-Epiambrosin Series. Tetrahedron 22, 3279 (1966).
70. Romo de Vivar, A. and J. Romo: The Constituents of *Helenium mexicanum* H. B. K. Chem. and Ind. 1959, 882.
71. — — Las lactonas del *Helenium mexicanum* H. B. K. Ciencia (México) 21, 33 (1961).
72. — — Mexicanin E. A Norsesquiterpenoid Lactone. J. Amer. Chem. Soc. 83, 2326 (1961).
73. Ruzicka, L.: The Isoprene Rule and the Biogenesis of Terpenic Compounds. Experientia 9, 357 (1953).
74. — History of the Isoprene Rule. Proc. Chem. Soc. (London) 1959, 341.
75. Sondheimer, F., S. Burstein and R. Mechoulam: Syntheses in the Cardiac Aglicone Field. III. The Conversion of a 14α- to a 14ß-Hydroxy Group in the Androstane Series. The Ultraviolet Spectra of Δ^{15}-Androsten-17-ones. J. Amer. Chem. Soc. 82, 3209 (1960).
76. Šorm, F. and L. Dolejš: Guaianolides and Germacranolides. In: E. Lederer (Edit.), Chimie des substances naturelles. Paris: Herman. 1966.
77. Šorm, F., M. Suchý and V. Herout: On Terpenes. C. The Structure of Ambrosin. Collect. Czech. Chem. Comm. 24, 1548 (1959).
78. Suchý, M., V. Herout and F. Šorm: On Terpenes. CLV. Structure of Damsine, a Sesquiterpenic Lactone from *Ambrosia maritima* L. Collect. Czech. Chem. Comm. 28, 2257 (1963).
79. Ungnade, H. E. and E. C. Hendley: The Bitter Product of *Helenium tenuifolium*. J. Amer. Chem. Soc. 70, 3921 (1948).
80. Ungnade, H. E., E. C. Hendley and W. Dunkel: Tenulin. II. Anhydrotenulin and Pyrotenulin. J. Amer. Chem. Soc. 72, 3818 (1950).

(Received, September 14, 1966.)

The Nonadrides.

By J. K. SUTHERLAND, London.

Contents.

I. Introduction.

The nonadrides are a small group of fungal metabolites characterised chemically by the presence of a nine-membered alicyclic ring and two five-ring anhydride functions. The first two members, glauconic and glaucanic acids, were isolated by WIJKMAN (*18*) in 1931 from a fungus described as *Penicillium glaucum* and in 1934 by YUILL (*19*) from a *P. purpurogenum* species. The third nonadride, byssochlamic acid, was obtained from *Byssochlamys fulva* by RAISTRICK and SMITH (*14*) in 1933. Although extensive chemical work was carried out on glauconic acid by WIJKMAN, KRAFT, SUTTER and their co-workers (*16, 17, 15, 12, 11*) and on byssochlamic acid by RAISTRICK's group (*14*) and by COOK, LOUDON and PATON (*7*), the structures of the metabolites were not elucidated until the early sixties when these problems were solved chemically by BARTON, SUTHERLAND and their collaborators (*1*) and X-ray crystallographically by ROBERTSON and SIM and their group (*8, 9*).

Biosynthetically the nonadrides appear to be closely related and to belong to the larger group of fungal metabolites which are derived from substituted citric acids (*1*).

II. Glauconic Acid.

Glauconic acid has been shown to have structure (1) and as the detailed chemical arguments leading to this conclusion have already been discussed (*1*) and confirmed by X-ray crystallography (*8*), the chemistry of the compound will be described in terms of (1). The early workers showed that glauconic acid contained an acylable hydroxyl group and titrated as a tetracarboxylic acid (*18*). However, most of their efforts were devoted to studying the two products obtained on heating glauconic acid above its melting point (201°). The volatile liquid was soon identified as α,β-diethylacrolein (2) by the reactions summarised in *Chart 1*. The α-ethylvaleric acid was characterised as the amide and shown (*15*) to be identical with an authentic specimen (*16*); and α,β-diethylacrylic acid was identified as the oxidation product of the aldehyde. Extensive

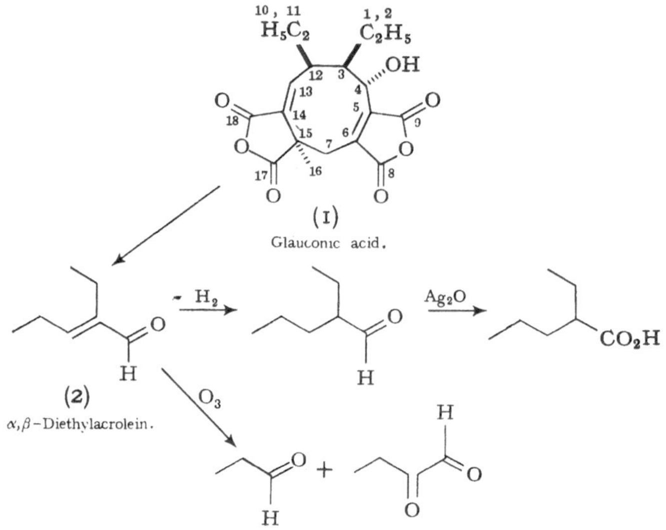

(1) Glauconic acid.

(2) α,β-Diethylacrolein.

Chart 1. Degradation of Glauconic Acid and of Diethylacrolein.

degradative work was carried out on the second product, glauconin (3), and this is collated in *Chart 2*. The isolation and identification by synthesis of the *meso*- and (±)-forms of the keto-acid (4) and of the three racemates of constitution (5) was a considerable technical achievement (*17*) and, coupled with the degradation of glauconin to the tricarboxylic acid (6) and its synthesis (*11*), established the carbon skeleton of glauconin except for the position of one carboxyl group. This was placed to give structure (7) for glauconin owing to some misleading experiments (*11*). Later, however, the structure (3), proposed prior to (7) and rejected, was shown to be correct (*5*) on the basis of ultraviolet, infrared, and nuclear magnetic resonance data and its synthesis (*4*) from the ester (8) which, by acid catalysis, was converted to the mixture of enol-lactones (9). Alkaline hydrolysis of the mixture gave glauconin from one component and the keto-acid (4) from the other.

References, pp. 148—149.

Knowledge of the structures of its two pyrolysis products is not of immediate use in the determination of the structure of glauconic acid since the carbon skeleton of glauconin is not present in (1). In fact the

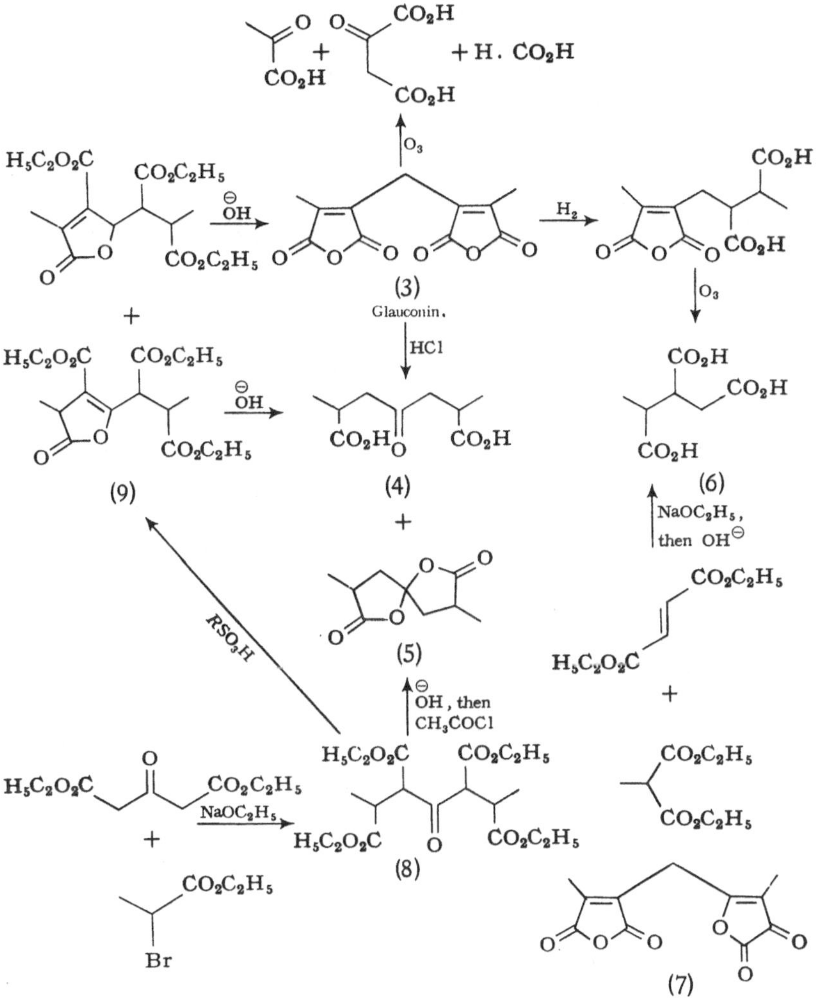

Chart 2. Degradation and Synthesis of Glauconin.

structure of glauconic acid was deduced (1) independently of the structures of the pyrolysis products and the mechanism of their formation became obvious only after structure (1) was known. The first step in their formation is a Cope rearrangement to give (9a) which, by a retro-aldol, or more probably by a retro-Prins reaction, accepting the stereochemistry shown, gives (10). In turn (10) can undergo a reversed Michael reaction leading to the two degradation products. The stereochemistry depicted for (9a) follows if it is accepted that glauconic acid reacts in the conformation (11) found by X-ray analysis of its m-iodobenzoate (8). It has not been possible to detect any of the intermediates postulated; presumably, the rate determining step is the initial Cope rearrangement.

When the acetate of glauconic acid is pyrolysed an isomer is formed (4). Infrared data showed that there was no change in functional groups but

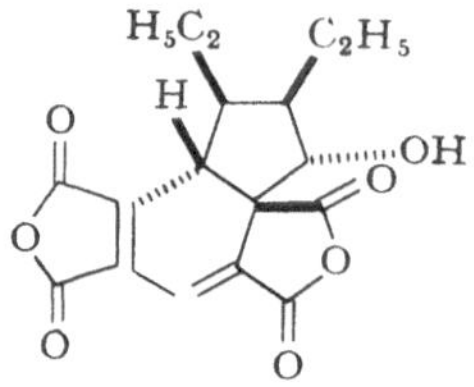

(9a)

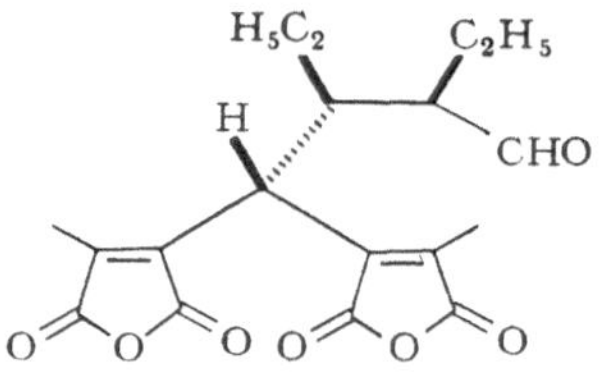

(10)

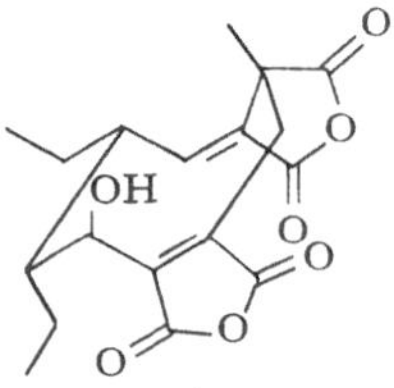

(11)

Glauconic acid.

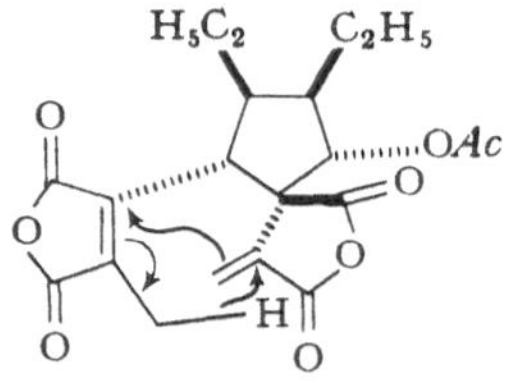

(12)

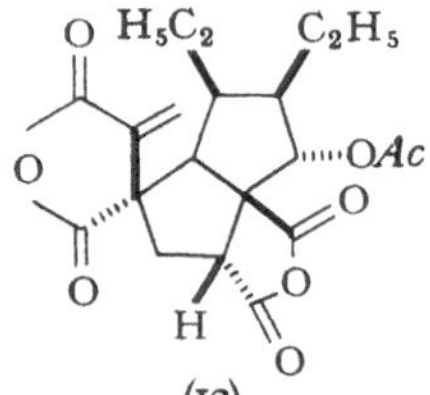

(13)

that their environment had altered. In the NMR spectrum the doublet for the single vinyl proton of glauconic acid acetate had disappeared and there were now present two vinyl protons at 3.26 and 3.95 τ, each split as a doublet ($J = 1 \, c \cdot sec^{-1}$). That these protons were present in a methylene group was confirmed by ozonolysis to give formaldehyde. The absence of the characteristic maleic anhydride UV absorption (λ_{max} ca. 255 mμ) and reduction of the isomer to a saturated dihydro compound showed that the isoacetate was bicyclic. The first stage of this reaction was formulated as giving the acetate (12) which then, by an 'ene' reaction, gave (13). A study of the hydrolysis of (13) supported this structure.

References, pp. 148—149.

In the degradation of glauconic acid and its derivatives reduction with zinc dust in acetic acid played a great role because of the selective action of this reagent in reducing only maleic anhydride functions. With glauconic acid itself a simple reduction followed by hydration and lactonisation took place (*17*) to give dihydroglauconic acid (*14*). The functional groups present were demonstrated by the preparation of a monomethyl

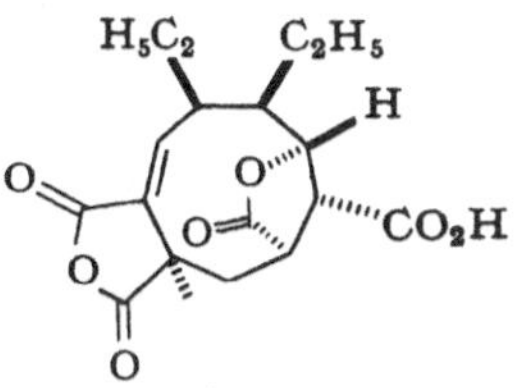

(14)
Dihydroglauconic acid.

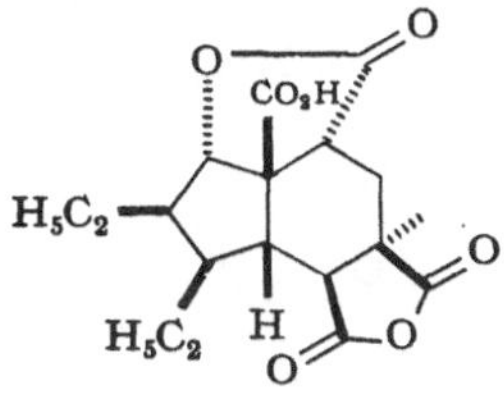

(15)

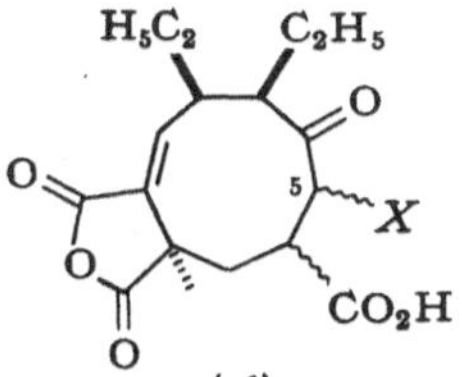

(16)
X=H
(18)
X=CO$_2$H

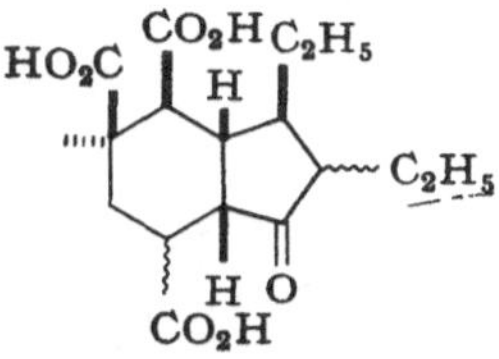

(17)

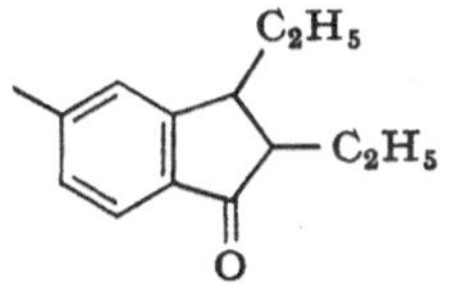

(19)

ester (*17*) with diazomethane and of a trimethyl ester (*11*) with dimethyl sulphate and alkali; the latter showed γ-lactone absorption in the IR. The retention of the itaconic anhydride in (*14*) followed from the UV absorption (λ_{max} 232 mμ, ε 8550) and the vinyl proton doublet at 3.2 τ (J = 12 c · sec^{-1}) (*4*).

Pyrolysis of the acid chloride of dihydroglauconic acid gave the saturated bicyclic compound (*15*) by a Michael addition (*4*). An addition of this type was observed during the reduction of glauconic acid ketone, obtained by oxidation of glauconic acid (*1*) with chromium trioxide in sulphuric acid-acetone (*4*). Two products were isolated from the reduction. The first one was assigned structure (*16*) since the itaconic anhydride residue was still intact [λ_{max} 227 mμ, ε 8600; ν_{max} 1830, 1770 cm^{-1}; singlet CH$_3$ at 8.59 τ, vinyl-H doublet at 3.08 τ (J = 10 c · sec^{-1})]. Spectral data also indicated the presence of a carbonyl group and a carboxyl whose presence was confirmed by the preparation of a mono-methyl ester. The second product (17) was recognised (by UV) as a saturated tricarboxylic keto-acid from which a trimethyl ester could be prepared and, by sublimation, a succinic anhydride. A band in the IR spectrum of (17) at 1740 cm^{-1} placed the ketonic group in a five-

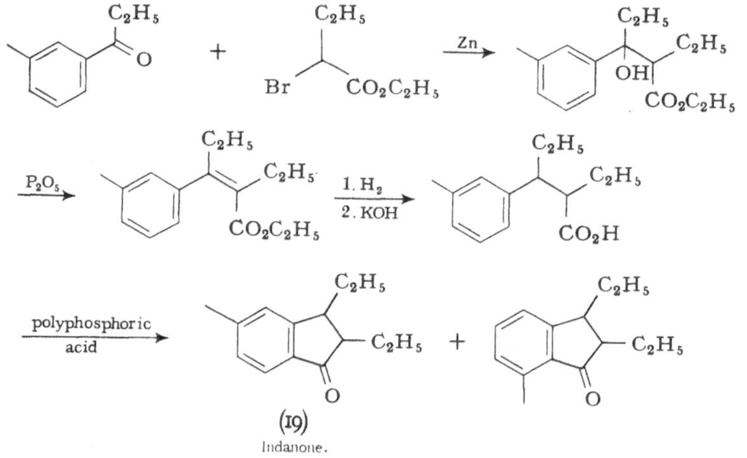

Chart 3. Synthesis of 5-Methyl-2,3-diethylindan-1-one.

membered ring. These two products can be accounted for readily by postulating the usual reduction of the maleic anhydride to the dihydro compound which is then hydrated to give (18). Decarboxylation of the β-keto acid yields (16) while (17) is formed by Michael addition of the enol from C$_{(5)}$ to the conjugated double bond. The decarboxylation to give (17) must be concommitant with the cyclisation or follow it since (16) does not cyclise under the reaction conditions. The carbon skeleton of (17) was proven by dehydrogenation to the indanone (19) which was synthesised by the scheme shown in Chart 3. The 2,4-dinitrophenyl-hydrazone of one of the indanones synthesised was shown to be identical to that of the indanone obtained by degradation.

Oxidative degradation of glauconic acid or its ketone also yielded useful information, giving three compounds (3). One was meso-diethyl succinic acid and the others were a pair of geometrical isomers, (20) and (21), of which (21) was further oxidised to meso-diethyl succinic acid. The two isomers were characterised as their dimethyl esters and typical spectroscopic data left no doubt that they both contained the itaconic anhydride residue. In (20) the vinyl proton appeared as a doublet (J = = 11 c · sec^{-1}) at 3.28 τ just as in glauconic acid and its derivatives, whereas in (21) this proton was no longer deshielded by the *cis* carbonyl group and appeared at 4.00 τ (J = 12 c · sec^{-1}). Thus, (20) is the immediate oxidation product and (21) is formed from it, or from some derivative, probably by addition of one of the carboxyl groups to the double bond to give a lactone which can then reverse the addition giving either (20) or (21).

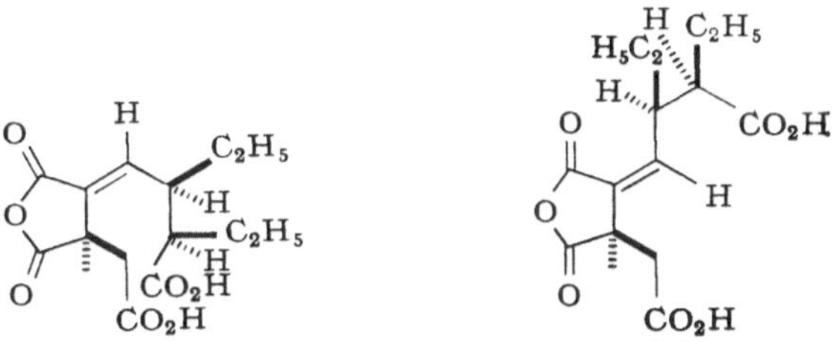

(20)   (21)

III. Glaucanic Acid.

The spectroscopic similarity of glauconic acid to glaucanic acid suggested that they were closely related and this was confirmed by zinc dust and acetic acid reduction of glauconic acid acetate (5, 3) to glaucanic acid (22) which can be further reduced but less rapidly than the acetate.

Glaucanic acid is much less reactive than glauconic acid and is inert both to pyrolysis and a variety of chemical reagents. However, it could be degraded by prolonged ozonolysis in acetic acid to give propane-1,2,2-tricarboxylic acid and (+)-α,β-*erythro*-diethylglutaric acid, the absolute configuration of which was determined as shown in *Chart 4* (3). The key compound in this correlation was the dimethylamide (23) which was converted into the half-ester (24). Homologation of (24) by the Arndt-Eistert procedure, followed by hydrolysis gave the enantiomeric diethylglutaric acid (25). The conversion of the dimethylamide to the half-ester was made possible by the development of an oxidative de-N-methylation (ozone) of (26) to (27) which by nitrosation and hydrolysis gave (24). These reactions enabled the dimethylamide group to be converted to carboxyl without hydrolysis of the methyl ester function.

The absolute configuration of the dimethylamide (23) was determined by its photochemical decarboxylation to (28). The absolute configuration of the corresponding (−)-acid had been determined previously but the acid had never been obtained optically pure. This was now done by a parallel series of experiments starting from (+)-diethylsuccinic acid of known absolute configuration which gave the enantiomeric dimethyl-amide (29). Oxidative de-N-methylation, nitrosation and hydrolysis yielded optically pure (—)-α-ethyl valeric acid.

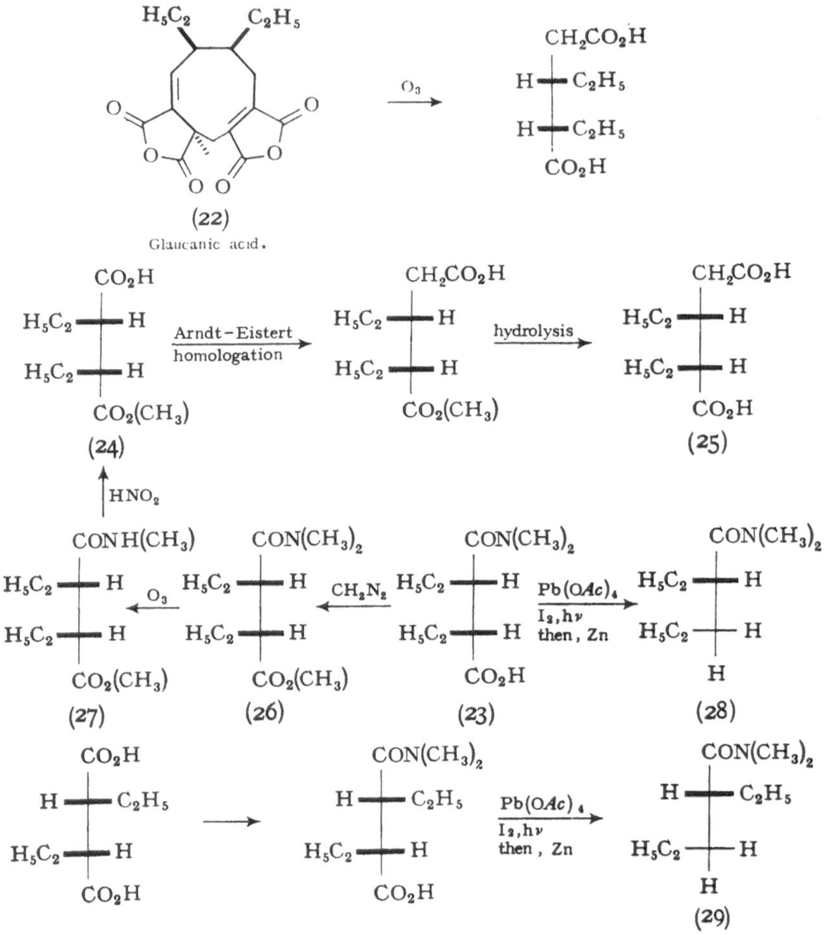

$$\text{(22)} \quad \xrightarrow{\;O_3\;} \quad \begin{array}{c} CH_2CO_2H \\ H{-}\!\!-C_2H_5 \\ H{-}\!\!-C_2H_5 \\ CO_2H \end{array}$$

(22)
Glaucanic acid.

$$\begin{array}{c} CO_2H \\ H_5C_2{-}\!\!-H \\ H_5C_2{-}\!\!-H \\ CO_2(CH_3) \end{array} \;\xrightarrow[\text{homologation}]{\text{Arndt--Eistert}}\; \begin{array}{c} CH_2CO_2H \\ H_5C_2{-}\!\!-H \\ H_5C_2{-}\!\!-H \\ CO_2(CH_3) \end{array} \;\xrightarrow{\text{hydrolysis}}\; \begin{array}{c} CH_2CO_2H \\ H_5C_2{-}\!\!-H \\ H_5C_2{-}\!\!-H \\ CO_2H \end{array}$$

(24)   (25)

$$\uparrow\;HNO_2$$

$$\begin{array}{c} CONH(CH_3) \\ H_5C_2{-}\!\!-H \\ H_5C_2{-}\!\!-H \\ CO_2(CH_3) \end{array} \;\xleftarrow{\;O_3\;}\; \begin{array}{c} CON(CH_3)_2 \\ H_5C_2{-}\!\!-H \\ H_5C_2{-}\!\!-H \\ CO_2(CH_3) \end{array} \;\xleftarrow{\;CH_2N_2\;}\; \begin{array}{c} CON(CH_3)_2 \\ H_5C_2{-}\!\!-H \\ H_5C_2{-}\!\!-H \\ CO_2H \end{array} \;\xrightarrow[\text{then, Zn}]{\substack{Pb(OAc)_4 \\ I_2,h\nu}}\; \begin{array}{c} CON(CH_3)_2 \\ H_5C_2{-}\!\!-H \\ H_5C_2{-}\!\!-H \\ H \end{array}$$

(27)   (26)   (23)   (28)

$$\begin{array}{c} CO_2H \\ H{-}\!\!-C_2H_5 \\ H_5C_2{-}\!\!-H \\ CO_2H \end{array} \;\longrightarrow\; \begin{array}{c} CON(CH_3)_2 \\ H{-}\!\!-C_2H_5 \\ H_5C_2{-}\!\!-H \\ CO_2H \end{array} \;\xrightarrow[\text{then, Zn}]{\substack{Pb(OAc)_4 \\ I_2,h\nu}}\; \begin{array}{c} CON(CH_3)_2 \\ H{-}\!\!-C_2H_5 \\ H_5C_2{-}\!\!-H \\ H \end{array}$$

(29)

Chart 4. The Absolute Configuration of Glaucanic Acid.

IV. Byssochlamic Acid.

X-ray analysis of the bis-p-bromophenylhydrazide of byssochlamic acid (9) first showed conclusively that byssochlamic acid must have structure (30), although this structure had been suspected previously on chemical and biogenetic grounds (1). The earlier workers (7, 14) had noted the chemical inertness of byssochlamic acid to electrophilic reagents

References, pp. 148—149.

and no real progress was made in its degradation until a modified Lossen rearrangement was developed (2) using the anhydride functions. This procedure and its probable mechanism is summarised in *Chart 5*. When byssochlamic acid tetrasodium salt was reacted with one mole of hydroxylamine hydrochloride and the mono-N-hydroxyimide degraded as shown, two crystalline ketones (31) and (32) were obtained. The ketone (31) was investigated extensively. Nitrosation gave the oximino ketone (33) which, with *p*-toluenesulphonyl chloride and alkali underwent a second order Beckman transformation yielding the nitrile acid (34). This open-chain compound was more susceptible to oxidation than any of its

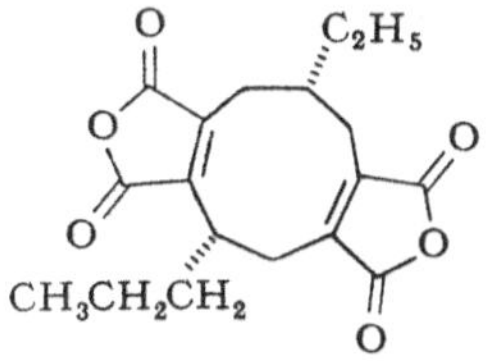

(30)
Byssochlamic acid.

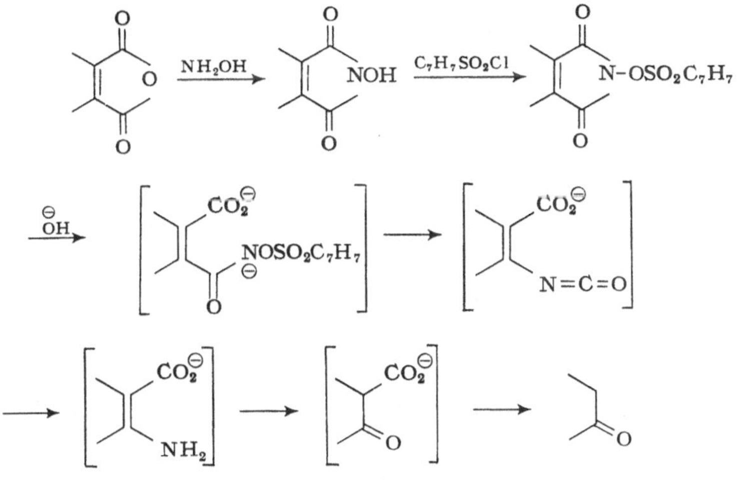

Chart 5. Mechanism of the Anhydride Degradation.

medium-ring precursors and on ozonolysis followed by oxidative work up gave (S)(—)-n-propylsuccinic acid (35) and the half-nitrile of β-ethyl-glutaric acid (36) *(Chart 6)*. The identification of these products reduces the possible structures for the ketone to two, viz. (31) and (37).

Accepting that the alkyl substituents in the ketone are *cis* as is known from the X-ray work to be the case in byssochlamic acid, then it follows that (31) and (37) would give rise to enantiomeric nitrile acids (36). Thus determination of the absolute configuration of the nitrile acid would then distinguish between the two formulae. To this end the racemic nitrile acid was synthesised and resolved to give the pure $(R)(-)$-isomer which was enantiomeric with that obtained from byssochlamic acid. The absolute configuration of the $(R)(-)$-isomer was determined by Hofmann

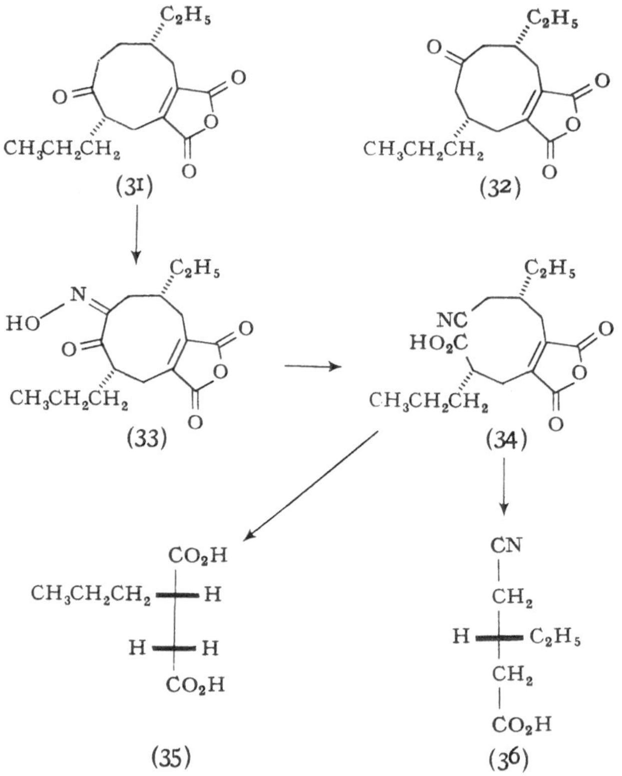

Chart 6. The Absolute Configuration of Byssochlamic Acid.

rearrangement of the derived amide, followed by oxidation of the amine to $(S)(-)$-ethylsuccinic acid of known absolute configuration. Accordingly (36), represents the absolute configuration of the nitrile acid obtained from byssochlamic acid and (31), not (37), must be the structure of the ketone.

However, this argument is open to the objection that epimerisation of the propyl group could have taken place in the formation of the ketone. If, in fact, this epimerisation had taken place then the structure of the ketone would be (37), with the ethyl group inverted, and it is the propyl group α to the anhydride function which would have been epimerised.

This is unlikely under the conditions applied, but to make certain byssochlamic acid was ozonized as the tetrasodium salt in aqueous solution. The reaction was very slow but did give a small yield of (S)(—)-n-propyl-

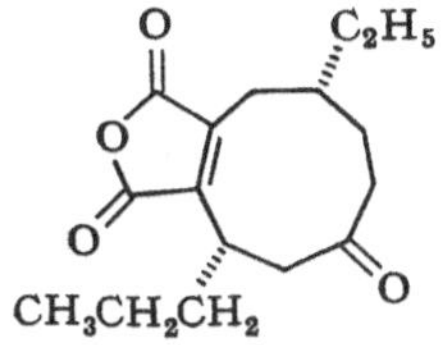

(37)

succinic acid confirming that no epimerisation had occurred on ketone formation and that the absolute configuration of byssochlamic acid is that shown in (30, p. 139).

The second ketone (32) could not be nitrosated but on oxidation with peroxytrifluoroacetic acid gave a crystalline, sharp-melting lactone.

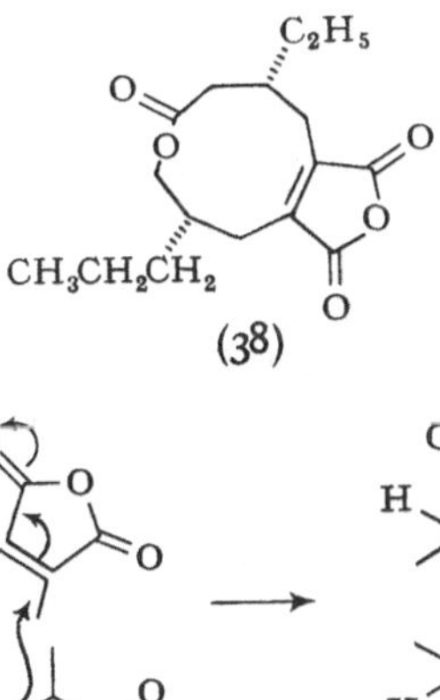

(38) 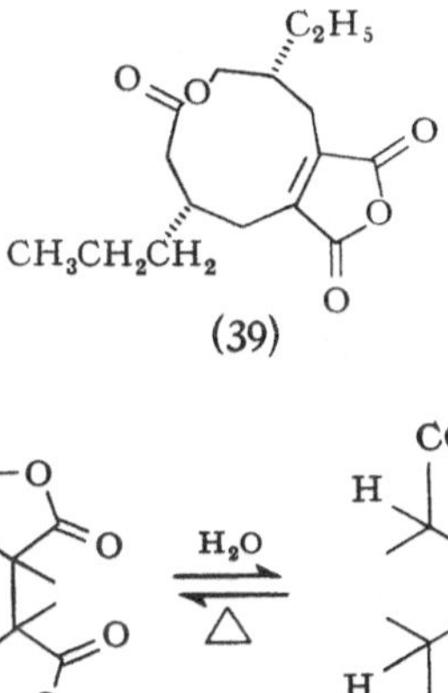   (39)

Chart 7. Part Structure of Dihydrobyssochlamic Acid.

Vigorous alkaline hydrolysis, methylation and acetylation yielded an acetoxy methyl ester which on ozonolysis gave β-ethylglutaric (42%), β-n-propyl glutaric (32%), ethylsuccinic (6%) and n-propylsuccinic (7%) acids as judged by gas-liquid chromatography of the methyl esters. The two glutaric acids were isolated and conclusively identified. These results are only consistent with the lactone being a mixture of the positional isomers (38) and (39). Double bond isomerism can be excluded by the absence of any vinyl proton absorption in the NMR spectrum of the acetoxy methyl ester.

Like glauconic acid, byssochlamic acid readily undergoes trans-annular reactions with the formation of bicyclic products. On zinc dust and acetic acid reduction, byssochlamic acid gives a dihydro compound (2, 14) which is saturated and hence bicyclic. Dissolution of the dihydro compound in alkali and re-acidification gives a dihydrohydrate which according to the IR spectrum is a succinic anhydride dicarboxylic acid. This is best interpreted as a reductive Michael addition as shown in *Chart 7*; and the properties of the compound fit in with this disposition

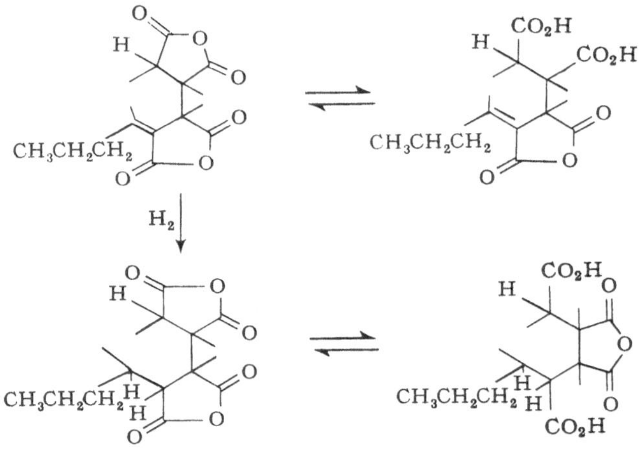

Chart 8. Part Structure of Isobyssochlamic Acid.

of functional groups, since, in general, α,α,α'-trisubstituted succinic acids exist in aqueous solution as the free acids, whereas $\alpha,\alpha,\alpha',\alpha'$-tetra-substituted succinics spontaneously form anhydrides under these conditions; thus anhydride formation by the two inner carboxyls would be expected on acidification of the tetra-anion. This interpretation of the reaction leads to four possible structures for the dihydro compound and it is not possible at present to choose among them.

A similar type of compound, isobyssochlamic acid, is obtained on heating byssochlamic acid with alumina (7) or, better, by heating in a sealed tube at 210° (2). Considering its UV spectrum (λ_{max} 236 mμ, ε 6300) the isomer contains an itaconic anhydride residue which is reduced on catalytic hydrogenation giving a saturated dihydro compound. Accepting again that tetra-substituted 1,2-dicarboxylic acids form anhydrides spontaneously on acidification of the salts, while trisub-stituted 1,2-dicarboxylic acids do not, the behaviour of both the isomer

References, pp. 148—149.

and its dihydro derivative in giving anhydride dicarboxylic acids on acidification of the salts leads to the part structures (2) shown in *Chart 8*. The propyl group is placed on the double bond since with N-bromo-succinimide isobyssochlamic acid gives a monobromo derivative (2) in

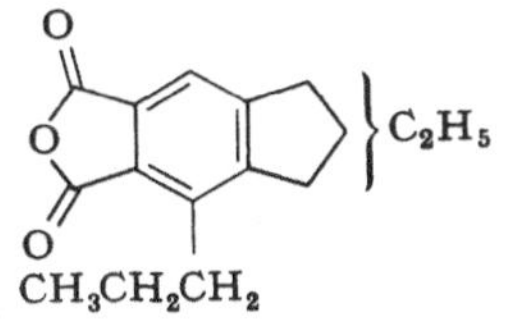

$$CH_3CH_2CH_2$$

(40)

which the CHBr proton absorps at 4.14 τ split as a symmetrical triplet (J = 4 c · sec^{-1}) indicating the presence of a freely rotating methylene group next to it. A platinum dehydrogenation product obtained from byssochlamic acid (7) has been shown to arise from the isomer and, by a combination of spectroscopic and chemical information, to have the part structure (40) (2). It is probably formed from isobyssochlamic acid by a skeletal rearrangement.

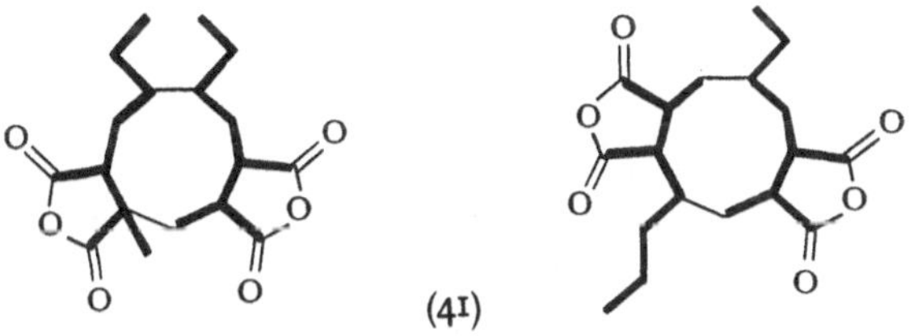

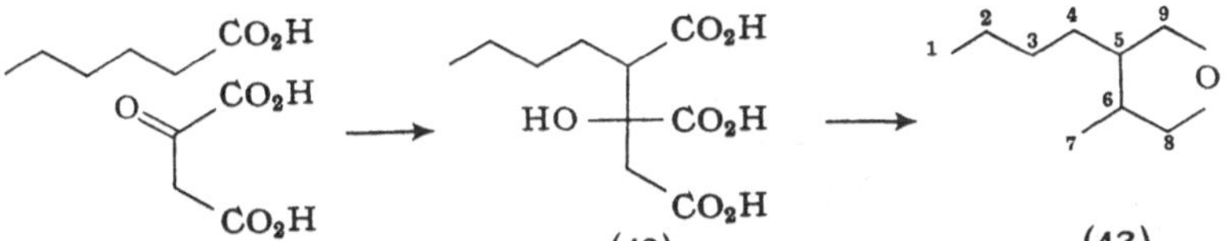

At present the structure of isobyssochlamic acid cannot be defined but some of its derivatives are being investigated by X-ray crystallo-graphy.

V. Biosynthesis.

On inspecting the structures of glaucanic and byssochlamic acids, it is striking that each contains two C_9-units of identical carbon skeleton (see 41). Accepting that these similarities have biogenetic significance, it was postulated (1) that the acids arose by dimerisation of a C_9-unit of carbon skeleton (42) which, in turn, resulted from the citric acid (43) by decarboxylation and elimination of water. The citric acid was presumed to arise by condensation of oxalacetic acid with hexanoic acid.

This biosynthetic scheme was investigated by BLOOMER, MOPPETT and SUTHERLAND (6) with reference to glauconic acid (11, p. 134), produced by *Penicillium purpurogenium*. It was expected that, in common with other fungal metabolites, it was derived ultimately from acetate; and, using [1-^{14}C]-acetate appreciable activity was incorporated. Glauconic acid was chosen because of the relative ease with which it could be degraded

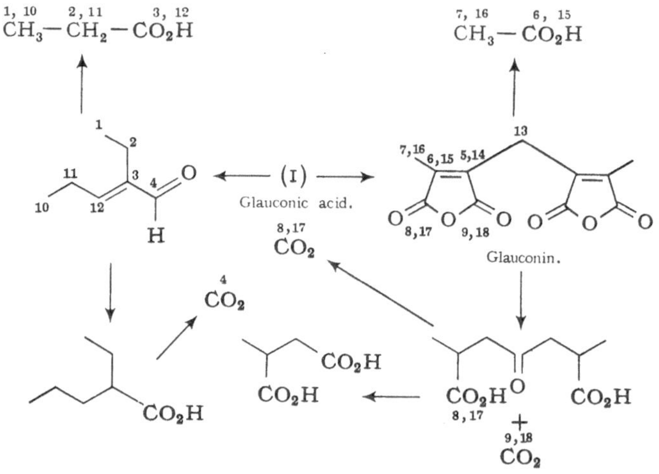

Chart 9. Radioactive Assay of Glauconic Acid Atoms.

and the individual carbon atoms isolated and their activity determined. The method of degradation is summarised in *Chart 9*. The acetic and propionic acids obtained from the reactions listed were further degraded stepwise to carbon dioxide. It will be realised that because of the symmetry of glauconin and the degradation of the diethylacrolein to two moles of propionic acid the activity values found for all of the carbons except $C_{(4)}$ and $C_{(13)}$ are averages for pairs of carbons originally present at different positions in glauconic acid. However, these pairs of atoms are at equivalent positions in the postulated C_9-precursor and if the two parts of the glauconic acid molecule arise from a single C_9-precursor, then the average values will be the true ones for each atom. In support of this assumption it has been shown that the activities of $C_{(4)}$ and $C_{(13)}$ from [1-^{14}C]-acetate derived glauconic acid are identical within experimental error.

References, pp. 148—149.

The full results for the distribution of activity in the [1-^{14}C]-acetate derived glauconic acid are summarised in (44); and (45) shows the activities found after [2-^{14}C]-acetate feeding. These patterns are consistent with the original scheme. The activity distribution in the C_6-chain agrees with it being formed from one acetyl and two malonyl units condensing by the well-established route of fatty acid biosynthesis to form an hexanoyl derivative. In the C_3-chain $C_{(6)}$ and $C_{(7)}$ are derived exclusively from the methyl of acetate while $C_{(8)}$ results both from carboxyl and methyl of acetate.

Further information on the origin of the C_3-chain (comprising $C_{(7)}$, $C_{(6)}$ and $C_{(8)}$) came from other incorporation experiments. Since all of the carbons of the C_3-chain appear, after pyrolysis, in glauconin, any preferential incorporation of activity into this fragment will raise the ratio

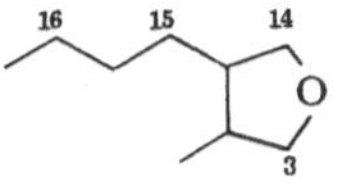

(44) 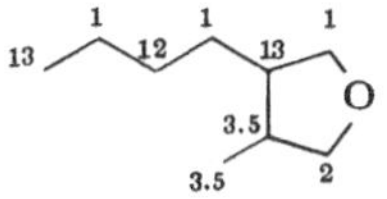   (45)

(49)

Glauconic acid = 100

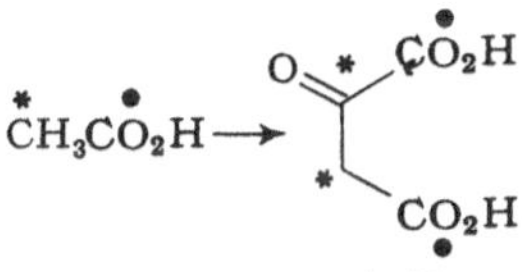

(46) 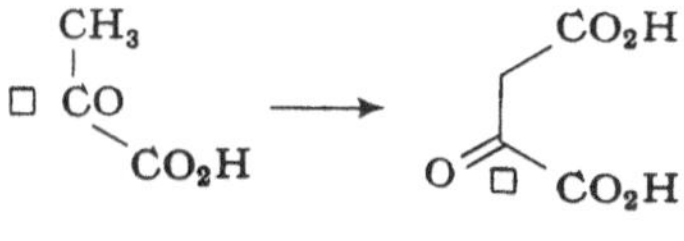   (47)

of activities glauconin: acrolein. It was first shown that both [1-^{14}C]- and [2-^{14}C]-acetate were more rapidly incorporated into the C_6-chain. When, however, [2-^{14}C]-pyruvate or [2-^{14}C]-glucose (which should be metabolised to [2-^{14}C]-pyruvate by the Embden-Myerhof pathway) were fed, the ratio rose from the 1.13 observed with [1-^{14}C]-acetate to 1.58. The pyruvate is known to be a good precursor for [1-^{14}C]-acetate but if the activity were being incorporated only via acetate, then the labelling pattern would be expected to be the same as that found with [1-^{14}C]-acetate. Degradation showed that 8% of the total activity was present at $C_{(6)}$ and 1.5% at $C_{(7)}$, whereas none was found at these positions in the [1-^{14}C]-acetate experiments. Thus pyruvate can be incorporated without degradation to acetate and without going through a symmetrical C_4-intermediate.

These results do not distinguish between the precursor being a C_3- or C_4-unit. The feeding of [2,3-^{14}C]-succinate did resolve this problem; since if the precursor were degraded to acetate or pyruvate before incorporation, labelling patterns similar to those already observed would be expected. The succinate was incorporated efficiently and 55% of the total activity was found at $C_{(6)}$ and $C_{(7)}$, in contrast to 14% with [2-^{14}C]-acetate and (calculated) 19% with pyruvate, strongly suggesting that the succinate was not degraded before incorporation.

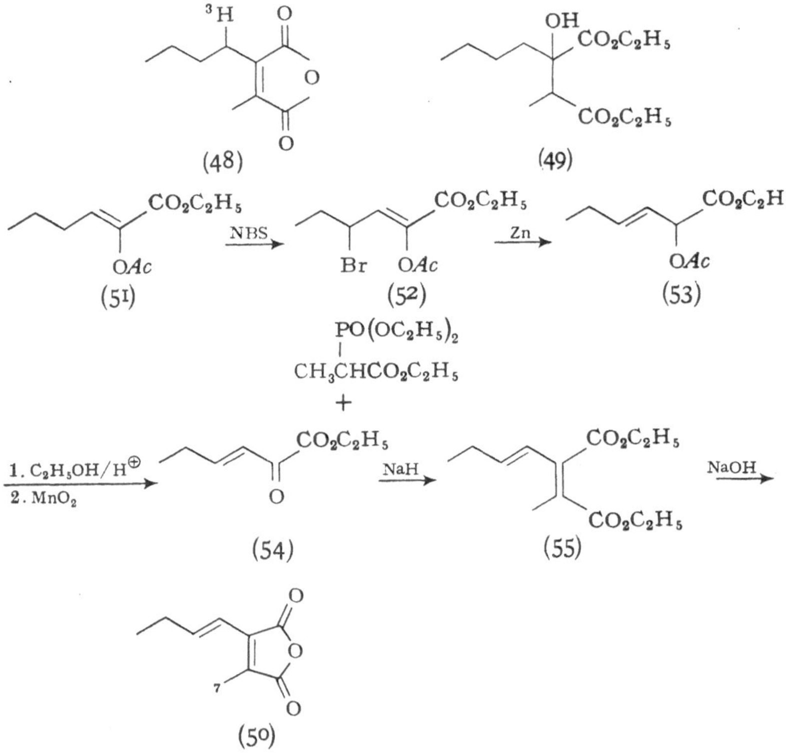

Chart 10. Synthesis of the Unsaturated Anhydride (50).

These results all point to oxaloacetate, $HOOC \cdot CH_2 \cdot CO \cdot COOH$, being the immediate precursor of the C_3-chain, since oxaloacetate with labelling pattern (46) can arise from operation of the citric acid cycle. In this, labelled acetate condenses with unlabelled oxaloacetate giving citric acid. Enzymic degradation of the citriate via aconitic, isocitric and α-ketoglutaric acids then leads to the symmetrically labelled succinate which is oxidized to fumaric and then malic acid and finally to oxalo-acetate labelled as in (46).

A second known route to oxaloacetate is the carboxylation of pyruvate giving the unsymmetrical labelling pattern (47). The appearance of some labelling at the methylene position is readily accountable by some of the oxaloacetate being recycled and giving the symmetrically labelled acid. These labelling patterns are consistent with the proposed bio-synthetic route but do not unambiguously prove it.

To gain more information on the latter stages of the process larger metabolites, which are not expected to participate in the general basal metabolism, were studied by MOPPETT and SUTHERLAND (13). The first C_9-unit prepared was (48) labelled with tritium as shown in the formula. Ethyl [3-³H]-2-ketohexanoate was obtained by ethanolysis of the corre-sponding enol acetate with [O-³H]-ethanol. A Reformatsky reaction between the ketohexanoate and ethyl α-bromopropionate gave the hydroxy diester (49) which by hydrolysis, dehydration and anhydride formation yielded (48). When the mold was grown in the presence of this compound radioactive glauconic acid was obtained (0.25% incor-poration). Pyrolysis to glauconin and diethylacrolein showed the activity ratio of the two fragments to be exactly one — a further support for glau-conic acid arising from two identical C_9-units.

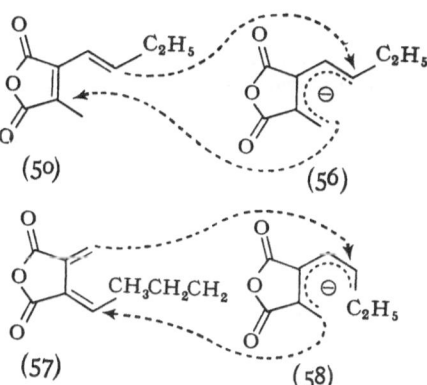

Chart 11. Dimerisation Mechanisms.

The precursor which undergoes dimerisation appears to be the un-saturated anhydride (50) which was synthesised (13) from the enol-acetate (51) as shown in *Chart 10*. Interaction of (51) and N-bromosuccinimide gave (52) which was reduced with zinc in acetic acid to the non-conju-gated ester (53). Ethanolysis of (53) gave the alcohol which on manganese dioxide oxidation yielded (54). In common with other α-ketoesters (54) reacted with the anion of the phosphonate to give stereospecifically (13) the maleic ester (55) which was hydrolysed to (49). Labelling of (50) at $C_{(7)}$ was achieved by alkylating the α-phosphonato-acetic ester with ¹⁴C-methyl iodide to give the labelling propionic phosphonate. When the mold was grown in the presence of (50), the glauconic acid isolated contained 51.5% of the total activity (97.5% of it at $C_{(7)}$ and $C_{(16)}$) and the glaucanic acid contained 4%.

In addition it has been shown that labelled glaucanic acid can be converted into labelled glauconic acid in the mold. Thus, there seems little doubt that (50) is the precursor of glaucanic acid. The relatively low incorporation in (48) suggests that this is not a true precursor. Maybe the incorporation observed is due to an exchange of free (48) with an enzyme-bound derivative.

The dimerisation of (50) to glaucanic acid is unique in biosynthetic reactions and could be an enzyme-induced electrocyclic addition between (50) (4π electrons) and its anion (56) (6π electrons) (*10*), which, by an endo transition state would give the required stereochemistry. Byssochlamic acid could arise in a similar way from the diene (57) and the anion (58), which is a stereoisomer of the anion (56) *(Chart 11)*.

References.

1. Baldwin, J. E., D. H. R. Barton, J. L. Bloomer, L. M. Jackman, L. Rodriguez-Hahn and J. K. Sutherland: The Constitutions of Glauconic, Glaucanic, and Byssochlamic Acids. Experientia **18**, 345 (1962).

2. Baldwin, J. E., D. H. R. Barton and J. K. Sutherland: The Nonadrides. IV. The Constitution and Stereochemistry of Byssochlamic Acid. J. Chem. Soc. (London) **1965**, 1787.

3. Barton, D. H. R., L. D. S. Godinho and J. K. Sutherland: The Nonadrides. III. The Absolute Configuration of Glauconic and Glaucanic Acids. J. Chem. Soc. (London) **1965**, 1779.

4. Barton, D. H. R., L. M. Jackman, L. Rodriguez-Hahn and J. K. Sutherland: The Nonadrides. II. The Constitutions of Glauconic and Glaucanic Acids. J. Chem. Soc. (London) **1965**, 1772.

5. Barton, D. H. R. and J. K. Sutherland: The Nonadrides. I. Introduction and General Survey. J. Chem. Soc. (London) **1965**, 1769.

6. Bloomer, J. L., C. E. Moppett and J. K. Sutherland: The Biosynthesis of Glauconic Acid. Chem. Commun. **1965**, 619.

7. Cook, J. W., J. D. Loudon and R. P. Paton; see R. P. Paton: Thesis, Univ. Glasgow. 1954.

8. Ferguson, G., J. M. Robertson and G. A. Sim: The Stereochemistry of Glauconic Acid. Proc. Chem. Soc. (London) **1962**, 385.

9. Hamor, T. A., I. C. Paul, J. M. Robertson and G. A. Sim: Fungal Metabolites. II. The Structure of Byssochlamic Acid: X-Ray Analysis of Byssochlamic Acid Bis-*p*-bromophenyl Hydrazide. J. Chem. Soc. (London) **1963**, 5502.

10. Hoffmann, R. and R. B. Woodward: Selection Rules for Concerted Cycloaddition Reactions. J. Amer. Chem. Soc. **87**, 2046 (1965).

11. Kraft, K.: Zur Konstitution der Glaukonsäuren. VI. Liebigs Ann. Chem. **530**, 20 (1937).

12. Kraft, K. und H. Porsch: Zur Konstitution der Glaukonsäuren. V. Untersuchungen am Glaukonin. Liebigs Ann. Chem. **527**, 168 (1936).

13. Moppett, C. E. and J. K. Sutherland: unpublished.

14. Raistrick, H. and G. Smith: Studies in the Biochemistry of Micro-organisms. XXXV. The Metabolic Products of *Byssochlamys fulva* Olliver and Smith. Biochem. J. **27**, 1814 (1933); and unpublished observations.

15. SUTTER, H., F. ROTTMAYR und H. PORSCH: Zur Konstitution der Glaukonsäuren. IV. Liebigs Ann. Chem. **521**, 189 (1936).

16. SUTTER, H. und N. WIJKMAN: Über einige neue, durch Schimmelpilze gebildete Substanzen. II. Zur Konstitution der Glaukonsäuren. Liebigs Ann. Chem. **505**, 248 (1933).

17. — — Zur Konstitution der Glaukonsäuren. III. Liebigs Ann. Chem. **519**, 97 (1935).

18. WIJKMAN, N.: Über einige neue, durch Schimmelpilze gebildete Substanzen. Liebigs Ann. Chem. **485**, 61 (1931).

19. YUILL, J. L.: The Acids Produced from Sugar by a *Penicillium* Parasitic upon *Aspergillus niger*. Biochem. J. **28**, 222 (1934).

(Received, September 29, 1966.)

Natürlich vorkommende Auronglykoside.

Von L. FARKAS und L. PALLOS, Budapest.

Herrn Professor Dr. H. WAGNER (Institut für pharmazeutische Arzneimittellehre der Universität München) danken wir für seine Mithilfe bei der Abfassung und für Durchsicht des Manuskriptes.

I. Einleitung.

Die Auronglykoside sind Polyhydroxy-benzal-cumaran-3-on-Derivate, die durch das Grundgerüst C_6—C_3—C_6 charakterisiert sind und damit zur Familie der Flavonoidverbindungen gehören. Einem Vorschlag von BATE-SMITH und GEISSMAN (*10*) folgend, wurde für diese Naturstoffe die Bezeichnung Aurone in die Fachliteratur eingeführt. Der Name deutet auf die schöne goldgelbe Farbe der Blüten hin, in denen diese Verbindungen vorkommen und bringt gleichzeitig die Verwandtschaft mit den Flavonen zum Ausdruck, die weniger intensiv gelb gefärbte Pigmente sind und im C_3-Molekülteil die gleiche Oxydationsstufe besitzen.

Zusammen mit den oft gleichzeitig auftretenden Chalconen gehören die Aurone und ihre Glykoside (*1*) zur Gruppe der Anthochlor-Pigmente.

Literaturverzeichnis: SS. 169—174.

Für diese ist kennzeichnend, daß ihre Farbe unter Einwirkung von schwachen Alkalien oder Ammoniak sich von Gelb nach Orangerot verschiebt (*90*, *73*). Die allein durch Carotinoide gelb gefärbten Blütenblätter zeigen diese Farbänderung nicht.

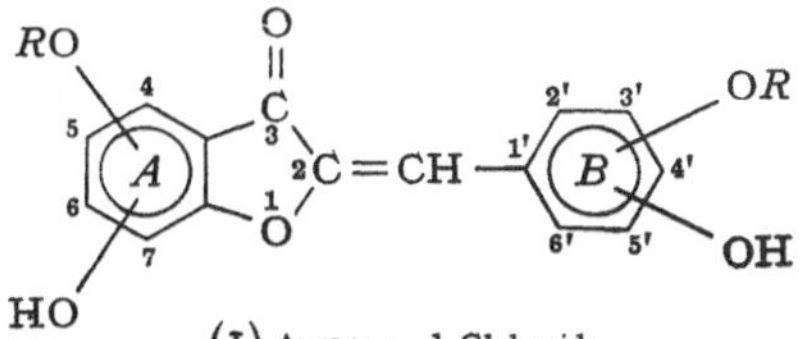

(I) Aurone und Glykoside.
R = H oder Glycosyl.

Das erste Auronglykosid, das Leptosin (*2*, S. 153), ein 3',4',6-Trihydroxy-7-methoxy-auron-6-glucosid und sein Aglucon, das Leptosidin (*3*) wurden im Jahre 1943 von Geissman (*44*) aus den Korbblüten von *Coreopsis grandiflora* Nutt. isoliert. Bis heute sind 8 weitere Auronglykoside aus verschiedenen Pflanzen gewonnen worden, wo sie in der Regel gemeinsam mit ihren 5 Aglykonen vorkommen. Bei einem einzigen Auron, dem im Kernholz des Rengas-Baumes vorkommenden Rengasin (*70*) (3',4',6-Trihydroxy-4-methoxy-auron) konnte das entsprechende Glykosid bisher noch nicht gefunden werden. Bemerkenswert ist, daß bisher die Auronglykoside am häufigsten in den Blütenblättern von Pflanzen der Compositen-Familie aufgefunden wurden.

Die Strukturaufklärung der Auronglykoside verdanken wir in erster Linie Geissman und Mitarb. Wertvolle Arbeiten auf diesem Gebiet haben außerdem Marini-Bettòlo, Shimokoriyama, Hattori, Seshadri, Chopin und Harborne geleistet. Während die erste Auronsynthese (Leptosin) von Geissman (*49*) durchgeführt wurde, haben sich in der Folgezeit Farkas und Pallos (*30*) um die Ausarbeitung allgemein anwendbarer Synthesen von Auronglykosiden und um den Strukturbeweis der natürlich vorkommenden Verbindungen bemüht.

Über die pharmakologische Wirkung der Auronglykoside ist bis heute noch nichts bekannt. Dagegen liegen bereits zahlreiche Untersuchungen über die Biosynthese der Aurone vor.

Die vorliegende Studie gibt einen Überblick über Vorkommen, Analytik und Chemie der bis heute bekannten Auronglykoside, wobei die Literatur bis zum Jahre 1966 berücksichtigt wurde.

II. Vorkommen und Isolierung der Auronglykoside.

Wie bereits erwähnt, finden sich die meisten Auronglykoside in den Vertretern der Compositen-Familie. In *Tabelle 1* (S. 152) sind sämtliche bisher als Auron-haltig erkannte Pflanzen und ihre botanische Zugehörigkeit aufgeführt. In den meisten Fällen ließen sich die Auronglykoside und ihre Aglykone aus den Blütenblättern isolieren. Eine Ausnahme bilden *Rhus cotinus* (*71*), der ein Auronglykosid in höchster Konzentration im Holzteil enthält, ferner Rengas (*70*) und *Citrus limonum* (*14*), wo die Verbindungen im Kernholz bzw. in der Fruchtschale angetroffen werden. Die Extraktion erfolgt aus den lufttrockenen oder frischen Blütenblättern im allgemeinen mit Alkohol, bei Zimmertemperatur. Während die Aglykone aus dem durch Eindampfen erhaltenen Rohextrakt mit Äther herausgelöst werden können, lassen sich die Glykoside in der Regel durch fraktionierte Kristallisation reinigen. In den letzten Jahren hat die Säulenchromatographie an verschiedenen Adsorbentien, besonders an Polyamiden für die Isolierung der Auronglykoside zunehmende Bedeutung erlangt.

Tabelle 1. Auron-haltige Pflanzen.

Vorkommen	Familie	Tribus
Coreopsis grandiflora Nutt....	Compositae	Heliantheae, Coreopsidinae
C. lanceolata L.	Compositae	Heliantheae, Coreopsidinae
C. saxicola Alexander........	Compositae	Heliantheae, Coreopsidinae
Cosmos sulphureus Cav.	Compositae	Heliantheae, Coreopsidinae
C. bipinnatus Cav.	Compositae	Heliantheae, Coreopsidinae
Coreopsis gigantea..........	Compositae	Heliantheae, Coreopsidinae
C. maritima Hook.	Compositae	Heliantheae, Coreopsidinae
C. tinctoria Nutt.	Compositae	Heliantheae, Coreopsidinae
Bidens laevis L.	Compositae	Heliantheae, Coreopsidinae
Dahlia variabilis Cav.	Compositae	Heliantheae, Coreopsidinae
Baeria chrysostoma F. u. M. .	Compositae	Heleniae
Viguiera multiflora Nutt....	Compositae	Heleniae
Helichrysum bracteatum Willd. (Vent.)	Compositae	
Chrysanthemum segetum L. ..	Compositae	
Butea frondosa.............	Leguminosae	
Antirrhinum majus L.	Scrophulariaceae	
Oxalis cernua Thumb.	Oxalidaceae	
Citrus limonum L.	Rutaceae	
Rhus cotinus Mill.	Anacardiaceae	Heliantheae, Coreopsidinae
Rengas	Melanorrhea	

Leptosin (2).

Die Verbindung wurde von Geissman (*44*) im Jahre 1943 aus den Blüten von *Coreopsis grandiflora* Nutt. isoliert. Die Strukturaufklärung erfolgte durch Darstellung von Derivaten und auf Grund biogenetischer Überlegungen. Die für Polyhydroxy-benzalcumaranone charakteristischen Farbreaktionen (*4, 31*), angestellt mit dem durch saure Hydrolyse erhaltenen Aglykon, zeigten bald, daß es sich hier nicht um ein Antho-

Literaturverzeichnis: SS. 169—174.

chlor-Pigment mit Chalkon-Struktur handeln konnte. Die Färbung der Verbindung sowie die Identität der nach der Diazomethan- und Dimethylsulfat-Methode erhaltenen Methyl-Produkte deuteten darauf hin, daß auch kein Flavanon vorliegen konnte. Der Analyse zufolge enthielt die Verbindung drei Hydroxylgruppen und eine Methoxygruppe, und die durch Oxydation mit Permanganat erhaltene Veratrumsäure bewies, daß sich zwei Hydroxylgruppen in den Stellungen 3' und 4' befinden.

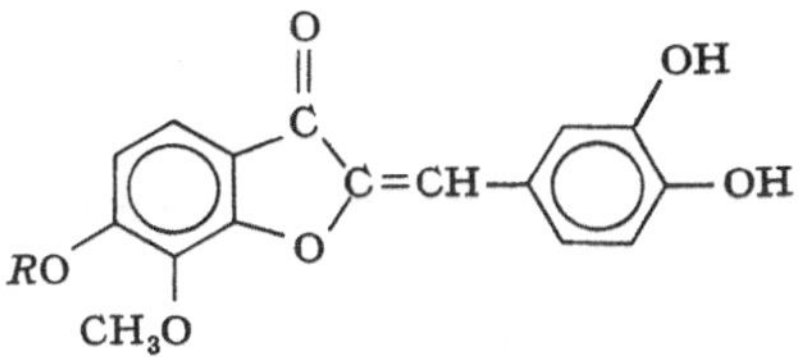

(2) $R = C_6H_{11}O_5$. Leptosin.

(3) $R = H$. Leptosidin.

Der restliche Teil dieses Auronmoleküls wurde vor allem an Hand von biogenetischen Analogieschlüssen strukturell sichergestellt. Schon nach kurzer Zeit gelang es dann, die Richtigkeit der aufgestellten Struktur durch Darstellung (45) des Leptosidin-trimethyläthers (vgl. 3) und später durch die Synthese des Glykosids (49) zu beweisen.

Aureusin (4).

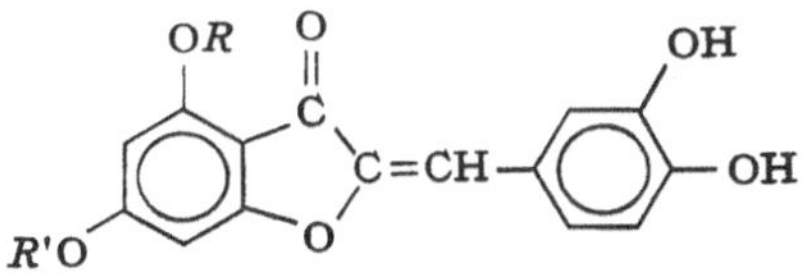

(4) $R' = C_6H_{11}O_5$. $R = H$. Aureusin.

(5) $R' = R = H$. Aureusidin.

(6) $R' = H$, $R = C_6H_{11}O_5$. Cernuosid.

GEISSMAN und Mitarb. (*100*) isolierten dieses Glykosid aus den Blütenblättern von *Antirrhinum majus* gemeinsam mit dem Aglykon Aureusidin (5). Die Struktur wurde analog der beim Leptosin angewandten Methode ermittelt, wobei erstmals auch die Absorptionsspektroskopie (*101*, *40*) mit Erfolg herangezogen wurde. Auf Grund der bathochromen UV-Verschiebung (*57*), die nur Aurone mit freier Hydroxylgruppe in $C_{(4)}$-Stellung auf Zusatz von Aluminiumchlorid zeigen, ließ sich eindeutig feststellen, daß der Zucker in 6-Stellung geknüpft war.

Cernuosid (6).

Im Jahre 1952 isolierten Lamonica und Marini-Bettòlo (75) die Verbindung aus den gelben Blüten einer vorwiegend im Mittelmeerraum heimischen Oxalidacee, *Oxalis cernua* Thumb. Zunächst schrieb man Cernuosid eine Chinon-Struktur zu, da Fernández und Pizarroso (32) aus *Oxalis purpurata* Jacq. ein Oxaloxanthin benanntes gelbes Pigment isoliert hatten, von dem Fieser (33) nachweisen konnte, daß es mit 2,5-Dihydroxy-3-tridecyl-1,4-benzochinon und dem schon früher von Kawamura (68) isolierten Rapanon (5) identisch ist. Später fand man jedoch, daß das Cernuosid und das mit diesem gemeinsam isolierte Aglykon zur Gruppe der Flavonoidverbindungen gehören müsse und daß das Aglykon mit Aureusidin (5) identisch ist. Die Struktur des Glykosids wurde im Jahre 1955 endgültig festgelegt (7), nachdem schon vorher die Stelle der Bindung mit dem Zuckeranteil geklärt worden war (40).

Sulfurein (7).

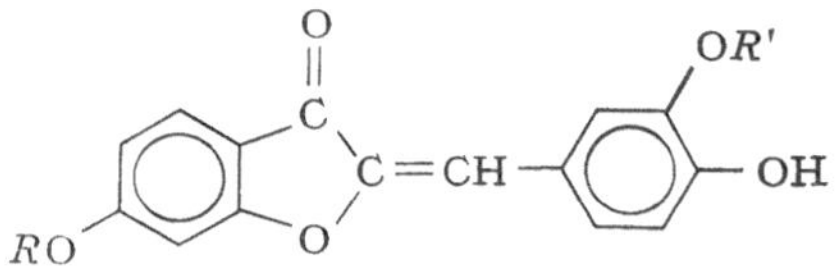

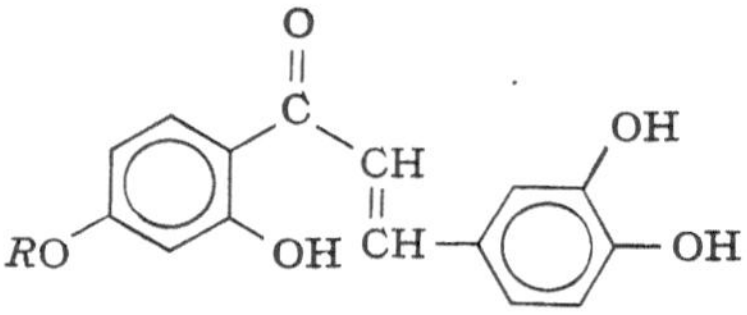

(7) $R = C_6H_{11}O_5$. $R' = H$. Sulfurein.   (10) $R = C_6H_{11}O_5$. Coreopsin.

(8) $R = R' = H$. Sulfuretin.

(9) $R = R' = C_6H_{11}O_5$. Palasitrin.

Shimokoriyama und Hattori (104) isolierten Sulfurein aus den Blütenblättern von *Cosmos sulphureus* und beobachteten, daß neben diesem Auronglykosid ein strukturanaloges Chalkonglykosid, das bekannte Coreopsin (10) vorkommt. Das Sulfurein lieferte durch saure Hydrolyse das bekannte Sulfuretin (8). Zuckerart und Verknüpfungsstelle des Zuckers mit dem Aglykon konnten durch Überführen des Coreopsins in das Sulfurein festgestellt werden.

Im gleichen Jahr isolierten Nordström und Swain (84) aus den gelben Blüten von *Dahlia variabilis* ein zweites „Sulfurein", dessen physikalische Konstanten von denen der von Shimokoriyama und Hattori (104) beschriebenen Verbindung abwichen, so daß Zweifel an der Richtigkeit der Struktur entstanden. Durch Vergleich der UV-Spektren konnten jedoch Geissman und Jurd (48) eindeutig die Richtigkeit der durch die japanischen Forscher beschriebenen Struktur erhärten.

Erhebliche Mengen des gleichen Sulfuretins und etwas weniger Sulfurein isolierten King und White (71) aus *Rhus cotinus*. Dieser Befund ist deshalb bemerkenswert, da bishin allein dem von Schmid (97) aus

Literaturverzeichnis: SS. 169—174.

dem gleichen Material isolierten 3,7,3',4'-Tetrahydroxy-flavon (Fisetin) die färberischen Eigenschaften des Fustin-Extraktes zugeschrieben worden waren.

Palasitrin (9).

Außer Sulfurein (7) existiert noch ein zweites natürliches Glykosid des Sulfuretins (8). PURI und SESHADRI (92) isolierten im Jahre 1955 aus *Butea frondosa* das Palasitrin (9), das nach dem indischen Namen der Pflanze, Palas, benannt wurde und das einzige bisher in der Natur vorkommende Aurondiglykosid, ein 6,3'-Diglykosid des Sulfuretins, darstellt. Die Zuckerstellung wurde durch erschöpfende Methylierung, nachfolgende Hydrolyse und Vergleich mit den in Frage kommenden Monomethyläthern des Aglykons bewiesen. Das daneben in *Butea frondosa* vorkommende strukturanaloge Chalkonglykosid, das schon früher ebenfalls von PURI und SESHADRI (91) isolierte Isobutrin (11) läßt sich nach Acetylierung, Bromierung und nachfolgender Behandlung mit alkoholischer Lauge in Palasitrin überführen.

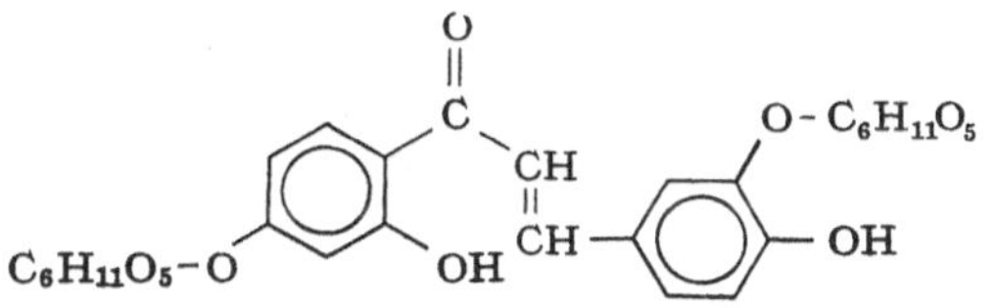

$$C_6H_{11}O_5-O \qquad OH \qquad \underset{\text{CH}}{\overset{\displaystyle O \atop \| \atop C}{}} \quad CH \quad O-C_6H_{11}O_5 \quad OH$$

(11)
Isobutrin.

Maritimein (12).

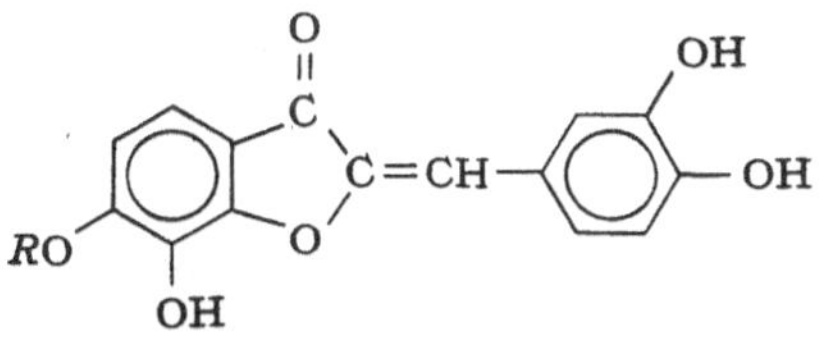

$$RO \qquad OH \qquad \overset{O}{\underset{O}{\overset{\|}{C}}} \quad C=CH \qquad OH \quad OH$$

(12) $R = C_6H_{11}O_5$. Maritimein.
(13) $R = H$. Maritimetin.

Diese Verbindung wurde von GEISSMAN und Mitarb. (42) aus *Coreopsis maritima* im Rahmen einer Gesamtanalyse der Pflanze isoliert. Auf Grund chromatographischer Untersuchungen und mit Hilfe der UV-Spektroskopie konnten sie Butein, Coreopsin, Sulfurein, Luteolin-7-glykosid und, außer der bis dahin in den *Coreopsis*-Arten noch nicht nachgewiesenen Chlorogensäure, Maritimein (12) und das entsprechende Chalkonglykosid, Marein (2',3',4',3,4-Pentahydroxychalkon-4'-glykosid), nachweisen bzw.

isolieren. Angeregt durch diese Ergebnisse unterzogen die Autoren auch *C. gigantea* einer erneuten Untersuchung, wobei sie Maritimein, Marein sowie die beiden entsprechenden Aglykone, Maritimetin (13) und Okanin isolieren konnten. Die Struktur von Maritimein und Marein wurde von Harborne und Geissman (60) aufgeklärt, indem sie das Marein durch Luftoxydation in Maritimein überführten und damit bewiesen, daß die Anordnung der Substituenten in beiden Verbindungen die gleiche ist. Durch Kochen mit äthanolischer Salzsäure liefert Marein das bekannte 3′,4′,7,8-Tetrahydroxyflavanon, so daß damit gleichzeitig die Struktur des Chalkons geklärt war. Die Zuckerstellung ergab sich aus dem UV-Spektrum des nach totaler Methylierung und nachfolgender Hydrolyse erhaltenen 6-Hydroxy-3′,4′,7-trimethoxy-aurons. Shimokoriyama (102) erhielt Marein und Maritimein auch aus *C. tinctoria* und fand daneben auch das Flavanon-isomere Flavanomarein. Schließlich konnten die beiden Aurone von Geissman und Steelink (50) auch aus *Chrysanthemum segetum* und von Shimokoriyama und Geissman (103) aus den Blütenblättern zweier nicht zur Untergruppe der Coreopsidinae gehörenden Compositen-Arten, nämlich aus *Viguiera multiflora* und *Baeria chrysostoma* isoliert werden.

Bractein (14).

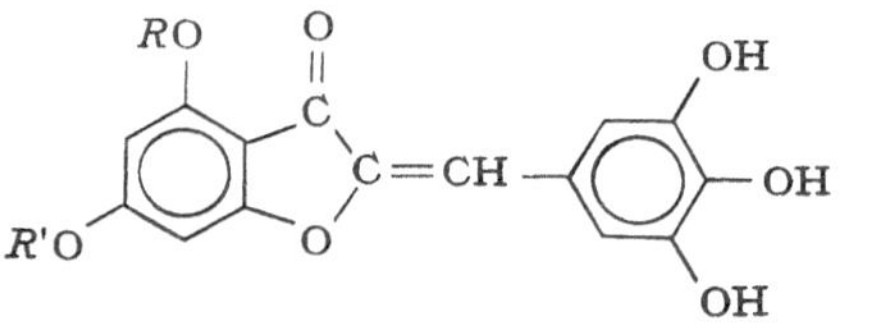

(14) $R = C_6H_{11}O_5 . R′ = H .$ Bractein.

(15) $R = R′ = H .$ Bracteatin.

(16) $R = H, R′ = C_6H_{11}O_5 .$ Bracteatin-6-glucosid.

Dieses neue Glykosid wurde von Hänsel und Mitarb. (56) im Jahre 1962 in den gelben Blüten der in Australien heimischen Compositen-Art *Helichrysum bracteatum* (Vent.) Willd. entdeckt, nachdem bereits Rosoll (95) 1884 auf Grund charakteristischer Farbreaktionen auf diesen neuartigen Farbstoff aufmerksam gemacht hatte und dieses, als Helichrysin benannte Pigment von Klein (73) in die Gruppe der Anthochlorverbindungen eingereiht worden war. Hänsel nannte das neue Glykosid Bractein und sein Aglykon Bracteatin. Für Bracteatin ermittelte er die Struktur 3′,4′,5′,4,6-Pentahydroxyauron (15) und konnte diese später auch auf synthetischem Wege beweisen (55). Das Bractein stellt das erste aus einer Compositen-Art isolierte Auronglykosid dar, dessen *A*-Ring in den Stellungen 4 und 6 Hydroxylgruppen trägt; es ist auch die erste

Literaturverzeichnis: SS. 169—174.

Verbindung dieses Typs, die im *B*-Ring drei Hydroxylgruppen in vicinaler Anordnung besitzt. Auf Grund spektroskopischer Untersuchungen nahm man an (*56*), daß der Zuckerrest in 4-Stellung geknüpft ist. In neuerer Zeit haben HÄNSEL und Mitarb. (*93*) auch das entsprechende Chalkonglykosid aus der Pflanze isoliert.

Aureusidin-6-rutinosid (**17**).

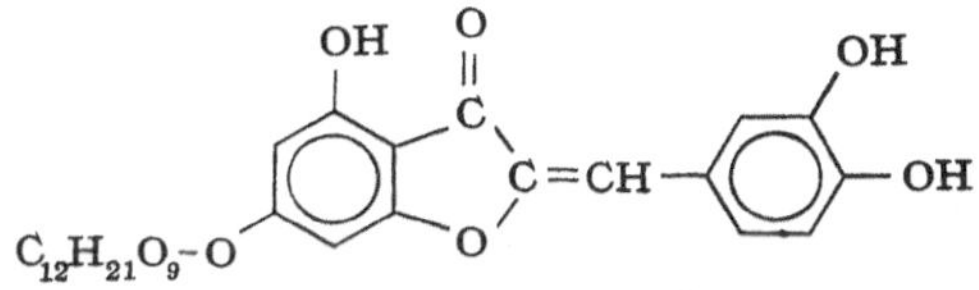

OH   O
        ||
        C

$C_{12}H_{21}O_9$- O

C=CH

OH

OH

(**17**)

Aureusidin - 6 - rutinosid.

CHOPIN und Mitarb. (*14*) isolierten (1963) aus dem Alkoholextrakt der Schalen von *Citrus limonum* ein neues Aureusidin-Glykosid, das bei saurer Hydrolyse neben dem Aglykon Glucose und Rhamnose lieferte. Durch Methylierung und Hydrolyse stellten sie fest, daß beide Zucker in Form eines Disaccharids in Stellung 6 gebunden sein müssen. Da unter den Begleitflavonoiden auch das in seiner Struktur geklärte Eryocitrin (3′,4′,5,7-Tetrahydroxy-flavanon-7-rutinosid) vorkommt, ein Flavanon mit analoger Hydroxylsubstitution, nahmen die genannten Autoren an, daß die Disaccharidkomponente auch in dem Auronglykosid Rutinose sein müsse. Die gleichzeitige Gegenwart eines entsprechenden Flavanonglykosids ließ schon CHOPIN vermuten, daß Aureusidin-6-rutinosid nicht genuin in der Pflanze enthalten sei, sondern ein Kunstprodukt darstelle, das sich während der Alkoholextraktion gebildet hatte. In der Tat läßt sich Eryocitrin unter den genannten Bedingungen zum entsprechenden Chalkon isomerisieren und dieses kann nach SHIMOKORIYAMA und HATTORI (*104*) spontan zum Auronrutinosid oxydiert werden. Dies erhielt eine Stütze durch die Beobachtung von CHOPIN und DELLAMONICA (*13*), daß das Eryocitrin zur Umwandlung in Aureusidin-6-rutinosid auch in neutralem Medium befähigt ist. Diese Reaktion bewies gleichzeitig, daß der Zuckeranteil Rutinose ist.

Bracteatin-6-glucosid (**16**).

Das neunte natürliche Auronglykosid, über dessen Isolierung HARBORNE (*59*) 1963 berichtet hat und dessen Identifizierung zusammen mit fünf anderen, in den Blüten von *Antirrhinum majus* vorkommenden Flavonen bzw. ihren Glykosiden erfolgte, ist das Bracteatin-6-glucosid. Während die Identität des Aglykons durch Vergleich mit synthetisch

Tabelle 2. Verzeichnis der bekannten Auronglykoside.

Benennung des Auron-glykosids	Struktur	Benennung des Aglykons	Struktur
Leptosin..........	$R_{3'} =$ OH, $R_{4'} =$ OH	Leptosidin	$R_{3'} = R_{4'} = R_6 =$ = OH
Leptosidin-6-gluco-sid	$R_6 =$ O-Glucose, $R_7 =$ OCH$_3$	3′,4′,6-Trihydroxy-7-methoxy-auron	$R_7 =$ OCH$_3$
Aureusin	$R_{3'} = R_4 = R_{4'} =$ = OH	Aureusidin	$R_{3'} = R_4 = R_{4'} =$ = $R_6 =$ OH
Aureusidin-6-gluco-sid	$R_6 =$ O-Glucose	3′,4′,4,6-Tetra-hydroxy-auron	
Cernuosid........	$R_{3'} = R_{4'} = R_6 =$ = OH	Aureusidin	$R_{3'} = R_4 = R_{4'} =$ = $R_6 =$ OH
Aureusidin-4-gluco-sid	$R_4 =$ O-Glucose	3′,4′,4,6-Tetra-hydroxy-auron	
Aureusidin-6-ruti-nosid..........	$R_{3'} = R_4 = R_{4'} =$ = OH $R_6 =$ O-Rutinose	Aureusidin 3′,4′,4,6-Tetra-hydroxy-auron	$R_{3'} = R_4 = R_{4'} =$ = $R_6 =$ OH
Sulfurein	$R_{3'} = R_{4'} =$ OH	Sulfuretin	$R_{3'} = R_{4'} = R_6 =$ = OH
Sulfuretin-6-gluco-sid	$R_6 =$ O-Glucose	3′,4′,6-Tri-hydroxy-auron	
Palasitrin	$R_{4'} =$ OH	Sulfuretin	$R_{3'} = R_{4'} = R_6 =$ = OH
Sulfuretin-3′,6-diglucosid	$R_{3'} = R_6 =$ O-Glucose	3′,4′,6-Trihydroxy-auron	
Maritimein........	$R_{3'} = R_{4'} = R_7 =$ = OH	Maritimetin	$R_{3'} = R_{4'} = R_6 =$ = OH
Maritimetin-6-glu-cosid	$R_6 =$ O-Glucose	3′,4′,6,7-Tetra-hydroxy-auron	$R_7 =$ OH
Bractein	$R_{3'} = R_{4'} =$ = $R_{5'} = R_6 =$ OH	Bracteatin	$R_{3'} = R_4 = R_{4'} =$ = $R_{5'} = R_6 =$ OH
Bracteatin-4-gluco-sid	$R_4 =$ O-Glucose	3′,4′,4,5′,6-Penta-hydroxy-auron	
Bracteatin-6-gluco-sid	$R_{3'} = R_4 = R_{4'} =$ = $R_{5'} =$ OH, $R_6 =$ O-Glucose	Bracteatin 3′,4,4′,5′,6-Penta-hydroxy-auron	$R_{3'} = R_4 = R_{4'} =$ = $R_{5'} = R_6 =$ OH
		Rengasin 3′,4′,6-Trihydroxy-4-methoxy-auron	$R_{3'} = R_{4'} = R_6 =$ = OH, $R_4 =$ = OCH$_3$

Literaturverzeichnis: SS. 169—174.

hergestelltem Bracteatin erbracht wurde, ergab sich die Zuckerstellung aus dem spektroskopischen und papierchromatographischen Verhalten des Glykosids.

In *Tabelle 2* sind die beschriebenen Auronglykoside und ihre Strukturen zusammengestellt.

III. Allgemein anwendbare Methoden zur Strukturaufklärung der Auronglykoside.

Bei der Strukturaufklärung der Auronglykoside haben von Anfang an Farbreaktionen und das papierchromatographische sowie spektroskopische Verhalten eine wichtige Rolle gespielt. Vorschriften zur Anfärbung natürlich vorkommender Aurone haben GEISSMAN (*38*), HARBORNE (*57a*, *58*), NIKONOV (*83*) und KING (*71*) ausgearbeitet. Die wichtigsten Farbreaktionen sind in den *Tabellen 3* und *4* zusammengefaßt.

Tabelle 3.
Farbreaktionen natürlicher Aurone auf Papier in Tageslicht.

Reagens	Farbe
Eisen-III-chlorid	blau
Ammoniak	orange
	orange-schwach rot
Aluminium-III-chlorid	blaßgelb
	orange
Natriumcarbonat	orange
	schwach rot
	purpur
Natriumborhydrid	farblos
p-Toluolsulfonsäure	rosa
	orange
Kaliumferricyanid.................................	blau
Essigsäureanhydrid und Schwefelsäure	orange
	purpur

Tabelle 4. Farbreaktionen natürlicher Aurone in UV-Licht.

Reagens	Farbe
Ammoniak	kräftig gelb
	oder grünlich-gelb
	gelblich-orange
	rötlich-orange
	braun
Aluminium-III-chlorid	grün fluorescierend
	grünlich-gelb
	hellbraun

160   L. Farkas und L. Pallos:

Unter den papierchromatographischen Arbeiten sind vor allem die systematischen Analysen von Roubalová (96) an natürlichen und synthetischen Auronen sowie die im Jahre 1956 beschriebenen Untersuchungen von Hattori und Mitarb. (62) über die Verteilung der Aurone und ihrer Glykoside in den verschiedenen Pflanzenteilen von Cosmos- und Bidens-Arten hervorzuheben. Tabelle 5, die hauptsächlich den Arbeiten von Geissman (38) und Harborne (57a, 58) entnommen ist, gibt das chromatographische Verhalten der natürlichen Aurone und Auronglykoside wieder.

Tabelle 5. Chromatographie natürlicher Aurone.

Lösungsmittel: Butanol-Eisessig-Wasser (4 : 1 : 5).

Benennung	R_f-Wert
Leptosin	0,51
Sulfurein	0,49
Aureusin	0,22
Cernuosid	0,49
Maritimein	0,42
Palasitrin	0,66*
Bractein	0,27
Aureusidin-6-rutinosid	0,35
Bracteatin-6-glucosid	0,06
Leptosidin	0,76
Sulfuretin	0,80
Aureusidin	0,61
Maritimetin	0,53
Bracteatin	0,32

* In phenol-gesättigtem Wasser.

Mit der Polarographie der Aurone haben sich Geissman und Harborne beschäftigt (41).

Die umfassendsten Angaben über die UV-Spektren der Aurone und ihrer Glykoside befinden sich bei Jurd (66). IR-spektroskopisch haben Hänsel und Langhammer (55) das natürliche und synthetische Bracteatin identifiziert und vor kurzem berichteten Batterham und Highet (11) über die NMR-Analyse des Aureusidins.

IV. Gegenseitige Umwandlung der Aurone und ihrer Glykoside in andere Flavonoide.

Nach der klassischen Methode von Kostanecki (69) können aus Chalkonen durch Bromierung und nachfolgende Behandlung mit alkoholischer Kalilauge Flavone erhalten werden. Diese Reaktion verläuft jedoch nur in wenigen Fällen quantitativ in dieser Richtung. Sehr häufig, und be-

Literaturverzeichnis: SS. 169—174.

sonders dann, wenn der aromatische Aldehyd substituiert ist, bilden sich neben den Flavonen auch isomere Benzalcumaranone. Trotz eingehender Untersuchungen konnten NADKARNI, WARRIAR und WHEELER (82) nicht eindeutig klären, von welchen Einflüssen die wechselweise Bildung von Flavonen und Auronen aus den Chalkondibromiden unter Basenkatalyse abhängig ist. Sie untersuchten den Temperatureinfluß auf diese Reaktion und beobachteten, daß sich mit verdünnten Laugen in der Kälte Flavone, in der Wärme Aurone bilden, während in Pyridin sowohl in der Kälte wie in der Wärme nur Flavone entstehen. Ob dabei auch die Methylsubstitution im Molekül eine Rolle spielt, konnten die genannten Autoren nicht entscheiden. Nach GOWAN, HAYDEN und WHEELER (52) entstehen bei höherer Temperatur in Gegenwart von weniger Lauge neben den Flavonen auch Aurone, während dies in der Kälte auch bei Anwendung eines Laugenüberschusses nicht der Fall ist. MARATHEY (77) sowie DONNELLY und Mitarb. (20) haben ebenfalls die Reaktionsbedingungen untersucht, bei denen Chalkondibromide in Gegenwart äthanolischer Lauge entstehen.

ALGAR und FLYNN (1) und gleichzeitig OYAMADA (87) unterwarfen die 2'-Hydroxy-methoxychalkone der Peroxydation und gelangten über die Dihydroflavonole, die MURAKAMI und IRIE (81) in einzelnen Fällen aus stark gekühlten Oxydationsgemischen isolieren konnten, zu Flavonolen. BARGELLINI und OLIVERIO (9) isolierten bei der alkalischen Oxydation von 2-Hydroxy-, 2-Hydroxy-4-methoxy-, 2-Hydroxy-3,4-dimethoxy- sowie 2-Hydroxy-3,4,5-trimethoxy-acetophenon nur Flavonole. Nach GEISSMAN und FUKUSHIMA (39) vermögen sich Aurone auf oxydativem Wege besonders dann zu bilden, wenn das Chalkon in 6'-Stellung eine Methoxylgruppe enthält. Diese Theorie wurde ergänzt durch die Angabe indischer Autoren (6), daß sowohl eine Methoxy- als auch eine Methyl-Substitution in 6'-Stellung für die Bildung von Auronen ausschlaggebend sei. Im Gegensatz dazu gelang es BARGELLINI (86), aus solchen Chalkonen nur Flavonole darzustellen. In Erweiterung dieser Befunde vermuten VENKATARAMAN und Mitarb. (2), daß bei der alkalischen Oxydation außer dem 6'-Substituenten auch die Hydroxylgruppe in $C_{(4)}$-Stellung eine Wirkung ausübt, und zwar in dem Sinne, daß sie den Effekt der 6'-Substitution aufhebt und die Oxydation eindeutig in Richtung Flavonol lenkt. Ein ähnlicher neutralisierter Effekt wurde beim 2-Hydroxychalkon (107), bei der Synthese des 2'-Hydroxy-5,7-dimethoxy-flavonols, sowie neuerdings auch im Falle der 3-Alkoxy-substituierten Chalkone (22) beobachtet. Auf Grund einer systematischen Studie über den Temperatureinfluß bei der Algar-Flynn-Oyamada-Reaktion kommen WHEELER und Mitarb. (89, 110) zu dem Schluß, daß die alkalische Oxydation der 2'-Hydroxy-chalkone in der Kälte vornehmlich Aurone und in der Wärme meist Flavonole liefert. Einige

Jahre später berichteten dieselben Autoren (*17*), daß sie bei der Oxydation der 2'-Hydroxy-3',4'-dimethoxy-chalkone einmal nur 2-Benzyl-2-hydroxy-cumaran-3-one oder 2-Arylbenzofuran-3-carbonsäuren, ein andermal beide Verbindungen isolieren konnten, wobei die Hydroxycumaranone als „hydratisierte" Derivate der Aurone aufgefaßt werden können. Zusammenfassend stellten sie fest, daß beim Fehlen des 6'-Methoxyl-Substituenten vornehmlich Flavonole und nur wenig Aurone bzw. ihre hydratisierte Analoge gebildet werden. Hydroxylgruppen in den Stellungen 2, 3, oder 4 sollen die Bildung der Flavonole begünstigen (*76*).

Im Zusammenhang mit den hydratisierten Auronen sind Untersuchungen von Gripenberg (*54*) erwähnenswert, in denen durch Kaliumpermanganat-Oxidation die entsprechenden 2-Benzyl-2-hydroxy-cumaranone und aus diesen durch Behandlung mit Schwefelsäure wieder die ursprünglichen Aurone gewonnen wurden. In letzter Zeit gewann Gripenberg in alkalischem Milieu aus Dihydroflavonolen hydratisierte Aurone, die mit einem sogenannten „Isoauron" (3-Benzylidencumaran-2-on) verunreinigt waren. Die Alkali-Umlagerung der Dihydroflavonole hatte bereits Chopin (*21*) untersucht. Chopin und Dural (*15*) behandelten 2'-Benzyloxy-chalkon-epoxyde in Essigsäure mit Natriumacetat und erhielten auf diese Weise α-Diketone, aus denen durch Ringschluß in Gegenwart von Mineralsäuren die entsprechenden 2-Benzyl-2-hydroxy-cumaranone hergestellt werden konnten. Hergert, Coad und Logan (*64*) erhielten mit Kalilauge sowohl aus den Methyläthern der Dihydroflavonole als auch aus dem Alkoxy-substituierten 2-Benzyl-2-hydroxy-cumaranon die entsprechenden Auron-methyläther.

Aus einem natürlichen hydratisierten Auron und aus 2-Benzyl-2,3',4',6-tetrahydroxycumaranon, erhalten aus einem Quebrachotannin-Extrakt, stellten King und Mitarb. (*72*) mit Schwefelsäure Sulfuretin (8) her.

Im Zusammenhang mit der Algar-Flynn-Oyamada-Reaktion sind Untersuchungen von Venturella und Bellino (*109*) sowie Molho (*79*) von Interesse, wonach die Aurone in alkalischem Medium mit Wasserstoffperoxyd zu substituierten Benzoe- oder Salicylsäuren oxydiert werden können.

Abschließend sei für diese Ringschlußreaktion vermerkt, daß zwar ihr Mechanismus auf Grund der Arbeiten von Wheeler (*110*) und Geissman (*43*) im Prinzip bekannt ist, daß es aber trotz vieler Versuche bisher nicht gelungen ist, die Reaktion in ihrer Gesetzmäßigkeit eindeutig zu erfassen.

Eine elegante Umwandlung von 2'-Hydroxy-chalkonglykosiden in die entsprechenden Auronglykoside beschrieben Shimokoriyama und Hattori (*105*). Nach ihren Versuchen erfolgte die Umwandlung des Chalkons in den *Cosmos*- oder *Coreopsis*-Blütenblättern in pH-Bereich

Literaturverzeichnis: SS. 169—174.

5—6, und es wurde angenommen, daß sie auf die Mitwirkung eines sogenannten „Chalkonase"-Enzyms zurückzuführen sei. Der eindeutige Beweis für ein solches Enzym konnte jedoch bisher nicht erbracht werden. Wesentlich wertvoller vom praktischen Gesichtspunkt ist jedoch die Darstellung von Auronglykosiden aus 2'-Hydroxy-chalkon-glykosiden durch Luftoxydation in Gegenwart von Natriumbicarbonat (*104*).

Aus der Reihe der Umwandlungsreaktionen der Aurone ist die Gewinnung von Flavonen und Flavonolen durch Erweiterung des fünfgliedrigen Furan-Ringes die wichtigste. WHEELER und Mitarb. (*35*) wurden 1952 auf diese interessante Reaktion aufmerksam, als sie 4'-Methoxy-auron in alkoholischer Kaliumcyanidlösung erhitzten und dabei in guter Ausbeute 4'-Methoxy-flavone erhielten. Später beschrieben sie den Mechanismus der Umwandlung (*34*) und teilten mit, daß diese Reaktion allgemein zur Synthese von Flavonen Anwendung finden kann. Nach WHEELER (*36*) können die Aurone auch durch Oxydation mit Wasserstoffperoxyd in alkalischem Medium in Flavonole übergeführt werden. Bei der Behandlung der Aurondibromide mit Kalilauge entstehen gleichfalls Flavonole (*3*). Nach GOWAN, PHILBIN und WHEELER (*53*) kann diese oxydative Ringerweiterung auch anders verlaufen. So bildet z. B. das 5,7-Dimethoxyauron bei der Behandlung mit Wasserstoffperoxyd im alkalischen Medium ein Epoxyd. Oxydiert man Aurone in Pyridin mit Natriumperoxyd, so bleibt die Zahl der Ringglieder unverändert. Lediglich aus dem Benzylidencumaranon bildet sich Benzoylcumaranon.

Durch Oxydation von Auronen mit Peroxyden in alkalischen Media entstehen Auron-epoxyde und Flavonole. Mit Natriumperoxyd bilden sich Aroylcumaran-3-one [GEOGHEGAN, O'SULLIVAN und PHILBIN (*51*)].

V. Synthese natürlich vorkommender Aurone und Auronglykoside.

1. Allgemeine Bemerkungen.

Über die Herstellung von Benzalcumaranon-Derivaten findet man in der Literatur verhältnismäßig wenige Angaben. Das gemeinsame Prinzip der klassischen Verfahren ist folgendes: Die entsprechend substituierten Cumaranone werden in der Regel in warmer alkoholischer Lösung mit Benzaldehyd-Derivaten in Gegenwart von wäßrigem Natriumhydroxyd (*31*, *3*) oder von Salzsäure (*4*) kondensiert. Piperidin (*19*) ist als Kondensationsmittel ebenfalls anwendbar. In Eisessig an Stelle von Alkohol und in Gegenwart von Salzsäure stellten SHRINER und GROSSER (*106*) und später GEISSMAN und HARBORNE (*40*) benzoylierte Aurone dar. Diese Modifikation der Claisen-Schmidt-Kondensation (*16*, *98*) war auch dann erfolgreich, wenn die Hydroxylgruppen nicht mit Acyl- oder Alkyl-gruppen geschützt waren. CLAISEN (*16*) hatte festgestellt, daß sich beim Erhitzen von Benzaldehyd, Aceton und Essigsäureanhydrid im Einschlußrohr Benzalaceton bildet. Später gelangen KALFF und ROBINSON (*67*) Kondensationen dieses Typs auch ohne Druck, allein durch Kochen mit Essigsäureanhydrid.

Benzalcumaranon-Verbindungen können aus Chalkondibromiden mit alkoholischer Lauge (*82*, *77*) oder mit Hilfe der Algar-Flynn-Oyamada-Reaktion (*1*, *87*) erhalten werden. Die Synthese von OKAYIMA (*85*), der o-Hydroxy-benzoyl-phenylacetylene mit alkoholischer Lauge in Benzalcumaranon-Derivate überführen konnte, sei lediglich wegen ihrer Außergewöhnlichkeit erwähnt.

Die einzige früher durchgeführte Auronglykosid-Synthese stammt von GEISSMAN und MOJÉ (*49*). Nach dieser liefern 6-Hydroxy-7-methoxy-cumaranon-(3) und Dibenzoyl-protocatechualdehyd bei der Kondensation mit Eisessig in Gegenwart von Salzsäure 3',4'-Dibenzoyl-leptosidin. Die einzige noch freie Hydroxylgruppe kann dann ohne Schwierigkeiten mit Acetobromglucose in Gegenwart von Kaliumhydroxyd umgesetzt werden. So entsteht 3',4'-Dibenzoyloxy-6-tetraacetyl-glykosidoxy-7-methoxy-benzalcumaran-3-on, aus dem durch Entbenzoylierung und Entacetylierung in methanolischer Ammoniaklösung Leptosin erhalten wird. Diese Synthese liefert schlechte Ausbeuten und läßt sich nicht verallgemeinern.

Erwähnenswert sind auch zwei halbsynthetische Verfahren, die Luftoxydation der Chalkonglykoside in alkalischem Medium (*104*) und die enzymatische Umwandlung (*105*). Brauchbare präparative Methoden standen noch aus. Zum Beweis der Struktur des Sulfureins haben FARKAS und PALLOS (*24*) durch Kondensation von 2,4-Dihydroxy-acetophenon-β-d-glykosid-4-tetraacetat und Protocatechualdehyd in alkalischem Medium Coreopsin [(2',4'-Dihydroxy-phenyl)-(3,4-dihydroxy-styryl)-keton-β-d-glykosid-(4')] hergestellt und damit auch die Struktur des Coreopsins bewiesen. Dieses Glykosid ließ sich nach der Methode von SHIMOKORYAMA und HATTORI (*104*) in Gegenwart von Natriumbicarbonat glatt in das Sulfurein überführen. Ein zur Synthese der Auronglykoside und Aurone allgemein anwendbares Verfahren haben FARKAS, PALLOS und PAÁL (*30*) ausgearbeitet. Sie wählten ein schon bei der Synthese von Flavonglykosiden im Zemplén-Institut angewandtes Verfahren. Nach diesem wird die Zuckerkomponente nicht an das partiell acylierte oder alkylierte Derivat des Aglykons, sondern an eine gut definierte und leicht zugängliche synthetische Vorstufe des Aglykons geknüpft und dann das so erhaltene Glykosid mit dem anderen Reaktionspartner umgesetzt. Im Falle der Auronglykoside lieferte die Darstellung der entsprechenden Hydroxycumaranon-glykoside und ihre Kondensation mit substituierten aromatischen Aldehyden in Gegenwart von Essigsäureanhydrid die

Literaturverzeichnis: SS. 169—174.

acetylierten Auronglykoside in guter Ausbeute. Nach Verseifung der acylierten Glykoside nach der Zemplén-Kunz-Methode (*112*) entstanden die freien Glykoside.

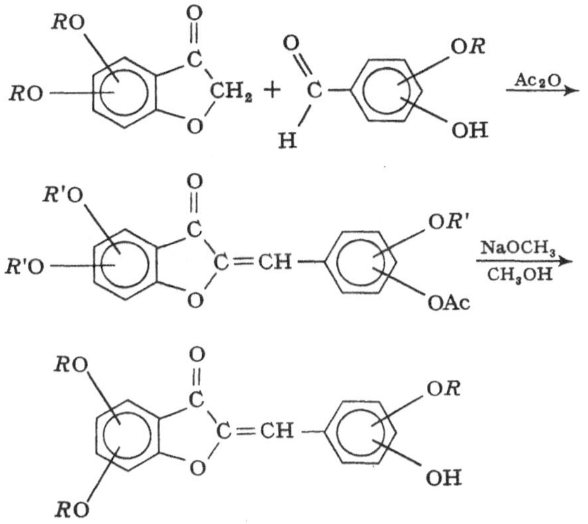

$R = $ H, Alkyl, Glykosyl- usw.
$R' = $ Acetyl-, Alkyl-, Tetraacetyl-glykosyl usw.

Die zu dieser Reaktion erforderlichen Hydroxy-cumaranon-glykoside erhält man, indem die partiell acylierten oder alkylierten Hydroxy-cumaranone mit α-Acetobromglucose (*8*) nach einer modifizierten Methode von MICHAEL (*78*) sowie KOENIGS und KNORR (*74*) in acetoniger Suspension durch Zugabe von verdünnter Natronlauge umgesetzt werden. Für die Kondensation der Hydroxy-cumaranon-glykoside mit den Aldehyd-Derivaten mußten nur noch die optimalen Bedingungen ermittelt werden (*88*). Die in der Literatur beschriebenen Methoden — meist Modifizierungen der Claisen-Schmidt-Kondensation (*16, 98*) — sind zur Herstellung von Auronglykosiden vor allem wegen der dort angewendeten pH-Verhältnisse nicht allgemein anwendbar. In einem mineralsauren Medium wird die Glykosidbindung gespalten, während in alkalischem Milieu die Cumaranon-Komponenten isomerisieren oder polymerisieren können. Außerdem findet in Alkali eine Disproportionierung bzw. Oxydation der aromatischen Hydroxy-aldehyde statt. Diese Probleme lassen sich umgehen, wenn man als Kondensationsmittel Essigsäureanhydrid verwendet und das Reaktionsgemisch 3—4 Stunden lang kocht. Nach Umkristallisieren des Rohproduktes aus Alkohol und Entacetylierung (*112*) können die Auronglykoside gewonnen werden. Entsprechend abgeändert eignet sich das Verfahren auch zur Synthese von Auronen.

166   L. Farkas und L. Pallos:

2. Anwendungsbeispiele.

Im folgenden sollen einige Beispiele für diese Synthese gegeben werden.

Rengasin (18).

King und Mitarb. (70) hatten festgestellt, daß das Rengasin aus dem Kernholz des zu den Melanorrhea-Arten gehörenden Rengas-Baumes ein 3′,4,4′-Trihydroxy-6-methoxy-auron (19) ist. Die durch Kondensation von Protocatechualdehyd mit 4-Hydroxy-6-methoxy-cumaran-3-on hergestellte Verbindung war mit dem Naturstoff nicht identisch. Erst die Darstellung anderer in Frage kommender Monomethyläther (19—21) des Aureusidins (5, S. 153) ergab (23), daß dem Rengasin in Wirklichkeit die Struktur des 3′,4′,6-Trihydroxy-4-methoxy-aurons (18) zukommt.

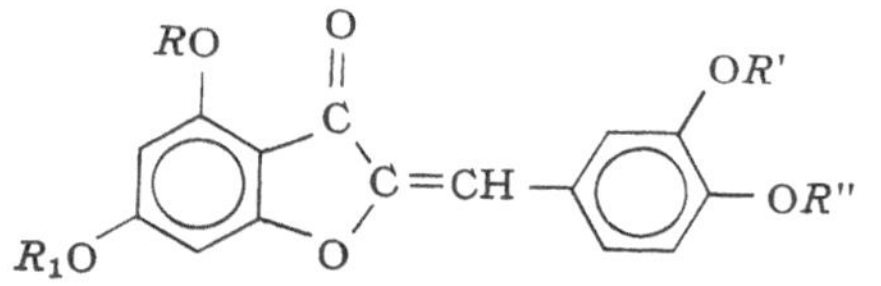

(19) $R_1=CH_3, R'=R''=R=H$.
(20) $R=R''=H, R'=CH_3, R_1=H$.
(21) $R=R'=H, R''=CH_3, R_1=H$.
(18) $R=CH_3, R_1=R'=R''=H$. Rengasin.

Sulfurein (7).

Dieses kann aus 6-Hydroxy-cumaran-3-on und α-Acetobromglucose in alkalisch-acetoniger Lösung über das 6-Hydroxy-cumaranon-6-β-d-glykosid-tetraacetat (22), Kondensation mit Protocatechualdehyd in Gegenwart von Essigsäureanhydrid und nachfolgende Verseifung des Hexaacetats erhalten werden (30).

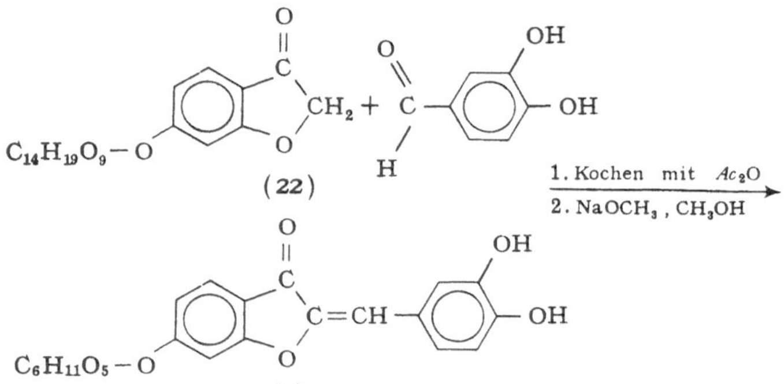

$C_{14}H_{19}O_9-O$ (22)

1. Kochen mit Ac_2O
2. $NaOCH_3$, CH_3OH

$C_6H_{11}O_5-O$ (7)

Literaturverzeichnis: SS. 169—174.

Palasitrin (9, S. 154).

Die Totalsynthese des Palasitrins (*25*) geht vom 6-Hydroxy-cuma-ranon-6-β-d-glykosid-tetraacetat (*22*) aus und führt über die Kondensation mit Protocatechualdehyd-3-glykosid zunächst zum Nonaacetat und durch dessen Hydrolyse (*112*) direkt zur gewünschten Verbindung. Das Protocatechualdehyd-3-glykosid ist nach der Methode von HELFERICH und PAPALAMBROU (*63*) in befriedigender Reinheit zugänglich.

Cernuosid (6, S. 153).

Die Reaktion von 4,6-Dihydroxy-cumaran-3-on mit Acetobrom-glucose führt zu einem Isomeren-Gemisch von 4- und 6-Glykosid-tetra-acetat. Auf Grund seiner günstigen Kristallisierbarkeit kann aus diesem Gemisch das 4-Glykosid-acetat isoliert werden (*27*), während sich von dem 6-Glykosid-Derivat selbst bei sorgfältiger Aufarbeitung nur Spuren isolieren lassen, die für die Weiterkondensation in der Regel nicht aus-reichen. Aus 4,6-Dihydroxy-cumaran-3-on-4-glykosid (*23*) oder seinem Tetraacetat und Protocatechualdehyd kann in guter Ausbeute Cernuosid-heptaacetat und nach dessen Verseifung (*112*) Cernuosid erhalten werden (*27*).

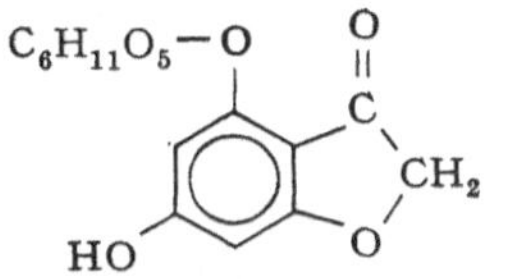

$C_6H_{11}O_5$—O

(*23*)
4,6 - Dihydroxy - cumaran - 3 - on - 4- glykosid.

Maritimein (12, S. 155) *und Leptosin* (2, S. 153).

Aus 6-Hydroxy-7-benzyloxy-cumaranon (*49*) bzw. 6-Hydroxy-7-methoxy-cumaranon (*108*) sind durch Kupplung mit α-Acetobrom-glucose die entsprechenden 6-Glykosid-tetraacetate zu gewinnen. Ihre Kondensation mit Protocatechualdehyd in Gegenwart von Essigsäure-anhydrid liefert das Acetat des Maritimeins (*28*) bzw. Leptosins (*29*). Im Falle des Maritimein-acetats führte die Entbenzylierung und Ent-acetylierung (*112*), bei Leptosinacetat die in gleicher Weise durch-geführte Entacetylierung zu den gewünschten freien Glykosiden.

Bractein (14, S. 156).

Das Bracteatin (*15*) wurde von HÄNSEL und Mitarb. (*55*) hergestellt, die Synthese des Glykosids jedoch im Laboratorium der Verfasser (*26*). Man erhält durch Kondensation des 4-Tetraacetyl-glykosidoxy-6-hydroxy-cumaran-3-ons (Acetat von **23**) mit Triacetyl-gallaldehyd (*37*) das Octa-acetyl-Derivat und durch Hydrolyse im Stickstoffstrom das äußerst oxydationsempfindliche freie Glykosid.

VI. Zur Genetik und Biosynthese der Aurone und ihrer Glykoside.

Die Aurone und ihre Glykoside sind im Pflanzenreich viel weniger verbreitet als die Flavone oder Anthocyan-Derivate und kommen nur in einigen Pflanzengruppen und dort nur in geringen Mengen vor. Mit Hilfe der papierchromatographischen und spektrophotometrischen Methoden ist es jedoch möglich geworden, diese Pigmente auch ohne Isolierung zu identifizieren. Für Biosynthese-Studien schien am besten *Antirrhinum majus* geeignet, da die Blüten mannigfaltige Farbtönungen zeigen und darüber hinaus auch farblose Blüten-Mutanten existieren. Aus diesen Blüten wurden bisher 40 Flavonoid-Komponenten isoliert, von denen etwa 25% identifiziert sind, während von den restlichen nur soviel bekannt ist, daß es sich bei diesen um Glykoside bereits bekannter Aglykone handelt.

Drei Gene scheinen die Bildung der Flavonoid-Pigmente zu steuern. Sie lassen sich nach ihrer Spezifität wie folgt charakterisieren: Das „M"-Gen ist zur Produktion der Pigmente allgemein notwendig. Das „P"-Gen steuert die Entstehung der Flavonole und Anthocyanidine, während das „Y"-Gen die Oxydationsstufe dieser Verbindungen sowie die Hydroxyl-Substitution im *B*-Ring regelt, aber dabei den Oxydationsgrad der Auron-Pigmente nicht beeinflußt (47). Die Biosynthese des Aureusidins in *Antirrhinum majus* wird durch den „Y"-Faktor gesteuert. Dominiert dieser, so bildet sich wenig Auron, und umgekehrt. Jorgensen und Geissman (65), die bei ihren Untersuchungen über die Wirkungsweise der „P"-, „M"- und „Y"-Faktoren auch Messungen des Anthocyan- und Aurongehaltes durchgeführt haben, fanden, daß je mehr genetische Faktoren zusammenwirken, sich um so weniger Auron-Pigmente und um so mehr Anthocyane bilden, und umgekehrt.

Auch Seikel (99) und Dayton (18) haben genetische Studien an den verschiedensten Genotypen von *Antirrhinum majus* durchgeführt, und bereits von Dayton stammt ein erster Biosynthesevorschlag für Aureusidin. Eine aufschlußreiche Zusammenfassung dieses Themas wird von Harborne und Sherratt (61) gegeben. Auf Grund dieser und der von Geissman (50) durchgeführten Untersuchungen sind einige, die Auronbiosynthese betreffenden Zusammenhänge aufgestellt worden. Nach diesen soll, ähnlich den Flavonoid-Verbindungen, auch die Bildung der Aurone von dem Robinsonschen (94) C_{15}-Vorläufer (24) ausgehen.

Von der Oxydationsstufe der C_3-Einheit hängt es ab, welcher Flavonoidtyp sich bildet. Wie parallel durchgeführte Anthocyan- und Auron-Bestimmungen zeigen, muß die Biosynthese des Aurons sehr bald vom Syntheseweg der übrigen Flavonoidverbindungen abzweigen. Das Vorhandensein von Chalkon-Intermediärprodukten zwischen der Präcursor-

und Auronphase ist zwar nachgewiesen worden, doch ist trotz der engen chemischen Verwandtschaft bisher kein einziges Gen bekannt, das die Oxydation von Chalkonen zu Auronen steuert. Immerhin läßt das von SHIMOKORIYAMA und HATTORI (*105*) isolierte, das in vitro Umwandlung katalysierende „Chalkonase"-Enzympräparat, darauf schließen, daß sich diese Umwandlung auch in vivo abspielen kann. Es ist sehr wahrscheinlich, daß die Hydroxylverteilung noch vor der Bildung des heterocyclischen Ringes stattfindet.

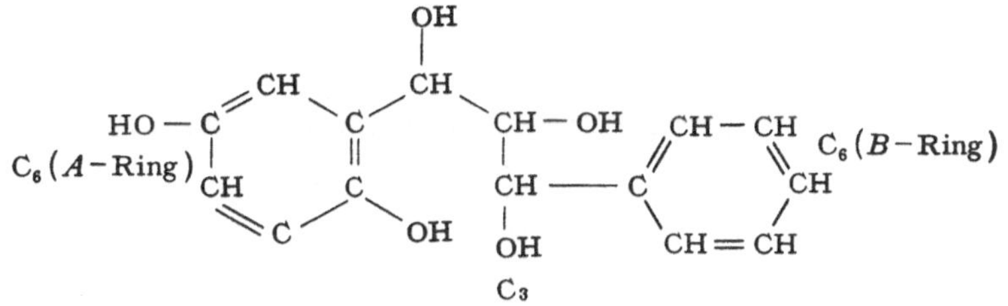

(24)

In letzter Zeit wurden von MONEY, DOUGLAS und SCOTT (*80*) synthetische Studien nach dem Biosynthese-Schema durchgeführt. Dabei sollten chemisches Verhalten und Synthese solcher Polypyrone untersucht werden, die den gleichen Vorläufer wie die Stilbene und Flavonoide (*12*) zu haben scheinen. Die genannten Autoren halten es für möglich, daß das synthetische Benzyliden-bis-pyron-epoxyd ein Vorläufer der Dihydroflavonole, Flavanone, Isoflavone und auch der Aurone ist. Eingehend hat kürzlich WONG (*111*) die Intermediärprodukte der Auron-Biosynthese untersucht. Im Zusammenhang mit den Auronglykosiden, die nach HARBORNE in allen nicht-albino Mutanten von *Antirrhinum majus* vorkommen, haben GEISSMAN und HINREINER (*46*) bereits im Jahre 1952 angenommen, daß die Glykosidierung eine der letzten Phasen der Biosynthese ist. Alle späteren Mitteilungen haben die Richtigkeit dieser Annahme erhärtet.

Literaturverzeichnis.

1. ALGAR, J. and J. P. FLYNN: New Synthesis of Flavonols. Proc. Roy. Irish Acad. **42 B,** 1 (1934).

2. ANAND, N., R. N. IYER and K. VENKATARAMAN: Synthetical Experiments in the Chromone Group. XXIII. Synthesis of Rhamnazin and a Synthesis of 3,4'-Dihydroxy-7-methoxyflavone. Proc. Indian Acad. Sci. **29 A,** 203 (1949).

3. AUWERS, K. v. und K. MÜLLER: Umwandlung von Benzal-cumaranonen in Flavonole. Ber. dtsch. chem. Ges. **41,** 4233 (1908).

4. AUWERS, K. v. und P. POHL: Über die Umwandlung von Benzalcumaranonen in Flavonole. Liebigs Ann. Chem. **405,** 243 (1914).

5. Asano, M. and K. Yamaguti: Hydroxyquinones. III. The Constitution and Synthesis of Rapanone, the Anthelmintic Principle of *Rapanea Maximowiczii* Koidz. J. Pharmac. Soc. Japan 60, 585 (1940).

6. Balakrishna, K. J., T. R. Seshadri and G. Viswanath: Nuclear Oxidation in Flavones and Related Compounds. XXV. Isomers of Pedicinin. Proc. Indian Acad. Sci. 30 A, 120 (1949).

7. Ballio, A. e G. B. Marini-Bettòlo: La costituzione del cernuoside, il pigmento giallo dei fiori di *Oxalis cernua* Thumb. Gazz. chim. ital. 85, 1319 (1955).

8. Bárczay-Martos, M. and F. Körösy: Preparation of Acetobrome-sugars. Nature 165, 369 (1950).

9. Bargellini, G. und A. Oliverio: Flavon-Derivate des 1,2,3,4-Tetraoxybenzols. Ber. dtsch. chem. Ges. 75, 2083 (1942).

10. Bate-Smith, E. C. and T. A. Geissman: Benzalcoumaranones. Nature 167, 688 (1954).

11. Batterham, T. J. and R. J. Highet: Nuclear Magnetic Resonance (NMR) Spectra of Flavonoids. Austral. J. Chem. 17, 428 (1964).

12. Birch, A. J.: Pathways in Biosynthesis. Proc. Chem. Soc. (London) 1962, 3.

13. Chopin, J. et G. Dellamonica: Formation de glycosides d'aurones par oxydation spontanée de l'ériocitrine, de l'hespéridine et de la naringine. C. R. hebd. séances Acad. Sci. 260, 5582 (1965).

14. Chopin, J., G. Dellamonica et P. Lebreton: Isolement d'un glucoside-6 de l'auréusidine à partir des extraits d'écorce de citron. C. R. hebd. séances Acad. Sci. 257, 534 (1963).

15. Chopin, J. et P. Durual: Synthèse et réactions des époxydes de benzyloxy-2′ chalcones. Bull. soc. chim. France 1965, 3350.

16. Claisen, L. und A. Claparede: Condensationen von Ketonen mit Aldehyden. Ber. dtsch. chem. Ges. 14, 2460 (1881).

17. Cummins, B., D. M. X. Donnelly, E. M. Philbin, J. Swirski, T. S. Wheeler and R. K. Wilson: New Results in the Algar-Flynn-Oyamada Oxidation of 2′-Hydroxychalcones. Chem. and Ind. 1960, 348.

18. Dayton, T. O.: The Inheritance of Flower Color Pigments. I. The Genus *Antirrhinum.* J. Genet. 54, 249 (1956).

19. Desai, R. B. and J. N. Ray: Experiments on the Synthesis of Cyanomaclurin and Related Substances. J. Indian Chem. Soc. 35, 83 (1958).

20. Donnelly, D. J., J. A. Donnelly, J. J. Murphy, E. M. Philbin and T. S. Wheeler: Steric Effects on the Cyclisation of Chalcone Dibromides. Chem. Commun. 1966, 351.

21. Durual, P., M. Chadenson et J. Chopin: Nouveaux exemples de transposition alcaline des hydroxy-3 flavanones. Bull. soc. chim. France 1964, 11.

22. Farkas, L., L. Hörhammer, H. Wagner, H. Rösler und R. Gurniak: Die Struktur des Jaceins und dessen Synthese aus dem Aglucon und Acetobromglucose. Chem. Ber. 97, 610 (1964).

23. Farkas, L., M. Nógrádi und L. Pallos: Aurone und Auronglucoside. VI. Die endgültige Konstitutionsaufklärung und die Synthese des Rengasins. Chem. Ber. 97, 1044 (1964).

24. Farkas, L. und L. Pallos: Endgültiger Strukturbeweis und Synthese des Coreopsins. Chem. Ber. 92, 1263 (1959).

25. — — Synthese des Palasitrins, eines Glucosids von *Butea frondosa.* Chem. Ber. 93, 1272 (1960).

26. — — Aurone und Auronglykoside. X. Synthese und endgültiger Strukturbeweis des Bracteins, eines Glucosids aus *Helichrysum bracteatum* (Vent.) Willd. Chem. Ber. 98, 2930 (1965).

27. FARKAS, L., L. PALLOS und GY. HIDASI: Synthese des Cernuosids und Aureusidins. Chem. Ber. 94, 2221 (1961).
28. FARKAS, L., L. PALLOS und M. NÓGRÁDI: Aurone und Auronglucoside. VIII.
Synthese des Maritimeins, eines Glucosids von Coreopsis maritima. Chem.
Ber. 98, 2103 (1965).
29. — — — Aurone und Auronglucoside. IX. Eine neue Synthese des Leptosins.
Acta Chim. Hung. 44, 341 (1965).
30. FARKAS, L., L. PALLOS und Z. PAÁL: Synthese und endgültige Strukturaufklärung des Sulfureins. Chem. Ber. 92, 2847 (1959).
31. FELIX, A. und P. FRIEDLÄNDER: Über indigoide Farbstoffe. VI. Aliphatisch-
aromatische Verbindungen. Monatsh. Chem. 31, 68 (1910).
32. FERNÁNDEZ, O. and A. PIZARROSO: Chemical Study of the Bulbs of Oxalis
purpurea var. jacquinii (Sonder). Farm. nueva (Madrid) 11, 1 (1946) [Chem.
Abstr. 42, 8888 (1948)].
33. FIESER, L. F. and E. M. CHAMBERLIN: Synthesis of Embelin, Rapanone and
Related Quinones by Peroxyde Alkylation. J. Amer. Chem. Soc. 70, 71
(1948).
34. FITZGERALD, D. M., J. F. O'SULLIVAN, E. M. PHILBIN and T. S. WHEELER:
Ring Expansion of 2-Benzylidenecoumaran-3-ones. A Synthesis of Flavones.
J. Chem. Soc. (London) 1955, 860.
35. FITZGERALD, D. M., E. M. PHILBIN and T. S. WHEELER: Ring Expansion
of 2-Benzylidenecoumaran-3-ones: A Synthesis of Flavones. Chem. and Ind.
1952, 130.
36. FITZMAURICE, W. E., W. I. O'SULLIVAN, E. M. PHILBIN, T. S. WHEELER and
T. A. GEISSMAN: Oxidative Ring Expansion of Aurones to Flavonols. Chem.
and Ind. 1955, 652.
37. FREUDENBERG, K. und H. H. HÜBNER: Oxyzimtalkohole und ihre Dehy-
drierungspolymerisate. Chem. Ber. 85, 1181 (1952).
38. GEISSMAN, T. A.: Anthocyanins, Chalcones, Aurones, Flavones and Related
Water-soluble Plant Pigments. In: K. PAECH and M. V. TRACEY (Edits.).
Modern Methods of Plant Analysis, Vol. III, p. 450. Berlin: Springer-Verlag.
1955.
39. GEISSMAN, T. A. and D. K. FUKUSHIMA: Flavonones and Related Compounds.
V. The Oxidation of 2'-Hydroxychalcones with Alkaline Hydrogen Peroxide.
J. Amer. Chem. Soc. 70, 1686 (1948).
40. GEISSMAN, T. A. and J. B. HARBORNE: Anthochlor Pigments. X. Aureusin and
Cernuoside. J. Amer. Chem. Soc. 77, 4622 (1955).
41. — — Anthochlor Pigments. XIII. The Ultraviolet Absorption Spectra of
Phenolic Plant Pigments. Polyhydroxyaurones. J. Amer. Chem. Soc. 78,
832 (1956).
42. GEISSMAN, T. A., J. B. HARBORNE and M. K. SEIKEL: Anthochlor Pigments.
XI. The Constituents of Coreopsis maritima, Reinvestigation of Coreopsis
gigantea. J. Amer. Chem. Soc. 78, 825 (1956).
43. — — — Chemistry of Aurones Isolated from Plants. Colloq. intern. centre
natl. recherche sci. (Paris) No. 64, 277 (1957) [Chem. Abstr. 54, 12094
(1960)].
44. GEISSMAN, T. A. and C. D. HEATON: Anthochlor Pigments. IV. The Pigments
of Coreopsis grandiflora Nutt. I. J. Amer. Chem. Soc. 65, 677 (1943).
45. — — Anthochlor Pigments. V. The Pigments of Coreopsis grandiflora Nutt. II.
J. Amer. Chem. Soc. 66, 486 (1944).
46. GEISSMAN, T. A. and E. HINREINER: Theories of the Biogenesis of Flavonoid
Compounds. Botan. Rev. 18, 77 (1952).

47. Geissman, T. A., E. C. Jorgensen and B. L. Johnson: The Chemistry of Flower Pigmentation in *Antirrhinum majus*. Color Genotypes. I. The Flavonoid Components of the Homozygous P, M, Y Color Types. Arch. Biochem. Biophys. **49**, 368 (1954).

48. Geissman, T. A. and L. Jurd: Anthochlor Pigments. IX. The Structure of the Aurone Pigment of *Cosmos sulfureus*, "Orange Flare" and "Yellow Flare". J. Amer. Chem. Soc. **76**, 4475 (1954).

49. Geissman, T. A. and W. Mojé: Anthochlor Pigments. VIII. The Pigments of *Coreopsis grandiflora* Nutt. III. J. Amer. Chem. Soc. **73**, 5765 (1951).

50. Geissman, T. A. and C. Steelink: Flavonoid Petal Constituents of Chrysanthemum segetum L. J. Organ. Chem. (USA) **22**, 946 (1957).

51. Geoghegan, M., W. I. O'Sullivan, E. M. Philbin and T. S. Wheeler: Flavonoid Epoxides. I. Oxidation of Aurones. Tetrahedron **22**, 3203 (1966).

52. Gowan, J. E., P. M. Hayden and T. S. Wheeler: A New Synthesis of Flavonols. J. Chem. Soc. (London) **1955**, 862.

53. Gowan, J. E., E. M. Philbin and T. S. Wheeler: Synthesis of γ-Pyrones. Sci. Proc. Roy. Dublin Soc. **27**, 185 (1956).

54. Gripenberg, J.: The Structure of some Alleged 3-Hydroxyflavanones. Acta Chem. Scand. **7**, 1323 (1953).

55. Hänsel, R. und L. Langhammer: *Helichrysum bracteatum:* Über die Identität von natürlichem Bracteatin mit synthetischem 4,6,3′,4′,5′-Pentahydroxyauron. Arch. Pharmaz. **296**, 619 (1963).

56. Hänsel, R., L. Langhammer and A. G. Albrecht: Ein neues Auronglykosid aus *Helichrysum bracteatum*. Tetrahedron Letters **1962**, 599.

57. Harborne, J. B.: Use of Alkali and Aluminium Chloride in the Spectral Study of Phenolic Plant Pigments. Chem. and Ind. **1954**, 1142.

57a. — The Chromatographic Identification of Anthocyanin Pigments. J. Chromatogr. **1**, 473 (1958).

58. — Chromatography of the Flavonoid Pigments. J. Chromatogr. **2**, 581 (1959).

59. — Plant Polyphenols. X. Flavone and Aurone Glycosides of *Antirrhinum*. Phytochem. **2**, 327 (1963).

60. Harborne, J. B. and T. A. Geissman: Anthochlor Pigments. XII. Maritimein and Marein. J. Amer. Chem. Soc. **78**, 829 (1956).

61. Harborne, J. B. and H. S. A. Sherratt: Identification of the Sugars of Anthocyanins. Biochem. J. **65**, 23 P (1957).

62. Hattori, S., M. Shimokoriyama et K. Oka: Localisation des pigments flavoniques dans *Cosmos sulphureus*. Bull. soc. chim. biol. (Paris) **38**, 557 (1956).

63. Helferich, B. und P. Papalambrou: Synthese des 3-(β-d-Glucosido)-protocatechualdehyds und seine fermentative Spaltbarkeit. Liebigs Ann. Chem. **551**, 242 (1942).

64. Hergert, H. L., P. Coad and A. V. Logan: The Methylation of Dihydroquercetin. J. Organ. Chem. (USA) **21**, 304 (1956).

65. Jorgensen, E. C. and T. A. Geissman: The Chemistry of Flower Pigmentation in *Antirrhinium majus*. II. Glycosides of PPmmYY, PPMMYY, ppmmYY and ppMMYY Color Genotypes. Arch. Biochem. Biophys. **54**, 72 (1955).

66. Jurd, L.: Spectral Properties of Flavonoid Compounds. In: T. A. Geissman (Edit.), The Chemistry of Flavonoid Compounds, p. 107. Oxford-London-New York-Paris: Pergamon Press Inc. 1962.

67. Kalff, J. and R. Robinson: A Synthesis of Datiscetin. J. Chem. Soc. (London) **127**, 1968 (1925).

68. Kawamura, S.: Constitution of Rhapontin. J. Pharmac. Soc. Japan **58**, 405 (1938).

69. KESSELKAUL, L. und S. v. KOSTANECKI: Zur Einwirkung des Benzaldehyds auf Chloracetopyrogallol. Ber. dtsch. chem. Ges. 29, 1886 (1896).

70. KING, F. E., T. J. KING and D. W. RUSTIDGE: The Chemistry of Extractives from Hardwoods. XXXIII. Extractives from *Melanorrhea* spp. (Rengas). J. Chem. Soc. (London) 1962, 1192.

71. KING, H. G. C. and T. WHITE: The Colouring Matter of *Rhus cotinus* Wood (Young Fustic). J. Chem. Soc. (London) 1961, 3538.

72. KING, H. G. C., T. WHITE and R. B. HUGHES: The Occurrence of 2-Benzyl-2-hydroxycoumaran-3-ones in Quebracho Tannin Extract. J. Chem. Soc. (London) 1961, 3234.

73. KLEIN, G.: Studien über das Anthochlor. I. und II. Sitzber. Akad. Wiss. Wien, Math.-naturw. Kl., Abt. I, 129, 341 (1920), 130, 247 (1921).

74. KOENIGS, W. und E. KNORR: Über einige Derivate des Traubenzuckers und der Galactose. Ber. dtsch. chem. Ges. 34, 957 (1901).

75. LAMONICA, R. e G. B. MARINI-BETTÒLO: Sul glucoside dell'*Oxalis cernua*. Ann. chim. (Roma) 42, 496 (1952).

76. MANI, R. and K. VENKATARAMAN: Synthetical Experiments in the Chromone Group. XXXIV. A Synthesis of Chrysosplenetin. J. Sci. Indust. Res. (India) 21 B, 477 (1962).

77. MARATHEY, M. G.: Benzopyrone Series. I. Formation of Benzylidenecoumaranones. J. Univ. Poona, Chem. Sect. 1, No. 2, 7 (1952).

78. MICHAEL, A.: Sur la synthèse du phénolglucoside et de l'orthoformylglucoside ou hélicine. C. R. hebd. séances Acad. Sci. 89, 355 (1879).

79. MOLHO, D.: Dégradations par l'eau oxygénée en milieu alcalin dans le domaine des dérivés flavoniques des coumarines et des benzal 2 ou 3 coumaranones. Bull. soc. chim. France 1956, 39.

80. MONEY, T., J. L. DOUGLAS and A. I. SCOTT: Biogenetic-Type Synthesis of Phenolic Compounds. J. Amer. Chem. Soc. 88, 624 (1966).

81. MURAKAMI, M. and T. IRIE: Synthesis of Flavonol and Dihydroflavonol. Proc. Imper. Acad. (Tokyo) 11, 229 (1935).

82. NADKARNI, S. M., A. M. WARRIAR and T. S. WHEELER: Chalkones: Reactivity of Some Aryl Alkoxystyryl Ketones and their Dihalides. J. Chem. Soc. (London) 1937, 1798.

83. NIKONOV, G. K.: Paper Chromatography of Natural Derivatives of α- and γ-Benzopyrone and Tanning Substances. Med. Prom. S. S. S. R. 12, No. 3, 16 (1959) [Chem. Abstr. 53, 14093 (1959)].

84. NORDSTRÖM, C. G. and T. SWAIN: Isolation of a Benzylidenecoumaranone (Aurone) from a Yellow Dahlia. Chem. and Ind. 1953, 823.

85. OKAYIMA, Y.: Acetylenic Compounds. X. Claisen Rearrangement and Fries Rearrangement. J. Pharm. Soc. Japan 80, 318 (1960).

86. OLIVERIO, A., G. B. MARINI-BETTÒLO and G. BARGELLINI: Flavonic Derivatives of 1,2,3,5-Tetrahydroxybenzene (Hydroxyphloroglucinol). Gazz. chim. ital. 78, 363 (1948) [Chem. Abstr. 43, 1772 (1949)].

87. OYAMADA, T.: A New General Method for the Synthesis of Flavonol Derivatives. J. Chem. Soc. Japan 55, 1256 (1934).

88. PALLOS, L. und Z. PAÁL: Unpublished.

89. PHILBIN, E. M., J. SWIRSKI and T. S. WHEELER: Action of Alkaline Hydrogen Peroxide on 2'-Hydroxychalcones. Chem. and Ind. 1956, 1018.

90. PRANTL, K.: Notiz über einen neuen Blütenfarbstoff. Bot. Ztg. 29, 425 (1871).

91. Puri, B. and T. R. Seshadri: Survey of the Anthoxanthins. III. Paper Chromatography of Some Flavanones and Chalcones and their Glycosides. Isolation and Constitution of Isobutyrin, a Glycoside of the Flowers of *Butea frondosa.* J. Sci. Indust. Res. (India) **13 B,** 321 (1954).

92. — — Survey of Anthoxanthins. IX. Isolation and Constitution of Palasitrin. J. Chem. Soc. (London) **1955,** 1589.

93. Rimpler, H. und R. Hänsel: Zwei neue Chalkonpigmente aus *Helichrysum bracteatum* (Vent.) Willd. Arch. Pharmaz. **298,** 838 (1965).

94. Robinson, R.: Formation of Anthocyanins in Plants. Nature **137,** 172 (1936).

95. Rosoll, A.: Beiträge zur Histochemie der Pflanze. I. Das Helichrysin. Sitzber. Akad. Wiss. Wien, Math.-naturw. Kl., Abt. I, **89,** 137 (1884).

96. Roubalová, D.: Paper Chromatography of Aurones (in Russian). Collect. Czech. Chem. Comm. **24,** 2166 (1959).

97. Schmid, J.: Über das Fisetin, den Farbstoff des Fisetholzes. Ber. dtsch. chem. Ges. **19,** 1734 (1886).

98. Schmidt, J. G.: Über die Einwirkung von Aceton auf Furfurol und auf Bittermandelöl bei Gegenwart von Alkalilauge. Ber. dtsch. chem. Ges. **14,** 1459 (1881).

99. Seikel, M. K.: The Chemistry of Flower Pigmentation in *Antirrhinum majus.* V. Pigments of Yellow *Antirrhinum majus,* Genotype ppmmyy. J. Amer. Chem. Soc. **77,** 5685 (1955).

100. Seikel, M. K. and T. A. Geissman: Anthochlor Pigments. VII. The Pigments of Yellow *Antirrhinum majus.* J. Amer. Chem. Soc. **72,** 5725 (1950).

101. Sherratt, H. S. A.: The Relationship between Anthocyanidins and Flavonols in Different Genotypes of *Antirrhinum majus.* J. Genet. **56,** 1 (1958).

102. Shimokoriyama, M.: Anthochlor Pigments of *Coreopsis tinctoria.* J. Amer. Chem. Soc. **79,** 214 (1957).

103. Shimokoriyama, M. and T. A. Geissman: Anthochlor Pigments. XIV. The Pigments of *Viguiera multiflora* (Nutt.) and *Baeria chrysostoma* (F. and M.). J. Organ. Chem. (USA) **25,** 1956 (1960).

104. Shimokoriyama, M. and S. Hattori: Anthochlor Pigments of *Cosmos sulphureus, Coreopsis lanceolata* and *C. saxicola.* J. Amer. Chem. Soc. **75,** 1900 (1953).

105. — — On a Probable Enzymatic Conversion of Hydroxychalcone Glycoside into Hydroxybenzalcoumaranone Glycoside. J. Amer. Chem. Soc. **75,** 2277 (1953).

106. Shriner, R. L. and F. Grosser: Coumaran Derivatives. IX. Synthesis of 3,4,6,3',4'-Pentahydroxy-2-benzylcoumaran. J. Amer. Chem. Soc. **64,** 382 (1942).

107. Simpson, T. H. and W. B. Whalley: 2'-Hydroxy- and 2'-Methoxy-flavanones. J. Chem. Soc. (London) **1955,** 166.

108. Späth, E. und H. Schmid: Synthese des Luvangetins. (LV. Mitt. über natürliche Cumarine.) Ber. dtsch. chem. Ges. **74,** 193 (1941).

109. Venturella, P. and A. Bellino: Oxidation of 2-Hydroxychalcones. II. The Relation between the Hydroxy Group Position in the Chalcone Nucleus and the Course of the Oxidative Algar-Flynn-Oyamada Reaction. Ann. chim. (Roma) **50,** 1510 (1960) [Chem. Abstr. **55,** 9412 (1961)].

110. Wheeler, T. S.: Unsolved Problems in Flavonoid Chemistry. Record Chem. Progr. (Kresge-Hooker Sci. Lib.) **18,** 133 (1957).

111. Wong, E.: An Intermediate in Aurone Biosynthesis. Chem. and Ind. **1966,** 598.

112. Zemplén, G. und A. Kunz: Über die Natriumverbindungen der Glucose und die Verseifung der acylierten Zucker. Ber. dtsch. chem. Ges. **56,** 1705 (1923).

(Eingelaufen am 22. November 1966.)

Recent Advances in the Chemistry of Hashish.

By R. MECHOULAM, Jerusalem, and Y. GAONI, Rehovoth.

With 10 Figures.

Contents.

I. Introduction.

Cannabis sativa L. (family Moraceae) is a dioecious plant of which two varieties exist, var. indica and var. non indica or typica. The flowering top of the female plant is covered with glandular hairs which secrete a resin. The resin formation ceases when the seeds mature and its function is to protect them during the ripening period. The cannabis resin is known as "hashish" in the Middle East and Europe and as "charas" in India. Frequently not only the resin but the whole flowering top is collected and used. It is then known as "ganja" in India, "kif" in North Africa, "dagga" in South Africa, "marihuana" in North America, "maconha" in Brazil, etc.

The resin has been used as a medicine and a psychotomimetic drug* since ancient times (*19*). Cannabis preparations are mentioned in Avesta, the sacred Book of Knowledge of the Zoroastrian faith (1000–600 B. C.). It has probably been known to the Assyrians (*34*). A Chinese treatise, about 2000 years old, records the use of cannabis as an anaesthetic in surgery (*59*). Herodotus reported its use as an intoxicant by the Scythians, but apparently the Greeks, Romans and Hebrews did not adopt it. The Arab invasions in the seventh century A. D. brought it to the Middle East and North Africa. In the Moslem world of today it is still the major illegal drug. It was introduced into European medicine by the physicians on the scientific commission which accompanied Napoleon's army in Egypt. Cannabis preparations were used in Europe for nearly a century in certain mental conditions and for their analgetic and sedative properties. In indigenous medicine it is still quite popular, especially in India (*29*). In both Ayurvedic (Hindu) and Tibbi (Mohammedan) systems of medicine it is used as a spasmolytic, hypnotic and analgetic, in mental conditions and to increase body resistance to severe physical stress.

As an illicit drug hashish and the other cannabis preparations are used widely almost all over the world. Hashish does not cause physiological addiction, although a certain psychological habituation probably exists. Though legally and through popular misconception it is usually grouped together with the opiates, it is not, as generally believed, used for its narcotic action but mainly for its psychotomimetic effects. It should therefore be considered as a member of the important group of materials which includes among others LSD, mescaline, etc. (*75*). Hashish intoxication (*19, 75, 80*) causes euphoria, motor excitation and hilarity followed by mental confusion, sometimes accompanied by depersonalization, hallucinations and depression. As with LSD the effects have a wavelike character. Reports are common of spatial and temporal distortion, increased sensitivity to sound and a feeling of a profound understanding of the meaning of things. McGlothlin (*75*) has concluded that "virtually all of the phenomena associated with LSD are, or can, also be produced with cannabis".

It is surprising that despite the widespread interest in psychochemistry, progress in both the chemical and medical aspects of a material known as long and used as widely as hashish has been rather slow. The reasons have been numerous – the lack of success until recently in the isolation and structure elucidation (*44*) of the pure major active prin-

---

* A psychotomimetic drug has been defined (*39*) as one which will "consistently produce changes in thought, perception and mood occuring alone or in concert without causing major disturbances of the autonomic nervous system or other serious disability".

ciple, Δ^1-tetrahydrocannabinol (Δ^1-THC)*, and hence its nonavailability, the lack of an animal test which parallels the activity in humans and the lack of controlled clinical experiments. The successful solution of many of the chemical problems in the last few years will undoubtedly lead to a parallel advance in the pharmacological and clinical aspects of the problem.

The biological activity of the natural cannabioides is given in Table 1, p. 203.

II. Literature on Hashish and Scope of the Present Review.

The analytical, botanical, pharmacological, clinical and sociological aspects of the hashish problem are not within the scope of this review. In the following bibliographies and reviews these aspects are presented.

A detailed bibliography on cannabis up to 1950 covers nearly 900 papers and books (14); a yearly supplement is published. A new bibliography with 1860 references has recently been compiled (15). The articles by Bouquet (19), Chopra and Chopra (29) and Watt and Breyer-Brandwijk (96) should be consulted for fascinating information on the botany of C. sativa, the various illegal preparations made from the resin, their use and misuse and their physiological and psychic effects. The Mayor's Report on Marihuana (New York, 1944) (74) deals mainly with sociological, clinical and pharmacological problems. This report refutes many of the fallacies generally connected with the use of marihuana, such as connection with crime, addiction etc. Critical reviews of recent psychological and psychiatric literature on the subject (75, 80) and one dealing mainly with the antibiotic properties of the resin and some of the pure constituents (60) have been published. In 1950 Loewe (73) summarized the detailed investigation of his group on the pharmacology of tetrahydrocannabinol analogs. For more recent work in pharmacology the papers by Boyd (20–22) and by Carlini (28, 84a) should be consulted. The lectures held in 1964 at a Ciba Foundation Symposium on hashish have been published (99); various aspects of hashish analysis, physiological action and drug dependence are presented**.

It is likewise impossible within the framework of this review to present and discuss fully the investigations on the chemistry of hashish reported over a period of 120 years. Much of the older work has been surveyed by Blatt (17). Adams (1) has summarized his own work in a Harvey Lecture and Todd (95) has reviewed the field up till 1946. Since then a few short reviews have appeared (39, 40, 53, 86). Grlić (53) has emphasized mainly the advances in analytical procedures. Korte (66) has summarized the work by the Bonn group. Papers by Farmilo (36), Lerner (71) and Claussen (30) deal with vapour phase chromatographic analysis of cannabis. The present article attempts to summarize the advances made in the chemistry of hashish since the early fifties; the older literature will be discussed only briefly.

At the end of this Review some of the physical, chemical and physiological properties of the natural cannabinoids are summarized. Spectral data, crystalline

---

* THC = tetrahydrocannabinol.

** Due to the non-availability of the natural active components, most of the pharmacological work reported at this symposium as well as that in previous articles is based on experiments done either with crude hashish extracts, whose content is known to vary widely, or with synthetic analogs of the active constituents, which may not possess the same activity as the natural active principles. These drawbacks have hampered biological work in this area to a considerable extent.

derivatives, melting points and optical rotations are presented in *Table 2* (p. 204). The biological activity of these compounds or their simple derivatives is described in *Table 1* (p. 203). The mass spectral curves of most of the natural cannabinoids have been reported (*23, 31, 32 a*). *Figures 1 to 10* (pp. 206–208) represent the IR curves of these compounds or their simple derivatives.

III. Numbering of the Cannabinoids.

We propose the term *cannabinoids* for the group of C_{21}-compounds typical of and present in *C. sativa*, as well as for their analogs and transformation products.

ADAMS (*1*) and others (*32, 41*) have followed the dibenzopyran nomenclature, while TODD (*48*) has used a numbering starting from the hetero-atom of the pyran ring. Cannabidiol type compounds have been numbered by ŠANTAVÝ (*60*) and by KORTE (*61*) as derivatives of diphenyl. Our group (*44, 46, 79*) as well as TAYLOR et al. (*93*) and HIVELY et al. (*57*) regard the cannabinoids as substituted monoterpenes. The latter numbering will be used in this review.

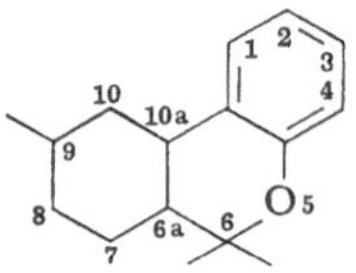

Dibenzopyran numbering.

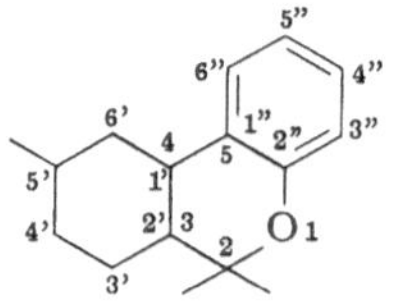

Numbering used by TODD

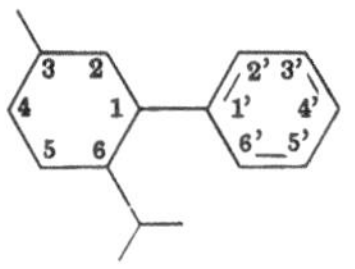

Diphenyl numbering.

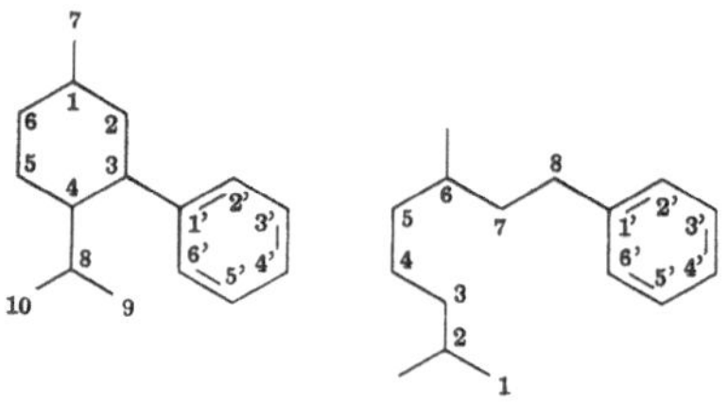

Monoterpene numbering.
(used in this review)

IV. An Outline of the Chemical Research until the Early Fifties.

Publications dealing with the chemistry of hashish have been appearing intermittently for nearly 120 years.

The first papers on the subject which we were able to locate are those by BOHLIG (*18*) (1840), DECOURTIVE (*37*) (1848) and SMITH and SMITH (*91, 92*) (1847). The latter authors have ascribed the physiological action to an alkali-insoluble resin obtained by alcoholic extraction of the plant. In 1899 WOOD (*100, 101*) purified the resin through distillation and after acetylation obtained crystalline cannabinol acetate. Work by CAHN (*24, 26, 27*) and BERGEL (*16*) on cannabinolactone (*1*), a degradation product, led to a partial formula for cannabinol, the full structure of which (*2*) was elucidated through synthesis by ADAMS (*3*) and by TODD (*49*) in 1940. During the following few years considerable progress was made. Cannabidiol was isolated (*7*) and its structure largely elucidated (*5*). A synthetic product, $\varDelta^3$-THC

References, pp. 208—213.

(3) (2, 49) was shown to be active. Acid treatment of the inactive cannabidiol gave active products (11, 5) the structure of one of which was deduced to be $\Delta^{1(6)}$-THC (4), without stereochemical assignment at $C_{(3)}$ and $C_{(4)}$. A group at the U. S. Bureau of Narcotics (98) was able to isolate a highly purified active fraction and, although no definite structure was put forward, it could be concluded that this active fraction was a THC isomer. A large number of analogs were synthesized, some of which were found to be many times more active than the different synthetic or semisynthetic THC isomers (9, 95 and ref. cited there). In spite of the considerable progress achieved the major objectives were not reached: The active constituent(s) were not obtained pure; their structures were not fully elucidated; the only natural cannabinoid whose structure was established was cannabinol. Probably because of the war, research on the chemistry of hashish was largely discontinued and the problems were not taken up for nearly two decades.

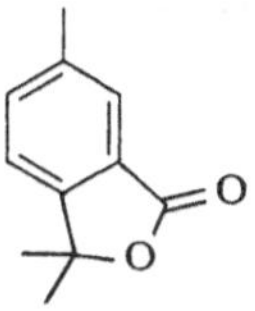

(1)
Cannabinolactone.

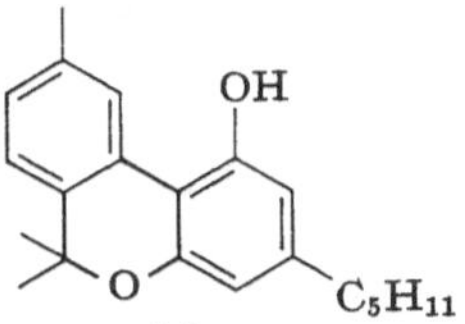

(2)
Cannabinol.

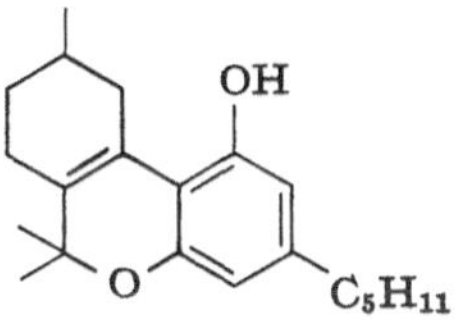

(3)
Δ^3- THC.

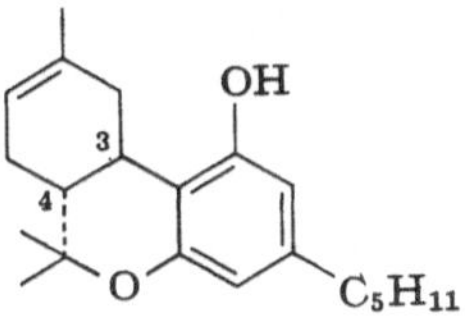

(4)
$\Delta^{1(6)}$ THC.

V. Isolation of Naturally Occurring Cannabinoids.

Most of the natural cannabinoids boil within the same temperature range and separation of the constituents of the active "red oil" by fractional distillation is experimentally worthwhile only if one of the components is present in relatively high concentration. Cannabidiol (5, p. 180) and cannabinol (2) were indeed obtained by distillation followed by the preparation of crystalline derivatives and hydrolysis back to the pure constituents (7, 12, 25, 100). WOLLNER et al. (98) have reported the isolation of a "THC" by fractional distillation of acetylated "red oil" followed by ammonolysis. The oil obtained cannot be considered pure as neither crystalline derivatives were obtained nor was any other criteria of purity reported. On the basis of the analytical results and high biological activity recorded it can be assumed that the oil was a highly concentrated THC mixture. The rotation given, $[\alpha]_D - 193°$, is intermediate between those of Δ^1-THC ($[\alpha]_D - 150°$) (6) and $\Delta^{1(6)}$-THC ($[\alpha]_D - 266°$) (4). Two previous preliminary reports on the isolation of active compounds have not been confirmed (*55, 83*). Šantavý (*60, 69*) and Schultz (*88*) have independently reported the isolation of cannabidiolic acid (7, p. 181) which was shown to be the major component of the acidic fraction of the petroleum ether extract of hemp. De Ropp (*38*) described the isolation of a THC. His method involved adsorption chromatography of a methanolic extract of the flowering tops of the plant, followed by partition chromatography on Celite using a N,N-dimethyl formamide/cyclohexane system, and high vacuum distillation. Although the material obtained was not crystalline, nor were crystalline derivatives reported, the compound was considered to be pure (on paper chromatographic evidence). Its infrared spectrum and other physical properties reported are similar to those of Δ^1-THC (6).

(5)
Cannabidiol.

(6)
^{1}- THC.

Mechoulam and Gaoni (*44, 46, 77*) have based their separation method on column chromatography.

A hexane extract of hashish was separated into acidic and neutral fractions. The latter was chromatographed on Florisil or acid-washed alumina. The following identified compounds were eluted (in order of increasing polarity): cannabidiol (5) (eluted with 5% ether in pentane), Δ^1-THC (6), cannabinol (2), cannabichromene (8) and cannabigerol (9) (eluted with 15% ether in pentane). Repeated chromatography was needed to effect full separation. When the fractions containing cannabidiol were chromatographed on acid-washed alumina containing 12% silver nitrate, a further component, cannabicyclol, could be separated (*47*). Cannabichromene and cannabigerol could likewise be separated by silver nitrate-alumina chromatography. Cannabidiol, m. p. 66–7°, cannabicyclol, m. p. 152–3° and cannabigerol, m. p. 51–3° were crystallized directly while Δ^1-THC and cannabichromene were further purified by the preparation of crystalline derivatives and hydrolysis to the parent compound. Δ^1-THC and cannabichromene gave crystalline 3,5-dinitrophenylurethanes, m. p. 115–6° and 106–7°, respectively. Chromatography of the esterified acidic fraction yielded cannabigerolic (11) cannabinolic (12) and cannabidiolic (7) acids (as esters). The column chromatographic separations were monitored by thin-layer chromatography.

The amounts of cannabinoids present in hashish were found to vary considerably from sample to sample. In one relatively fresh hashish sole (10–15 months old) the following amounts were present (*78*): cannabidiol 4%; Δ^1-THC 0.4%; cannabinol 1.2%; cannabichromene 0.1%; cannabigerol 0.3%; cannabicyclol 0.1%; cannabidiolic acid 3.25%; cannabigerolic acid 0.5%; and cannabinolic acid 0.25%.

References, pp. 208—213.

A number of additional compounds are observed upon thin-layer chromatography, especially in fractions which are more polar than cannabigerol, but no pure compounds have as yet been isolated.

KORTE (62, 66) has used adsorption chromatography on aluminum oxide to remove colored impurities, followed by countercurrent distribution. Crystalline cannabidiol and cannabinol were obtained. Further distributions led to the separation of three components which were initially named THC I, II and III, present in the ratio of 70 : 30 : 1. THC I proved to be identical with Δ^1-THC described by GAONI and MECHOULAM (44). THC II has been renamed cannabichromene (33). THC III, m. p. 146° [originally (64) reported 125–128°], might be identical with cannabicyclol. Another crystalline compound isolated (10), m. p. 128°, is probably an artefact as it is a racemate. The labile Δ^1-THC acid (13) has likewise been isolated by countercurrent distribution (62).

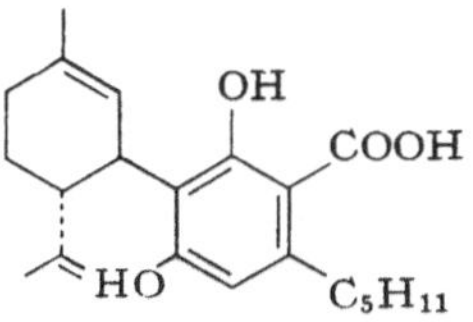

(7)
Cannabidiolic acid.

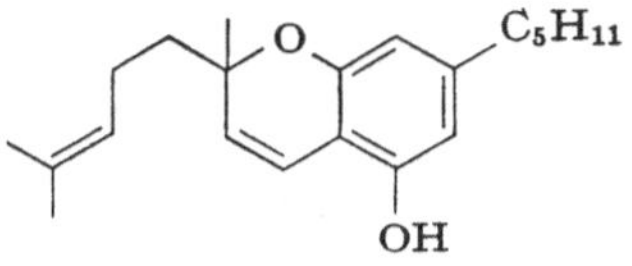

(8)
Cannabichromene.

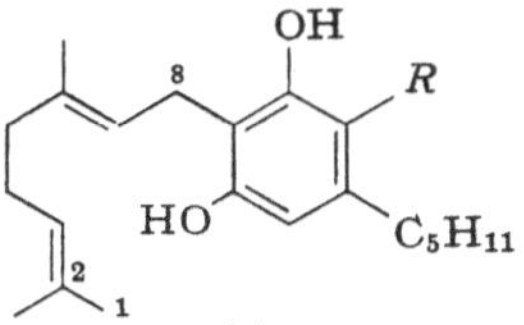

(9)

(11)

$R = H.$ Cannabigerol.

$R = COOH.$ Cannabigerolic acid.

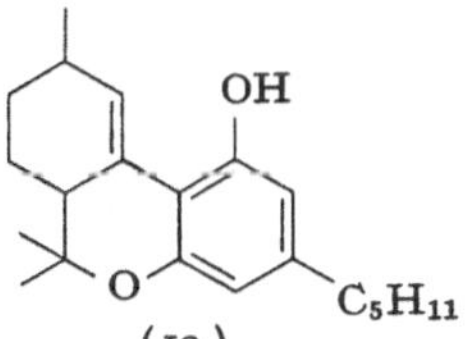

(10)

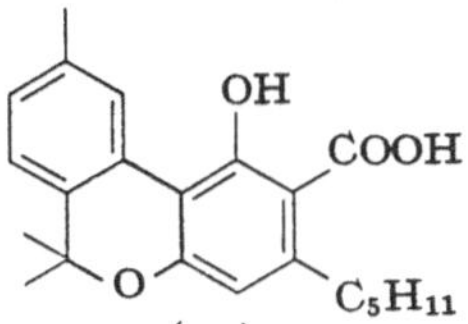

(12)
Cannabinolic acid.

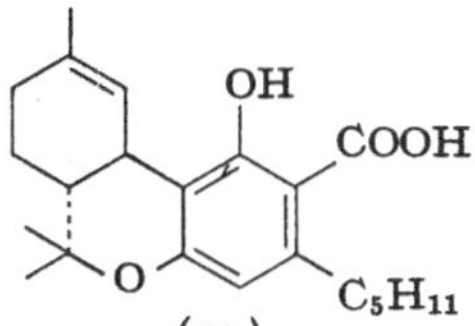

(13)
Δ^1– THC acid.

In Middle Eastern hashish, German hemp, South African dagga and Indian ganja the only THC is the Δ^1 isomer (6) (47); from some marihuana samples, however, $\Delta^{1(6)}$-THC (4) has been isolated by HIVELY et al. (57). Chromatography on silicic acid of a petroleum ether extract of the flowering tops and leaves of *C. sativa* from Maryland and rechromatography on silicic acid-silver nitrate gave cannabinol, followed by Δ^1-THC, $\Delta^{1\,(6)}$-THC and cannabidiol. De Ropp's method (*38*), described on p. 180, also gave a good separation. The ratio of the Δ^1 to the $\Delta^{1\,(6)}$ isomer is reported to be 10 : 1.

Todd (*58*) has described the isolation of cannabol, $C_{21}H_{30}O_2$ (*p*-phenylazobenzoate, m. p. 117–8°). Covello (*35*) has isolated a crystalline substance, m. p. 129–133°, possessing a highly inebriating effect in the dog. No further work has been reported on these two substances.

Numerous non-cannabinoid terpenes (*82*), phenolic compounds (*56*, *82a*), sugars (*56*), nitrogen bases (cholin, trigonellin) (*84*), etc. have been isolated from or detected in *C. sativa*.

The amount of cannabinoids in the plant seems to depend both on climatic factors and on the botanical variety. The female inflorescences of both *C. sativa* var. indica and var. non indica grown in Germany under identical agricultural conditions produce the same cannabinoids albeit in different proportions (*65*, *66*). In the var. indica the ratio of cannabidiol to tetrahydrocannabinol was 1 : 1, while in the var. non indica it was 2.5 : 1. Only minor differences have been noted (*87*) in the content of cannabidiol and cannabidiolic acid in plants grown from seeds from different countries. However, plants grown at Zagreb from seeds from various geographical locations yielded resins with different chemical characteristics (*52*).

It is generally accepted that the resin of *C. sativa* grown in temperate climates contains more cannabidiolic acid and probably cannabidiol and less THC and cannabinol than the resin formed in subtropical or tropical regions. Grlić (*51*, *53*, *54*) has based his classification of cannabis samples on these differences. Thus, samples in which cannabidiolic acid is predominant are classified as "unripe" while samples containing mostly THC and cannabinol are referred to as "ripe" or "overripe", respectively. These terms are meant to indicate biogenetic evolution and hence geographic origin and are not a measure of seasonal variation of chemical content.

The isolation of numerous new cannabinoids in recent years and improved methods of analysis should make possible more precise determination of the chemical differences due to botanical variation, climatic conditions and geographic location.

VI. Structural Elucidation.

Cannabinol (2, p. 179).

Cannabinol was the first cannabinoid whose structure was fully elucidated [see (*24*, *26*) for degradative evidence and (*3*, *49*) for total syntheses].

Cannabidiol (5, p. 180).

Structure (**14**) (without stereochemical assignments) was originally suggested for cannabidiol (*1*). On catalytic reduction two moles of hydrogen were absorbed, showing that two double bonds were present. Pyrolysis with pyridine hydrochloride caused cleavage into *p*-cymene and olivetol (**15**) *(Chart 1)*. These two compounds account for all 21

carbons present in cannabidiol. The position of the linkage between these
residues was established by conversion of cannabidiol into cannabinol
by ring closure with acid followed by dehydrogenation. The terminal
position of one of the double bonds was established by ozonization of
cannabidiol to give formaldehyde. The position of the double bond in
the alicyclic ring, however, was based partially on negative evidence.

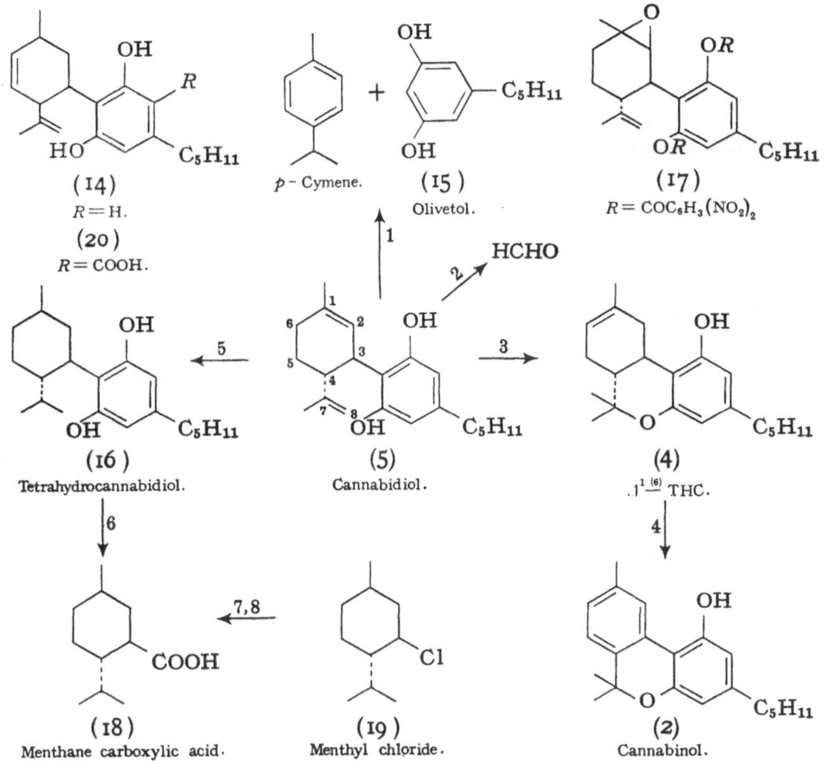

Chart 1. Conversions of Cannabidiol.

Reagents: 1) Δ, pyridine hydrochloride; 2) O_3; 3) p-toluene sulfonic acid; 4) S;
5) H_2/PtO_2; 6) $KMnO_4/CH_3COCH_3$; 7) Mg; 8) CO_2.

The ultraviolet spectrum of cannabidiol showed that the double bond
in the terpene ring was conjugated neither to the terminal methylene
group nor to the aromatic nucleus. Furthermore, it did not isomerize
to a conjugated position to the aromatic ring on either basic or acidic
treatment. By a process of elimination the double bond was assumed
to be at the Δ^5 position. Although TODD (95) in 1946 had already expressed
doubt as to the validity of this assignment, the correct position was
deduced only in 1963, mainly by NMR measurements (79). Structure

(14) was eliminated by the presence of only one olefinic proton in the NMR spectrum. The chemical shift of the $C_{(3)}$ proton, at 3.85 ppm, was considered too low for an ordinary benzylic proton and it was assumed that the additional deshielding contribution arose from an adjacent olefinic center, namely Δ^1. Support for this assignment was obtained from the NMR spectra of tetrahydrocannabidiol (16) and the mono-epoxide (17). In (16) the chemical shift of the $C_{(3)}$ proton is at 3.0 ppm while in (17) it is at 3.24 ppm. These differences could be explained only by assuming that the double bond occupied the Δ^1 and not the $\Delta^{1(6)}$ position.

The relative stereochemistry of the two asymmetric centers $C_{(3)}$ and $C_{(4)}$ was deduced to be *trans* from an analysis of the coupling constants of the protons at these centers and by degradative evidence (79). Adams (1) had shown that oxidation of tetrahydrocannabidiol (16) with potassium permanganate in acetone gave a menthane carboxylic acid (18), the anilide of which was identical with the anilide of the acid obtained from menthyl chloride (19) through a Grignard reaction with carbon dioxide. Attempted equilibration of the methyl ester of (18) showed (79) that the carbomethoxyl group was equatorial and *trans* to the isopropyl group. Assuming that the relatively mild oxidation of tetrahydrocanna-bidiol caused no inversion at $C_{(3)}$, the stereochemistry of cannabidiol should be identical with that of menthane carboxylic acid, i. e. *trans*.

The large coupling constant (J = 11 cps) between the $C_{(3)}$ and $C_{(4)}$ protons in both (5) and (17) indicates that the dihedral angle between them is that of two *trans* diaxial hydrogens, confirming the assignment based on degradative evidence.

The same conclusion has been reached (85) by analysis of optical rotation data of compounds reported previously. It was shown recently (45), however, that some of the compounds whose rotations were compared were actually mixtures.

Cannabidiolic acid (7, p. 181).

Cannabidiolic acid was first assigned structure (20) (67, 68, 89) mainly on the basis of its conversion through decarboxylation to cannabidiol which at that time was considered to be (14). The modification of the structure of cannabidiol to (5) (79) necessitated a similar revision of the structure of cannabidiolic acid to (7) (77). The position of the carboxyl group in (7) was established by the infrared (68, 89) and NMR (77) spectra. In the infrared a peak at 1698 cm^{-1} was assigned to an aromatic carboxyl group. In the NMR spectrum only one aromatic proton was observed, thus conclusively determining the position of the carboxyl group.

Cannabigerol (9, p. 181) (46).

Cannabigerol has two hydrogens more than cannabidiol (5), but the same number of double bonds. Hence, it possesses one ring less than (5). Optical inactivity suggested that the two asymmetric centers present in (5) are absent in cannabigerol. The NMR spectrum indicates that the aromatic protons are identical, that two protons (those at $C_{(8)}$) are strongly deshielded and split by a single adjacent proton and that an isopropylidene group is present. The ultraviolet spectrum is identical with that of cannabidiol and indicates that the double bonds are conjugated neither to each other nor to the aromatic ring. Structure (9) has been corroborated by synthesis (see p. 196).

Δ^1-THC (6, p. 180).

The structure of this major psychotomimetic principle was elucidated in 1964 by GAONI and MECHOULAM (44). The carbon skeleton was determined by conversion to cannabinol (2, p. 179) on dehydrogenation. The position of the double bond and the stereochemistry at the asymmetric centers were deduced from NMR measurements and by synthetic studies.

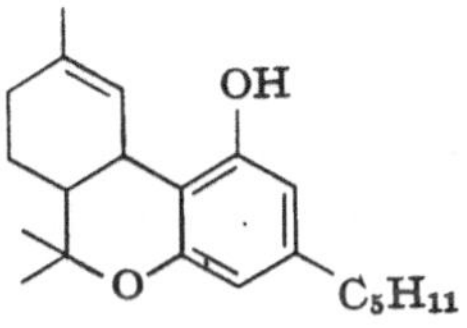

(21)
Δ^1– 3,4 – cis –THC.

The NMR spectrum of (6) shows the presence of one olefinic proton, two methyl groups α to an oxygen atom and one olefinic methyl group. These observations place the double bond in the Δ^1 or $\Delta^{1(6)}$ position. A comparison of the chemical shifts of the $C_{(2)}$ and $C_{(3)}$ protons in Δ^1-THC with those of cannabidiol (5) shows that the olefinic proton in Δ^1-THC is strongly deshielded as compared with that in cannabidiol, while the reverse relationship is observed with regard to the $C_{(3)}$ proton. This effect is due to the conformation of the two compounds. In cannabidiol the free rotating aromatic ring has been assumed to be in the same plane as the $C_{(3)}$ proton, while in Δ^1-THC the additional ring places the aromatic portion of the molecule in (or nearly in) the same plane as the olefinic proton at $C_{(2)}$, thus causing the observed deshielding effects. These observations place the double bond in the Δ^1 position in (6) and eliminate the alternative $\Delta^{1(6)}$ position. In $\Delta^{1(6)}$-THC (4) (45, 57, 93) the olefinic proton is not deshielded by the aromatic ring. Conversion of cannabidiol into Δ^1-THC confirmed the structure suggested and established its stereochemistry. Corroboration of the stereochemical assignments can be found in later synthetic work. In Δ^1-3,4-*trans*-THC

(6) the $C_{(3)}$ proton is at 3.14 ppm while in the synthetic *cis* isomer (**21**) (*93*) it is at 3.59. Models show that in the *cis* compound the $C_{(3)}$ proton is more deshielded by the aromatic ring than in the *trans* isomer.

Cannabigerolic Acid (11, p. 181) (*77*), Cannabinolic Acid (12) (*77*) and Δ^1-THC Acid (13) (*62*).

The structures of these acids were determined by comparison of their NMR spectra with those of the corresponding neutral cannabinoids, cannabigerol (**9**) cannabinol (**2**, p. 179) and Δ^1-THC (**6**) and, in the case of (**11**) and (**12**), by decarboxylation to (**9**) and (**2**). The NMR spectra (see Table 2, p. 204) indicate that the major difference between the acids (or their methyl esters) and the respective neutral cannabinoids is the presence of one aromatic proton only in each cannabinoid acid. A certain deshielding effect on the benzylic methylene of the amyl side-chain is likewise observed. It is due to the neighboring carboxyl or carbomethoxyl group. In cannabinol and Δ^1-THC the two aromatic positions are not equivalent and hence two different acids could be derived in each case. Actually only a single cannabinolic acid and a single Δ^1-THC acid have been observed so far. In both instances the position of the carboxyl group was shown to be adjacent to the free phenolic group on the basis of the strong deshielding of the phenolic proton caused by hydrogen bonding.

Cannabichromene (8, p. 181) (*33, 42*).

Cannabichromene is of a different structural type than the other cannabinoids. Its ultraviolet spectrum shows conjugation of a double bond with the aromatic ring, while the NMR spectrum indicates that a) two of the olefinic protons are not flanked by additional hydrogen atoms (sharp AB pattern), b) an isopropylidene grouping is present and c) one of the methyl groups is α to an oxygen while two others are olefinic.

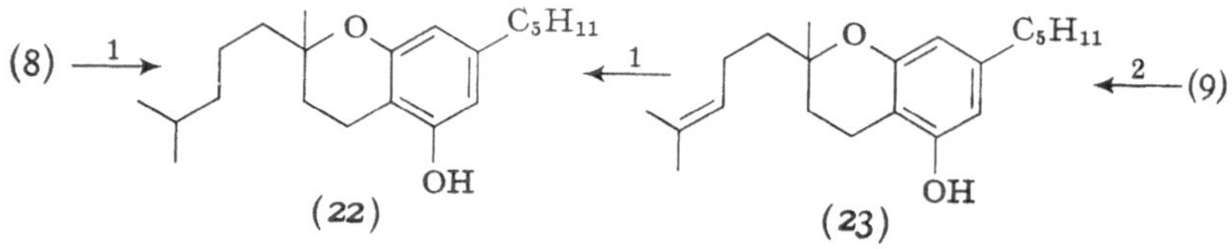

Reagents: 1) H_2/Pd-C; 2) *p*-toluene sulfonic acid in benzene.

These data and the molecular weight (mass spectrum) suggested structure (**8**), which was corroborated (*42*) by correlation with cannabigerol (**9**). Catalytic hydrogenation of cannabichromene gave (**22**) whose *dl* form was obtained from (**9**) by cyclization to (**23**) followed by hydrogenation.

References, pp. 208—213.

$\Delta^{1\,(6)}$-THC (4, p. 179).

In 1941 ADAMS (5) showed that cannabidiol (5) cyclized to $\Delta^{1\,(6)}$-THC (4, without stereochemistry) on refluxing with p-toluene sulfonic acid in benzene. Later work based on comparison of optical rotations (85), mass spectra (23), chemical evidence and NMR data (45, 57) as well as total synthesis (41, 15a, 93) has confirmed the suggested structure and established the *trans* stereochemistry at $C_{(3)}$ and $C_{(4)}$. The NMR evidence has been discussed in connection with the structure of Δ^1-THC (p. 185). Dehydrogenation to cannabinol and the mass spectrum (23) have established the carbon skeleton. Hydrogenation of the double bond gave the two possible $C_{(1)}$ isomers. The same hexahydrocannabinols, although in different ratios, were obtained from Δ^1-3,4-*trans*-THC, thereby showing the stereochemical relationship of the Δ^1 and $\Delta^{1\,(6)}$ isomers (45). $\Delta^{1\,(6)}$-THC isolated from marihuana is identical with the semisynthetic $\Delta^{1\,(6)}$-THC (57).

Added in Proof.

LERNER et al. (71a) have found that Δ^1-THC and $\Delta^{1(6)}$-THC can be separated well on VPC with a 2% OV-17 (a methyl phenyl silicone) column. Using a smoking machine the same authors (71a) have shown that the ratio of Δ^1-THC to $\Delta^{1(6)}$-THC (99.6 : 0.4) in a marihuana sample remains essentially unchanged on smoking; however, in cigarette tobacco impregnated with "red oil" the ratio on smoking decreased from 97 : 3 to 91 : 9. It is yet to be determined whether this change is due to partial isomerization of Δ^1-THC to $\Delta^{1(6)}$-THC or to preferential oxidation of Δ^1-THC.

Professor C. J. MIRAS of the School of Medicine, University of Athens, has informed us that he has found no conversion of Δ^1-THC to $\Delta^{1(6)}$-THC when Greek tobacco impregnated with Δ^1-THC was smoked by a machine.

Cannabicyclol (47).

The structure of this minor component, m. p. 152–3°, which is probably identical with "THC III", m. p. 146°, (65) has not yet been established. The molecular weight (mass spectrum) and elementary analysis indicate the composition $C_{21}H_{30}O_2$. The ultraviolet spectrum is identical with that of cannabidiol, which is typical for the olivetol moiety. The NMR spectrum shows a) two nonequivalent aromatic protons, b) four methyl groups, none of which is olefinic, but at least one is α to an oxygen atom and one is apparently the terminal methyl group of the pentyl side-chain, and c) no olefinic protons. Apparently cannabicyclol has no double bond. Consequently, the elemental composition requires a tetracyclic structure. Structure (24) can be visualized as a working hypothesis.

C_5H_{11}

OH

(24)

VII. Absolute Configuration.

The absolute configuration of the natural cannabinoids, except that of cannabichromene which is unknown, is D, relative to $D(+)$-glyceraldehyde. The key to this assignment is the degradative correlation of cannabidiol (5) with $D(—)$-menthane carboxylic acid (18, p. 183) derived from $D(—)$- menthol as described on p. 183. Adams (8) has reported that the anilide of (18), m. p. 152°, prepared from $D(—)$-menthol does not depress the melting point of the anilide of (18) prepared by degradation of cannabidiol. Unfortunately, the rotations of these compounds were not reported. Recently (78) this work has been repeated. $D(—)$-Menthane carboxylic acid, m. p. 65–6°, $[\alpha]_D — 44°$, was obtained both from $D(—)$-menthol and from cannabidiol, thus defining the absolute stereochemistry. The structures of all cannabinoids in this review are presented with their correct absolute stereochemistry.

VIII. Syntheses.

Two syntheses of cannabinol (2) reported by the groups of Adams and Todd are presented in *Charts 2* and *3*. For a further synthesis by Adams see (3).

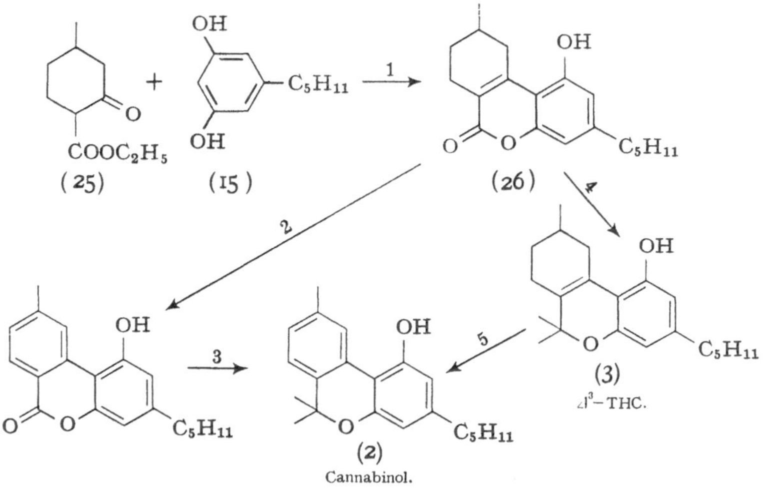

Chart 2. Synthesis of Δ^3-THC and Cannabinol.

Adams et al. (2 and ref. cited there). Reagents: 1) POCl$_3$; 2) S; 3) or 4) CH$_3$MgI.
Ghosh, Todd and Wilkinson (48, 49). Reagents: 1) H$_2$SO$_4$; 4) CH$_3$MgI; 5) Pd/charcoal.

The synthetic route shown in Chart 2 is of particular interest because the intermediate THC (3) was found to cause the characteristic effects

References, pp. 208—213.

of hashish in animals and in man. A series of closely related synthetic compounds was then prepared by this procedure using the same keto ester (25), with homologs of olivetol (9, 95 and ref. cited there). A definite relationship between the length and branching of the side-chain and hashish activity was established: Activity in the n-alkyl series rises to a maximum at n-hexyl and then falls off again; branching increases the potency remarkably. Nitrogen analogs have likewise been prepared by this procedure using 4-carbethoxy-N-methyl-3-piperidone hydrochloride (27) instead of the keto ester (25). The compounds obtained [$R = C_5H_{11}$, —CH_3, —$CH(CH_3)(CH_2)_4CH_3$, etc.] showed physiological activity comparable to that of the natural THC-s (82 b).

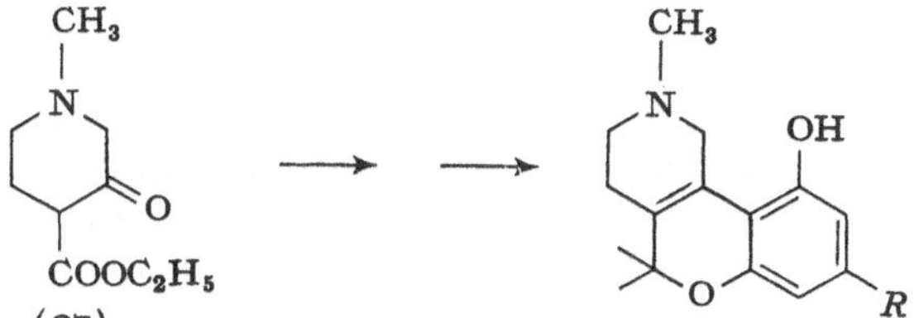

The synthetic method presented in Chart 3 has not been used as much as the previous ones because the Δ^3-THC (3) is accompanied by a number of inactive by-products which are difficult to remove.

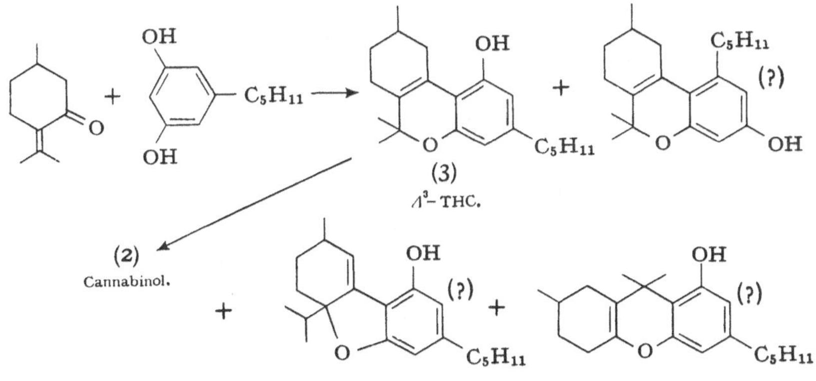

Chart 3. Synthesis of Δ^3-THC and Cannabinol.

ADAMS et al. (13). Reagent: $POCl_3$.
TODD's group (50, 70). Reagent: HCOOH.

The l form of (3) has been shown to be 11–15 times as active as the d form (70).

The reactions leading to Δ^3-THC have recently been reinvestigated (*32, 64–66*). Repetition of the reaction scheme reported by Adams (*2*) (Chart 2) led to a mixture which could be separated by countercurrent distribution into three crystalline isomers, viz.: (**3**), m. p. 62–3°, (**10**, p. 181), m. p. 128°, and an isomer, m. p. 86–7°, present in a ratio 100 : 10 : 0.5, with an overall yield of 95%. The isomer (**10**) is identical with a compound considered to be an artefact obtained (*64*) from the resin of German hemp (see p. 181).

A reinvestigation of the condensation of olivetol with pulegone, according to Todd (*50, 70*) (Chart 3) was reported (*66*) to give a mixture of two compounds, which could not be separated by countercurrent distribution. They were reported to be different (TLC data only) from the three isomers (**3**), (**10**) and isomer m. p. 86° described above. These observations contradict those of Todd (*70*), who has obtained crystalline menthoxyacetates of d-Δ^3-THC prepared by both methods (Charts 2 and 3) and has shown identity by direct comparison (mixed melting point and identical UV). The Δ^3-THC prepared through both routes has also been converted into cannabinol by dehydrogenation. In view of this discrepancy the suggested non-identity of the products prepared via both routes has to be further substantiated.

A different approach to the synthesis of a $\Delta^{1(6)}$-THC (**4**, p. 179) through a Diels-Alder reaction of an appropriately substituted cinnamic acid with isoprene has been initiated by Adams (*4, 6*). o,o'-Dihydroxy-cinnamic acid did not react with isoprene (*4*); however, the dimethyl ether of 2,6-dihydroxy-4-methyl- (or -4-amyl-) cinnamic acid condensed readily giving (**28**). Subsequent hydrolysis to the phenolic analog, however, failed (*4, 6*).

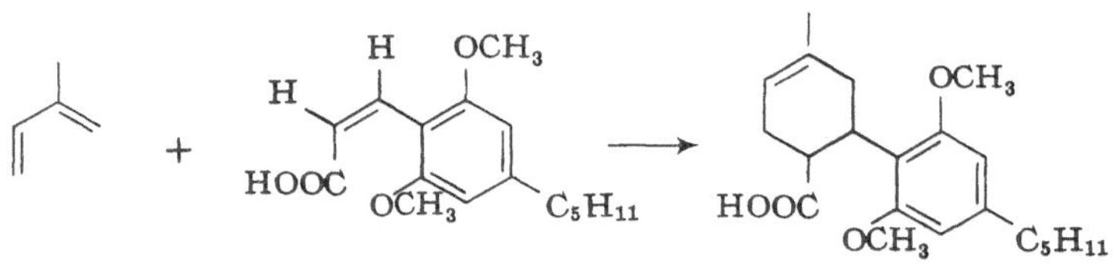

(28)

A variation on this theme was the projected attempt (*10*) to react an appropriately substituted coumarin with isoprene to give directly (**29**).

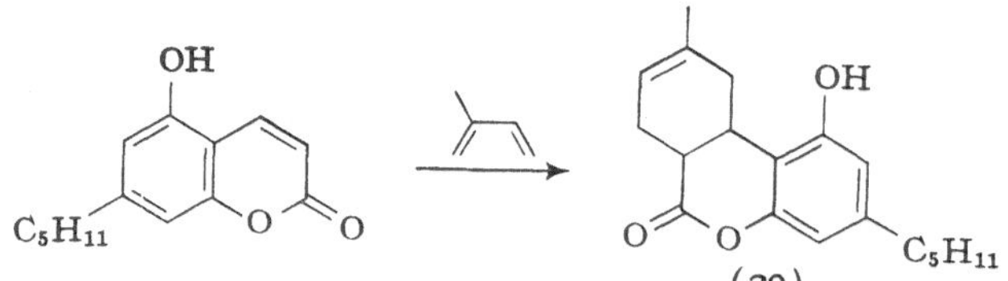

(29)

Model experiments with coumarin have not been successful and no further work by Adams' group along these lines has been reported. This approach has been taken up by Taylor and Strojny (94). They have based their synthetic modification on the fact that introduction of an electronegative group in conjugation with a double bond increases its dienophilic activity in the Diels-Alder reaction. 3-Carbethoxycoumarin reacted with isoprene to give (30). This lacton was converted to (31) and (32) by the sequence described in *Chart 4*.

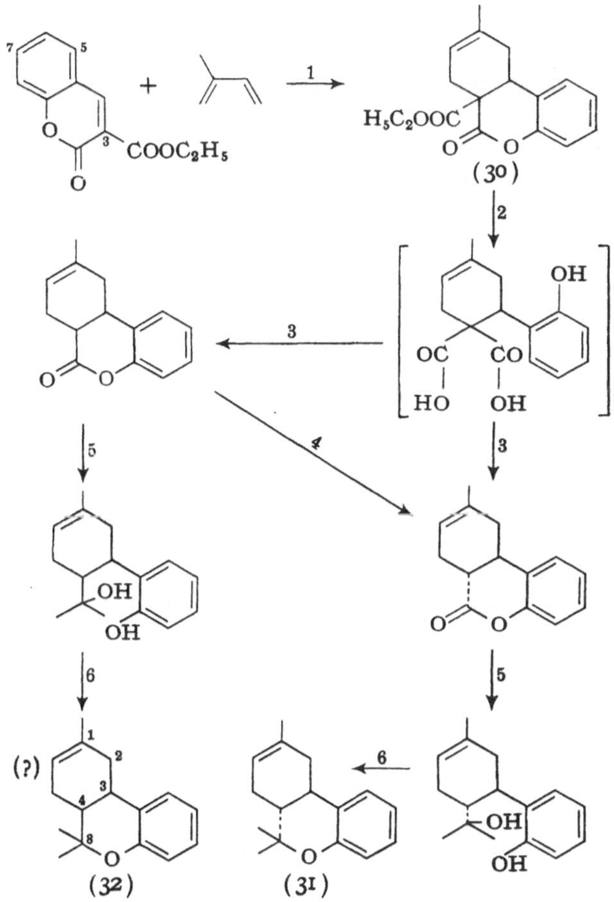

Chart 4. Synthesis of Model Compounds Related to THC, by Taylor et al. (94).

Reagents: 1) Δ; 2) NaOH/C$_2$H$_5$OH then H$^+$; 3) Δ; 4) NH$_4$OAc, Δ; 5) CH$_3$MgI; 6) p-toluene sulfonic acid.

The structure of the assumed *cis* isomer (32), which is reported to be a mixture, requires confirmation. Dreiding models show that the cyclization during the last step of the reaction can take place at the $C_{(1)}$ as well as the $C_{(8)}$ carbons and hence compounds such as (33), (34), (35), etc. might be expected.

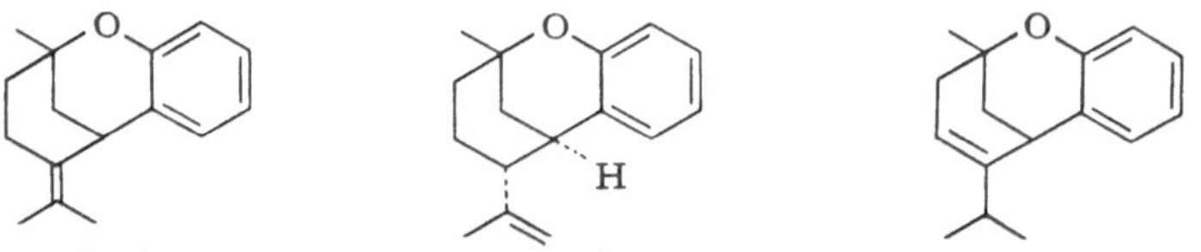

(33)   (34)   (35)

KORTE (63) has reported the synthesis of an isomer of dimethoxy-cannabidiol using a modified Diels-Alder approach *(Chart 5)*. The stereochemistry of the asymmetric centers has not been determined. The double bond occupies the "unnatural" $\Delta^{1\,(6)}$ position.

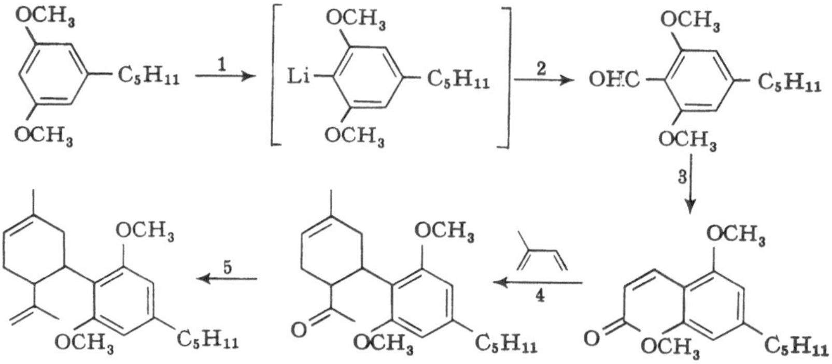

Chart 5. Synthesis of an Isomer of Dimethoxy-cannabidiol, by KORTE et al. (63).

Reagents: 1) C_4H_9Li; 2) N-methylformanilide; 3) aldol condensation with $(CH_3)_2CO$; 4) Δ; 5) $(C_6H_5)_3P=CH_2$.

The first total synthesis of *dl*-cannabidiol (5) and of *dl*-Δ^1-THC (6) was reported by MECHOULAM and GAONI (76) in 1965 *(Chart 6)*. As ADAMS (5) has described the preparation of $\Delta^{1\,(6)}$-THC (4) from canna-bidiol (5), this synthesis is also a route to *dl*-(4). The synthesis is pat-terned along the suggested biogenetic pathway (see p. 201). The cycli-zation step involves a *trans-cis* isomerization of a double bond, which formally does not participate in the reaction. It is possible that the hypothetical geranyl derivative (36) isomerizes through internal return to the linalyl derivative (37) which undergoes cyclization. Compound (38) is also formed. The demethylation step is of interest. The lability of cannabidiol precluded the demethylation with strong acidic or basic reagents. The use of dry Grignard compound is rather un-

References, pp. 208—213.

usual and has apparently not been employed previously in total syntheses of natural products. *dl*-Cannabidiol is best characterized as its ditosylate, m. p. 138–140°; the ditosylate of natural cannabidiol melts at 81–83°. *dl*-Cannabidiol (5) on mild acid treatment, gives *dl*-Δ¹-3,4-*trans*-THC (6) (3,5-dinitrophenylurethane, m. p. 202–3°). The overall yield in this synthesis is 2%.

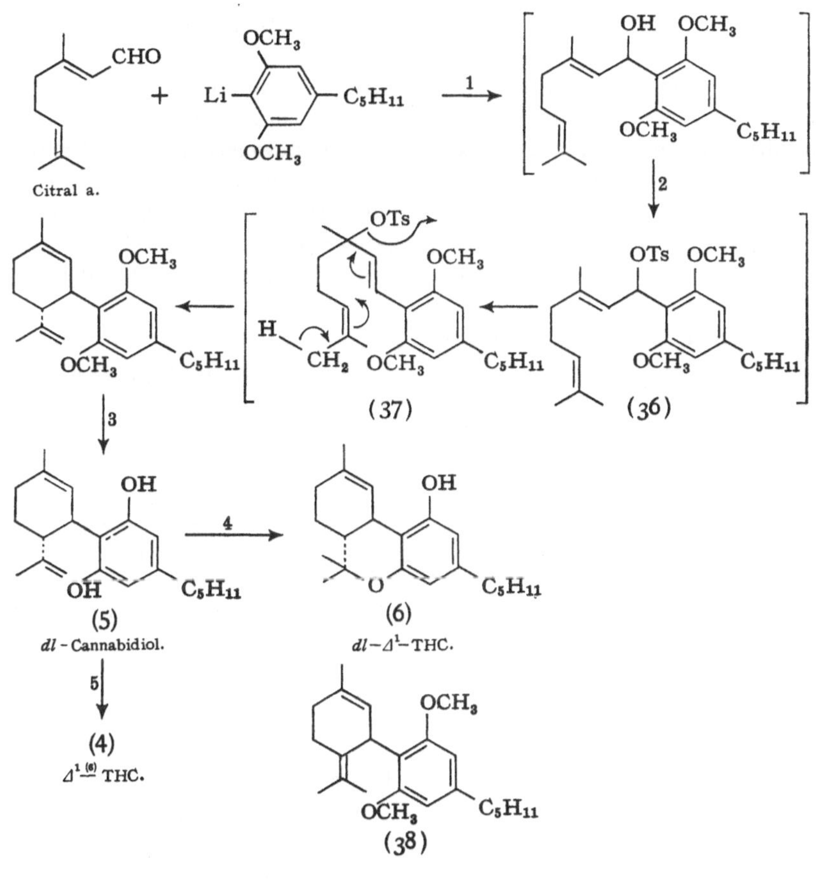

Chart 6. Synthesis of *dl*-Cannabidiol and *dl*-Δ¹-THC, by MECHOULAM and GAONI (76).

Reagents: 1) room temp.; 2) TsCl, pyridine; 3) CH₃MgI, 160°; 4) HCl; 5) *p*-toluene sulfonic acid.

TAYLOR, LENARD and SHVO (93) have independently employed a similar approach leading to *dl*-Δ¹ ⁽⁶⁾-*trans*-(4) and *dl*-Δ¹-*cis*-THC (21) as well as to compound (39) *(Chart 7)*. The Δ¹ ⁽⁶⁾-*trans* isomer is obtained in 10–20% yield. In the original publication (39) was considered to be dl-$\Delta^{1\,(6)}$-cis-THC (39 a), but this structure has since been modified (43). Some chromatographic fractions were reported to contain (by NMR) the dl-Δ^1-3,4-trans isomer but its vapour phase chromatography (vpc) separation (column, 10% GE-SE 30 on Diatoport S) was frustrated by isomerization to the $\Delta^{1\,(6)}$ isomer. In view of this facile isomerization it was suggested "that the physiological effects attendant upon smoking of hashish ascribed to the Δ^1 isomer may in actuality be due to the Δ^6 isomer". This assumption is not correct for on a different column (SE-30 on Chromosorb-W up to 300°) no such isomerization occurs (43). Several instances of isomerization of terpenes on vpc have been reported. The resulting products are those experimentally produced on acid isomerization. Zubyk and Conner (102) have shown that such reactions are due to acidic catalysis by the column support. Hashish is usually smoked in cigarettes or narghilas, which are probably chemically neutral and hence it is doubtfull whether they can cause the isomerization of the Δ^1 to the $\Delta^{1\,(6)}$ isomer.

A minor experimental modification of the synthetic procedure of Taylor et al. leads to the dl-Δ^1-3,4-trans isomer in 20% yield (47) (Chart 7).

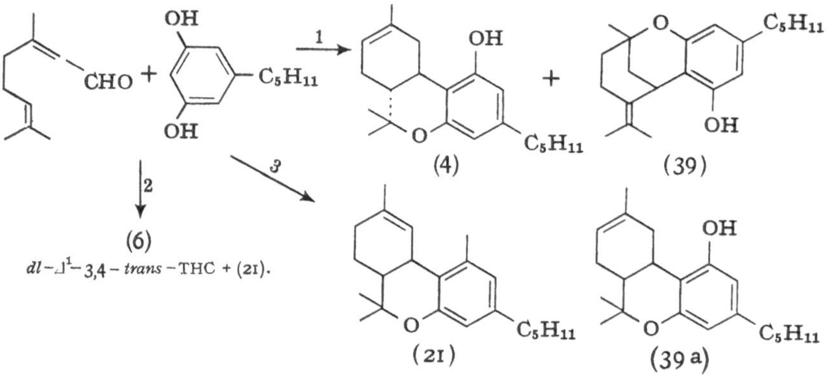

Chart 7. Synthesis of dl-Δ^1-THC and dl-$\Delta^{1\,(6)}$-THC.

Taylor et al. (93). Reagents: 1) 10% BF$_3$ in C$_6$H$_6$; 3) HCl in C$_2$H$_5$OH. Gaoni et al. (47). Reagent: 2) 1% BF$_3$ in CH$_2$Cl$_2$.

Fahrenholtz, Lurie and Kierstead (41) have reported a total synthesis of dl-Δ^1- and dl-$\Delta^{1\,(6)}$-3,4-trans-THC based on a method basically different from those employed by the other authors *(Chart 8)*. The overall yield of this somewhat lengthy synthesis is low.

KORTE et al. *(61)* have reported a second total synthesis of *dl*-cannabidiol *(Chart 9)* based on a modification of the method previously used by the same group for the preparation of an isomer of dimethoxycannabidiol (see p. 192). The overall yield (based on olivetol dimethyl ether) was 9.5%.

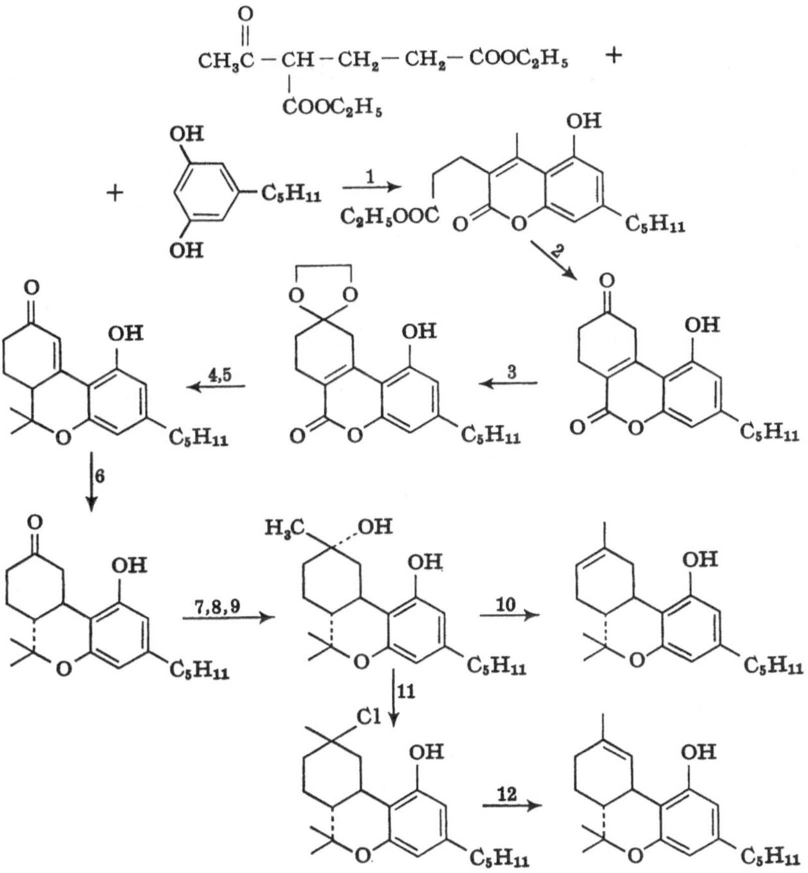

Chart 8. Synthesis of *dl*-Δ^1-THC and *dl*-$\Delta^{1(6)}$-THC, by FAHRENHOLTZ et al. *(41)*.

Reagents: 1) $POCl_3$; 2) NaH, in $(CH_3)_2SO$; 3) $(CH_2OH)_2$, H^+; 4) CH_3MgI; 5) H^+; 6) Li, liq. NH_3; 7) dihydropyran, H^+; 8) CH_3MgI; 9) H^+; 10) *p*-toluene sulfonic acid in benzene; 11) Lucas reagent; 12) NaH in tetrahydrofuran.

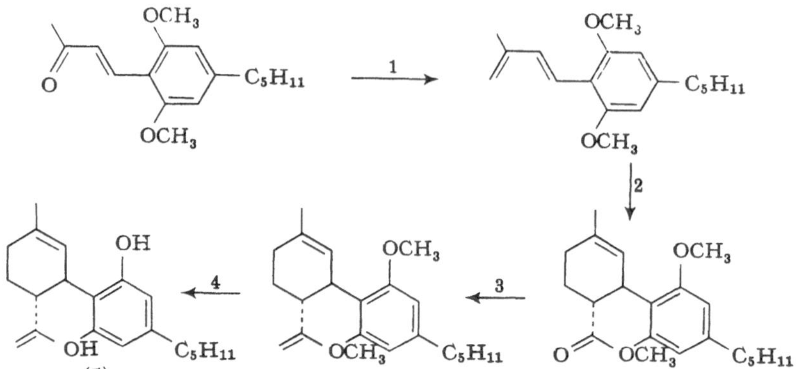

(5)
dl - Cannabidiol.

Chart 9. Synthesis of *dl*-Cannabidiol, by KORTE et al. (*61*).

Reagents: 1) $(C_6H_5)_3P{=}CH_2$; 2) Diels-Alder reaction with $CH_2{=}CH{-}CO{-}CH_3$; 3) $(C_6H_5)_3P{=}CH_2$; 4) CH_3MgI as in Chart 6.

Cannabigerol has been synthesized by GAONI and MECHOULAM (*46*) by heating geraniol with olivetol in decaline.

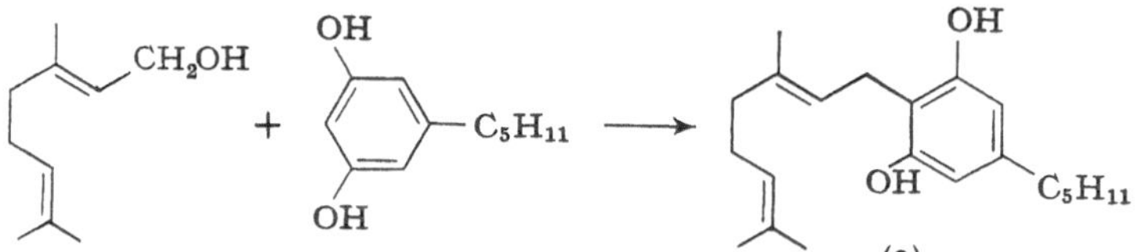

(9)
Cannabigerol.

Added in Proof.

Recently, PETRZILKA et al. (*82c*) have synthesized (—)-cannabidiol (5) in 25% yield by condensation of (+)-*trans-p*-menthadien-2,8-ol-1 with olivetol:

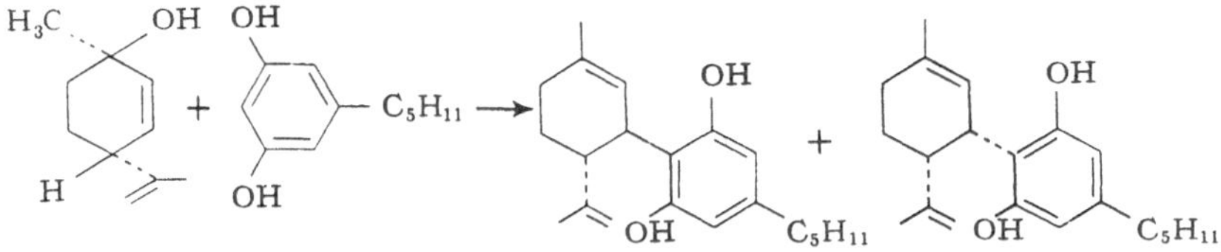

(5) *cis* - Cannabidiol.

Reagent: $(CH_3)_2N{-}CH[OCH_2{-}C(CH_3)_3]_2$.

MECHOULAM et al. (*75a*) have completed a stereo-specific synthesis of (—)-$\Delta^{1(6)}$-THC (4) in 38% yield and of (—)-Δ^{1}-THC (6) in 21% yield:

References, pp. 208—213.

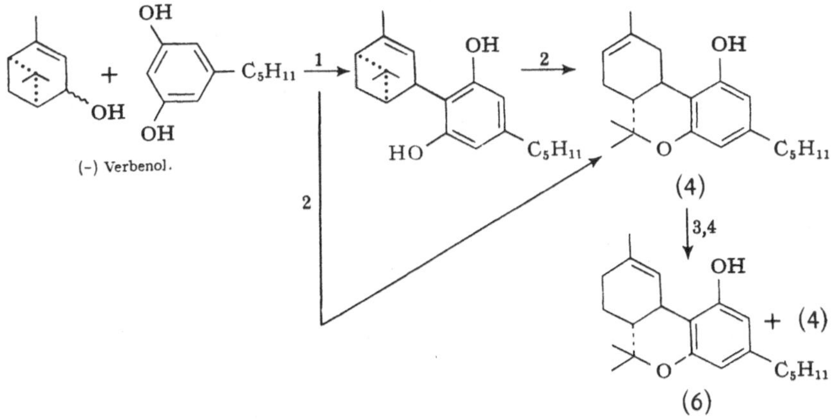

(-) Verbenol.

(4)

+ (4)

(6)

Reagents: 1) p-toluenesulfonic acid; 2) BF_3-etherate; 3) $HCl/ZnCl_2$, — 15°;
4) NaH in tetrahydrofuran.

$(+)$-$\Delta^{1(6)}$-THC prepared from $(+)$-verbenol by the same route has been shown
to be inactive in the ataxia test at the ED_{50} level of the natural isomer $(75a)$.

IX. Chemical Transformations in the Cannabinoid Series.

As most of the naturally occurring cannabinoids were isolated and
their structures elucidated only recently the chemical transformations
in this series have not yet been thoroughly investigated. Some of the
typical reactions are summarized below.

1. Cleavage Reactions.

ADAMS (8) has reported that cannabidiol (5) is split into olivetol
$(15, $ p. 183$)$ and p-cymene by pyridine hydrochloride at 220°. Boiling
dihydrocannabidiol dimethyl ether in benzene with p-toluene sulfonic
acid gives olivetol dimethyl ether (5). From a mechanistic view-point
this cleavage can be considered a reversed Friedel-Crafts reaction.

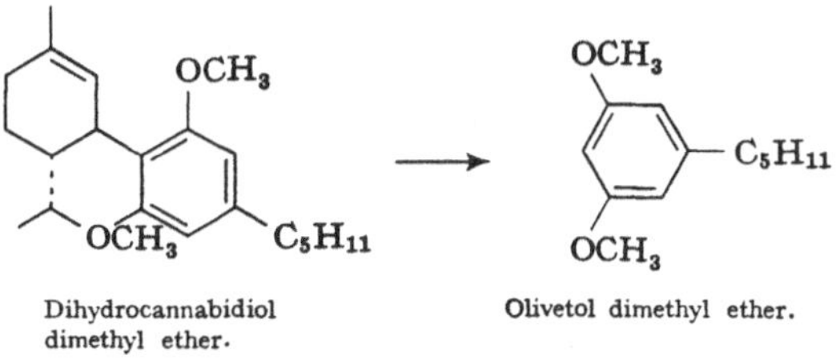

Dihydrocannabidiol
dimethyl ether.

Olivetol dimethyl ether.

Reagent: p-toluene sulfonic acid.

2. The Δ^1 to $\Delta^{1(6)}$ Double Bond Migration.

Δ^1-3,4-*trans*-THC (6) isomerizes with ease to $\Delta^{1(6)}$-3,4-*trans*-THC (4) when boiled with *p*-toluene sulfonic acid in benzene (45, 57, 93). In the *cis* series this double bond migration does not take place under these conditions (43). This difference has been attributed to steric strain. In the *trans* series the olefinic $C_{(2)}$ hydrogen is hindered to a considerable extent; the Δ^1 double bond tends to isomerize to the $\Delta^{1(6)}$ position thus reducing the non-bonded interactions. In the *cis* series such a driving force is apparently small.

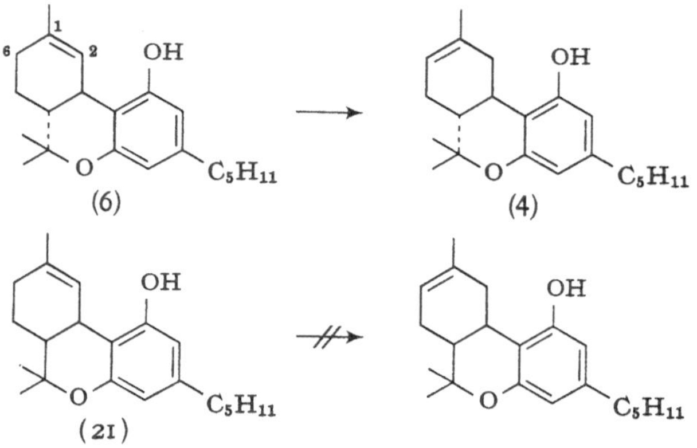

Reagent: *p*-toluene sulfonic acid in benzene.

3. Addition of Alcohols to Double Bonds and Internal Ether Formation.

Boiling cannabidiol (5) with dilute hydrochloric acid in ethanol gives a mixture of compounds *(Chart 10)* which has been partially separated (45) into 1-ethoxy-hexahydrocannabinol, m. p. 141–2° (40), 8-ethoxy-iso-hexahydrocannabinol, m. p. 86–7° (41)*, Δ^1-THC (6), $\Delta^{1(6)}$-THC (4) and Δ^8-*iso*-THC (42). The latter compound is a minor component of the mixture. It is apparent from this reaction that addition to both double bonds of either ethanol or, intramolecularly, of a phenol can take place. In the *trans* series, (5 being *trans*) cyclization takes place predominantly at $C_{(8)}$. In the *cis* series the situation is more complicated. *cis*-Cannabidiol is yet unknown and consequently an exact comparison cannot be made. However, in Taylor's total cannabinoid synthesis (93) (p. 194), Δ^1-3,4-*cis*-THC (21) which is undoubtedly formed from *cis*-cannabidiol, is obtained in high yield, while Δ^8-3,4-*cis*-iso-THC (43) is not observed. It seems that under relatively mild acidic conditions cyclization occurs likewise predominantly at $C_{(8)}$. More drastic acidic treatment, however, converts (21) into $\Delta^{4(8)}$-iso-THC (39) (43).

* In the original publication (45) this compound was considered to be the isomeric 1-ethoxy-hexahydrocannabinol (40); this has been shown to be incorrect (47).

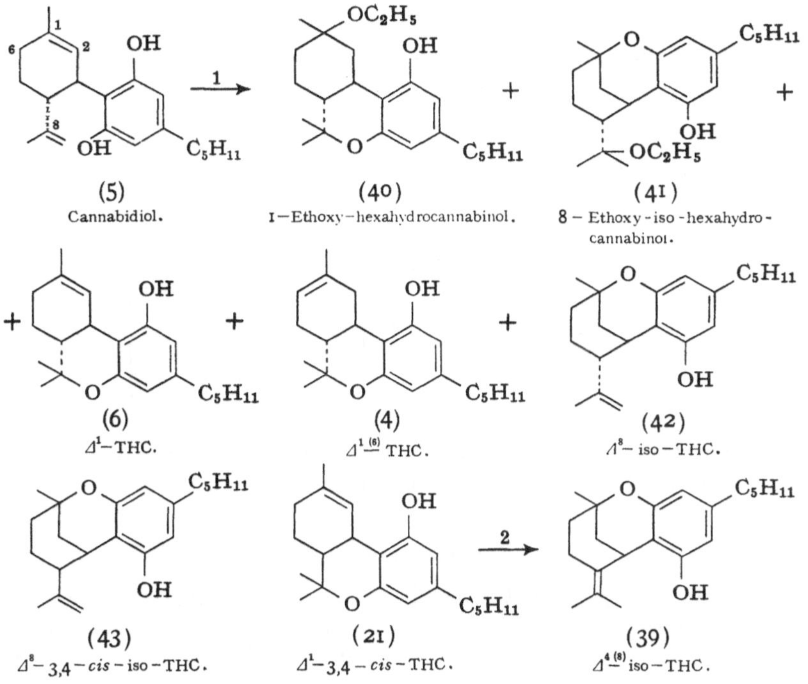

(5)
Cannabidiol.

(40)
I—Ethoxy—hexahydrocannabinol.

(41)
8 — Ethoxy-iso-hexahydro-cannabinol.

(6)
Δ^1—THC.

(4)
$\Delta^{1\,(6)}$ THC.

(42)
Δ^8—iso—THC.

(43)
Δ^8—3,4—cis—iso—THC.

(21)
Δ^1—3,4—cis—THC.

(39)
$\Delta^{4\,(8)}$ iso—THC.

Chart 10.

Reagents: 1) dil. HCl/C_2H_5OH; 2) p-toluene sulfonic acid in benzene.

4. Cyclizations.

Two cyclizations forming C—C bonds have been reported. Cannabigerol (9) on acid treatment gives (22) and (44) (42, 47). The stereochemistry of (44) is *trans*.

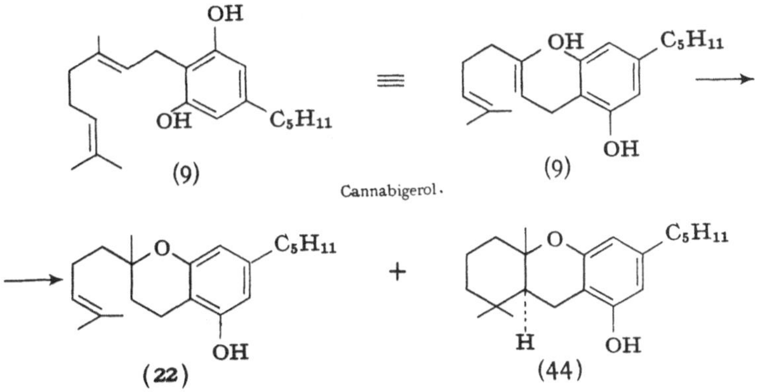

(9)

(9)
Cannabigerol.

(22)

(44)

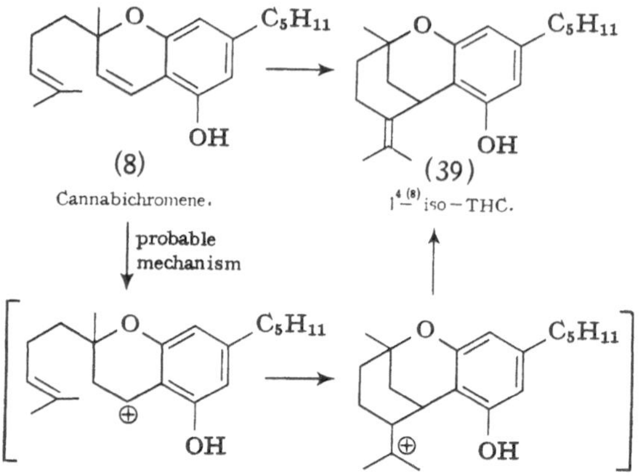

(8)   (39)
Cannabichromene.   1⁴⁽⁸⁾iso−THC.

Cannabichromene (8) on boiling with *p*-toluene sulfonic acid in benzene
gives (39) (*43*) probably by the mechanism just indicated.

5. Aromatic Substitution.

Δ^1-THC (6) reacts with 3,5-dinitrophenyl isocyanate to give the
expected urethane (45) accompanied by the amide (45a), m. p. 142–4°
(*44, 47*). The structure of (45 a) has been deduced on the basis of its NMR
spectrum, elemental analysis and infrared spectrum. The formation of
(45 a) is unusual and seems to be the first case of a comparable reaction
of a phenol.

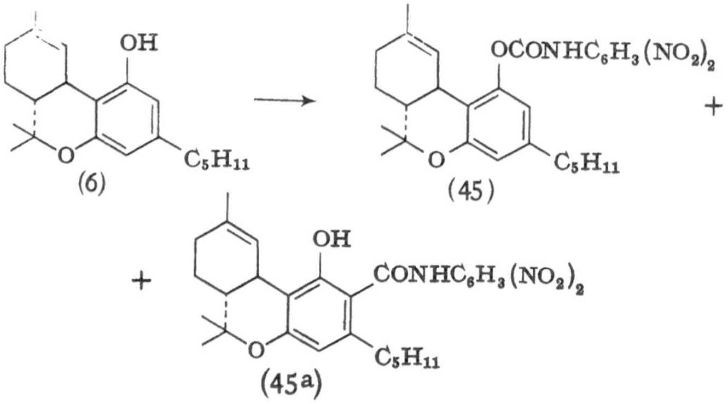

Reagent: $3,5\text{-}(NO_2)_2C_6H_3\text{—NCO}$.

6. Aromatization.

Both Δ^1- and $\Delta^{1(6)}$-THC give cannabinol (2) on dehydrogenation in
almost quantitative yield (*11, 44*). $\Delta^{4(8)}$-iso-THC (39) likewise gives

References, pp. 208—213.

cannabinol (2) although in lower yields (43). While unexpected, this reaction is, on mechanistic grounds, not exceptional.

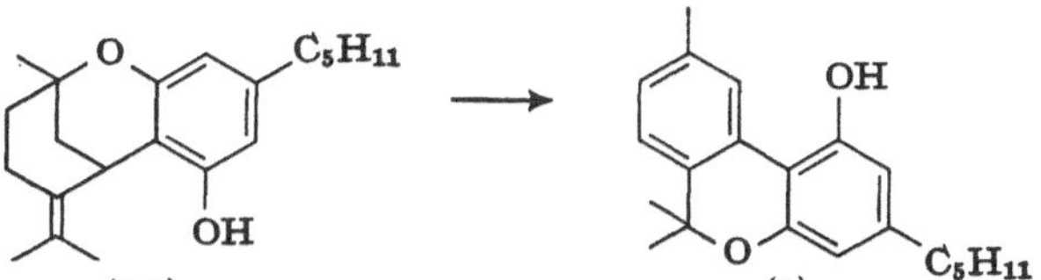

(39)
$\varDelta^{1-(8)}$ iso $-$ THC.

(2)
Cannabinol.

Reagent: S.

X. Biogenesis.

In 1942 TODD (90, 95) suggested that the cannabinoids originated in the plant from a condensation of a terpene derivative (a hypothetical menthatriene) with olivetol. The initial cannabidiol-type compound could then cyclize to a THC and to cannabinol. The isolation of cannabidiolic acid has necessitated a modification of this suggestion. SCHULTZ (87) has brought evidence that cannabidiol is formed from cannabidiolic acid.

Recent advances in the elucidation of the biogenesis of terpenes and steroids and the isolation of numerous new cannabinoids have made possible the presentation of TODD's scheme in modern terms (42, 46, 77, 97) (Chart 11). Condensation of geranyl pyrophosphate (46) with olivetol (15) or olivetolic acid (47) (or an open chain precursor of these compounds) will lead to cannabigerol (9) or cannabigerolic acid (11). A direct cyclization of cannabigerol to a cannabidiol-type compound is not possible as cannabidiol is in a stage of oxidation higher than that of cannabigerol. If it is assumed that cannabigerol undergoes oxidation at $C_{(8)}$, a position which is both allylic and benzylic and hence easily oxidizable, the postulated compound obtained, 8-hydroxycannabigerol (48), can cyclize to either cannabidiol (5) or cannabichromene (8). The cyclization to the former can take place only if the configuration of the double bond is inverted. This is possible by allylic rearrangement, followed by cyclization in which the double bond reverts to its original position. The cyclization of cannabidiol to $\varDelta^1$-THC (6) and dehydrogenation of the latter to cannabinol (2) are straightforward. It is indeed possible that the dehydrogenation reaction is not necessarily an enzymatic process, for an increase in the amount of cannabinol in hashish has been reported on prolonged storage (72).

The presence of two parallel lines of compounds in hashish, one derived from olivetol and the other from olivetolcarboxylic acid poses a problem of biogenesis. With the data on hand it is impossible to

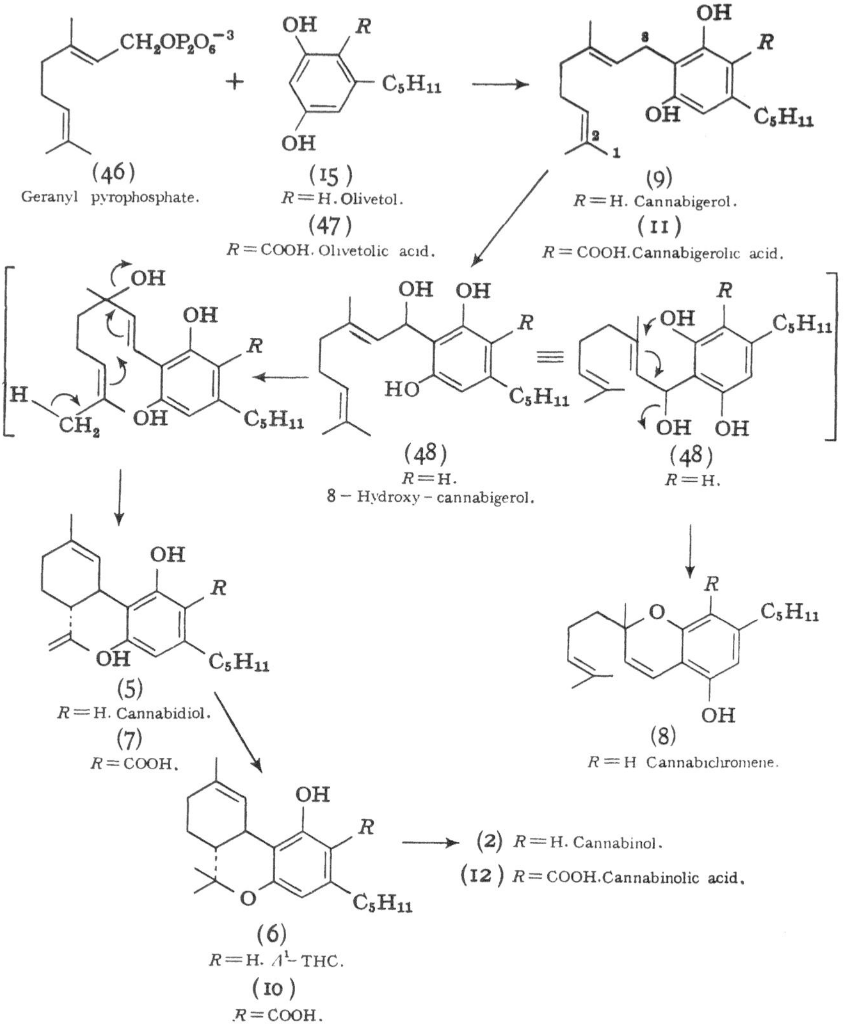

Chart 11. Assumed Cannabinoid Biogenesis.

decide whether cannabigerol and cannabigerolic acid are formed independently, each being the starting point of a chain of compounds, or whether the olivetolic acid chain is the only one and the other compounds are formed by decarboxylation in the plant. No experimental biosynthetic work in this field has as yet been published.

Some inconclusive preliminary results with labelled compounds appear in a Thesis (*81*).

References, pp. 208—213.

XI. Tables.

Table 1. Biological Activity of the Natural Cannabinoids.

Compound	Activity	References
Cannabinol (2).................	no specific activity	—
Cannabidiol (5)	antibiotic (gram + bacteria)	(60, 89)
Cannabigerol (9)	antibiotic (gram + bacteria)	(77)
Δ^1-Tetrahydrocannabinol (6)	ataxia (dog), psychotomimetic (humans, effective dose 3—5 mg) analgetic (mice, rabbits)	(16a, 44, 47)
$\Delta^{1(6)}$-Tetrahydrocannabinol (4) ...	ataxia (dog), psychotomimetic (humans) analgetic (mice, rabbits)	(1, 16a, 73)
Cannabichromene (8)	sedation and ataxia (dog)*	(42)
Cannabicyclol (24)	unknown	—
Cannabidiolic acid (7)	sedative, antibiotic (gram + bacteria)	(60, 88)
Cannabigerolic acid (11)	antibiotic (gram + bacteria)	(77)
Cannabinolic acid (12).........	unknown	—
Δ^1-THC acid (13)	unknown	—

* Cannabichromene does not possess psychotomimetic activity in humans (KORTE, private communication).

Table 2. Cannabinoids, Some Physical Properties and Derivatives.

Constituent	UV Spectrum[a]	NMR Spectrum[b]	Solid derivatives	References[c]
Cannabigerol (9) $C_{21}H_{32}O_2$ m. p. 51-3°	272 (1,100) 280 (1,050)	0.95 (aliphatic CH₃); 1.60, 1.69, 1.82 (olefinic CH₃); 3.35 (2) (d, J = 7.5 cps) (C₈—H) 4.90–5.35 (2) (m) (olefinic H); 6.18 (2) (s) (aromatic H) (in CCl₄)		(46, 77)
Cannabidiol (5) $C_{21}H_{30}O_2$ m. p. 66-7° [α]D — 125°	212 (37,150) 273 (1,100) 280 (1,050)	0.88 (aliphatic CH₃); 1.68, 1.80 (olefinic CH₃); 3.81 (1) (d, br) (C₃—H); 4.58, 4.66 (2) (>C=CH₂); 5.59 (1) (s, br) (C₂—H); 6.22 (2) (s) (aromatic H). (in CDCl₃)	bis-3,5-dinitrobenzoate m. p. 106-7° (7) ditosylate m. p. 81-3° (76)	[α]D, m. p. (12) UV (64) NMR (79)
Δ¹-Tetrahydrocannabinol (6) $C_{21}H_{30}O_2$ [α]D — 150°	277 (1,640)[e] 282 (1,550)	0.88 (aliphatic CH₃); 1.08, 1.38 (CH₃ α to O); 1.68 (olefinic CH₃); 3.14 (1) (d, br) (C₃—H); 6.35 (1) (s, br) (C₂—H); 6.00 (1) (d, J = 2 cps), 6.18 (1) (d, J = 2 cps) (aromatic H) (in CCl₄)	3,5-dinitrophenylurethane, m. p. 115-6°	(44)
Δ¹⁽⁶⁾-Tetrahydrocannabinol (4) $C_{21}H_{30}O_2$ [α]D — 265°	275 (1,260) 282 (1,320)	0.88 (aliphatic CH₃); 1.08, 1.32 (CH₃ α to O); 1.68 (olefinic CH₃); 3.18 (1) (d, br) (C₃—H); 5.35 (1) (s, br) (C₆—H); 5.90 (1) (d, J = 2 cps), 6.13 (1) (d, J = 2 cps) (aromatic H). (in CCl₄)	unknown	[α]D (5) UV, NMR (45)
Cannabichromene[d] (8) $C_{21}H_{30}O_2$ m. p. 144-6°[d] (?)	228 (25,100) 280 (8,900)	0.87 (aliphatic CH₃); 1.32 (CH₃ α to O); 1.58, 1.62 (olefinic CH₃); 5.05 (1) (t, br) (olefinic H); 5.44, 6.60 (2) (AB quartet, J_AB = 10 cps) (olefinic H); 5.97, 6.15 (2) (s) (aromatic H). (in CCl₄)	3,5-dinitrophenylurethane, m. p. 106-7°	UV, NMR derivative (12) m. p. (33)[d]

Compound	UV[a]	NMR[b]	Derivatives	Ref[c]
Cannabinol (2) $C_{21}H_{26}O_2$ m. p. 75–6°	220 (35,600) 285 (18,000)	0.88 (aliphatic CH_3); 1.58 (6) (s) (CH_3 α to O); 2.35 (3) (s) (aromatic CH_3); 6.25 (1) (d, J = 2 cps), 6.40 (1) (d, J = 2 cps), 7.10 (2) (s), 8.20 (1) (s) (aromatic H). (in CCl_4)	m. p. (12) UV (64) NMR (77)	
Cannabigerolic acid (11) as methyl ester	223 (22,650) 271 (12,580) 302 (4,080)	0.91 (aliphatic CH_3); 1.55, 1.66, 1.80 (olefinic CH_3); 3.37 (2) (d, J = 7 cps) (C_8—H); 3.92 (3) (s) (—$COOCH_3$); 4.8–5.2 (2) (m) (olefinic H); 6.15 (1) (s) (aromatic H). (in CCl_4)	cannabigerolic acid methyl ester, *bis*-3,5-dinitrobenzoate, m. p. 110–2°	(77)
Cannabidiolic acid (7) as methyl ester	224 (23,100) 271 (12,000) 304 (4,530)	0.88 (aliphatic CH_3); 1.70, 1.82 (olefinic CH_3); 3.88 (3) (s) (—$COOCH_3$); 4.00–4.30 (1) (br) (C_3—H); 4.45, 4.55 (2) (s) (>C=CH_2); 5.60 (1) (s) (C_2—H); 6.25 (1) (s) (aromatic H). (in $CDCl_3$)	cannabidiolic acid di-acetate, m. p. 126–8° (68), cannabidioloic acid methyl ester, mono-3,5-dinitrobenzoate, m. p. 173–5°	(77)
Cannabinolic acid (12) as methyl ester m. p. 86–7°	220 (sh) (28,300) 266 (41,300) 295 (sh) (11,200) 326 (6,100)	0.88 (aliphatic CH_3); 1.50 (6) (s) (CH_3 α to O); 2.32 (3) (s) (aromatic CH_3); 3.85 (3) (s) (—$COOCH_3$); 6.22 (1) (s), 7.00 (2) (s), 8.50 (1) (s) (aromatic H). (in CCl_4)		(77)
Δ¹-Tetrahydrocannabinolic acid (13) [α]D — 206.8°	223 (1,795) 278 (1,426) 308 (4,642) (in cyclohexane)	6.14 (1) (s) (aromatic H); 6.46 (1) (s, br) (C_2—H)		(62)
Cannabicyclol (24) m. p. 152–3°	275 (1,240) 282 (1,270)	0.80, 0.90 (aliphatic CH_3); 1.38 (aliphatic and α to O CH_3); 3.12 (1) (d, br) (C_3—H); 6.18 (1) (d, J = 2 cps), 6.33 (1) (d, J = 2 cps) (aromatic H). (in $CDCl_3$)		(47)

a In ethanol, unless otherwise stated. Values given in mμ (ε).

b Values given in p. p. m. relative to $(CH_3)_4$ Si as internal standard. Number in parentheses denotes number of protons, determined by integration of areas. Letters in parentheses denote singlet (s); doublet (d); triplet (t); quartet (q); broad (br); and multiplet (m).

c The references indicate the source of the data and are not intended to assign priorities.

d Two [α]D have been reported: [α]D + 3.4° (33) and [α]D — 9° (42). A purified sample (47) has shown [α]D — 0.9°. It is possible that cannabichromene is a racemate. The m. p. reported (33) has not been confirmed and may be due to an impurity.

e Revised data, slightly different from those published in (44).

XII. Spectral Curves.

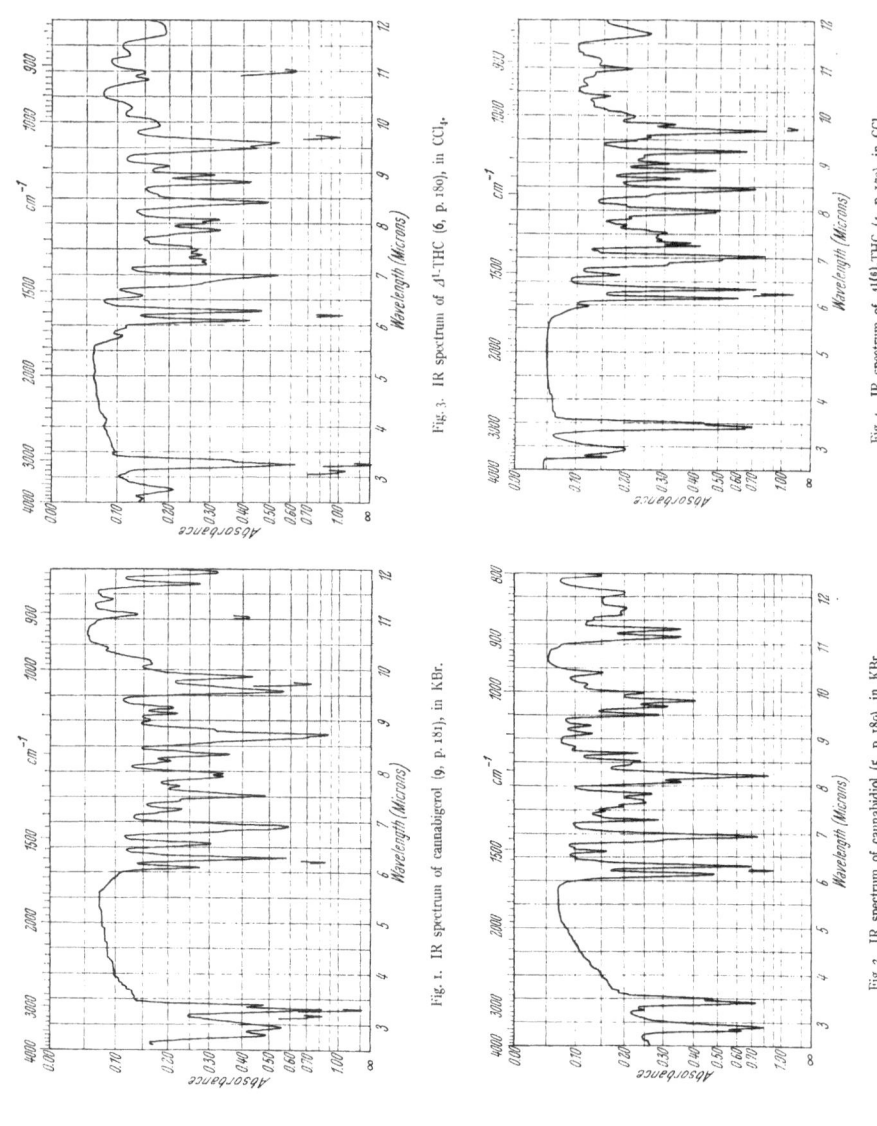

Fig. 3. IR spectrum of Δ¹-THC (6, p. 180), in CCl₄.

Fig. 4. IR spectrum of Δ¹⁽⁶⁾-THC (4, p. 170), in CCl₄.

Fig. 1. IR spectrum of cannabigerol (9, p. 181), in KBr.

Fig. 2. IR spectrum of cannabidiol (5, p. 180), in KBr.

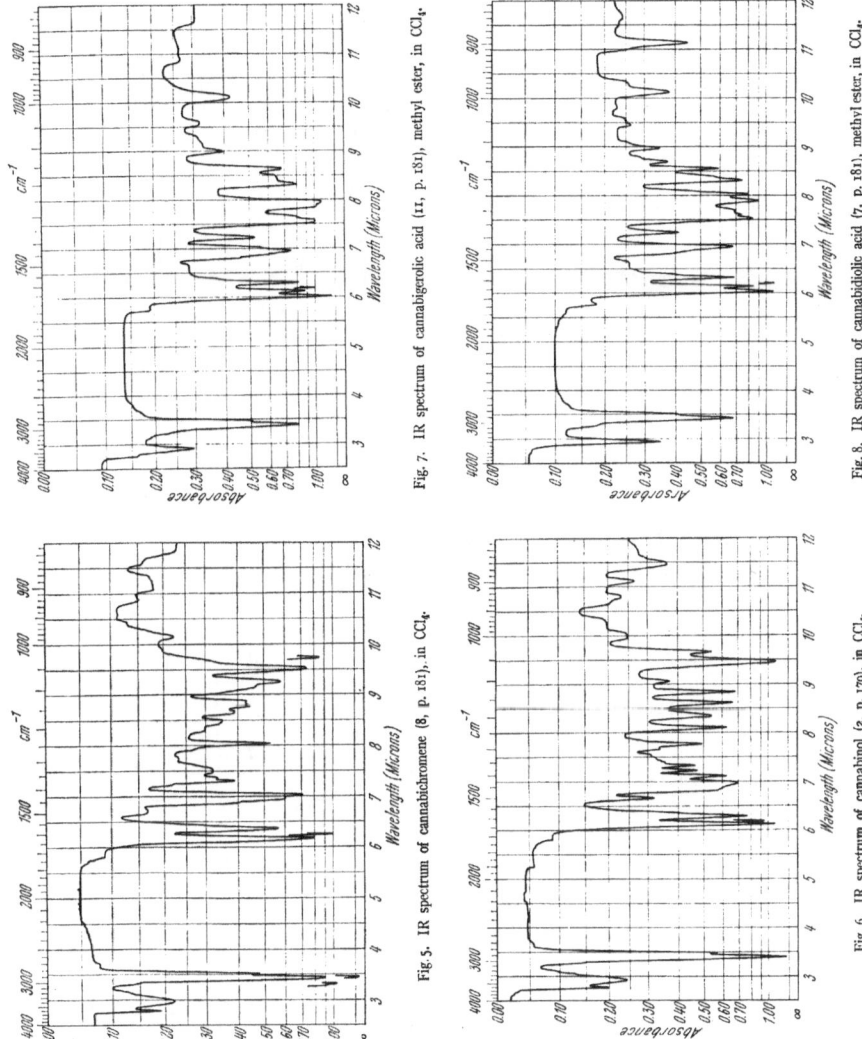

Fig. 7. IR spectrum of cannabigerolic acid (11, p. 181), methyl ester, in CCl₄.

Fig. 8. IR spectrum of cannabidiolic acid (7, p. 181), methyl ester, in CCl₄.

Fig. 5. IR spectrum of cannabichromene (8, p. 181), in CCl₄.

Fig. 6. IR spectrum of cannabinol (2, p. 179), in CCl₄.

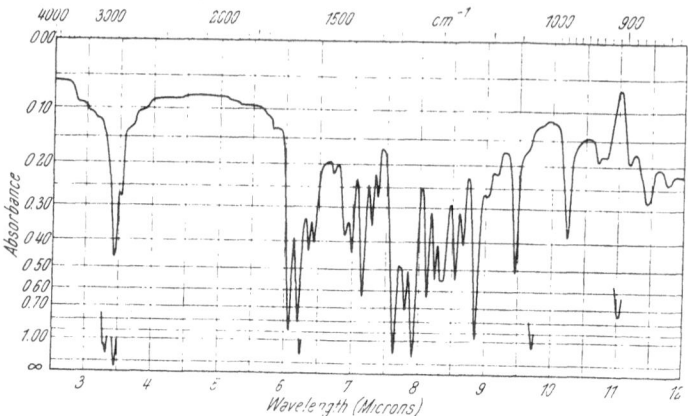

Fig. 9. IR spectrum of cannabinolic acid (12, p. 181), methyl ester, in CCl$_4$.

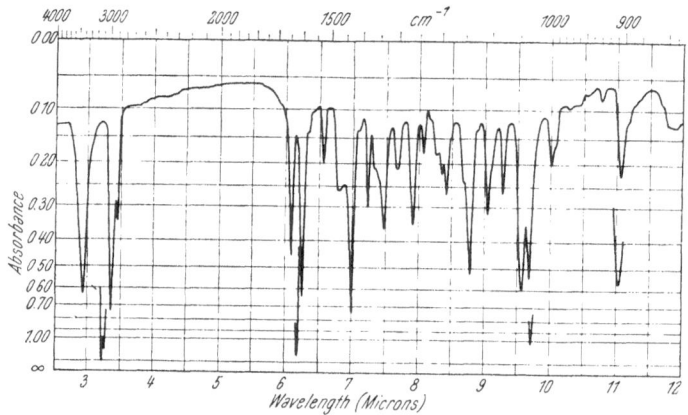

Fig. 10. IR spectrum of cannabicyclol (24, p. 187), in KBr.

References.

1. ADAMS, R.: Marihuana. Harvey Lect. 37, 168 (1941–1942).
2. ADAMS, R. and B. R. BAKER: Structure of Cannabidiol. VII. A Method of Synthesis of a Tetrahydrocannabinol which Possesses Marihuana Activity. J. Amer. Chem. Soc. 62, 2405 (1940).
3. ADAMS, R., B. R. BAKER and R. B. WEARN: Structure of Cannabinol. III. Synthesis of Cannabinol, 1-Hydroxy-3-n-amyl-6,6,9-trimethyl-6-dibenzopyran. J. Amer. Chem. Soc. 62, 2204 (1940).
4. ADAMS, R. and T. E. BOCKSTAHLER: Preparation and Reactions of o-Hydroxy-cinnamic Acids and Esters. J. Amer. Chem. Soc. 74, 5346 (1952).
5. ADAMS, R., C. K. CAIN, W. D. McPHEE and R. B. WEARN: Structure of Cannabidiol. XII. Isomerization to Tetrahydrocannabinols. J. Amer. Chem. Soc. 63, 2209 (1941).

6. ADAMS, R. and R. B. CARLIN: The Addition of Dienes to certain Di-*o*-methoxy-substituted Cinnamic Acids. II. J. Amer. Chem. Soc. **65**, 360 (1943).

7. ADAMS, R., M. HUNT and J. H. CLARK: Structure of Cannabidiol, a Product Isolated from the Marihuana Extract of Minnesota Wild Hemp. I. J. Amer. Chem. Soc. **62**, 196 (1940).

8. — — — Structure of Cannabidiol. III. Reduction and Cleavage. J. Amer. Chem. Soc. **62**, 735 (1940).

9. ADAMS, R., S. MacKENZIE, Jr. and S. LOEWE: Tetrahydrocannabinol Homologs with Doubly Branched Alkyl Groups in the 3-Position. XVIII. J. Amer. Chem. Soc. **70**, 664 (1948).

10. ADAMS, R., W. D. McPHEE, R. B. CARLIN and Z. W. WICKS: The Addition of Dienes to Coumarin and to Certain Substituted Cinnamic Acids. I. J. Amer. Chem. Soc. **65**, 356 (1943).

11. ADAMS, R., D. C. PEASE, C. K. CAIN and J. H. CLARK: Structure of Cannabidiol. VI. Isomerization of Cannabidiol to Tetrahydrocannabinol, a Physiologically Active Product. Conversion of Cannabidiol to Cannabinol. J. Amer. Chem. Soc. **62**, 2402 (1940).

12. ADAMS, R., D. C. PEASE and J. H. CLARK: Isolation of Cannabinol, Cannabidiol and Quebrachitol from Red Oil of Minnesota Wild Hemp. J. Amer. Chem. Soc. **62**, 2194 (1940).

13. ADAMS, R., C. M. SMITH and S. LOEWE: Tetrahydrocannabinol Homologues and Analogues with Marihuana Activity. X. J. Amer. Chem. Soc. **63**, 1973 (1941).

14. Anonymous: Bibliography, *Cannabis sativa*. Bull. Narcotics **3**, No. 1, 59; No. 2, 42 (1951). A yearly supplement is published in the same journal.

15. Anonymous: The Question of Cannabis. Cannabis Bibliography. U. N. Econ. Soc. Council, Comm. Narcotic Drugs Document E-CN 7-479, 1965.

16. BERGEL, F. und K. VÖGELE: Synthese des Cannabino-lactons und isomerer Verbindungen. Zur Konstitution des Cannabinols. II. Liebigs Ann. Chem. **493**, 250 (1932).

16a. BICHER, H. I., J. KRUPNIK and S. ELIASH: Pharmacological Properties of $\Delta^{1(6)}$-Tetrahydrocannabinol. Proc. Intern. Congr. Pharmacol. Brazil, 1966, p. 114, and private commun.

17. BLATT, A. H.: A Critical Survey of the Literature Dealing with the Chemical Constituents of *Cannabis sativa*. J. Washington Acad. Sci. **28**, 465 (1938).

18. BOHLIG, J. F.: *Cannabis sativa* und *Urtica dioica* chemisch analysiert. Jahrb. prakt. Pharm. **1840**, 1.

19. BOUQUET, R. J.: Cannabis. Bull. Narcotics **2**, No. 4, 14 (1950); **3**, No. 3, 22 (1951).

20. BOYD, E. S. and D. A. MERITT: Effects of a Tetrahydrocannabinol Derivative on some Motor Systems in the Cat. Arch. Int. Pharmacodyn. Therapeut. **153**, 1 (1965).

21. — — Effects of Barbiturates and a Tetrahydrocannabinol Derivative on Recovery Cycles of Medial Lemniscus, Thalamus and Reticular Formation in the Cat. J. Pharmacol. Exp. Therapeut. **151**, 376 (1966).

22. — — Effect of Thiopental and a Tetrahydrocannabinol Derivative on Arousal and Recruiting in the Cat. J. Pharmacol. Exp. Therapeut. **149**, 138 (1965).

23. BUDZIKIEWICZ, H., R. T. APLIN, D. A. LIGHTNER, C. DJERASSI, R. MECHOULAM und Y. GAONI: Massenspektroscopische Untersuchung der Inhaltsstoffe von Haschisch. Tetrahedron **21**, 1881 (1965).

24. CAHN, R. S.: *Cannabis indica* Resin. I. The Constitution of Nitrocannabinolactone (Oxycannabin). J. Chem. Soc. (London) **1930**, 986.

25. Cahn, R. S.: Cannabis indica Resin. II. J. Chem. Soc. (London) 1931, 630.
26. — Cannabis indica Resin. III. The Constitution of Cannabinol. J. Chem. Soc. (London) 1932, 1342.
27. — Cannabis indica Resin. IV. The Synthesis of some 2 : 2-Dimethyldibenzo-pyrans, and Confirmation of the Structure of Cannabinol. J. Chem. Soc. (London) 1933, 1400.
28. Carlini, E. A. and C. Kramer: Effects of Cannabis sativa (Marihuana) on Maze Performance of the Rat. Psychopharmacol. 7, 175 (1965).
29. Chopra, I. C. and R. N. Chopra: The Use of Cannabis Drugs in India. Bull. Narcotics 9, No. 1, 4 (1957).
30. Claussen, U., W. Borger und F. Korte: Gaschromatographische Analyse der Inhaltsstoffe des Hanfes. Liebigs Ann. Chem. 693, 158 (1966).
31. Claussen, U., H.-W. Fehlhaber und F. Korte: Massenspektrometrische Bestimmung von Haschisch-Inhaltsstoffen. II. Tetrahedron 22, 3535 (1966).
32. Claussen, U. und F. Korte: Die Struktur zweier synthetischer Isomeren des Tetrahydrocannabinols. Z. Naturforsch. 21 b, 594 (1966).
32 a. — — Massenspektrometrische Bestimmung von Haschisch-Inhaltsstoffen. Tetrahedron, Suppl. No. 7, 89 (1966).
33. Claussen, U., F. v. Spulak und F. Korte: Cannabichromen, ein neuer Haschisch-Inhaltsstoff. Tetrahedron 22, 1477 (1966).
34. Contenau, J.: La divination chez les Assyriens et les Babyloniens, p. 49. Paris: Payot. 1940.
35. Covello, M.: Ricerche chimiche e farmacologiche sulla Cannabis indica coltivata in Italia. II. Degradazione dell'attività biologica della droga in rapporto all'in-vecchiamento e separazione chromatografica delle frazioni attive dagli estratti alcoolico ed etereo. Il farmaco, scienza e tecnica 3, 8 (1948).
36. Davis, T. W. M., C. G. Farmilo and M. Osadchuk: Identification and Origin Determinations of Cannabis by Gas and Paper Chromatography. Analyt. Chemistry 35, 751 (1963).
37. Decourtive, E.: Note sur le Haschisch. C. R. hebd. séances Acad. Sci. 26, 509 (1848).
38. De Ropp, R. S.: Chromatographic Separation of the Phenolic Compounds of Cannabis sativa. J. Amer. Pharmac. Ass., Sci. Ed. 49, 756 (1960).
39. Downing, D. F.: The Chemistry of the Psychotomimetic Substances. Quart. Rev. (Chem. Soc. London) 16, 133 (1962).
40. — Psychotomimetic Compounds. In: M. Gordon (Edit.), Psychopharmacological Agents, p. 555. New York: Academic Press. 1964.
41. Fahrenholtz, K. E., M. Lurie and R. W. Kierstead: Total Synthesis of dl-Δ^9-Tetrahydrocannabinol and of dl-Δ^8-Tetrahydrocannabinol, Racemates of Active Constituents of Marihuana. J. Amer. Chem. Soc. 88, 2079 (1966).
42. Gaoni, Y. and R. Mechoulam: Hashish. VII. Cannabichromene, a New Active Principle in Hashish. Chem. Commun. 1966, 20.
43. — — Hashish. IX. Concerning the Isomerization of Δ^1- to $\Delta^{1(6)}$-Tetrahydro-cannabinol. J. Amer. Chem. Soc. 88, 5673 (1966).
44. — — Hashish. III. The Isolation, Structure and Partial Synthesis of an Active Constituent of Hashish. J. Amer. Chem. Soc. 86, 1646 (1964).
45. — — Hashish. VI. The Isomerization of Cannabidiol to Tetrahydrocanna-binols. Tetrahedron 22, 1481 (1966).
46. — — Hashish. II. The Structure and Synthesis of Cannabigerol, a New Hashish Constituent. Proc. Chem. Soc. (London) 1964, 82.
47. — — unpublished.

48. GHOSH, R., A. R. TODD and S. WILKINSON: Cannabis indica. IV. The Synthesis of Some Tetrahydrodibenzopyran Derivatives. J. Chem. Soc. (London) 1940, 1121.

49. — — — Cannabis indica. V. The Synthesis of Cannabinol. J. Chem. Soc. (London) 1940, 1393.

50. GHOSH, R., A. R. TODD and D. C. WRIGHT: Cannabis indica. VI. The Condensation of Pulegone with Alkyl Resorcinols. A New Synthesis of Cannabinol and of a Product with Hashish Activity. J. Chem. Soc. (London) 1941, 137.

51. GRLIĆ, LJ.: A Comparative Study on Some Chemical and Biological Characteristics of Various Samples of Cannabis Resin. Bull. Narcotics 14, No. 3, 43 (1962).

52. — A Study of Some Chemical Characteristics of the Resin from Experimentally Grown Cannabis of Various Origins. U. N. Secr. Publ. Ser. ST. SOA/SER. S/10 (1964).

53. — Recent Advances in the Chemical Research of Cannabis. Bull. Narcotics 16, No. 4, 29 (1964).

54. GRLIĆ, LJ. and A. ANDREC: The Content of Acid Fraction in Cannabis Resin of Various Age and Provenance. Experientia 17, 325 (1961).

55. HAAGEN-SMIT, A. J., C. Z. WAWRA, J. B. KOEPFLI, G. A. ALLES, G. A. FEIGEN and A. N. PRATER: A Physiologically Active Principle from Cannabis sativa (Marihuana). Science 91, 602 (1940).

56. HEGNAUER, R.: Chemotaxonomie der Pflanzen, Band 3, S. 350. Basel: Birkhäuser Verl. 1964.

57. HIVELY, R. L., W. A. MOSHER and F. W. HOFFMANN: Isolation of trans-Δ^6-Tetrahydrocannabinol from Marijuana. J. Amer. Chem. Soc. 88, 1832 (1966).

58. JACOB, A. and A. R. TODD: Cannabidiol and Cannabol, Constituents of Cannabis indica Resin. Nature 145, 350 (1940).

59. JULIEN, S.: Substance anesthésique employée en Chine dans le commencement du IIIème siècle de notre ère pour paralyser momentanément la sensibilité. C. R. hebd. séances Acad. Sci. 28, 195 (1849).

60. KABELIK, J., Z. KREJČI and F. ŠANTAVÝ: Cannabis as a Medicament. Bull. Narcotics 12, No. 3, 5 (1960).

61. KORTE, F., E. DLUGOSCH und U. CLAUSSEN: Synthese von DL-Cannabidiol und seinem Methylhomologen. Liebigs Ann. Chem. 693, 165 (1966).

62. KORTE, F., M. HAAG and U. CLAUSSEN: Tetrahydrocannabinolcarboxylic Acid, a Component of Hashish. Angew. Chem. (Internat. Ed.) 4, 872 (1965).

63. KORTE, F., E. HACKEL und H. SIEPER: Synthese eines Cannabidioldimethyläthers. Liebigs Ann. Chem. 685, 122 (1965).

64. KORTE, F. und H. SIEPER: Isolierung von Hashish-Inhaltsstoffen aus Cannabis sativa non indica. Liebigs Ann. Chem. 630, 71 (1960).

65. — — Untersuchung von Hashish-Inhaltsstoffen durch Dünnschichtchromatographie. J. Chromatogr. 13, 90 (1964).

66. KORTE, F., H. SIEPER and S. TIRA: New Results on Hashish-specific Constituents. Bull. Narcotics 17, No. 1, 35 (1965).

67. KREJČI, Z., M. HORÁK and F. ŠANTAVÝ: Constitution of the Cannabidiolic Acid and of an Acid of m. p. 133°, Isolated from Cannabis sativa. Acta Univ. Palackianae Olomuc. 16, 9 (1958).

68. — — — Hanf (Cannabis sativa) — antibiotisches Heilmittel. III. Isolierung und Konstitution zweier aus C. sativa gewonnener Säuren. Pharmazie 14, 349 (1959).

69. KREJČI, Z. and F. ŠANTAVÝ: Isolation of Other Substances from the Leaves of the Indian Hemp (Cannabis sativa L. var. indica). Acta Univ. Palackianea Olomuc. 6, 59 (1955).

70. LEAF, G., A. R. TODD and S. WILKINSON: *Cannabis indica.* IX. The Isolation of 3′ : 4′ : 5′ : 6′-Tetrahydrodibenzopyran Derivatives from Pulegone-Orcinol and Pulegone-Olivetol Condensation Products. Synthesis of *d*-Tetrahydrocannabinol. J. Chem. Soc. (London) **1942**, 185.

71. LERNER, M.: Marihuana: Tetrahydrocannabinol and Related Compounds. Science **140**, 175 (1963).

71a. LERNER, M. and J. T. ZEFFERT: Determination of Tetrahydrocannabinol Isomers in Marihuana and Hashish. Bull. Narcotics (in press, 1967).

72. LEVINE, J.: Origin of Cannabinol. J. Amer. Chem. Soc. **66**, 1868 (1944).

73. LOEWE, S.: Cannabiswirkstoffe und Pharmakologie der Cannabinole. Arch. exp. Pathol. Pharmakol. **211**, 175 (1950).

74. Mayor's Committee on Marihuana: The Marihuana Problem in the City of New York, Sociological, Medical, Psychological and Pharmacological Studies. Lancaster, Penna.: Cattel Press. 1944.

75. McGLOTHLIN, W. H.: Hallucinogenic Drugs: A Perspective with Special Reference to Peyote and Cannabis. Santa Monica, Calif.: Rand Corp. 1964.

75a. MECHOULAM, R., P. BRAUN and Y. GAONI: A Stereospecific Synthesis of (—)-Δ¹- and (—)-Δ¹⁽⁶⁾-Tetrahydrocannabinols. J. Amer. Chem. Soc. (in press, 1967).

76. MECHOULAM, R. and Y. GAONI: Hashish. V. A Total Synthesis of *dl*-Δ¹-Tetrahydrocannabinol, the Active Component of Hashish. J. Amer. Chem. Soc. **87**, 3273 (1965).

77. — — Hashish. IV. The Isolation and Structure of Cannabinolic, Cannabidiolic and Cannabigerolic Acids. Tetrahedron **21**, 1223 (1965).

78. — — Hashish. X. The Absolute Configuration of Δ¹-Tetrahydrocannabinol, the Major Active Constituent of Hashish. Tetrahedron Letters **1967**, 1109.

79. MECHOULAM, R. and Y. SHVO: Hashish. I. The Structure of Cannabidiol. Tetrahedron **19**, 2073 (1963).

80. MURPHY, H. B. M.: The Cannabis Habit, a Review of Recent Psychiatric Literature. Bull. Narcotics **15**, No. 1, 15 (1963).

81. NI, R. R.-J.: Studies on the Biosynthesis of Cannabinol and Cannabidiol in *Cannabis sativa.* Dissert. Abstr. **25**, 1576 (1964).

82. NIGAM, M. C., K. L. HANDA, I. C. NIGAM and L. LEVI: Essential Oils and Their Constituents. XXIX. The Essential Oil of Marihuana: Composition of Genuine Indian *Cannabis sativa* L. Canad. J. Chem. **43**, 3372 (1965).

82a. OBATA, Y. and Y. ISHIKAWA: Isolation of a Gibbs-positive Compound from Japanese Hemp. Agr. Biol. Chem. (Japan) **30**, 619 (1966).

82b. PARS, H. G., F. E. GRANCHELLI, J. K. KELLER and R. K. RAZDAN: Physiologically Active Nitrogen Analogs of Tetrahydrocannabinols. Tetrahydrobenzopyrano[3,4-d]pyridines. J. Amer. Chem. Soc. **88**, 3664 (1966).

82c. PETRZILKA, T., W. HAEFLIGER, C. SIKEMEIER, G. OHLOFF und A. ESCHENMOSER: Synthese und Chiralität des (—)-Cannabidiols. Helv. Chim. Acta **50**, 719 (1967).

83. POWELL, G., M. R. SALMON, T. H. BEMBRY and R. P. WALTON: The Active Principle of Marihuana. Science **93**, 522 (1941).

84. SALEMINK, C. A., E. VEEN und W. A. DE KLOET: Über die basischen Inhaltsstoffe von *Cannabis sativa.* Planta Med. **13**, 211 (1965).

84a. SALUSTIANO, J., K. HOSHINO and E. A. CARLINI: Effects of *Cannabis sativa* and Chlorpromazine on Mice as Measured by Two Methods Used for Evaluation of Tranqquilizing Agents. Med. Pharmacol. Exp. **15**, 153 (1966).

85. ŠANTAVÝ, F.: Notes on the Structures of Cannabidiol Compounds. Acta Univ. Palackianae Olomuc. **35**, 5 (1964).

86. SCHULTZ, O.-E.: Der gegenwärtige Stand der Cannabis-Forschung. Planta Medica 12, 371 (1964).
87. SCHULTZ, O.-E. und G. HAFFNER: Zur Frage der Biosynthese der Cannabinole. Arch. Pharmaz. 293, 1 (1960).
88. — — Zur Kenntnis eines sedativen Wirkstoffes aus deutschem Faserhanf (Cannabis sativa). I. Arch. Pharm. 291, 391 (1958).
89. — — Zur Kenntnis eines sedativen und antibakteriellen Wirkstoffes aus dem deutschen Faserhanf (Cannabis sativa). Z. Naturforsch. 14b, 98 (1959).
90. SIMONSEN, J. and A. R. TODD: The Essential Oil from Egyptian Hashish. J. Chem. Soc. (London) 1942, 188.
91. SMITH, T. and H. SMITH: On the Resin of Indian Hemp. Pharmac. J. and Trans. 6, 127 (1847).
92. — — Process for Preparing Cannabine or Hemp Resin. Pharmac. J. and Trans. 6, 171 (1847).
93. TAYLOR, E. C., K. LENARD and Y. SHVO: Active Constituents of Hashish. Synthesis of dl-Δ^6-3,4-trans-Tetrahydrocannabinol. J. Amer. Chem. Soc. 88, 367 (1966).
94. TAYLOR, E. C. and E. J. STROJNY: The Synthesis of Some Model Compounds Related to Tetrahydrocannabinol. J. Amer. Chem. Soc. 82, 5198 (1960).
95. TODD, A. R.: Hashish. Experientia 2, 55 (1946).
96. WATT, J. M. and M. G. BREYER-BRANDWIJK: The Medicinal and Poisonous Plants of Southern and Eastern Africa. Edinburgh and London: Livingstone Ltd. 1962.
97. WHALLEY, W. B.: Some Structural and Biogenetic Relationships of Plant Phenolics. In: W. D. Ollis (Edit.), Chemistry of Natural Phenolic Compounds, p. 38. New York: Pergamon Press. 1961.
98. WOLLNER, H. J., J. R. MATCHETT, J. LEVINE and S. LOEWE: Isolation of a Physiologically Active Tetrahydrocannabinol from Cannabis sativa Resin. J. Amer. Chem. Soc. 64, 26 (1942).
99. WOLSTENHOLME, G. E. W. and J. KNIGHT (Edit.): Hashish: Its Chemistry and Pharmacology. Ciba Found. Study Group No. 21. London: Churchill. 1965.
100. WOOD, T. B., W. T. N. SPIVEY and T. H. EASTERFIELD: Cannabinol. I. J. Chem. Soc. (London) 75, 20 (1899).
101. — — — Charas, the Resin of Indian Hemp. J. Chem. Soc. (London) 69, 539 (1896).
102. ZUBYK, W. J. and A. Z. CONNER: Analysis of Terpene Hydrocarbons and Related Compounds by Gas Chromatography. Analyt. Chemistry 32, 912 (1960)

(Received, September 26, 1966.)

The Toxic Peptides of *Amanita Phalloides*.

By THEODOR WIELAND, Frankfurt am Main.

With 9 Figures.

Contents.

I. Introduction.

Fatal mushroom poisoning is almost exclusively attributable to members of the genus *Amanita*. In the United States, it is *Amanita verna*, the so called "destroying angel", which along with another poisonous species, *A. tenuifolia*, appears to play the major part in lethal mushroom poisoning (*1*, *2*). Quite recently we were able to demonstrate the presence of the two main amanita toxins in a European sample of *A. verna* (*73*). Besides these mushrooms some relatively rare *Galerina* species also contain the amanita toxins (*42*). In Central Europe the predominant culprit is the greenish *Amanita phalloides (Fig. 1)*. This mushroom, also known as the "deadly agaric", is frequently confused with the delicious field mushroom *Agaricus campestris* or with the yellow

References, pp. 246—250.

Amanita mappa (*citrina*), which contains no toxic peptides, but the relatively untoxic bufotenine (5-hydroxy-N-dimethyl-tryptamine) (*67*, *43*), a base occurring in toads and recently isolated also from different plants, among them *Piptadenia peregrina* (*12*), *P. macrocarpa* and *P. excelsa* (*20*),

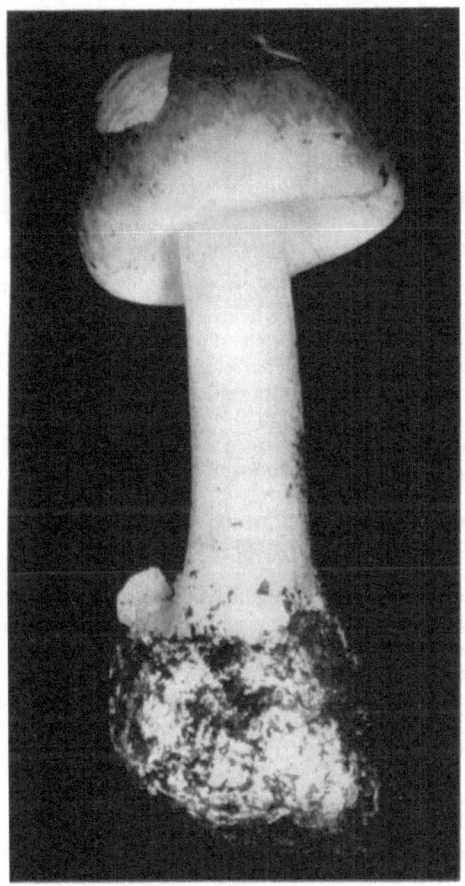

Fig. 1. *Amanita phalloides*, young mushroom (height about 8 cm).

Photo B. Haccius.

Desmodium pulchellum (*19*) or from Epena, a drug from a woody South American Leguminosa (*28*).

The poisonous amanitae are indistinguishable in taste from the edible mushrooms. The latent period of several hours from the time of consumption until the first appearance of toxic effects also markedly increases the danger, because during this time the toxins are completely absorbed from the gastrointestinal tract. Their action is primarily on the liver, and death occurs in hepatic coma. We do not yet know any specific therapeutic agent. Therefore, the understanding of the chemical structures and of the mechanism of action of the toxins is of great importance from the therapeutic standpoint. Over and above the specific toxicologic aspects, however, their action also is of much interest from the point of view of the structure and the function of the liver.

II. Isolation of the Toxic Ingredients.

Attempts to obtain the lethal poison of *A. phalloides* in a pure state started at the beginning of the last century. The following summary, taken from RAAB (*36*) up to 1932, gives the historically interested reader the chronological sequence of the major developments.

1793–1808 PAULET (*32*): The toxin of *A. phalloides* is soluble in water and alcohol and is not destroyed by drying or by heating in boiling water. Animal experiments with variously treated extracts from the mushroom.

1811 VAUQUELIN (*48*): The toxic substance is a "fatty matter".

1826 LETELLIER (*26*): The amanita poison is identical with that of the toadstool and has the properties of an alkaloid ("amanitine").

1866 BOUDIER (*5*): First systematic study of botanically well determined mushrooms, establishing that the toadstool poison is definitely different from that of amanita.

1877 ORÉ (*30*, *31*): The toxic principle (called "phalloidin") is said to be very similar to, or even identical with, strychnine.

1891 KOBERT (*22*): Freshly expressed juices from *A. phalloides* possess strong hemolytic activity. The unknown hemolysin is called "phallin". It is highly sensitive to heat, alkali and acids and is insoluble in organic solvents. It is said to be the active principle in human poisoning.

1893 SEIBERT (*41*): Mushrooms definitely toxic for various animals may lack hemolysin which, accordingly, is not the active principle in the typical poisoning.

1909 KOBERT (*23*): Hemolysin is present in *A. phalloides* in greatly varying amounts and may even be absent. As many as three toxins are supposed to occur in *A. phalloides*.

1906–1913 FORD et al. (*16*): *A. phalloides* contains two toxic substances: a thermolabile hemolysin identical with KOBERT's phallin, said to be a sulfur-containing glucoside, and another toxin, not an alkaloid, called "amanitatoxin". Neither can be obtained in pure form.

1932 RAAB (*36*): Introduction of the guinea pig test. Approximately 100-fold concentration of "amanitatoxin", principally by precipitation with heavy metal salts. Impurities removed by precipitation with lead acetate. Poisoned animals lose weight rapidly.

1934–1935 RENZ (*40*), RAAB and RENZ (*37*), KIMMIG (*21*): Use of chromatographic adsorption. The toxin is difficult to dialyze and can be extracted from water with butanol.

1937 LYNEN and U. WIELAND (*27*): The poison is precipitable from aqueous solution by ammonium sulfate. It consists of two components. The more rapidly acting one is obtained in crystalline form and designated as phalloidin, and data are given on its chemical nature. The other component, slower acting but more toxic, was brought to 50% purity. Indications for a third component are reported.

References, pp. 246—250.

1941 H. WIELAND and HALLERMAYER (*50*): The highly potent component is obtained in crystalline form and called amanitin.

When work was resumed following World War II, new methods of fractionation such as countercurrent distribution, paper chromatography and paper electrophoresis were available. Hence it was soon possible to demonstrate that amanitin consists of two crystalline components, a neutral α- and an acid β-component. The β-substance also proved to be toxic. On paper electrophoresis it travels to the anode (*86*). Subsequently, one of numerous combinations of solvents tried (the upper phase of a system of 20 vol. of methylethyl ketone, 2 vol. of acetone and 5 vol. of water) allowed paper-chromatographic separation of the phytotoxins and thus rapid analysis of mushroom juices (*74*). The toxins give, with cinnamic aldehyde, in an atmosphere rich in hydrochloric acid vapors, blue and bluish-purple colors. Still another compound, γ-amanitin, present in very small amounts, was detected with the aid of this procedure; it was recently isolated and also proved to be toxic (*58*). This new compound persistently accompanies phalloidin in regular paper chromatograms; its separation and micropreparation became first possible by impregnating the paper with borate buffer. Later on, a chromatographic procedure on silica gel and Sephadex led to a crystalline preparation (*82*) in a more convenient manner. Furthermore, a fifth toxin, phalloin, was crystallized from *A. phalloides*; in its rapid action it was comparable with phalloidin (*65*).

When investigating the contents of the mother liquors additional peptidic components could be discovered, which form a minor part of the complex mixture and appear in the paper chromatogram only after previous concentration and purification. Five of them have been isolated in pure state and characterized up till now: Phallisin (*17*) and phallin B (*80*), rapidly acting toxins of the phalloin-type; δ-amanitin belonging chemically and in its slow action to the amanitin group (*3*); amanin (*17*, *3*) which gives a blue color with cinnamic aldehyde/HCl like phalloidin but acts as slowly as the amanitins; finally, a substance of the chemical behavior of amanitin but showing no toxicity and which therefore was named amanullin (*6*). The variety of components is demonstrated in the schematic paper chromatogram in *Figure 2*: In the interest of clarity, the substances are represented separately in two main groups, the phalloins (left) and the amanitins (right) with amanin and amanullin between them.

As to the amounts of the individual toxins contained in amanita, only tentative conclusions can be drawn from the yields of the isolated substances or from toxicological data. More reliable is the spectrophotometric assay after paperchromatographic separation and elution of the compounds, particularly when these procedures are applied to the greatly pre-purified and quantitatively prepared "primary material". Such assays yield 10 mg of phalloidin, 8 mg of α-amanitin, 5 mg of β-amanitin, 0.5 mg of γ-amanitin and traces of phalloin per 100 g of fresh mushroom. The concentrations of the other toxins in Fig. 2 are too low

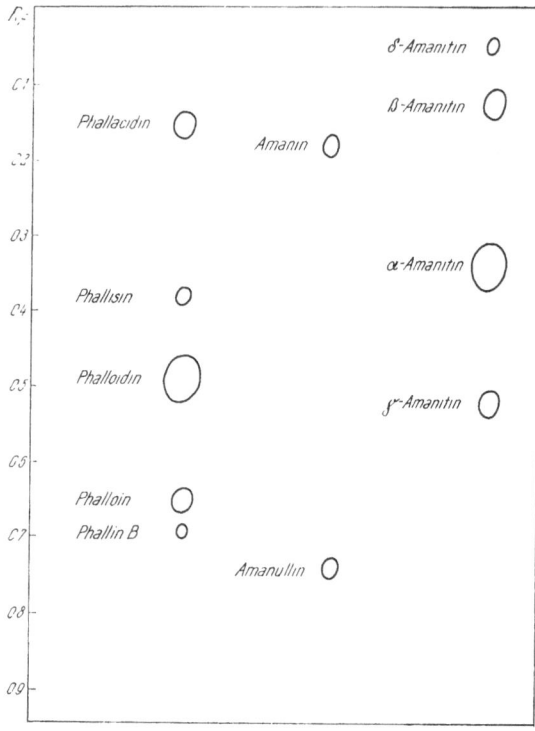

Fig. 2. Schematic representation of descending paper chromatograms (butanone : acetone : water 30 : 3 : 5 vol.) of the identified components of *Amanita phalloides*. Size of spots indicates roughly the concentration of substance in the mushroom extract.

to be determined in this way. In a mixture of the above ratio, the α- and β-amanitins are largely responsible for the toxic effect; since their lethal dose is about 0.2 mg/kg, the toxin content of a mushroom weighing 50 g may be sufficient to kill a human being.

For the isolation of the various toxic components of *A. phalloides*, the following general procedure has been employed.

Fresh mushrooms are placed in methanol and expressed after a few days. The extracts are filtered and evaporated to a syrup in vacuo. On addition of 5 to 10 vol. of methanol much inorganic salt and other insoluble material precipitates and is removed by centrifugation. Methanol is again added in the cold to the supernatant, thus giving a further precipatate, which is removed as above. The procedure is repeated until the methanolic solution remains clear at — 10°. The residue

References, pp. 246—250.

obtained after evaporation can now be subjected to a multistep chromatographic process, described below. For obtaining the so called "primary material", the residue is taken up with water, then lead acetate is added until marked precipitation no longer occurs; the filtrate, freed of lead by cautious precipitation of excess sulfuric acid and centrifugation of $PbSO_4$, is evaporated in vacuo to a small volume. The pH of the solution must be maintained at about 4 by adding small amounts of NaOH from time to time. The bulk of the toxins is precipitated by saturation with ammonium sulfate, the rest is extracted from the solution with isopropanol, evaporation of the alcohol, taking up the residue in a small volume of water and saturating with ammonium sulfate. In this manner (66) almost all the toxin is obtained as primary material (I resp. II) containing 10 to 20% toxins along with ammonium sulfate and mainly acidic impurities. The isolation of the different peptides is carried out by partition chromatography, after removing the inorganic salt by treating with methanol in the cold as described above.

Before applying the different chromatographic processes (39) the material, either as a whole or in the form of primary material, is divided into a lipophilic and a more hydrophilic fraction by treating 100 g with 0.5 l of a mixture of butanone: acetone : water (20 : 6 : 5 vol.). Two layers are formed. The upper phase, which contains the more lipophilic substances, is removed and the lower dark-brown liquid is shaken 5–10 times with an equal volume of the ternary mixture in a separatory funnel. The combined upper phases are evaporated, the residue is triturated with acetone. The insoluble major part (A) consists mostly of the toxins, whereas the soluble part contains untoxic lipophilic substances with some cyclic peptides among them. Part A is resolved by partition chromatography in the mentioned solvent on Sephadex G-25 (10 g on a column of 16 × 120 cm). The components appear in about 15 l solvent successively according to their R_F-values (Fig. 2, p. 218). The hydrophilic phase after evaporation is taken up in water and can be resolved by (adsorption-) chromatography on a similar column of Sephadex G-25, swollen in water, with water as an eluent (20 g substance/5 kg Sephadex, eluted with about 30 l). Here among others a clean separation of the toxins of the phalloidin group from those of the amanitin group takes place, the former being eluted with 20–25 l, the latter with 25–30 l of water. For further purification and isolation of the individual toxins the reader is referred to the original papers, esp. to (70 a).

The amanita toxins are soluble in methanol, liquid ammonia, pyridin and more or less soluble in water, ethanol and water-containing isopropanol or butanol. They are insoluble in weakly polar organic solvents.

III. Chemical Structure of the Toxic Peptides.

The amanita toxins can be divided into two main groups: the phalloidin group, mainly comprising phalloidin, phallacidin, phalloin, phallisin (and phallin B), and the amanitin group consisting of α-, β-, γ-amanitin, and amanin (see Fig. 2, p. 218). The basic difference between these toxin groups lies in some chemical features and in the rapidity of their action. The members of the phalloidin group act quickly, and at higher dose levels death of mice or rats occurs within 1 or 2 hours. In contrast, the action of the amanitin group is delayed, so that even with very high doses it is not possible to reduce the lethal interval to less than 15 hours. The toxicity is, however, just the opposite: α-Amanitin is 10–20 times

Table 1. Some Common and Individual Properties of Identified Toxins of *Amanita phalloides*.

Characteristics	Phalloidin Group					Amanitin Group			
	Rapid action; can cause death after 1 hour. Cyclic heptapeptides containing a γ-hydroxylated leucine					Slow action; death occurs only after 15 hours. Cyclic octapeptides containing a γ-hydroxylated isoleucine			
	PHD	PHN	PHC	PHS	PHB	α-AMA	β-AMA	γ-AMA	AMN
Toxicity in mouse (LD$_{50}$, mg/kg)	2	2	2.5	2.5	15	0.2	0.3	0.15	0.4
Indole derivative	Try	Try	Try	Try	Try	Otry	Otry	Otry	Try
Color reactions: with cinnamic aldehyde —HCl	blue	blue	blue	blue	blue	violet	violet	violet	blue
With diazotized sulfanilic acid..........	weak yellow	weak yellow	weak yellow	weak yellow	weak yellow	red	red	red	weak yellow
With Fe^{3+}-containing conc. sulfuric acid ...	blue	blue	blue	blue	blue	green	green	green	blue
Reduction of Ag(NH$_3$)$_2$$^+$	—	—	—	—	—	+	+	+	
Max. of UV-absorption (mμ)	292	292	292	292	292	302	302	302	285

Abbreviations: PHD = phalloidin, PHN = phalloin, PHC = phallacidin, PHS = phallisin, PHB = phallin B, AMA = amanitin, AMN = amanin, Try = tryptophan, Otry = 6-hydroxytryptophan.

more toxic than phalloidin and is therefore the major poisonous constituent of *A. phalloides.*

Both of these groups are cyclopeptides which have in common a
sulfur containing bridge formed by coupling of cysteine sulfur with the indole
nucleus of a tryptophan, and a γ-hydroxylated amino acid. They differ
as summarized in *Table 1.*

A. Phalloidin Group.

1. Phalloidin (9 a, p. 225).

Phalloidin is the best known of all the amanita toxins. By hydro-
lyzing the crystalline toxin with 20% hydrochloric acid or 25% sulfuric
acid at 100°, as is customary for cleavage of polypeptides, LYNEN and
U. WIELAND (*27*) obtained alanine (1), and H. WIELAND and WITKOP (*51*)
obtained L-α-oxytryptophan (2), more correctly termed β-oxindolyl-
α-alanine because of its lactam structure, and L-cysteine (3). They also
found L-allohydroxyproline (4), the diastereoisomer of the hydroxyproline usually occurring in nature, in which carboxyl and hydroxyl
groups are situated on opposite sides of the plane of the 5-membered ring.
The alanine was identified as the L-form.

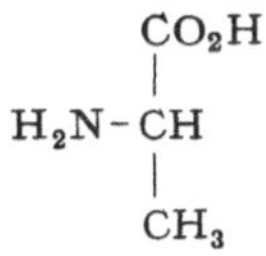

(1)
Alanine.

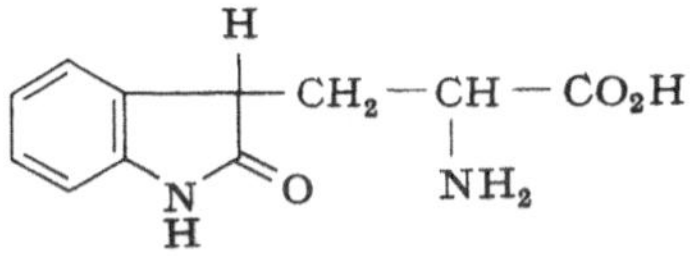

(2)
L-α-Oxytryptophan.

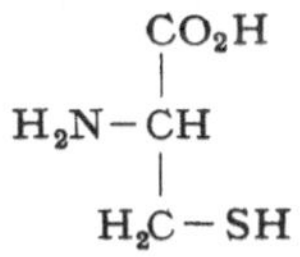

(3)
L-Cysteine.

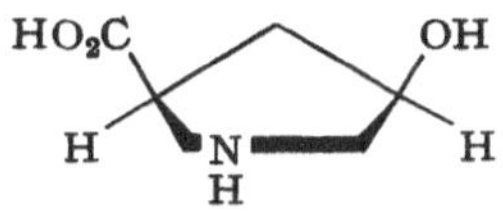

(4)
L-Allohydroxyproline.

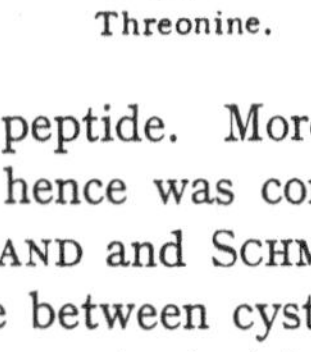

(5)
Threonine.

Thus the toxin is a rather unusual peptide. Moreover, it has no
terminal amino or carboxyl groups and hence was considered to be a
cyclopeptide. In later studies, TH. WIELAND and SCHMIDT (*74*) demonstrated the peculiar nature of the linkage between cysteine and tryptophan, previously suggested by NEUBERGER et al. (*7*) on the basis of
spectroscopic observations. The oxindolylalanine building block, when
isolated, exhibits maximum absorption at 250 mμ (indicating a N-acyl-
aniline rather than an indole spectrum), whereas in phalloidin the greater

wave-length maximum (292 mμ) points to an indole conjugation. On the basis of these observations and of the absence of the nitroprusside reaction even after reduction, a linkage of the cysteine sulfur to position 2 of the tryptophan ring was postulated. Indeed, reductive desulfuration by boiling with Raney nickel converted phalloidin to a nontoxic, sulfur-free cyclopeptide, desthiophalloidin (**10 a**, p. 225) whose hydrolyzate yielded tryptophan instead of oxindolylalanine (**2**) and showed a higher alanine content (**74**). These findings tended to substantiate a thioamide structure. In the same investigations, threonine (**5**), which had not been isolated by earlier investigators, was disclosed by paper chromatography as another building block of the phalloidin hydrolyzate. This constituent belongs to the D-series (**78**).

Further evidence for the presence of a thioamide structure came from synthesis of model substances. In various indole derivatives, including tryptophan itself, it is possible to introduce sulfur into position 2 by reaction with disulfurdichloride, ClSSCl, and sulfenic acid chlorides, RSCl (**85**). Such thioamides, e. g., 2-S-methylindolyl acetic acid (**8**), are very similar to phalloidin in their absorption characteristics. By comparing the extinction of the model compound ($\varepsilon = 12270$, mol. weight 221.3) with that of phalloidin, a molecular weight of 890 could be calculated for phalloidin *(Fig. 3)*.

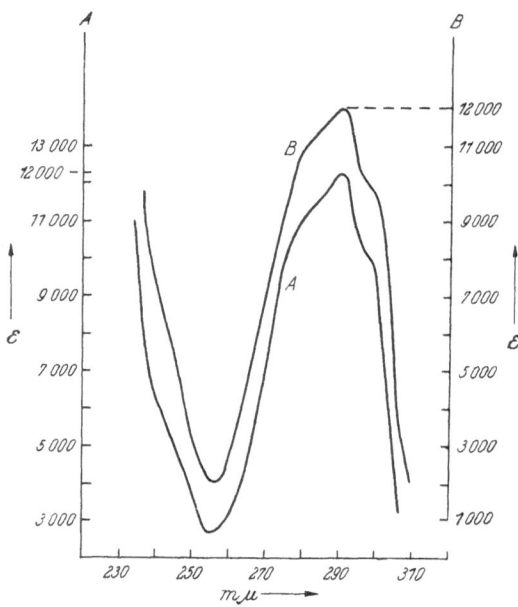

Fig. 3. Ultraviolet spectra of equimolecular amounts of 2-S-methylindolylacetic acid (8) = A (left ε-scale), and phalloidin = B (right ε-scale), in methanol. [From: Helv. Chim. Acta **44**, 919 (1961).]

References, pp. 246—250.

In further studies of phalloidin hydrolyzates paper chromatography proved inadequate, whereas cooled paper electrophoresis at elevated voltages was extremely helpful (*68*). With this method, a sixth hitherto unknown peptide building block was discovered (*77*). In the freshly prepared acid hydrolyzate a ninhydrine-positive cationic compound was demonstrable which is rapidly converted to a neutral α-amino acid by dilute alkali and reconverted into the cationic state by mineral acids. This suggested an α-amino-γ- (or δ-) hydroxy acid structure, travelling in an electric field to the cathode as an amino lactone. The compound was eluted as the lactone hydrochloride, crystallized and identified as α-amino-γ-hydroxy-γ-hydroxymethyl n-valeric acid (γ,δ-dihydroxyleucine, 6). In the original publication (*77*) it was assumed that the leucine derivative is present in the unsaturated state (δ-hydroxyleucenine) when bound in the phalloidin molecule. Later on it could be shown (*78*), that the amino acid exists already as γ,δ-dihydroxyleucine (6) in the cyclic peptide. Simultaneously it has been shown by an enzymatic method that the aspartic acid (7), which is formed by oxidation of the new amino acid with periodate and hypoiodite belongs to the L-series. For stereochemistry and synthesis of the different γ-hydroxy-amino acids of the toxins see pp. 239—241.

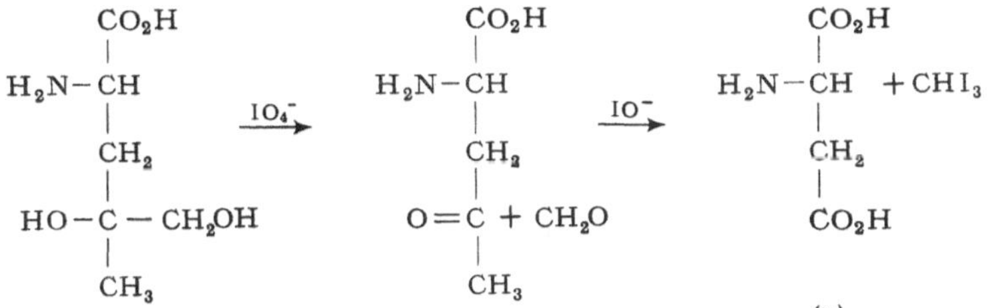

(6)
γ, δ – Dihydroxyleucine.

(7)
Aspartic acid.

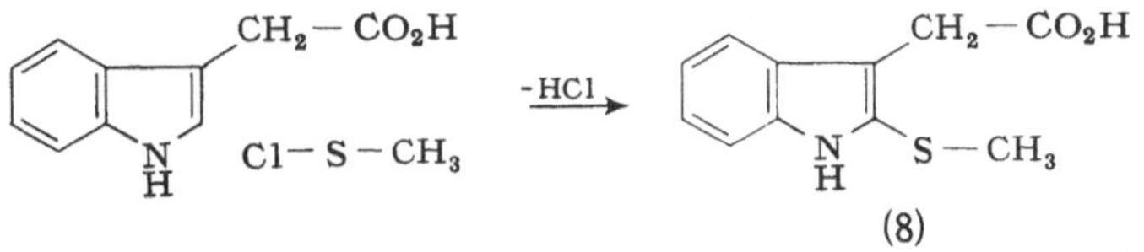

(8)
2 – S – Methylindolyl acetic acid.

With complete information on the component amino acid building stones and their molar ratios at hand, it has become possible to establish the formula of phalloidin as follows: 1 cysteine, 2 alanines, 1 allo-hydroxy-proline, 1 threonine, 1 oxindolylalanine, 1 γ,δ-dihydroxyleucine, minus 7 H_2O (seven peptide linkages), minus 1 H_2O (one thioether linkage), corresponding to the molecular formula $C_{35}H_{45}N_8O_{11}S$—$8\,H_2O$.

Since the air-dry peptide sample contains 5 moles of water rather stably bound, the molecular weight must be augmented to 878 (determined from spectroscopic evidence: 890). On this basis the analytical values for all elements agree well with the calculated ones.

The first step in elucidating the amino acid sequence was the hydrogenolytic substitution with Raney-nickel of sulfur by two hydrogen atoms, yielding the non-toxic desthiophalloidin (10 a) *(Chart 1)*. In a second step the peptide ring was opened. Selective splitting of the peptide ring took place, when (10 a) was heated with dilute acids for a short time. The peptide bond between the carboxyl of a γ-hydroxyamino acid and nitrogen was disconnected under formation of the lactone of desthio-seco-phalloidin (11 a). The reaction presumably occurs via a carbonium ion intermediate:

$$
\begin{array}{ccc}
\overset{H}{\underset{|}{-C}}-CH_2\;OH & \overset{H}{\underset{|}{-C}}-CH_2 & \overset{H}{\underset{|}{-C}}-CH_2\;\;CH_3 \\
| \quad\quad \diagdown C-CH_3 & | \quad\quad \diagdown C^{\oplus}CH_3 & | \quad\quad\quad C \\
C=O \quad\quad CH_2OH & C=O \quad\quad CH_2OH & O=C-O \diagup \quad\diagdown CH_2OH \\
| & +H^{\oplus} & | & \\
NH & (-H_2O) & \text{NH} & +H_2O & \overset{\oplus}{NH_3} \\
| & & | & & | \\
-\underset{H}{\overset{|}{C}}-CH_3 & -\underset{H}{\overset{|}{C}}-CH_3 & -\underset{|}{C}-CH_3 \\
& & H
\end{array}
$$

The principle of this reaction has been used by us (64) to develop a protecting group in peptide chemistry: the γ-hydroxyisocaproyl residue is introduced by heating a sodium salt of an amino acid with isocaprolactone in imidazole. It is removed at pH 1–2 and 100° in 20 minutes. An even milder amide bond splitting, namely by incubation with 0.5 N acetic acid at 20° for 10 minutes, is possible if the side-chain bears a γ,δ-epoxide ring (70):

$$
\begin{array}{cc}
\overset{H}{\underset{|}{-C}}-CH_2\;\;CH_3 & \overset{H}{\underset{|}{-C}}-CH_2\;\;CH_3 \\
| \quad\quad \diagdown C^{\oplus} & | \quad\quad\quad C \\
C=O \quad C-OH & O=C-O\diagup\quad\diagdown CH_2OH \\
| \quad\quad\quad CH_2 & +H_2O & \\
\text{NH} & \overset{+H^+}{\longrightarrow} & \overset{\oplus}{NH_3} \\
| & & | \\
-\underset{H}{\overset{|}{C}}-CH_3 & -\underset{H}{\overset{|}{C}}-CH_3
\end{array}
$$

In order to establish the sequence of amino acids in the heptapeptide the Edman degradation method was applied (76, 77). It yielded successively the phenylthiohydantoin derivatives (PTH) of alanine, D-threonine, alanine (from cysteine), allo-hydroxyproline, alanine and' tryptophan, thus disclosing the structure of phalloidin (9 a).

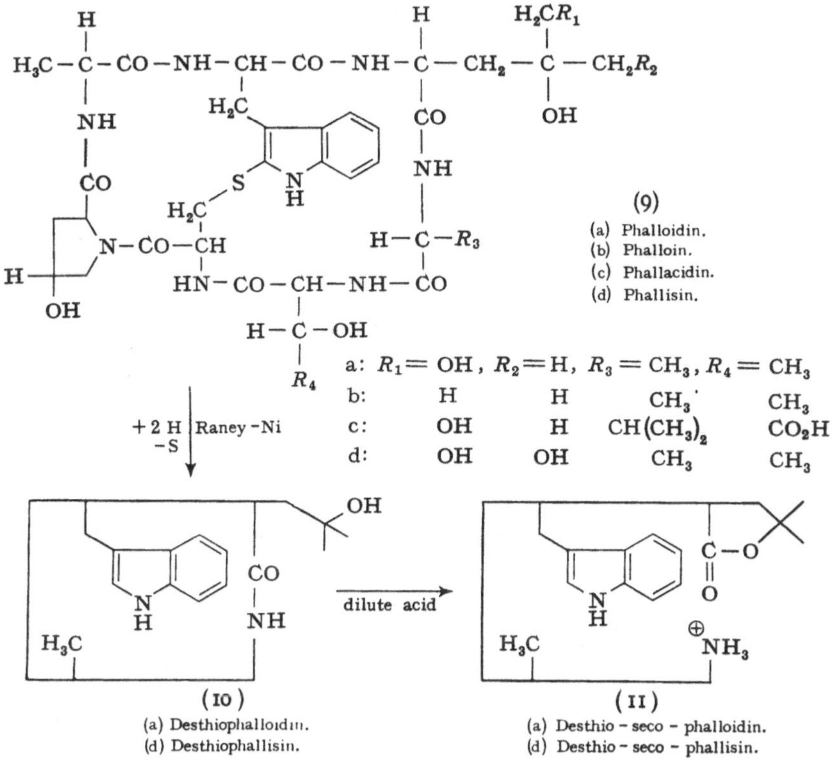

(9)
(a) Phalloidin.
(b) Phalloin.
(c) Phallacidin.
(d) Phallisin.

a: $R_1 =$ OH, $R_2 =$ H, $R_3 =$ CH$_3$, $R_4 =$ CH$_3$
b: H   H   CH$_3$'   CH$_3$
c: OH   H   CH(CH$_3$)$_2$ CO$_2$H
d: OH   OH   CH$_3$ CH$_3$

(10)
(a) Desthiophalloidin.
(d) Desthiophallisin.

(11)
(a) Desthio – seco – phalloidin.
(d) Desthio – seco – phallisin.

Chart 1.

2. *Phalloin* (9 b).

Phalloin is chemically closely related to phalloidin. Both of them give the same characteristic color reactions and have the same ultraviolet spectrum (see Table 1, p. 220) (65). The close relationship of the two peptides is further indicated by their amino acid composition. The two-dimensional paper chromatogram of a phalloin hydrolyzate also shows alanine, threonine, cysteine, allo-hydroxyproline and oxindolylalanine (66). In addition to the amino acids enumerated above, the sulfuric acid hydrolyzate of phalloin contains a ninhydrine-positive amino lactone which, on electrophoresis, travels to the cathode somewhat faster than the analogous building stone of phalloidin. It turned out to be the γ-lactone of γ-hydroxyleucine (12). Therefore, the sole difference between phalloidin (9 a) and phalloin (9 b) consists in the lack of one O-atom in the δ-position of the branched amino acid in phalloin (66, 76).

γ-Hydroxyleucine, a substance already known by synthesis (11), had not been found in nature. Quite recently DÖLLING succeeded in the author's laboratory in discovering this amino acid in a HCl-hydrolyzate of gelatin (57).

$$
\begin{array}{cccc}
\mathrm{CO_2H} & \mathrm{CO_2H} & \mathrm{CO_2H} & \mathrm{CO_2H} \\
| & | & | & | \\
\mathrm{H_2N-CH} & \mathrm{HC-NH_2} & \mathrm{HC-NH_2} & \mathrm{H_2N-CH} \\
| & | & | & | \\
\mathrm{CH_2} & \mathrm{HC-OH} & \mathrm{HO-CH} & \mathrm{HC-CH_3} \\
| & | & | & | \\
\mathrm{HO-C-CH_3} & \mathrm{CO_2H} & \mathrm{CO_2H} & \mathrm{CH_3} \\
| & & & \\
\mathrm{CH_3} & & & \\
(12) & (13a) & (13b) & (14) \\
\text{γ-Hydroxyleucine.} & & &
\end{array}
$$

3. Phallacidin (9 c).

In paper electropherograms of sufficiently concentrated hydrophilic fractions an anodic migrating component was visible which after total purification exhibited the same UV-spectrum (Fig. 3, p. 222) as phalloidin or phalloin. The toxin has not yet been obtained in crystalline state. Considering its acid character an acidic amino acid was expected in the total hydrolyzate. After hydrolysis with 25% sulfuric acid, however, no acidic substance could be detected, but after using 6 N hydrochloric acid the presence of two acidic ninhydrin-positive substances became evident (75). At pH 1.9 they travelled to the cathode with different speeds, the faster one being about 10 times higher in concentration. They agreed in electrophoretic behavior with the two diastereoisomeric β-hydroxyaspartic acids (13 a, 13 b). In phallacidin (9 c) only the faster moving erythro-compound (13 a) seems to be present; a synthetic sample of (13 a) was isomerized to a mixture of the diastereoisomeric hydroxyaspartic acids (13 a) and (13 b) (ca. 9 : 1), when treated under hydrolysis conditions with hydrochloric acid. On heating with 25% sulfuric acid the hydroxyaspartic acids are decomposed by splitting off a molecule of water, carbon dioxide and ammonia. Therefore, no acidic amino acid could be observed in the H_2SO_4-hydrolyzate of phallacidin. Comparison of a crystalline sample from phallacidin hydrolyzate with authentic specimens revealed the identity of the natural compound with D-erythro-β-hydroxyaspartic acid (13 a). The isolation of the L-isomer from tryptic casein hydrolyzate has been reported some years ago (44).

An additional deviation from phalloidin and phalloin is the occurrence of L-valine (14) instead of alanine in the molecule of phallacidin. Edman degradation in an analogous manner as before led to formula (9 c).

4. *Phallisin* (9 d, p. 225).

This toxin is one of the minor components of *A. phalloides*. It could only be purified by repeated chromatographic procedures on Sephadex G-25 with water and on silica gel layers with methanol-ethylacetate (4 : 1) [GEBERT (*17*)], but it has not yet crystallized. On hydrolysis it yielded the same amino acids as phalloidin or phalloin, the sole difference being the nature of a hydroxylated amino acid, whose easily formed amino lactone travelled more slowly to the cathode in paper electrophoresis at pH 6.5 than the aminolactones of (6) or (12). In thin layer chromatography on silica gel the new compound moved more slowly too, thus showing a more hydrophilic character, i. e. one additional OH-group. Having in mind a biogenetical relationship, also here a leucine skeleton was assumed which could contain three OH groups. The decision between the two possible trihydroxy side-chains of β,γ,δ-trihydroxyleucine (15) or of γ,δ,δ'-trihydroxyleucine (16) could be made by periodate oxidation of desthiophallisin (10 d, p. 225). In the former case, oxidation was to give peptide-bound aminomalonic semialdehyde (15 a) (which would decompose on acid hydrolysis of the oxidized cyclopeptide), while in the latter case aspartic acid (16 a) should be formed.

$$-NH-\underset{\underset{CO}{|}}{\overset{\overset{H}{|}}{C}}-\underset{\underset{CH_3}{|}}{\overset{\overset{OH}{|}}{\underset{\underset{H}{|}}{C}}}-\overset{OH}{\overset{|}{C}}-CH_2OH \xrightarrow{IO_4^-} -NH-\underset{\underset{CO}{|}}{\overset{\overset{H}{|}}{C}}-\overset{\overset{H}{|}}{C}=O \quad CH_3CO_2H$$

(15)
β,γ,δ-Trihydroxyleucine.

(15 a) + + CO_2H

$$-NH-\underset{\underset{CO}{|}}{\overset{\overset{H}{|}}{C}}-CH_2-\underset{\underset{H_2COH}{|}}{\overset{\overset{OH}{|}}{C}}-CH_2OH \xrightarrow{IO_4^-} -NH-\underset{\underset{CO}{|}}{\overset{\overset{H}{|}}{C}}-CH_2-CO_2H$$

(16)
γ,δ,δ'-Trihydroxyleucine.

(16 a) + 2 CH_2O

The occurrence of aspartic acid in the acid hydrolyzate of oxidized desthiophallisin proved unambiguously the structure of the new amino acid as γ,δ,δ'-trihydroxyleucine (16). Edman degradation of desthioseco-phallisin (11 d) led to the amino acid sequence of formula (9 d) for phallisin (*17*).

15*

5. Phallin B.

Phallin B is also a minor constituent of *A. phalloides* extracts and shows lipophilic properties. It was isolated by repeated chromatography of the fast travelling toxin fractions on alumina, silica gel and Sephadex G-50 in different solvents by DE VRIES in 1963 (*80*). Phallin B shows a typical phalloidin spectrum and also the blue color reaction with cinnamic aldehyde-HCl of the phalloidin group. On acid hydrolysis the following amino acids could be identified by two-dimensional paper electrophoresis chromatography: cysteine, γ-hydroxyleucine, phenylalanine, proline, threonine and alanine (probably twofold molar amount). The toxin seems to be a phalloin-like cyclopeptide in which allo-hydroxyproline is substituted by proline, and valine by phenylalanine.

6. Phalloidin Derivatives Obtained by Chemical Reactions.

The bicyclic shape gives to the molecules of the toxins in the phalloidin group a relatively stable conformation whose side-chains are apt to suffer several well-defined chemical attacks without alteration of the geometry of the peptide ring. Destruction of the architecture, however, can be achieved by heating the toxic molecules with Raney-nickel (formation of the desthiocompounds, 10 a, d, p. 225) or by treating them with acid under not too energetic conditions. On standing in 50% trifluoroacetic acid for several hours at room temperature, selective opening of the peptide ring takes place as a consequence of the lactonization tendency of the γ-hydroxylated amino acid building stones. The seco compounds (17) thus formed are as nontoxic as the desthio compounds; alteration of the shape of the molecules apparently causes loss of toxicity.

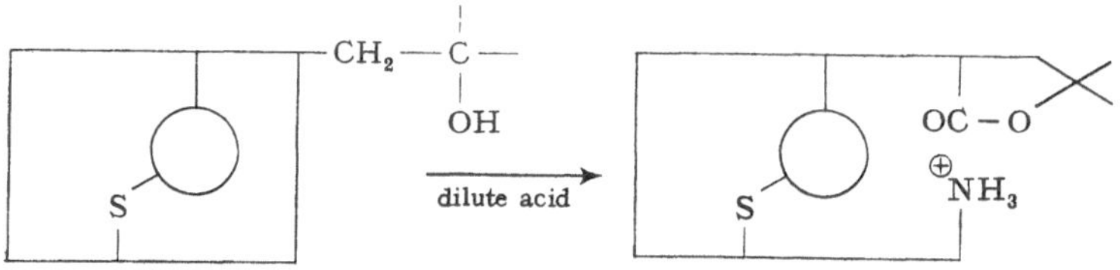

The circles in (9) and (17) stand for the skatyl side-chain of tryptophan, connected with sulfur of cysteine in 2-position.

On the other hand, chemical manipulations at the periphery of the intact molecules do not destroy toxicity. Thus, by treating phalloidin with acetic anhydride under reflux a crystalline triacetyl derivative could be obtained (27) which showed full toxicity. We know now that in this instance the alcoholic hydroxyl groups of allo-hydroxyproline, threonine, and the δ-OH of γ,δ-dihydroxyleucine have been esterified. The possibility, however, exists that in vivo the acetyl residues are split off enzymatically giving rise to the formation of the poison proper. The same holds for the toxic effect of "ketophalloidin" (18) *(Chart 2)*, the oxidation product of phalloidin by periodate *(78)* which shows a toxicity of the same range. Ketophalloidin can be hydrogenated to the toxic desmethylphalloin (19) by means of NaBH$_4$ *(70)*, and it cannot be excluded that the same mechanism is operating in the liver of the treated animal. By using NaBT$_4$ a rather strongly labeled phalloidin-like poison (19 T) could be prepared *(35)* which was useful in following the fate of the toxin in rats.

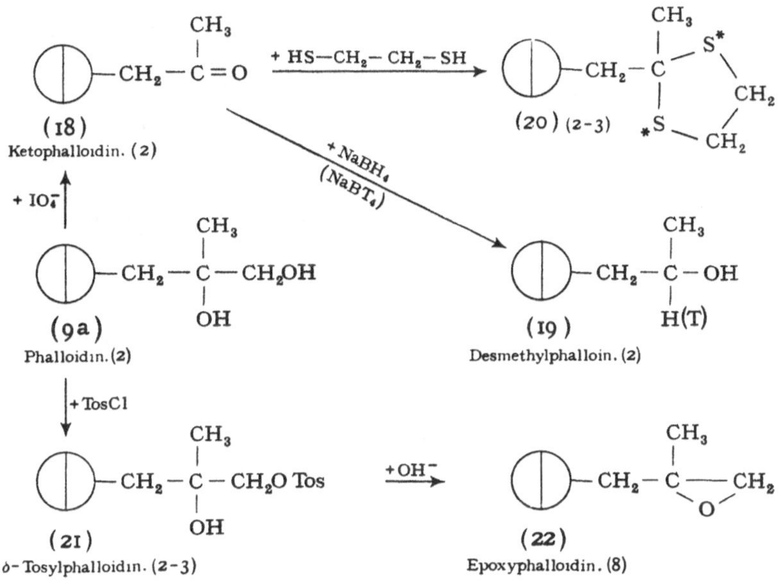

Chart 2.

⊙ is the symbol for the bicyclic system; numbers in parantheses: LD$_{50}$ in mg/kg in the white mouse; T = tritium.

Two years ago, REHBINDER prepared in the author's laboratory a ^{35}S-containing, likewise toxic dithiolane (20) by reacting (18) with ethylenedithiol-^{35}S, and δ-tosylphalloidin (21) from phalloidin (9 a) with tosylchloride *(70)*. The latter was converted to epoxyphalloidin (22) by weak bases. The extreme readiness for hydrolysis of the adjacent peptide bond in (22) has been mentioned on p. 224. Recently, the oxime of (18)

and its tosylhydrazone have been prepared and found to be toxic. Therefore, the presence of a γ-hydroxy group in a side-chain of phalloidins as
a prerequisite for the toxicity does not seem to be justified.

B. Amanitin Group.

1. α- and β-Amanitin (29, 30, p. 233).

α- and β-Amanitin give after acid hydrolysis identical phero-chromato-
grams of amino acids: glycine (23), isoleucine (24), aspartic acid (25),
hydroxyproline (26), cysteic acid (27), and an aminolactone (62). The
nature of hydroxyproline (in contrast to the allo-compound in the phalloidin group) could easily be clarified by paper electrophoresis at pH 1.9,
in which the allo-form travels faster to the cathode (69). Cysteic acid
is set free during hydrolysis, although intact α-amanitin is a neutral
substance. β-Amanitin has an acidic function, the β-carboxy group of
aspartic acid, which could be converted to its amide with ammonia using
the mixed anhydride method (56). The amide of β-amanitin is identical
with α-amanitin. To the lactonizing amino acid, the formula of β-methyl-
γ,δ-dihydroxyleucine had been ascribed earlier (62), but from a reinvestigation using mass spectrometry (33) and comparison with a synthetic
sample (18) has resulted the structure of one of the four diastereoisomeric
γ,δ-dihydroxyisoleucines (28).

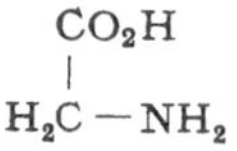

(23)

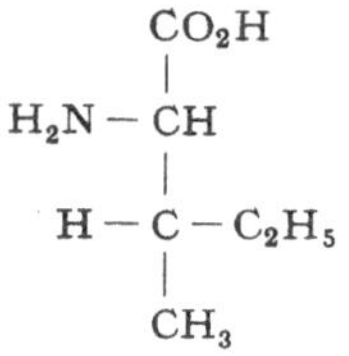

(24)

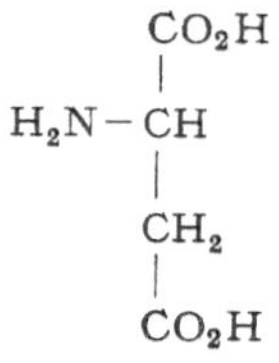

(25)

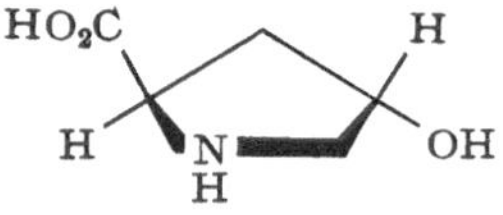

(26)
Hydroxyproline.

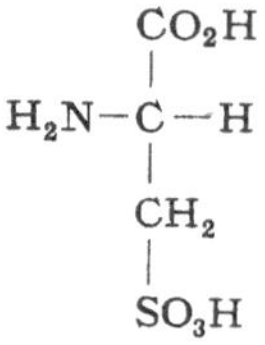

(27)
Cysteic acid.

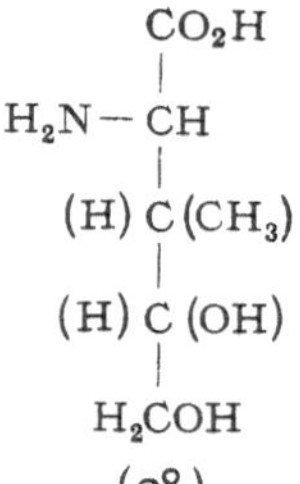

(28)
γ,δ-Dihydroxyisoleucine.

The amanitins also contain sulfur, which is located between an indole
nucleus and the periphery. They reduce silver ions in ammoniacal solu-

tion. The UV absorption maximum (302 mμ) is shifted by alkali to higher wavelengths *(Fig. 4)*; furthermore, the amanitins give azo dyes with

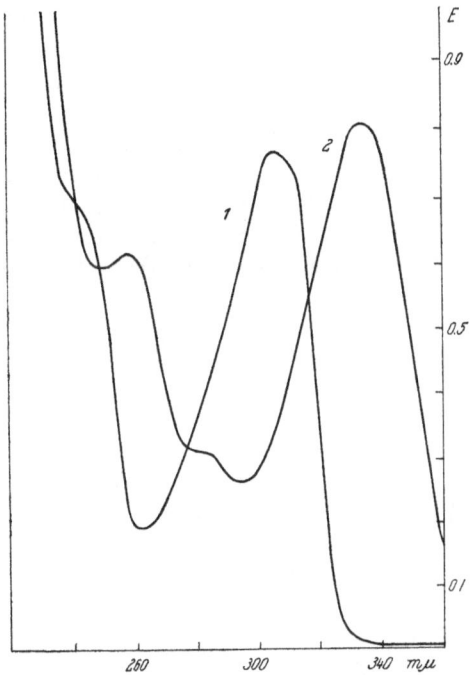

Fig. 4. Ultraviolet spectrum of 0.60 mg of α-amanitin in 10 ml of water, 1: neutral and 2: shortly after addition of one drop of 0.1 N NaOH.

diazonium salts (Pauly reaction), thus showing the presence of a phenolic hydroxyl group. This can be methylated (in α-amanitin) by diazomethane yielding an equally toxic methyl-*x*-amanitin (**29** — CH$_3$) which shows a nearly identical UV-spectrum. The indole system of the amanitins is extremely unstable towards hydrolysis by strong acids. In this case the S-containing part is converted into cysteic acid (**27**) by a hydrolysis of the sulfur bridge in which the indole part is set free as (peptide bound) 6-hydroxyindole. Thus, a localization of the bridge-head, like in the phalloidin series, was not possible.

The different steps in the elucidation of the structure of *α*-amanitin (**29**) are shown below using the finally established formula (**60**). The first substantial progress was made by treating the toxin with Raney-nickel. This gave a non-toxic sulfur-free compound, desthio-*x*-amanitin, which exhibited UV-absorption very similar to that of 6-hydroxytryptophan *(Fig. 5)*.

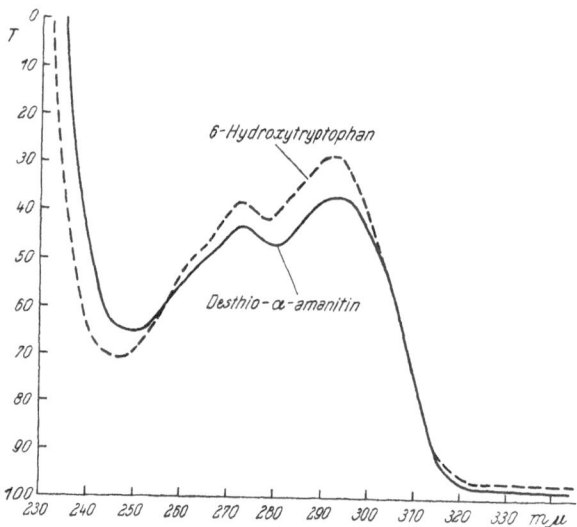

Fig. 5. Ultraviolet spectra of desthio-α-amanitin (1) and 6-hydroxytryptophan (2), in water. [From: Pure and Appl. Chem. 9, 145 (1964).]

Hence, it follows that the phenolic hydroxyl is attached to the 6-position. The suspected 6-hydroxytryptophan of desthioamanitin, however, is not stable enough to survive the partial hydrolysis now necessary as the next step. To stabilize the indole system hydrogenation over platinum was carried out. In this reaction not only saturation of the aromatic rings takes place but also elimination of the phenolic hydroxyl group. A peptide of octahydrotryptophan is obtained, perhydro-desthioamanitin, which migrates electrophoretically to the cathode on account of its basicity. Then the selective opening of the peptide ring with weak acids gave the octapeptide seco-perhydro-desthioamanitin (33).

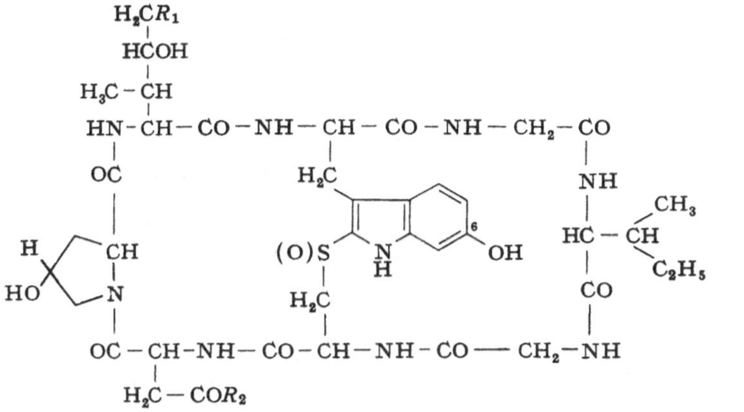

(29)   : α-Amanitin, R_1=OH, R_2=NH$_2$
(29 — CH$_3$) : Methyl-α-amanitin, OCH$_3$ in 6-position of indole
(30)   : β-Amanitin, R_1=OH, R_2=OH
(31)   : γ-Amanitin, R_1=H, R_2=NH$_2$
(32)   : Amanin, H in 6-position of indole in (29)

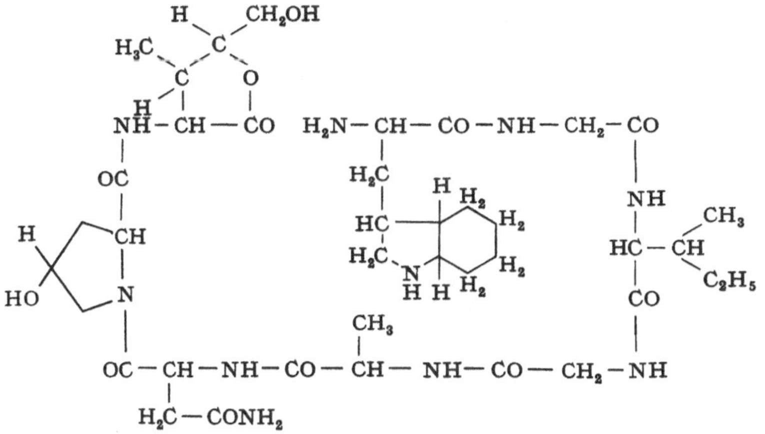

(33)

seco − Perhydro − desthioamanitin.

Total hydrolysis of (33) yielded the following amino acids: glycine, isoleucine, alanine (formed from cysteine by treatment with Raney-nickel), hydroxyproline, aspartic acid, octahydrotryptophan and γ,δ-dihydroxyisoleucine (lactone of **28**). None of these amino acids disappeared on incubation with D-amino acid oxidase, thus suggesting that all of them belong to the L-series. Edman degradation gave the PTH-derivatives of the amino acids in a sequence which led to formula (**33**) for the peptide analyzed.

The sole structural elements not directly proved by the reactions described above are the 2-position and the thioether bridge. We tried to furnish evidence by comparison of the UV-spectra of amanitin and synthetic model substances. GRIMM (*61*) obtained, e. g., the thioether (**34**) by reacting 2-methyl-6-methoxyindole with ethylsulfenylchloride *(Chart 3)*.

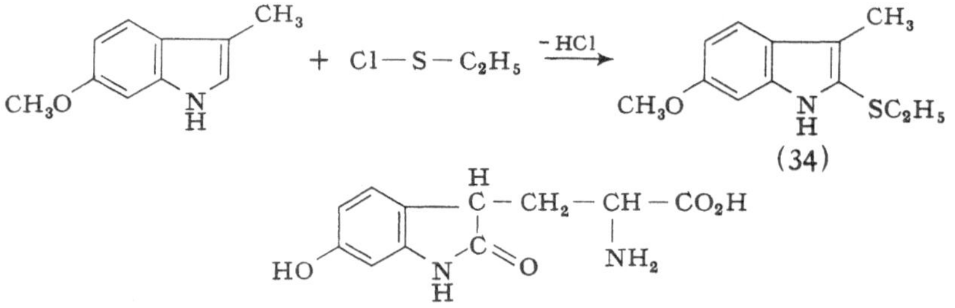

Chart 3.

These substances show UV-spectra, which resemble that of methyl-α-amanitin (**29** — CH_3) in the position of λ_{max} but not in the shape of the peak. All other positions of the sulfur bond, however, were unlikely; recently, it was possible to detect by refined chromatographic methods 6-hydroxy-2-oxindolyltryptophan (**35**) among the products of acid hydrolysis of α-amanitin (*17*). This points to a simultaneous "nucleophilic" hydrolysis of the thioether bridge, which takes place exclusively in phalloidin (p. 221), and has proved the 2-position of sulfur at the indole ring.

Added in Proof.

According to experiments of H. FAULSTICH in the author's laboratory, a sulfoxide structure for the amanitins and for amanin is very probable.

Allomethyl-α-amanitin (34).

The discrepancy of UV-curves of model thioethers (**34**) and methyl-α-amanitin (**29** — CH_3, p. 233) was cleared up by an incidental observation during experiments to desulfurize methyl-α-amanitin with Raney-nickel. When boiling the reactants in methanol of low water content, among other products a well crystalline, still sulfur-containing product, allomethyl-α-amanitin (AMA), was formed which upon hydrolysis yielded the same amino acids as (**29** — CH₃) but cysteine instead of cysteic acid, and differed from it in several respects as summarized in *Table 2*. Most noticeable was the appearance of fine structure in the UV-spectrum of AMA which very much resembles that of the model thioether *(Fig. 6)*.

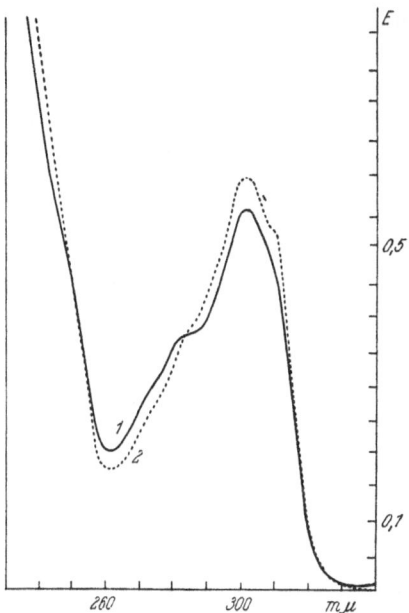

Fig. 6. Ultraviolet spectra of thioether (**34**) (1), and allomethyl-α-amanitin (**2**), in methanol.

This now can be explained, according to experiments of FAULSTICH (see note on p. 234) by assuming a reduction of a sulfoxide structure to a thioether chromophore. The small differences of UV-curves seem due to differences in the environment of the indole part. Earlier an alteration of the shape of the UV-curve had been observed in solvents which possess the capability of breaking hydrogen bridges (dimethyl-sulfoxide, dimethyl-formamide or tetramethylurea) (*53*).

Table 2. Properties of Methyl-α-amanitin (**29** — CH₃) and Allomethyl-α-amanitin (AMA).

	R_F in solvent of Fig. 2 α-amanitin = 1	$[\alpha]_D$	λ_{max} (methanol)	$\varepsilon \cdot 10^{-3}$	soluble in water	LD_{50}/kg (mouse)
(**29**—CH₃)	1.43	+ 120°	302 mμ 308 mμ	16.1 15.6	good	290 μg
AMA	1.90	+ 47°	298 mμ	10.2	slight	500 μg

2. γ-Amanitin (31, p. 233).

A minor component which gives a blue-violet color reaction with cinnamic aldehyde—HCl and which permanently accompanies phalloidin in chromatograms was named γ-amanitin (58). Its separation was only possible by using as a support paper or cellulose powder impregnated with borate, because phalloidin is thus retarded as a consequence of complex formation by its γ,δ-dihydroxylated side-chain. In γ-amanitin whose travelling velocity is not reduced, the presence of a monohydroxylated amino acid was anticipated instead. Recently, it was possible to isolate the crystalline toxin in a more convenient chromatographic way (82) and to obtain a crystalline sample of its aminolactone-hydrochloride from the hydrolyzate. By mass spectrometry (33) and by comparison with a synthetic specimen the compound turned out to be the lactone of one of the diastereoisomeric γ-hydroxyisoleucines (36). All other features and constituents being identical with those of α-amanitin, we assume with high probability analogous structures for the amanitines.

δ-Amanitin, mentioned on p. 217 and in Fig. 2, p. 218, seems to be the carboxylic acid (aspartic acid) of which γ-amanitin is the amide.

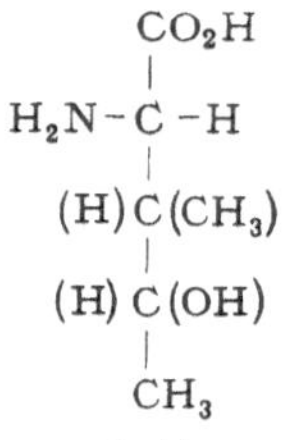

$$CO_2H$$
$$|$$
$$H_2N-C-H$$
$$|$$
$$(H)C(CH_3)$$
$$|$$
$$(H)C(OH)$$
$$|$$
$$CH_3$$
$$(36)$$

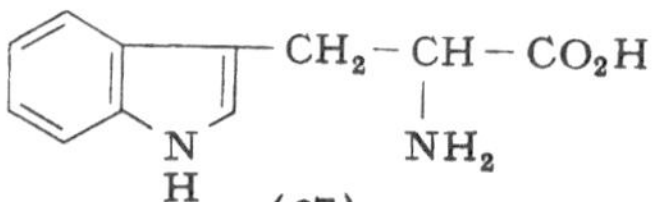

$$CH_2-CH-CO_2H$$
$$|$$
$$NH_2$$
(37)

3. Amanin (32, p. 233).

Amanin, the poison with the amanitin-like toxic action and with the color reaction of phalloidin has been obtained in pure state by repeated, different chromatographic processess (3, 17). It yields on acid hydrolysis almost the same amino acids as α-amanitin, the sole difference being the formation of β-oxindolylalanine (2, p. 221) plus tryptophan (37) (besides cysteine and cysteic acid), thus pointing to a phalloidin-like chromophore. Desthioamanin, produced by treatment of amanin with Raney-nickel, exhibits – like the phalloidin product – the indole-spectrum of tryptophan. According to these observations amanin should be an amanitin with the indole moiety of phalloidin; but, unfortunately, its UV-spectrum does not correspond to that of phalloidin (Fig. 7). This discrepancy can only arise from differences in the chromophoric systems, as described for the methylamanitines on p. 234. In amanin

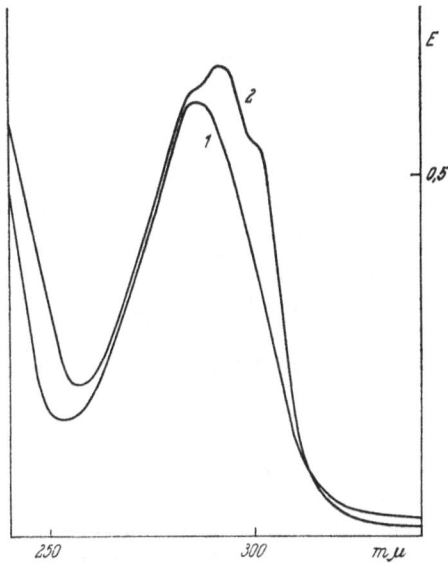

Fig. 7. Ultraviolet spectra of amanin (1) and phalloidin (2), in water.

the imino group of the indole nucleus may be situated next to a sulfoxide oxygen causing a hydrogen bond. In accordance with this suggestion the shape of the UV-peak is clearly different in methanol or in a hydro-

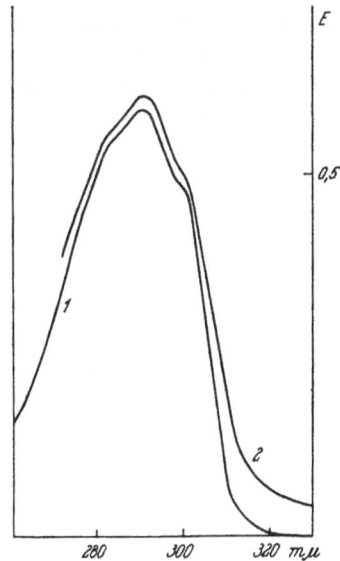

Fig. 8. Ultraviolet spectra of **phalloidin in water** (1) and amanin in tetramethylurea (2).

gen bond breaking solvent such as tetramethylurea. The spectrum of amanin in tetramethylurea is almost identical with that of phalloidin (in water) *(Fig. 8)*, and hence the true character of the compound is now evident.

Amanin was converted to an open desthio-seco peptide which could be analyzed by the Edman degradation method (*17*). These experiments led to the structural formula (**32**, p. 233) for amanin which differs from α-amanitin only by the absence of the OH-group from the 6-position in the indole nucleus. On account of the analogous spectrophotometric behavior, the molecular conformations are also presumed to be identical.

4. Amanullin.

An unknown, fast moving substance, violet with cinnamic aldehyde-HCl ("X-Körper"), was observed very early in paper chromatograms of lipophilic fractions of mushroom extracts, but its isolation became possible only after collecting suitable material from separation procedures over several years. By repeated, troublesome chromatographic processes, BUKU (*6, 70a*) finally succeeded in obtaining about 50 mg of the substance in an homogenous state. It was named amanullin, because of the lack of toxicity. On hydrolysis with hydrochloric acid the following amino acids were formed: cysteic acid, aspartic acid, isoleucine, hydroxyproline and glycine. Quantitative amino acid analysis revealed a ratio of Asp_1, Ile_2, Gly_2, $Hypro_1$. Thus a very close relationship in composition with the amanitines appeared. The sole difference is the absence of a γ-hydroxylated amino acid; instead of γ-hydroxy- or of γ,δ-dihydroxyisoleucine, a second molecule of isoleucine is present, as if during biosynthesis of amanitines the hydroxylation of a side-chain had been omitted. Since the ultraviolet spectra are identical, a similarity also in the molecular shapes seems probable. The amino acid sequence has not yet been determined. If also in this respect analogy should appear, the toxic effect in the amanitin field would depend not only on the presence of an intact peptide ring and of a thioether bridge, but also on the existence of a γ-hydroxylated side-chain.

As in the phalloidin series (p. 224), also in the amanitines the toxicity disappears when the sulfur bridge is hydrogenated (desthio compounds) or the peptide bond adjacent to the γ-hydroxylated amino acid is disconnected (seco compounds).

IV. Synthetic Experiments.

A. Amino Acids.

The toxins of *A. phalloides* contain some previously unknown amino acids, whose synthesis has been elaborated in the author's laboratory for the first time. Some other constituents have been described or found in nature earlier. They could be synthesized by new methods or by varying some older ones and will also be mentioned in the following pages.

L-*allo-Hydroxyproline* (4, p. 221).

This imino acid was synthesized (*81*) in medium yield but relatively simply, using a method already described for the D,L-compound (*25*). The steps were *(Chart 4)*: reaction of sodium N-carbobenzyloxy-glycine ethylester with diethylfumarate to give, after decarboxylation and stereospecific catalytic hydrogenation D,L-N-carboethoxy-allohydroxy-proline ethylester (38). The parent D,L-acid (39), obtained by partial alcaline hydrolysis gave, on treatment with brucine in butanone, a crystalline salt containing only the L-antipode. Removal of the carbo-ethoxy group could be achieved by boiling with 5 N hydrochloric acid for several hours.

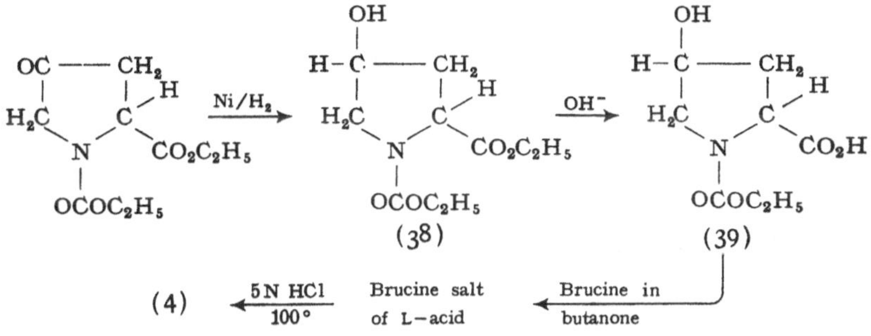

$$
\begin{array}{ccc}
\underset{\substack{| \\ OCOC_2H_5}}{\overset{\substack{OC - CH_2 \\ | \quad\quad |\;H \\ H_2C \;\; C \\ N \;\; CO_2C_2H_5}}{}} & \xrightarrow{Ni/H_2} & \underset{\substack{| \\ OCOC_2H_5 \\ (38)}}{\overset{\substack{OH \\ | \\ H - C - CH_2 \\ | \quad\quad |\;H \\ H_2C \;\; C \\ N \;\; CO_2C_2H_5}}{}} \xrightarrow{OH^-} \underset{\substack{| \\ OCOC_2H_5 \\ (39)}}{\overset{\substack{OH \\ | \\ H - C - CH_2 \\ | \quad\quad |\;H \\ H_2C \;\; C \\ N \;\; CO_2H}}{}}
\end{array}
$$

$$(4) \xleftarrow[100°]{5\,N\ HCl} \text{Brucine salt of L—acid} \xleftarrow{\text{Brucine in butanone}} $$

Chart 4.

L-*erythro-γ,δ-Dihydroxyleucine* (6 B, p. 240).

Of the four possible diastereoisomeric γ,δ-dihydroxyleucines, the L-erythro compound proved identical with the amino acid isolated from phalloidin hydrolyzate (*78*). The mixture of all of them (A + B) was synthesized *(Chart 5)* by reacting methallyl-acetamidomalonic ester (40) with bromine in chloroform solution (*83*). The main product, the hydrobromide of a bromomethyl-dihydrooxazin (41) was hydrolyzed by boiling with silver sulfate in aqueous sulfuric acid to give the mixture of aminolactones A and B. The hydrochlorides of A and B could be separated from each other by crystallization from alcohol with ether and the racemates A and B were resolved into their antipodes by recrystallization of their di-*p*-tolyltartrates (*63*).

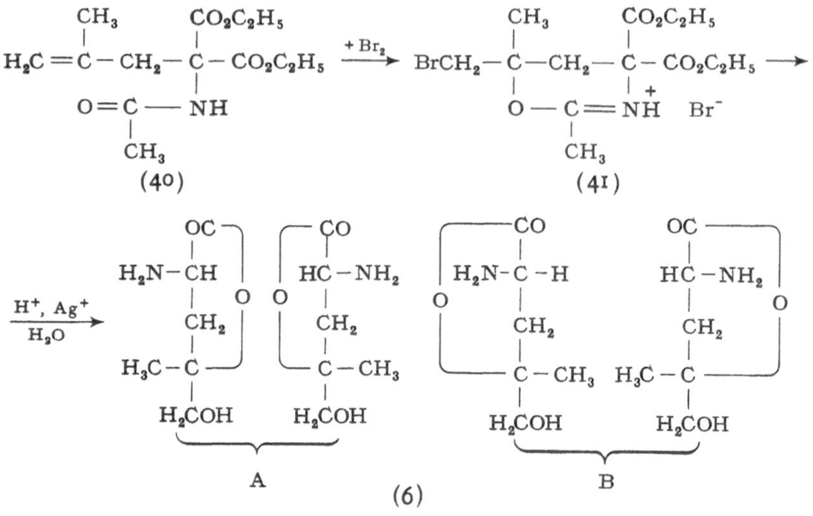

$$\underset{\substack{| \\ O=C \underset{|}{\overset{|}{}} NH \\ | \\ CH_3 \\ (40)}}{\overset{\displaystyle CH_3 \qquad CO_2C_2H_5}{H_2C=C-CH_2-C-CO_2C_2H_5}} \xrightarrow{+Br_2} \underset{\substack{| \\ O-C=NH \;\; Br^- \\ | \\ CH_3 \\ (41)}}{\overset{\displaystyle CH_3 \qquad CO_2C_2H_5}{BrCH_2-C-CH_2-C-CO_2C_2H_5}} \longrightarrow$$

$$\xrightarrow[H_2O]{H^+,\,Ag^+}$$

Chart showing structures A and B, compound (6):

OC — | H₂N–CH | CH₂ | H₃C–C — | H₂COH and CO | HC–NH₂ | CH₂ | C–CH₃ | H₂COH (A)

CO— | H₂N–C–H | CH₂ | C–CH₃ | H₂COH and OC— | HC–NH₂ | CH₂ | H₃C–C— | H₂COH (B)

(6)

Chart 5.

L-γ-*Hydroxyleucine* (**12**, p. 226).

The racemate of γ-hydroxyleucine has been synthesized as early as 1948 (**11**) by refluxing diethyl methallylacetamidomalonate (**40**) with aqueous mineral acids. After hydrolysis and decarboxylation the amino-lactone is formed by intramolecular addition of the carboxyl oxygen to the double bond of the α-aminopentenoic acid (**42**).

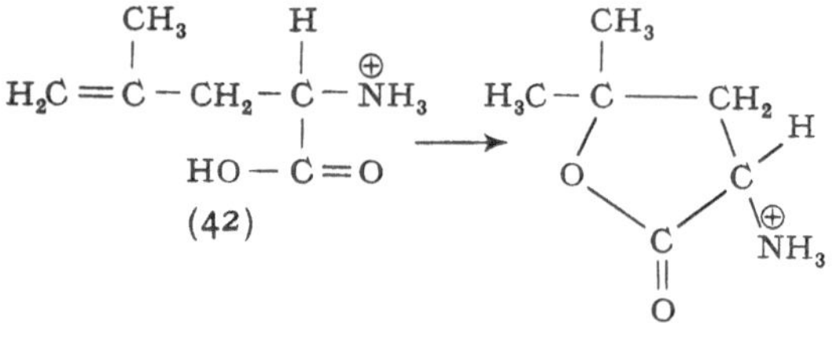

$$\underset{\substack{| \\ HO-C=O \\ (42)}}{\overset{\displaystyle CH_3 \qquad\quad H}{H_2C=C-CH_2-\overset{\displaystyle |}{\underset{\displaystyle |}{C}}-\overset{\oplus}{N}H_3}} \longrightarrow$$

(lactone structure with CH_3, CH_2, O, $C=O$, $\overset{\oplus}{N}H_3$)

The optically active compound (**12**) has been obtained by photo-chlorination (**24**) of L-leucine in 12 N hydrochloric acid by DÖLLING and FAULSTICH (**10**) in the author's laboratory. In this reaction the γ-chloro-amino acids initially formed are hydrolyzed rapidly to give the corresponding γ-lactones. It is of interest, that not only in leucine (where the γ-position has tertiary character) but also in other long-chain amino acids the γ-CH is preferentially substituted by chlorine. Thus, it also was possible to synthesize L-γ-hydroxyisoleucine (**36**) in a relatively simple manner.

References, pp. 246—250.

L-γ-*Hydroxyisoleucines* (36, p. 236).

This amino acid exists in γ-amanitin most probably as one of the four possible L-diastereoisomers (a–d). The allo-isomers c and d (derived from allo-isoleucine) can be excluded, because on photochlorination of allo-

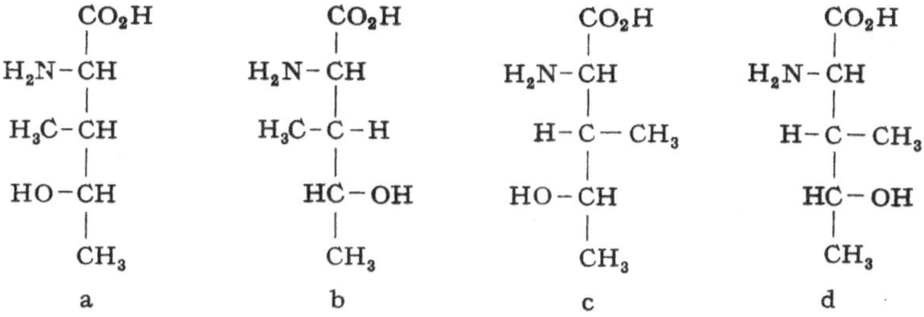

a	b	c	d
CO_2H	CO_2H	CO_2H	CO_2H
H_2N-CH	H_2N-CH	H_2N-CH	H_2N-CH
H_3C-CH	H_3C-C-H	$H-C-CH_3$	$H-C-CH_3$
$HO-CH$	$HC-OH$	$HO-CH$	$HC-OH$
CH_3	CH_3	CH_3	CH_3

(36)

isoleucine aminolactones are formed which are different from the natural product in paper electrophoresis. A decision, however, whether the γ-amanitin component has structure a or b has not yet been made. A mixture of the eight diastereoisomeric D,L-compounds is formed (*10*) by acidic hydrolysis of ethyl 2-acetamido-3-methyl-Δ^4-pentenoate (43) (*18*), according to the mechanism described on p. 240 for the hydrolysis of the aminomalonic derivative (40).

D,L-γ,δ-*Dihydroxyisoleucines* (28).

One diastereoisomer among the possible four L-compounds (28, a–d) is a constituent of α- and β-amanitines. Presumably the natural product

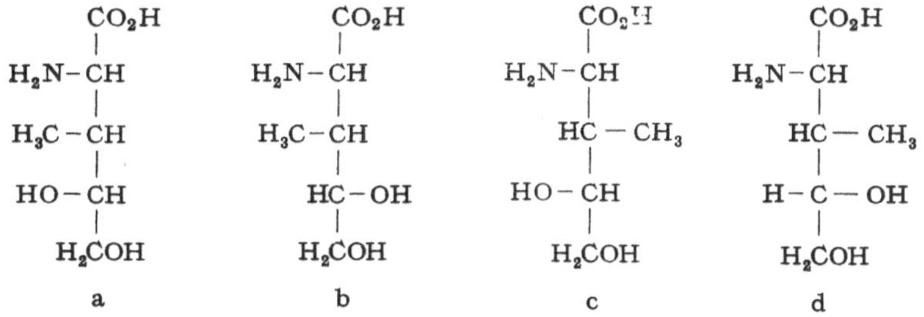

a	b	c	d
CO_2H	CO_2H	CO_2H	CO_2H
H_2N-CH	H_2N-CH	H_2N-CH	H_2N-CH
H_3C-CH	H_3C-CH	$HC-CH_3$	$HC-CH_3$
$HO-CH$	$HC-OH$	$HO-CH$	$H-C-OH$
H_2COH	H_2COH	H_2COH	H_2COH

(28)

belongs to the isoleucine series (a, b) also in this instance.* A mixture of the racemates of a–d was obtained synthetically by GEORGI (*18*) *(Chart 6)*:

---

* *Added in Proof.* Quite recently M. HASAN and P. PFAENDER in the author's laboratory proved, mainly by NMR analysis, that (28 b) is the structure of the natural product.

Bromine was added to ethyl 2-acetamido-3-methyl-Δ^4-pentenoate (43), obtained by addition of vinyl-magnesium bromide to ethylidenemalonic ester (8), nitrosation and reduction under acetylating conditions. Ag^+-assisted hydrolysis with sulfuric acid gave the γ-lactones of the D,L-compounds, whose crystalline hydrochlorides proved to be a mixture of all four species by NMR spectroscopy (four doublets of the four different CH_3-protons at about $\tau = 8.9$). The natural product exhibits only one doublet of methyl protons.

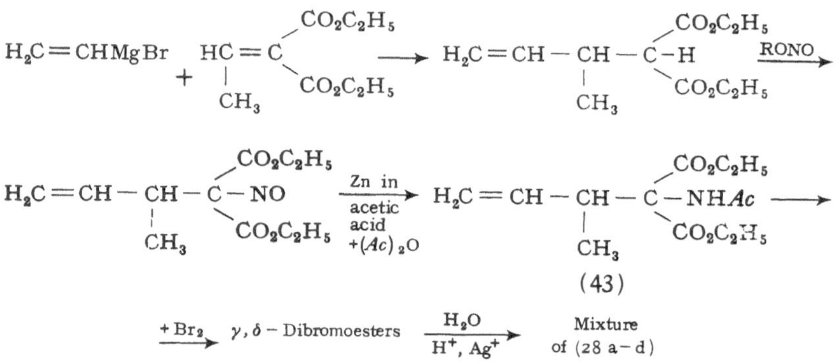

Chart 6.

2-Thioindole Compounds.

It was found in 1953 (85) that sulfur can be introduced into the 2-position of indoles containing a substituent in 3-position by reaction with appropriate S-chlorides (S_2Cl_2, $RSCl$) *(Chart 7)*. The 2-thio deri-

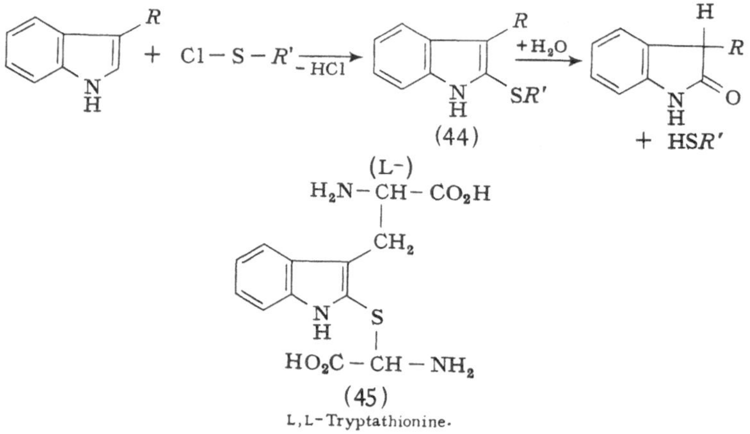

Chart 7.

vatives (44) are hydrolyzed with mineral acids under formation of oxindoles (most convenient synthesis of oxindolyl-alanine) (2, p. 221) (84).

Compounds of structure (44) showed a UV-spectrum of exactly the same shape as the phalloidines, thus proving the correct structure for the chromophoric system. By variation of residues R and R' numerous peptides related to phalloin have been synthesized (59, 72), among them the tetrafunctional bridge amino acid L,L-*tryptathionine* (45) (46). Introduction of several thio compounds, e. g., N-carbobenzyloxycysteine, into the 2-position of 6-methoxy-N-carbobenzyloxy-tryptophan yielded some model thioethers (34, p. 234) of the presumed amanitin-type (61) [OCH₃ in 6-position of (44) and (45)].

B. Cyclopeptides (54).

In order to find the conditions for the cyclization of appropriate peptides as a last step in the total synthesis of phalloin-like substances, experiments have been made to recyclize different seco compounds. All attempts to tie again the ruptured peptide bonds of the (ninhydrin-positive) seco compounds (17, p. 228) of phalloidin (9 a, p. 225) or phalloin (9 b) have not been successful. A ninhydrin-negative product, formed from (17) in molten imidazol as a condensing agent (49) and presumably having the bicyclic structure, was nontoxic. Furthermore, the bicyclus (46) obtained synthetically by Sarges (45) using ethoxyacetylene for forming the last peptide bond (at the arrow), showed no toxicity. The compound, it is true, does not contain a γ-hydroxylated amino acid like the natural toxins, but besides this feature, a stereochemical complication seems to intervene.

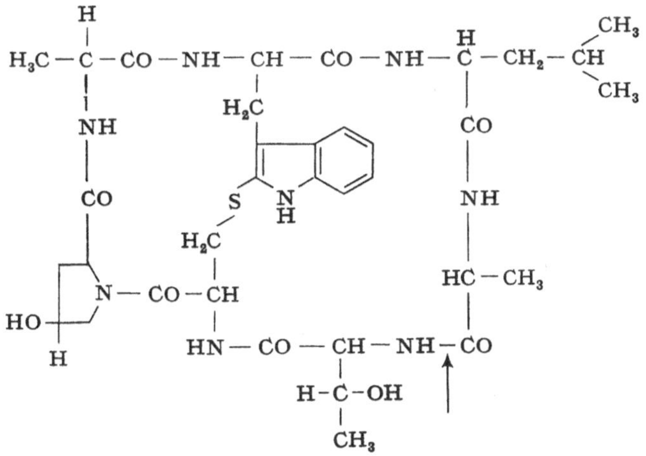

(46)

Models show that bridged cyclic compounds of the phalloin type may exist in two atropisomeric forms, A and B, because the bridge is thick enough to prevent it from swinging through the cyclopeptide ring *(Fig. 9)*.

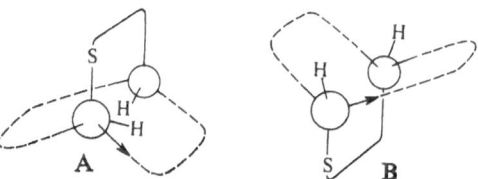

Fig. 9. Two possible atropisomers of phalloidines. [From: Liebigs Ann. Chem. **657**, 225 (1962).]

The isomers A and B may be markedly different in energy and entropy, and in a thermodynamically controlled synthesis only one of them would be expected to arise. This idea is supported by results obtained in cyclization experiments (*71*) with seco-ketophalloidin (**47**). The latter was obtained by glycol oxidation of seco-phalloidin (**17**, p. 228) with periodate (*78*).

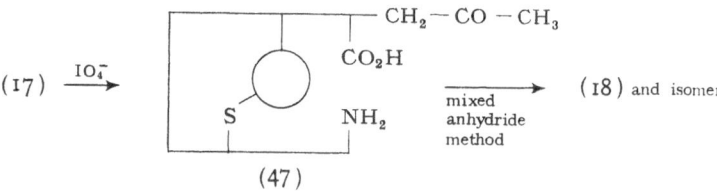

In trying to obtain a cyclic peptide from (**47**), rapidly acting reagents of peptide synthesis can be applied, whereas in similar compounds with γ-hydroxylated side-chains, under drastic conditions of condensation lactonization will always take place at the cost of forming the peptide bond. For cyclization the method of mixed anhydrides (*55*) was tried in the form recommended for the synthesis of cyclopeptides (*4*). Here, among other polar compounds, two neutral substances were formed which could be differentiated by paper chromatography. One of them was identical with ketophalloidin (**18**) in its color reactions, R_F-values and toxicity. The other substance, travelling a little faster, had the same amino acid composition but was nontoxic. This "isomer" is formed exclusively when cyclization of (**47**) is carried out using the slowly acting thionyl-bis-imidazole (**47**) as suggested for the synthesis of peptides (*79*).

V. Toxicological Remarks.

The actual intoxication by the individual phytotoxins manifests itself by morphologic changes of the parenchymatous organs, notably

the liver which is attacked first. The action of phalloidines and amanitines seems to be similar at first glance, but there are great differences
in their toxicities and in the onset and the sites of their effects: α-ama-
nitin is 10–20 times more toxic than phalloidin, as has been pointed out
on p. 219. Many more experiments have been conducted with the latter
toxin.

The features first observable biochemically point to a reaction of
phalloidin with cell and tissue structures of the liver. The microsomes
show an especially marked affinity to this toxin.

This has been shown in experiments with a radioactively labeled toxic
phalloidin derivative, which was used, instead of the original toxin, in the
perfused liver preparation (*38*). After subsequent homogenization of the
liver and differential centrifugation, radioactivity was found in cell nuclei
and debris, mitochondria and microsomes. The activity can be washed
out more-or-less easily by re-suspending the material in fresh buffer
and centrifugation. Only the microsome fraction retains a constant
amount of activity after the second washing. This corresponds well with
the results of HULTIN et al. (*9*), who found that incorporation of amino
acids into the protein of the liver microsome fraction is inhibited 1–2 hours
after the animals have been poisoned with phalloidin.

In this and other effects on liver metabolism (*52*) phalloidin acts only if it
has been administered to the intact animal. This observation led HULTIN et al.
(*9*) and recently FIUME (*13*) to the assumption that phalloidin is not toxic in
itself, but is converted into a toxic compound by metabolizing enzymes of the
liver. New-born rats endure a manifold lethal dose of the venom without being
killed (*13*). Apparently, the toxic effect is absent as long as the system
of drug-metabolizing enzymes is very rudimentary. If these enzymes are
damaged by carbon tetrachloride (*15*) or other liver poisons, the toxicity
of phalloidin is also reduced. In spite of all of these observations, there
is no consistent evidence available for a toxification of phalloidin in the
liver.

Further insight into the mechanism of the toxic action was obtained
by electron microscopy (*14, 29*) which confirmed that only the cells of
the liver are changed by phalloidin. The earliest ultrastructural changes,
observable as early as 15 minutes after phalloidin injection are, dilatation
of the endoplasmatic reticulum (which occasionally forms very large
vacuoles), swelling of mitochondria and deposition of fat droplets. The
alterations of the reticulum offer a plausible explanation of the reduced
protein synthesis (*9*) in the liver of intoxicated animals and are also
consistent with the observation of a preferred adsorption of radioactive
toxin by the microsomal elements in the liver (*38*).

The mechanism of the action of the amanitins is distinctly different
from that of the phalloins. As mentioned above, the evidence of intoxi-

cation appears considerably later. Then primarily the nuclei are affected, as demonstrated by recent electron micrographs of Fiume and Laschi (*14*). Without any damage to the endoplasmatic reticulum, 15 hours after application of α-amanitin to rats the nucleoli of liver cells begin to fragmentate. Cytoplasmatic lesions follow the nuclear ones, presumably as a consequence of the nuclear damage. The cytotoxic effects of the amanitins are not restricted to liver cells but cytological changes are observed also in the kidney.

As in the phalloidin field, also in that of the amanitins an insight into the mechanism of action on a molecular basis has not yet been gained. One might expect that such insight would be of general interest in cell physiology.

References.

1. Block, S. S., R. L. Stephens and W. A. Murrill: The Amanita Toxins in Mushrooms. Agric. Food Chem. **3**, 584 (1955).
2. Block, S. S., R. L. Stephens, A. Barretto and W. A. Murrill: Chemical Identification of the Amanita Toxin in Mushrooms. Science **121**, 505 (1955).
3. Boehringer, H.: Ein neuer Giftstoff aus den Extrakten des grünen Knollenblätterpilzes. Diplomarb., Univ. Frankfurt a. M. 1964.
4. Boissonnas, R. A. et I. Schumann: Sur la cyclisation des polypeptides. Helv. Chim. Acta **35**, 2229 (1952).
5. Boudier, E.: Des champignons, au point de vue de leurs caractères usuels chimiques et toxicologiques. Arch. hyg. publ. et méd. lég. **1867**, 27.
6. Buku, A.: Isolierung und Untersuchung von seltenen Stoffen aus *Amanita phalloides*. Diplomarb., Univ. Frankfurt a. M. 1965.
7. Cornforth, J. W., C. E. Dalgliesh and A. Neuberger: β-3-Oxindolylalanine (Hydroxytryptophan). 2. Spectroscopic and Chromotographic Properties. Biochem. J. **48**, 598 (1951).
8. Cuvigny, T. et H. Normant: Synthèse de composés γ-éthyléniques à partir des magnésiens vinyliques. Bull. soc. chim. France **1961**, 2423.
9. Decken, A. von der, H. Löw und T. Hultin: Über die primären Wirkungen von Phalloidin in Leberzellen. Biochem. Z. **332**, 503 (1960).
10. Dölling, J. und H. Faulstich: Synthese von γ-Hydroxyaminosäuren durch Photochlorierung. Dissert., J. Dölling, Univ. Frankfurt a. M. 1966/67.
11. Fillman, J. and N. F. Albertson: Amino Acid Intermediates: α-Amino-γ-lactones. J. Amer. Chem. Soc. **70**, 171 (1948).
12. Fish, M. S., N. M. Johnson and E. C. Horning: Piptadenia Alkaloids. Indole Bases of *P. peregrina* (L.) Benth. and Related Species. J. Amer. Chem. Soc. **77**, 5892 (1955).
13. Fiume, L.: Mechanism of Action of Phalloidin. Lancet **1965**, I, 1284.
14. Fiume, L. e R. Laschi: Lesioni ultrastrutturali prodotte nelle cellule parenchimali epatiche dalla falloidina e dalla α-amanitina. Sperimentale **115**, 288 (1965).
15. Floersheim, G. L.: Schutzwirkung hepatotoxischer Stoffe gegen letale Dosen eines Toxins aus *Amanita phalloides* (Phalloidin). Biochem. Pharmacol. **15**, 1589 (1966).

16. FORD, W. W. et al.: 15 references as cited in RAAB *(36).*

17. GEBERT, U.: Postdoctoral Work, Inst. Organ. Chem., Univ. Frankfurt a. M. 1965/66 (to be published).

18. GEORGI, V. und TH. WIELAND: Über die Inhaltsstoffe des grünen Knollenblätterpilzes, XXIX. Synthese von γ,δ-Dihydroxyisoleucin, der lactonisierenden Aminosäure des α- und β-Amanitins. Liebigs Ann. Chem. 700, 149 (1966).

19. GHOSAL, S. and B. MUKHERJEE: Alkaloids of *Desmodium pulchellum* Benth. ex Baker. Chem. and Ind. 1964, 1800.

20. IACOBUCCI, G. A. and E. A. RÚVEDA: Bases Derived from Tryptamine in Argentine *Piptadenia* Species. Phytochem. 3, 465 (1964).

21. KIMMIG, J.: Versuche zur Isolierung von Amanitatoxin. Dissert., Univ. München. 1935.

22. KOBERT, R.: Über Pilzvergiftung. St. Petersb. med. Wschr. 16, 463 (1891).

23. — Lehrbuch der Intoxikationen. II, SS. 624, 763. Stuttgart: Enke Verl. 1909.

24. KOLLONITSCH, J., A. ROSEGAY and G. DOLDOURAS: Reactions in Strong Acids. II. New Concept in Amino Acid Chemistry: C-Derivatization of Amino Acids. J. Amer. Chem. Soc. 86, 1857 (1964).

25. KUHN, R. und G. OSSWALD: Neue Synthese von β-Pyrrolidonen; Darstellung von DL-γ-Oxoprolin, DL-*allo*-Hydroxyprolin und 4-Äthoxypyrrolcarbonsäure-(2). Chem. Ber. 89, 1423 (1956).

26. LETELLIER, J. B.: Sur les propriétés alimentaires médicales et vénéneuses des champignons qui croissent aux environs de Paris. Thèse, Univ. Paris. 1826.

27. LYNEN, F. und U. WIELAND: Über die Giftstoffe des Knollenblätterpilzes, IV. Liebigs Ann. Chem. 533, 93 (1938).

28. MARINI-BETTÒLO, G. B., F. DELLE MONACHE and E. BIOCCA: Hallucinogenic Drugs from Amazonia. II. Epena, from the Yanoama Tribe in the Rio Negro and Upper Orinoco Basins. Ann. chim. (Roma) 54, 1179 (1964) [Chem. Abstr. 62, 8114 (1965)].

29. MILLER, F. and O. WIELAND: Elektronenmikroskopische Untersuchungen zum Wirkungsmechanismus des Knollenblätterpilzgiftes Phalloidin (im Druck).

30. ORÉ, J.: Recherches expérimentales sur l'empoisonnement par l'agaric bulbeux. Bull. Acad. méd. Paris (sér. 2) 6, 350 (1877).

31. — Recherches expérimentales sur l'empoisonnement par l'agaric bulbeux. Arch. physiol. norm. path. (sér. 2) 4, 274 (1877).

32. PAULET, G.: Traité des champignons, Vol. II. Paris. 1808.

33. PFAENDER, P. und TH. WIELAND: Über die Inhaltsstoffe des grünen Knollenblätterpilzes, XXVII. Massenspektrometrische Strukturanalyse natürlicher und synthetischer γ-Hydroxyaminosäurelactone. Liebigs Ann. Chem. 700, 126 (1966).

34. — — Reaction of Methyl-α-amanitin on Raney-Nickel. (To be published in Liebigs Ann. Chem.)

35. PUCHINGER, H.: Radioaktive Markierung eines Phalloidinderivats. Diplomarb., Univ. Frankfurt a. M. 1965.

36. RAAB, H. A.: Beiträge zur Kenntnis des Giftstoffs der Knollenblätterpilze. Z. physiol. Chem. 207, 157 (1932).

37. RAAB, H. A. und J. RENZ: Beiträge zur Kenntnis der Giftstoffe der Amanitaarten. Z. physiol. Chem. 216, 224 (1933).

38. REHBINDER, D., G. LÖFFLER, O. WIELAND und TH. WIELAND: Studien über den Mechanismus der Giftwirkung des Phalloidins mit radioaktiv markierten Giftstoffen. Z. physiol. Chem. 331, 132 (1963).

39. REMPEL, D.: Die Verwendung von Sephadex zur chromatographischen Trennung der Inhaltsstoffe von *Amanita phalloides* und *Amanita mappa.* Dissert., Univ. Frankfurt a. M. 1966.

40. RENZ, J.: Über das Amanitatoxin. Z. physiol. Chem. **230**, 245 (1934).

41. SEIBERT, J.: Beiträge zur Toxikologie der *Amanita phalloides*. Dissert., Univ. München. 1893.

42. TYLER, V. E., Jr., L. R. BRADY, R. G. BENEDICT, J. M. KHANNA and M. H. MALONE: Chromatographic and Pharmacologic Evaluation of Some Toxic *Galerina* Species. Lloydia **26**, 154 (1963).

43. TYLER, V. E., Jr. and D. GROEGER: Amanita Alkaloids. II. *Amanita citrina* and *Amanita porphyria*. Planta Med. **12**, 397 (1964).

44. SALLACH, H. J. and M. L. KORNGUTH: The Natural Occurrence of β-Hydroxyaspartic Acid. Biochim. Biophys. Acta **34**, 582 (1959).

45. SARGES, R.: Synthese eines phalloinähnlichen Peptids. Dissert., Univ. Frankfurt a. M. 1962.

46. SCHMIDT, K. H.: Beiträge zur Synthese heterodeter Cyclopeptide mit Thioätherbindung und zur Synthese des Tryptathionins. Dissert., Univ. Frankfurt a. M. 1963.

47. STAAB, H. A. und K. WENDEL: Reaktionen von N,N'-Thionyl-diimidazol. Angew. Chem. **73**, 26 (1961).

48. VAUQUELIN, M.: Expériments sur les champignons. Ann. chimie **85**, 5 (1811).

49. VOGELER, K.: Peptidsynthese mit γ-Lactonen und Alkylestern unter Imidazolkatalyse. Dissert., Univ. Frankfurt a. M. 1963.

50. WIELAND, H. und R. HALLERMAYER: Über die Giftstoffe des Knollenblätterpilzes, VI. Amanitin, das Hauptgift des Knollenblätterpilzes. Liebigs Ann. Chem. **548**, 1 (1941).

51. WIELAND, H. und B. WITKOP: Über die Giftstoffe des Knollenblätterpilzes, V. Zur Konstitution des Phalloidins. Liebigs Ann. Chem. **543**, 171 (1940).

52. WIELAND, O.: Changes in Liver Metabolism Induced by the Poisons of *Amanita phalloides*. Clin. Chem. **2**, 323 (1965).

53. WIELAND, TH.: Peptides of *Amanita phalloides*. Pure Appl. Chem. **9**, 145 (1964).

54. — Cyclisierung von Peptiden, besonders an Thioäthern. In: Peptides, Proc. 6th Europ. Sympos., Athens, 1963. New York and London: Pergamon Press. 1965.

55. WIELAND, TH. und H. BERNHARD: Über Peptidsynthesen, III. Verwendung von Anhydriden aus N-acylierten Aminosäuren und Derivaten anorganischer Säuren. Liebigs Ann. Chem. **572**, 190 (1951).

56. WIELAND, TH. und W. BOEHRINGER: Über die Giftstoffe des grünen Knollenblätterpilzes, XIX. Umwandlung von β-Amanitin in α-Amanitin. Liebigs Ann. Chem. **635**, 178 (1960).

57. WIELAND, TH. und J. DÖLLING: Die Entdeckung von γ-Hydroxyleucin im Hydrolysat von Gelatine. Naturwiss. **53**, 526 (1966).

58. WIELAND, TH. und CH. DUDENSING: Über die Giftstoffe des grünen Knollenblätterpilzes, XI. γ-Amanitin, eine weitere Giftkomponente. Liebigs Ann. Chem. **600**, 156 (1956).

59. WIELAND, TH., K. FRETER und E. GROSS: Versuche zur Synthese phalloinähnlicher Cyclopeptide. Liebigs Ann. Chem. **626**, 154 (1959).

60. WIELAND, TH. und U. GEBERT: Über die Inhaltsstoffe des grünen Knollenblätterpilzes, XXX. Die Strukturen der Amanitine. Liebigs Ann. Chem. **700**, 157 (1966).

61. WIELAND, TH. und D. GRIMM: Indole zum Vergleich mit Amanitagiften, V. Thioäthersynthesen in der Oxindol-, Indolin- und Indolreihe. Chem. Ber. **98**, 1727 (1965).

62. WIELAND, TH. und A. HÖFER: Über die Giftstoffe des grünen Knollenblätterpilzes, XVI. Die Bausteine des α-Amanitins. Liebigs Ann. Chem. **619**, 35 (1958).

63. WIELAND, TH. und H. KRANTZ: Die vier isomeren γ-Lactone des γ,δ-Dihydroxyleucins. Chem. Ber. **91**, 2619 (1958).

64. WIELAND, TH., CH. LAMPERSTORFER und CH. BIRR: Über Peptidsynthesen, 32. Der γ-Hydroxyisocaproyl- und der 3-Nitrophenoxycarbonylrest als Schutzgruppen. Makromol. Chem. **92**, 277 (1966).

65. WIELAND, TH. und K. MANNES: Über die Giftstoffe des grünen Knollenblätterpilzes, XIII. Phalloin, ein weiteres Toxin. Angew. Chem. **69**, 389 (1957).

66. WIELAND, TH., K. MANNES und A. SCHÖPF: Über die Giftstoffe des grünen Knollenblätterpilzes, XV. Die Konstitution des Phalloins. Liebigs Ann. Chem. **617**, 152 (1958).

67. WIELAND, TH., W. MOTZEL und H. MERZ: Über das Vorkommen von Bufotenin im gelben Knollenblätterpilz. Liebigs Ann. Chem. **581**, 10 (1953).

68. WIELAND, TH. und G. PFLEIDERER: Analytische und mikropräparative Trägerelektrophorese mit höheren Spannungen. Angew. Chem. **67**, 257 (1955).

69. — — Neuere Anwendungen der Hochspannungselektrophorese. Angew. Chem. **69**, 199 (1957).

70. WIELAND, TH. und D. REHBINDER: Über die Giftstoffe des grünen Knollenblätterpilzes, XXIII. ^{35}S-Markierung und chemische Umwandlungen an einer Seitenkette des Phalloidins. Liebigs Ann. Chem. **670**, 149 (1963).

70 a. WIELAND, TH., D. REMPEL, U. GEBERT, A. BUKU und H. BOEHRINGER: Über die Inhaltsstoffe des grünen Knollenblätterpilzes, XXXII. Chromatographische Auftrennung der Gesamtgifte und Isolierung der neuen Nebentoxine Amanin und Phallisin sowie des ungiftigen Amanullins. Liebigs Ann. Chem. **704**, 226 (1967).

71. WIELAND, TH. und I. SANGL: Synthese von Ketophalloidin durch Recyclisierung der Secoverbindung. Liebigs Ann. Chem. **671**, 160 (1964).

72. WIELAND, TH. und R. SARGES: Über Peptidsynthesen, XXV. Zur Synthese von Tryptathioninpeptiden. Liebigs Ann. Chem. **658**, 181 (1962).

73. WIELAND, TH., H. SCHIEFER und U. GEBERT: Giftstoffe von *Amanita verna.* Naturwiss. **53**, 39 (1966).

74. WIELAND, TH. und G. SCHMIDT: Über die Giftstoffe des Knollenblätterpilzes, VIII. Liebigs Ann. Chem. **577**, 215 (1952).

75. WIELAND, TH. und H. W. SCHNABEL: Über die Giftstoffe des grünen Knollenblätterpilzes, XXI. Die Konstitution des Phallacidins. Liebigs Ann. Chem. **657**, 218 (1962).

76. — — Über die Giftstoffe des grünen Knollenblätterpilzes, XXII. Neue Sequenzanalyse von Phalloidin und Phalloin. Liebigs Ann. Chem. **657**, 225 (1962).

77. WIELAND, TH. und W. SCHÖN: Über die Giftstoffe des grünen Knollenblätterpilzes, X. Die Konstitution des Phalloidins. Liebigs Ann. Chem. **593**, 157 (1955).

78. WIELAND, TH. und A. SCHÖPF: Über die Giftstoffe des grünen Knollenblätterpilzes, XVIII. Ergänzungen zur Phalloidinformel; Ketophalloidin. Liebigs Ann. Chem. **626**, 174 (1959).

79. WIELAND, TH. und K. VOGELER: Verwendung von Thionyl-diimidazol zur Peptidsynthese. Angew. Chem. **73**, 435 (1961).

80. WIELAND, TH. und J. X. DE VRIES: Über die Inhaltsstoffe des grünen Knollenblätterpilzes, XXXI. Phallin A, ein untoxischer und Phallin B, ein toxischer Bestandteil der lipophilen Extraktfraktionen. Liebigs Ann. Chem. **700**, 174 (1966).

81. WIELAND, TH. und H. WEHRT: Darstellung von Allohydroxy-L-prolin. Chem. Ber. **92**, 2106 (1959).

82. WIELAND, TH., H. WEHRT und J. DÖLLING: Über die Inhaltstoffe des grünen Knollenblätterpilzes, XXVI. Die Bausteine des γ-Amanitins. Liebigs Ann. Chem. **700**, 120 (1966).

83. WIELAND, TH. und O. WEIBERG: Synthese des δ-Hydroxyleuceninhydrats. Liebigs Ann. Chem. **607**, 168 (1957).

84. WIELAND, TH., O. WEIBERG, W. DILGER und E. FISCHER: Synthese des 2-Oxy- und des 2-Thiotryptophans. Liebigs Ann. Chem. **592**, 69 (1955).

85. WIELAND, TH., O. WEIBERG, E. FISCHER und G. HÖRLEIN: Darstellung schwefelhaltiger Indolderivate. Liebigs Ann. Chem. **587**, 146 (1954).

86. WIELAND, TH., L. WIRTH und E. FISCHER: Über die Giftstoffe des Knollenblätterpilzes, VII. β-Amanitin, eine dritte Komponente des Knollenblätterpilzgiftes. Liebigs Ann. Chem. **564**, 152 (1949).

(Received, October 6, 1966.)

Die Prolamine.

Von E. WALDSCHMIDT-LEITZ und H. KLING, Heiligenberg, Baden.

Mit 2 Abbildungen.

Inhaltsübersicht.

I. Einleitung.

Die Prolamine sind Bestandteile des Eiweißes allein von Getreidesamen, während Albumine und Globuline auch in den Samen vieler anderer Pflanzen verbreitet sind. Die Proteine der Getreidesamen wurden von OSBORNE (*48*) zuerst in grundlegenden Arbeiten untersucht und in vier Gruppen eingeteilt (*Tabelle 1*). Die Einteilung beruht auf der Löslichkeit der einzelnen Eiweißfraktionen: Die Albumine sind danach in

Tabelle 1. Zusammensetzung des Gersteneiweißes nach OSBORNE (*46*).

Proteinfraktion	% vom Gesamtprotein	Stickstoffgehalt (%)
Albumin	4	16,5
Globulin	31	18,1
Prolamin (Hordein)	36	19,2
Glutenin	29	16,2

salzfreiem Wasser, die Globuline in verdünnter Neutralsalzlösung, die Prolamine in hochprozentigem Äthylalkohol, die Glutenine nur in Alkali löslich. Die Bezeichnung „Prolamine" wurde von Osborne wegen ihres hohen Gehalts an Prolin und an Glutaminsäure gewählt.

Die einzelnen Eiweißfraktionen der Getreideproteine sind nicht einheitlich. Sie stellen nach neueren Untersuchungsergebnissen (Chromatographie, Elektrophorese, Sedimentationsanalyse usw.) Gemische nahe verwandter Komponenten dar.

II. Vorkommen.

Das Vorkommen der Prolamine ist auf die Samen der Getreidearten, die ja zu den Gräsern gehören, beschränkt, in den gewöhnlichen Grassamen selbst finden sich aber keine Prolamine (73). Sie kommen mit den Gluteninen zusammen im Endosperm und in der Aleuronschicht der Getreidesamen vor. Prolamine und Glutenine stellen Reservestoffe für die Ernährung des sich entwickelnden Keimlings dar; sie scheinen im Gegensatz zu den Albuminen, zu denen zahlreiche Enzyme gerechnet werden, keine physiologische Funktion zu besitzen.

Tabelle 2. Proteingehalt von Getreidesamen.

	Gesamt-protein %	davon (%)				Literatur
		Albumin	Globulin	Prolamin	Glutenin	
Weizen (Triticum vulgare)	10—15	3—5	6—10	40—50	30—40	(1, 32, 50, 47)
Roggen (Secale cereale)......	9—14	5—10	5—10	30—50	30—50	(32, 47)
Gerste (Hordeum vulgare)	10—16	3—4	10—20	35—45	35—45	(32, 10, 47)
Hafer (Avena sativa)	8—14	1	80	10—15	5	(32, 47, 15, 41)
Mais (Zea mays)	7—13	Spuren	5—6	50—55	30—45	(32, 47, 37)
Hirse (Panicum miliaceum)...	7—16	10—11		57	30	(24, 68)
Reis (Oryza sativa)	8—10	Spuren			85—90	(32, 47, 28, 49)

Von den Prolaminen sind bis heute das Hordein der Gerste, das Gliadin des Weizens und das Zein aus Mais am eingehendsten untersucht worden, während über das Gliadin des Roggens, das Avenin des

Literaturverzeichnis: SS. 264—268.

Hafers und das Panicin der Hirse nur wenige Untersuchungsergebnisse vorliegen.

Prolamine und Glutenine bilden den Hauptbestandteil der im gewöhnlichen Getreidemehl vorkommenden Eiweißstoffe (*10, 32*). Der Gehalt an Prolamin ist aber, wie *Tabelle 2* zeigt, in den verschiedenen Getreidesamen ein sehr unterschiedlicher.

Gliadin und Glutenin bilden im Weizen- und Roggenmehl zusammen das sogenannte Klebereiweiß, auch Gluten genannt, auf dessen physikalischer Konsistenz nach dem Anteigen mit Wasser die Möglichkeit beruht, Brot zu backen. Hafer enthält nur wenig Glutenin, Reismehl praktisch kein Prolamin und nur Glutenin; diese Mehlarten sind daher zur Brotbereitung nicht tauglich.

III. Löslichkeitsverhalten, Isolierung.

Die Isolierung der Prolamine erfolgt im allgemeinen durch Extraktion mit hochprozentigem Äthylalkohol (*48*). Für Hordein, Gliadin und Avenin verwendet man dabei 70—75%igen Alkohol, während Zein und Panicin einen höheren Alkoholgehalt erfordern (85—90%).

Je nach ihrer verschiedenen Löslichkeit in wasserhaltigem Alkohol können die Prolamine aus ihren Lösungen entweder durch Alkohol (Hordein, Gliadin, Avenin) oder durch Wasser (Zein, Panicin) ausgefällt werden. Die Prolamine lösen sich ferner in verschiedenen anderen Alkoholen wie Methyl-, Propyl- und Benzylalkohol; diese Lösungsmittel sind allerdings zu ihrer Isolierung weniger geeignet. Eine Lösung der Prolamine ist auch durch verdünnte Säuren und Basen möglich, diese betrifft indessen auch die anderen begleitenden Samenproteine. Weitere mehr oder weniger spezifische Extraktions- und Lösungsmittel sind Harnstoff, Glycerin, Dimethylformamid, Detergentien u. a. (*14, 44, 66, 23, 71, 57, 90, 64*).

IV. Molekulargewichte.

Zur Frage des Molekulargewichts der drei wichtigsten Prolamine, Weizengliadin, Hordein und Zein, sind zahlreiche Untersuchungen unternommen worden, mittels Sedimentations- und Diffusionsanalyse (*65, 45, 89, 42, 38*) und mittels Osmosemessungen (*19*). So hat man für Gliadin Werte von 27500 bzw. 40000 (*65, 45, 89, 42*), für Hordein einen solchen von 27500 (*55*) und für Zein von 51000 (*88, 22*) ermittelt.

Es hat sich indessen, worauf schon die unterschiedlichen Werte für Gliadin hinweisen, gezeigt, daß bei einigen Prolaminen Komponenten verschiedenen Molekulargewichts unterschieden werden können. So wird für das sogenannte α-Gliadin 200000 (*30*) bzw. 100000 (*3*), für das β-Gliadin 42000 (*30*) bzw. 31000 (*3*) und für das γ-Gliadin 47000 (*30*) bzw. 31000 (*3*) als Molekulargewicht angegeben. Diese unterschiedlichen

Werte haben sich allerdings nur bei der Anwendung verschiedenartiger Bestimmungsmethoden ergeben: Im ersteren Fall (*30*) wurden die Werte aus unmittelbaren Sedimentations- und Diffusionsmessungen errechnet, während im zweiten Fall (*3*) die Ergebnisse nach vorhergehender Anwendung eines Molekularsiebes (Sephadex) erhalten wurden. Auch für Zein wird angegeben, daß es polydispers sei (*22*); im Gegensatz dazu haben Molekulargewichtsuntersuchungen am Hordein gezeigt, daß dieses Prolamin monodispers zu sein scheint (*55*).

V. Elektrophoretische Analyse.

Im Gegensatz zu den Molekulargewichtsuntersuchungen haben elektrophoretische Messungen gezeigt, daß in allen bisher untersuchten Fällen die Prolamine wie auch die meisten tierischen und pflanzlichen Proteine Mischungen von Komponenten verschiedener Ladung darstellen. So hat man zuerst durch elektrophoretische Analyse festgestellt, daß das Hordein der Gerste aus 5 Komponenten besteht (*7, 75*), später hat man deren 7 unterschieden (*82*). Im Gliadin aus Weizen sind durch elektrophoretische Analyse 7 Komponenten aufgefunden worden (*85*), durch Elektrophorese an Stärkegel konnten sogar bis zu 30 verschiedene Komponenten unterschieden werden (*31, 90, 21*). Beim Gliadin des Roggens findet man 5 ladungsverschiedene Einzelkomponenten (*85*), Zein (*61*) setzt sich aus 6, Avenin (*87*) aus 4 elektrophoretisch unterscheidbaren Komponenten zusammen.

An dieser Stelle sei angeführt, daß sich bei Zein auch mit den Methoden der fraktionierten Fällung aus der Lösung desselben in Methylcellosolve verschiedene Komponenten gewinnen ließen (*25*). Diese Methoden wie auch die der Elektrophorese versagen beim Panicin aus Rispen- und Mohrenhirse infolge der zu geringen Löslichkeit dieser Prolamine auch in höherprozentigem Alkohol (*85*).

Die verschiedene Ladung der einzelnen Prolaminkomponenten, die bei der Elektrophorese ihrer Lösungen zutage tritt, erlaubt auch eine präparative Trennung derselben. So ist beim Hordein die präparative Darstellung der elektrophoretisch einheitlichen sogenannten ε-Komponente gelungen (*78*), deren Einheitlichkeit auch durch Anwendung der Methoden der Immunoelektrophorese bestätigt werden konnte*.

Eine verschiedene Ladung der Einzelkomponenten des Hordeins konnte auch durch Ionenaustausch an Carboxymethylcellulose festgestellt werden, es gelang damit eine Auftrennung in 3 Fraktionen (*54*).

Wie später gezeigt werden wird, ist als Hauptursache der unterschiedlichen Ladung der Einzelkomponenten der Prolamine ihre ver-

---

* Nach Versuchen von Dr. J. DAUSSANT, Institut Pasteur, Paris (Privatmitteilung von Prof. Dr. P. GRABAR, Paris).
Literaturverzeichnis: SS. 264—268.

schiedene Aminosäurezusammensetzung, nämlich ihr jeweils verschiedener Gehalt an sauren und basischen Bausteinen, ermittelt worden.

VI. Bausteinanalyse.

Da die Bausteinanalysen aus älterer Zeit unzureichend waren, kann auf ihre Anführung hier verzichtet werden. Erste mit den modernen Methoden der Papierchromatographie ausgeführte Analysen, die noch qualitativer Natur waren, sind von BISERTE und SCRIBAN (63) durchgeführt worden; die modernen quantitativen Analysen wurden mittels Chromatographie an Kunstharzsäulen durchgeführt. Quantitative Bausteinanalysen liegen heute für fast alle Prolamine vor, für Weizengliadin (85), Roggengliadin (85), Hordein (75), Zein (85), Avenin (87) und Panicin (85); ihre Ergebnisse sind in *Tabelle 3* zusammengestellt, sie beziehen sich in allen Fällen auf die natürlichen Komponentengemische der einzelnen Prolamine.

Tabelle 3. Bausteinanalyse der Getreideprolamine.
(Die Zahlen bedeuten % vom Gesamtstickstoff.)

Aminosäure	(85) Gliadin (Weizen)	(85) Gliadin (Roggen)	(75) Hordein (Gerste)	(87) Avenin (Hafer)	(85) Zein (Mais)	(85) Panicin (Hirse)
Glykokoll	1,5	1,6	1,2	1,3	1,3	1,6
Alanin	1,7	1,5	2,0	3,0	10,0	10,3
Valin	2,9	2,9	2,6	7,4	3,3	3,2
Leucin	4,8	3,7	4,6	6,4	14,2	7,3
Isoleucin	2,5	2,1	2,4	1,0	2,8	2,7
Serin	3,9	3,5	3,2	2,5	4,8	4,6
Threonin	1,2	1,7	1,3	1,5	2,1	1,7
Cystin (1/2)	1,2	2,0	1,3	1,3	0,4	0,9
Methionin	0,7	0,7	0,7	0,7	0,6	2,4
Phenylalanin	3,5	3,2	4,6	5,2	3,8	3,2
Tyrosin	1,2	0,6	1,8	1,3	2,8	2,0
Prolin	9,3	12,7	17,7	7,0	8,1	6,3
Asparaginsäure	1,3	1,3	1,0	1,3	3,4	2,1
Glutaminsäure	24,1	23,0	27,0	25,0	15,3	13,7
Lysin	0,8	1,1	0,6	0,9	0,0	0,1
Arginin	5,5	3,9	3,9	6,8	3,0	2,8
Histidin	4,5	2,5	1,4	6,1	1,7	3,9
Ammoniak	26,2	28,4	19,7	21,3	18,9	25,5
	96,8	96,4	97,0	100,0	96,5	94,3

Am Beispiel des Hordeins hat sich zuerst ergeben, daß die elektrophoretisch unterscheidbaren Einzelkomponenten der Prolamine sich ihrer verschiedenen Ladung entsprechend vor allem durch einen verschiedenen Gehalt an sauren und basischen Aminosäuren unterscheiden

(75). Dies wird in der *Tabelle 4* am Beispiel der Zusammensetzung der Komponenten mit extremer Wanderungsgeschwindigkeit, also mit maximalem bzw. minimalem Gehalt an basischen und sauren Aminosäuren, für Gliadin aus Weizen und Roggen (85) sowie für Hordein (72) und Avenin (87) belegt.

Tabelle 4. Bausteinanalysen der Komponenten mit extremer Ladung von Weizengliadin, Roggengliadin, Hordein und Avenin.
(Die Zahlen bedeuten % vom Gesamtstickstoff.)

Aminosäure	(85) Weizengliadin		(85) Roggengliadin		(72) Hordein		(87) Avenin	
	α	ζ	α	ε	α	ε	α	δ
Glykokoll	2,3	1,2	3,2	1,3	2,3	0,7	3,0	6,2
Alanin	2,2	0,9	2,9	1,1	2,4	0,8	4,4	5,2
Valin	3,1	1,3	2,8	1,9	3,4	0,9	5,7	4,2
Leucin	5,1	3,1	4,3	2,8	5,2	2,5	7,5	5,7
Isoleucin	2,3	1,6	2,8	1,8	2,4	1,2	2,4	1,6
Serin	4,5	3,6	3,3	3,1	3,1	2,2	2,4	4,0
Threonin	1,8	1,2	3,2	1,2	1,8	0,7	1,6	2,2
Cystin (1/2)	0,8	0,3	2,2	0,9	1,9	0,0	0,9	1,1
Methionin	0,9	1,0	1,2	0,5	0,7	0,6	0,2	0,9
Phenylalanin	2,5	4,0	3,0	3,6	3,2	4,5	2,8	4,5
Tyrosin	1,2	0,7	0,8	0,6	1,7	1,2	1,3	1,5
Prolin	10,4	12,1	9,9	11,3	14,1	20,8	6,8	3,9
Asparaginsäure	1,6	0,7	0,9	1,3	1,7	0,6	1,6	2,2
Glutaminsäure	21,6	25,5	21,7	23,7	19,8	21,2	18,0	20,7
Lysin	1,5	0,5	1,4	0,7	1,6	0,4	2,5	1,2
Arginin	6,9	2,1	6,6	3,8	7,2	3,9	10,9	3,1
Histidin	4,8	2,9	3,0	2,2	2,9	1,8	6,9	4,0
Ammoniak	25,3	33,0	25,4	34,0	18,6	29,9	21,0	26,3
	98,8	95,7	98,6	95,8	94,0	93,9	99,9	98,5

Daß der isoelektrische Punkt der Prolamine im annähernd neutralen Bereich liegt, z. B. für Weizengliadin bei pH = 6,5 (67, 16), und zwar trotz der großen am Aufbau beteiligten Menge Glutaminsäure, ist darauf zurückzuführen, daß die freien Carboxylgruppen dieser Aminosäure bei der Hydrolyse in etwa entsprechender Menge Ammoniak ergeben, daß also die Glutaminsäure großenteils als Amid vorliegt.

Die in der Tabelle 3 angeführten Ergebnisse zeigen ferner, daß das Gliadin aus Weizen von dem des Roggens, trotz der sonstigen Ähnlichkeit der beiden Prolamine der Zusammensetzung nach, mit Sicherheit zu unterscheiden ist; dies geht z. B. aus den erheblichen Unterschieden im Gehalt an Cystin, an Prolin und an Arginin und Histidin hervor. Für die Unterscheidbarkeit der beiden Gliadine ist weiterhin kennzeichnend, daß auch Komponenten ähnlicher Beweglichkeit aus Weizengliadin einerseits und Roggengliadin anderseits, z. B. β-Gliadin aus Weizen

Literaturverzeichnis: SS. 264—268.

(apparente Beweglichkeit = 4,7) und α-Gliadin aus Roggen (apparente Beweglichkeit = 4,6) sich in ihrer Zusammensetzung erheblich unterscheiden (85): Das erstere enthält beispielsweise viel mehr Prolin, mehr Glutaminsäure und weniger Cystin. Erwähnenswert erscheint ferner, daß auch das Panicin aus Hirse ebenso wie das Zein aus Mais kein oder fast kein Lysin enthält, daß es also nicht nur im Löslichkeitsverhalten, sondern auch in dieser Beziehung dem Zein verwandt und von den übrigen Getreideprolaminen unterschieden ist.

Der Vergleich der Zusammensetzung der untersuchten Getreideprolamine insgesamt bestätigt die alte, wenn auch noch auf unzureichende Bausteinanalysen sich gründende Anschauung von Osborne (48), wonach Beziehungen zwischen der systematischen Verwandtschaft der Getreidearten und der Zusammensetzung ihrer Prolamine bestehen. Unter den untersuchten Prolaminen sind nämlich ihrer Zusammensetzung, vornehmlich ihrem Gehalte an Glutaminsäure nach, zwei Gruppen zu unterscheiden, die Prolamine aus Weizen, Roggen, Gerste und Hafer mit hohem und die Prolamine aus Mais und Hirse mit viel geringerem Glutaminsäuregehalt. Der Vergleich des Gehaltes an Prolin andererseits läßt für das Haferprolamin eine gewisse Übergangsstellung erkennen. Die Löslichkeit in hochprozentigem Äthylalkohol scheint aber nicht, wie vermutet wurde, zum Gehalt an Prolin in Beziehung zu stehen, da sie für die prolinärmeren Prolamine größer gefunden wird.

VII. Endständige Aminosäuren.

Frühere Untersuchungen zur Bestimmung der aminoendständigen Aminosäuren, nach der Methode von Sanger (60) mit Dinitrofluorbenzol durchgeführt, hatten zu einander widersprechenden Ergebnissen geführt. So wurde für Weizengliadin in einem Fall Phenylalanin (17, 18), in einer zweiten Untersuchung (58) Glutaminsäure, Threonin, Phenylalanin, Leucin, in einer dritten (56) Histidin als aminoendständig ermittelt, für Roggengliadin einerseits Phenylalanin und Glutaminsäure (17, 18), andererseits Glutaminsäure, Threonin, Phenylalanin und Leucin (58), für Zein ausschließlich Phenylalanin angegeben (17, 18).

Demgegenüber fanden Waldschmidt-Leitz und Mitarbeiter (12, 85, 87) eine bemerkenswerte Übereinstimmung sowohl der amino- als auch der carboxylendständigen Bausteine für die meisten untersuchten Prolamine, nämlich die aus Weizen, Roggen, Gerste, Mais, Hafer und Hirse. Die Bestimmung der aminoendständigen Aminosäuren erfolgte ebenfalls nach der Sangerschen Methode (60), die der carboxylendständigen nach der Carboxypeptidasemethode von Waldschmidt-Leitz und Gauss (76).

Ähnlich der Einteilung in Gruppen auf Grund der Anteile bei den Bausteinanalysen sind also auch hinsichtlich der carboxylendständigen

Bausteine zwei Gruppen von Prolaminen zu unterscheiden, nämlich
solche mit Glutaminsäure und solche mit Asparaginsäure am Carboxyl-
ende, während als aminoendständige Aminosäuren, mit Ausnahme des
Zeins, wie *Tabelle 5* zeigt, bei allen Prolaminen übereinstimmend Glutaminsäure und Serin gefunden werden.

Tabelle 5. Endgruppen der Prolamine.

Prolamin	Aminoendstandig	Carboxylendstandig	Literatur
Weizengliadin ..	Glu, Ser, (Thr, Ala, Val)	Glu, (Asp, Leu)	(85)
Roggengliadin .	Glu, Ser, (Thr, Ala)	Glu, (Asp, Leu)	(85)
Hordein	Glu, Ser	Asp, (Leu, Ala, Val)	(12)
Panicin........	Glu, Ser, (Thr)	Asp, (Leu, Ala)	(85)
Zein	Ser, Thr	Asp, (Leu, Ala)	(85)
Avenin	Glu, Ser, (Thr)	Asp	(87)

Die Tatsache, daß bei fast allen untersuchten Prolaminen am Amino-
ende Glutaminsäure und Serin zusammen beobachtet werden, hat am
Beispiel des Hordeins eine Erklärung dahin gefunden, daß hier ein großer
Teil, mindestens aber die Hälfte der endständigen Glutaminsäure mit
der γ-Carboxylgruppe, vermutlich esterartig, an Serin gebunden vorliegt (86); dies gilt auch für das Weizengliadin (26, 36).

Diese Befunde insgesamt legen die Schlußfolgerung nahe, daß sich
das Molekül der untersuchten Prolamine aus mehreren, mindestens aber
aus zwei Peptidketten zusammensetzt; demgegenüber deuten Untersuchungsergebnisse aus neuester Zeit darauf hin, daß im Weizengliadin
nur eine einzige Peptidkette vorliege (4, 20). Weitere Befunde bezüglich
der Sekundär- und Tertiärstruktur der Prolamine, auch Versuche zur
Analyse der Aminosäuresequenzen liegen bisher in der Literatur nicht vor.

VIII. Veränderlichkeit
des Verhältnisses der Einzelkomponenten in Prolaminen.

Mit dieser Fragestellung sind eingehendere Untersuchungen am
Hordein, am Weizen- und Roggengliadin sowie am Avenin durchgeführt
worden; sie betreffen den Einfluß der Getreidesorte, des Standorts und
des Anbaujahres auf die Komponentenzusammensetzung. So haben
Versuche über einen Zeitraum von sechs aufeinanderfolgenden Anbau-
jahren bei Hordein zu der Feststellung geführt, daß das Mengenverhältnis
der elektrophoretisch unterscheidbaren Einzelkomponenten in weiten
Grenzen schwanken kann, man findet es abhängig von der Gerstensorte,
vom Standort und vom Anbaujahr (79, 81, 83, 84, 77). In den meisten
Fällen ist dabei der Sorteneinfluß der überwiegende (vgl. *Abb. 1* und *2*
Literaturverzeichnis: SS. 264—268.

sowie *Tabellen 6—9)*. Ein Einfluß einer Stickstoffdüngung in verschiedenem Ausmaß, der am Beispiel von Gerste *(81)* und von Weizen *(85)* untersucht wurde, ist dagegen kaum erkennbar.

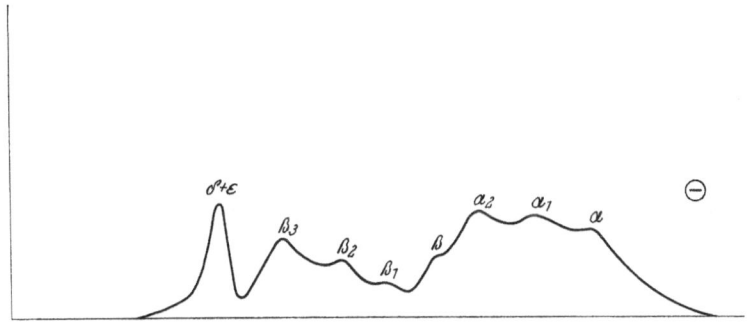

Abb. 1. Elektropherogramm von Hordein (Gerstensorte Haisa II, Anbaujahr 1959).

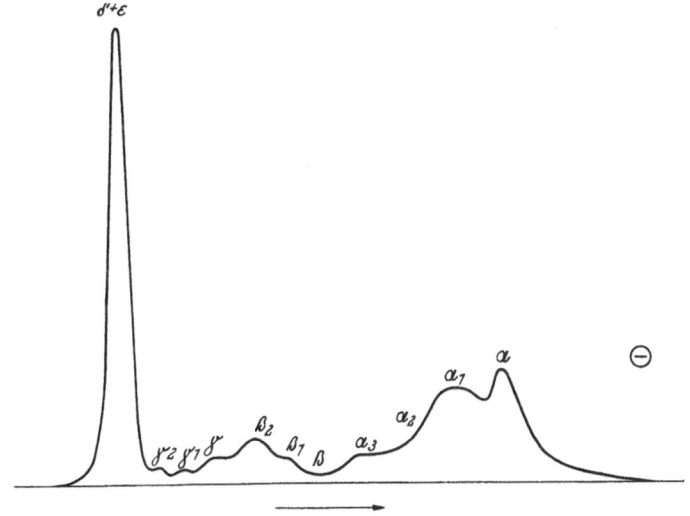

Abb. 2. Elektropherogramm von Hordein (Gerstensorte Ackermanns Donaria, Anbaujahr 1959).

Auch beim Avenin aus Hafer *(87)* ist eine ausgeprägte Sortenabhängigkeit des Komponentenverhältnisses zu bemerken, während eine solche für das Komponentenverhältnis bei Weizen- und Roggengliadin viel weniger ausgeprägt hervortritt *(85)*. In den *Tabellen 10* und *11*

17*

finden sich Beispiele für die Schwankungen im Komponentenverhältnis
bei Avenin bzw. Weizengliadin.

Tabelle 6. Sortenabhängigkeit des Anteils an α-Hordein.
(Anbaujahre 1959—1964) (77).
(+ = Abhängigkeit vorhanden, — = Abhängigkeit nicht ausgeprägt.)

Standort	1959	1960	1961	1962	1963	1964
1	+	+	+	+	—	+
2	—	+	+	+	+	±
3	—	+	+	+	+	+
4	—	+	+	+	+	+

Tabelle 7. Sortenabhängigkeit des Anteils an δ,ε-Hordein.
(Anbaujahre 1959—1964) (77).

Standort	1959	1960	1961	1962	1963	1964
1	+	+	+	+	+	+
2	+	+	+	+	+	+
3	+	+	+	+	+	+
4		+	+	+	+	+

Tabelle 8. Standortabhängigkeit des Anteils an α-Hordein (77).

Anbaujahr	Sorte						
	1	2	3	4	5	6	7
1959	+	—	—	+	+		
1960	—	+	—	—	—	—	—
1961	—	—	+	+	+	—	+
1962	—	+	—	+	—	+	—
1963	—	+	+	+	+	+	+
1964	+	—	—	+	+	—	+

Tabelle 9. Standortabhängigkeit des Anteils an δ,ε-Hordein (77).

Anbaujahr	Sorte						
	1	2	3	4	5	6	7
1959	+	+	+	+	+		
1960	—	+	—	+	—	+	+
1961	—	—	—	—	+	—	+
1962	—	—	—	+	+	+	+
1963	+	+	—	+	+	—	+
1964	+	—	+	—	+	+	+

Literaturverzeichnis: SS. 264—268.

Tabelle 10. Mengenverhältnis der Komponenten
im Avenin verschiedener Hafersorten (87).
(Die Zahlen bedeuten % des Gesamtavenins.)

Sorte	Komponente			
	α	β	γ	δ
1	18	15	53	14
2	20	17	52	11
3	27	16	50	7
4	18	27	47	8
5	27	15	52	6
6	25	18	51	6
7	24	15	48	13
8	24	13	46	17
9	22	14	49	15
10	24	15	51	10

Tabelle 11. Komponentenverhältnis im Gliadin verschiedener Weizensorten (85).

Sorte	Anbaujahr	Komponente					
		α	β	γ	δ	ε	ζ
1	1958	40	20	11	19	6	4
	1959	38	23	13	16	6	4
2	1958	30	22	13	23	6	6
	1959	30	21	13	23	7	6
3	1958	31	16	10	21	7	15
	1959	37	16	10	19	4	14
4	1958	36	19	9	21	4	11
	1959	33	22	15	19	1	10
5	1958	41	15	10	25	2	7
	1959	43	15	10	23	2	7

IX. Reifungsprozeß, Lagerung, Keimung.

Besondere Untersuchungen haben am Beispiel der Gerste gezeigt, daß im unreifen Samenkorn alle durch ihre Löslichkeit unterschiedenen Proteinfraktionen, also auch das Hordein, schon im Anfangsstadium der Kornbildung vorhanden sind; der Gehalt an der alkohollöslichen Proteinfraktion steigt dabei zugleich mit dem Gesamteiweißgehalt an (11). Bei Weizen andererseits soll das Gliadin erst als letztes der Weizenproteine ausgebildet werden (43).

Über die Entwicklung der Komponentenzusammensetzung des Hordeins während der Reifung der Gerste liegen einige vorläufige Untersuchungsergebnisse vor (33). Sie besagen, daß schon in ganz unreifem Zustand alle in der reifen Gerste nachweisbaren Einzelkomponenten des

Hordeins vorhanden sind, doch ändert sich im Verlauf des Reifungsprozesses das Mengenverhältnis der Komponenten, wenn auch die Verschiebung desselben meist keine erhebliche ist.

Auch bei der Lagerung* reifer Gerste sind größere Veränderungen im Komponentenverhältnis des Hordeins beobachtet worden. Die Tendenz dieser Veränderungen, die an insgesamt 6 Gerstensorten geprüft wurde, geht in der überwiegenden Anzahl der Beispiele dahin, daß eine Zunahme der glutaminsäure- und prolinreichsten Anteile, insbesondere der β- und δ-, ε-Komponenten auf Kosten der an Glutaminsäure und Prolin ärmeren α-Komponente erfolgt (35) (*Tabelle 12*).

Tabelle 12. Veränderung des Komponentenverhältnisses im Hordein nach halbjähriger Lagerung.

($+$ = Zunahme, $-$ = Abnahme, $\pm$ = im wesentlichen unverändert.)

Sorte	Komponente			
	α	β	γ	$\delta + \varepsilon$
1	$-$	$+$	$\pm$	$\pm$
2	$-$	$+$	$\pm$	$+$
3	$\pm$	$+$	$\pm$	$\pm$
4	$-$	$+$	$\pm$	$\pm$
5	$-$	$+$	$-$	$+$
6	$-$	$+$	$\pm$	$+$

Verschiebungen im Komponentenverhältnis des Hordeins treten auch bei der Keimung der Gerste ein (7, 8, 62); sie erfolgen in dem Sinn, daß vorwiegend die glutaminsäure- und prolinreichsten Komponenten abgebaut werden (79).

Aus alledem geht hervor, daß nicht nur bei der Reifung und Keimung des Samenkorns, sondern auch während der Lagerung desselben Umsetzungen der Eiweißkomponenten erfolgen.

X. Beziehungen zu Anthocyanogenen.

Von allen näher untersuchten Getreidesamen enthält nur das Gerstenkorn Anthocyane bzw. deren Vorstufen, die Anthocyanogene oder Leukoanthocyane (53). Während sich fertig ausgebildete Anthocyane in manchen Gerstensorten in den Spelzen vorfinden und deren Färbung verursachen (70), finden sich die farblosen Vorstufen in der Aleuronschicht des Samenkorns und in viel geringerer Menge auch im Endo-

---

* Nach Beobachtungen von Urion und Mitarb. (69) tritt bei der Lagerung der Gerste eine Abnahme ihres Prolamingehalts ein.

Literaturverzeichnis: SS. 264—268.

sperm (*39*, *13*, *2*, *53*). Aus den Leukoanthocyanen entstehen beim Erhitzen mit Säure die entsprechenden Farbstoffe, nämlich hauptsächlich Delphinidin und Cyanidin, sie können auf diesem Wege nachgewiesen und kolorimetrisch bestimmt werden (*2*).

Bei der Fraktionierung der Eiweißinhaltsstoffe der Gerste begleiten die Anthocyanogene das Hordein. Es ist dabei verschiedentlich die Frage untersucht worden, ob sie chemisch oder nur adsorptiv an das Hordein gebunden sind. Die bisher gewonnenen Erfahrungen sprechen dafür, daß keine chemische Bindung zwischen Hordein und Leukoanthocyanen besteht. Einerseits ist es nämlich möglich, durch Vorextraktion mit stark verdünntem Alkohol den größten Teil der Anthocyanogene von der Hauptmenge des Hordeins zu trennen (*52*), wobei allerdings gewisse Anteile des Hordeins, nämlich vor allem die α- und β-Komponenten mit in Lösung gehen; anderseits hat es sich gezeigt, daß auch schon durch Dialyse in Essigsäure die Farbstoffvorstufen zum größten Teil entfernt werden (*34*). Beim Vorliegen einer chemischen Bindung könnte also eine solche nur mit niedermolekularen Begleitstoffen als Partner vorliegen. Eine solche Deutung erscheint auch deshalb unwahrscheinlich, weil, wie man gefunden hat, die Anthocyanogene in allen elektrophoretisch unterscheidbaren Komponenten des Hordeins sich finden, und zwar vornehmlich in den δ- und ε-Komponenten (*34*), während ja anderseits beim Verfahren der Extraktion mit stark verdünntem Alkohol in den Lösungen sich die Anthocyanogene mit den α- und β-Komponenten vergesellschaftet finden.

Die mit Hordein in der Gerste assoziierten Anthocyanogene, die auch bei der Keimung der Gerste das Hordein bzw. höhermolekulare Abbauprodukte desselben in das Malz begleiten, spielen eine bedeutsame Rolle bei der Bildung der sogenannten reversiblen Kältetrübung im Bier, einer in der Brauereitechnik sehr unerwünschten Erscheinung (*59*, *9*). Die Bildung dieser Kältetrübung ist nämlich zurückzuführen auf das Zusammenwirken von Anthocyanogenen und Tanninsubstanzen mit höhermolekularen Abbauprodukten des Hordeins (*40*, *74*), unterstützt durch die Mitwirkung von Schwermetallspuren. Als Peptidpartner scheinen insbesondere Abkömmlinge der δ- und ε-Komponenten des Hordeins zu fungieren, wie dies aus zahlreichen Bausteinanalysen von Kältetrübungen hervorgeht (*80*).

XI. Ernährungsphysiologische Bedeutung von Prolaminen.

Wie die Zusammenstellung der Bausteinanalysen in Tabelle 3 (S. 255) erkennen läßt, ist die Zusammensetzung der Prolamine eine sehr einseitige; dies zeigt sich z. B. in ihrem ungewöhnlich hohen Glutaminsäuregehalt, der bei Hordein fast die Hälfte des Aminosäurestickstoffs bestreitet. Eine gewisse biologische Minderwertigkeit von Prolaminen ist schon lange bekannt, sie gilt nämlich für das Zein (*29*), das durch das Fehlen sowohl von Lysin wie von Tryptophan, zweier essentieller Aminosäuren gekennzeichnet ist, und gilt ebenso nach den neueren Ergebnissen für das Panicin der Hirse (*85*). Auch die anderen Prolamine sind durch einen sehr niedrigen Gehalt an den beiden Aminosäuren gekennzeichnet. Für die tierische Ernährung sind also die Prolamine als unzureichend anzusehen, während andererseits ihr hoher Gehalt an Glutaminsäure sie für die Ernährung des pflanzlichen Embryos besonders günstig erscheinen läßt, da ja die Glutaminsäure im Pflanzenreich mittels ihrer Umwandlung in Glutamin die Funktion eines Stickstoffspeichers besitzt. Es werden daher bei der Keimung des Embryos, wie früher angeführt, die Prolamine in erster Linie mobilisiert.

Wie neuere Untersuchungen ergeben haben, sind Prolamine auch für die Entstehung eines Nährmittelschadens bei der tierischen Ernährung verantwortlich. Es handelt sich um die sogenannte Cöliakie, einen bei Säuglingen bei der Ernährung mit Weizenmehl auftretenden Schaden (*6*). Wie in einer Anzahl immunologischer Untersuchungen beobachtet wurde, enthalten die Seren solcher Säuglinge Antikörper gegen Weizengliadin (*27, 51, 5*). Ein solcher Nährmittelschaden tritt nicht nur bei der Verwendung von Weizenmehl auf, sondern auch mit Gerstenmehl. Es hat sich hier gezeigt, daß die die Cöliakie hervorrufende Eigenschaft des Gerstenmehls auf Hordein zurückzuführen ist, daß aber nicht das Gesamthordein, sondern nur eine oder einige wenige Komponenten desselben dafür verantwortlich zu sein scheinen*. Bei der Gabe von Mehl aus einer an α-Hordein sehr armen marokkanischen Gerstensorte trat bei cöliakie-verdächtigen Säuglingen keine Schädigung ein, während das Auftreten von Cöliakie nach der Gabe von Mehl aus einer α-reichen, aber an den δ- und ε-Komponenten armen Gerstensorte beobachtet wurde.

Literaturverzeichnis.

1. BAILEY, C. H.: The Chemistry of Wheat Flower. New York: Chemical Catalog Co. 1925.
2. BATE-SMITH, E. C.: Leuco-anthocyanins. Detection and Identification of Anthocyanidins Formed from Leuco-anthocyanins in Plant Tissues. Biochem. J. **58**, 122 (1954).
3. BECKWITH, A. C., H. C. NIELSEN, J. S. WALL and F. R. HUEBNER: Isolation and Characterisation of High-molecular Weight Protein From Wheat Gliadin. Cereal Chem. **43**, 14 (1966).
4. BECKWITH, A. C. and J. S. WALL: Reversible Reduction and Reoxidation of the Disulfide Bond in Wheat Gliadin. Arch. Biochem. Biophys. **112**, 16 (1965).
5. BERGER, E., A. BÜRGIN-WOLFF und E. FREUDENBERG: Diagnostische Bewertung des Nachweises von Gliadin-Antikörpern bei Cöliakie. Klin. Wschr. **42**, 788 (1964).

* Privatmitteilung von Prof. Dr. E. FREUDENBERG, Basel.

6. BERGER, E. und E. FREUDENBERG: Ist die Cöliakie durch Allergie gegen Gliadin bedingt? Schweiz. med. Wschr. **93**, 549 (1963).

7. BISERTE, G. et R. SCRIBAN: Les protides de l'orge. Bull. soc. chim. biol. (Paris) **32**, 959 (1950).

8. — — Les protides du malt. Bull. soc. chim. biol. (Paris) **33**, 114 (1951).

9. — — Proc. E. B. C. Congr. Nizza, S. 48 (1953).

10. BISHOP, L. R.: The Composition and Quantitative Estimation of Barley Proteins. J. Inst. Brew. **34**, 101 (1928).

11. — The Composition and Determination of Barley Proteins. III. Fourth Report on Barley Proteins. The Proteins of Barley During Development and Storage of Mature Grain. J. Inst. Brew. **36**, 336 (1930).

12. BRUTSCHECK, H.: Zur Kenntnis des Hordeins, dem Prolamin der Gerste. Dissert. München. 1958.

13. CHAZE, J.: Sur la présence de pigments anthocyaniques ou de composés oxyflavoniques dans les grains d'aleurone de certaines Graminées. C. R. hebd. séances Acad. Sci. **196**, 952 (1933).

14. COOK, W. H.: Preparation and Heat Denaturation of the Gluten Proteins. Canad. J. Res. **5**, 389 (1931).

15. CSONKA, F. A.: Studies on Glutelins. III. The Glutelin of Oats *(Avena sativa)*. J. Biol. Chem. **75**, 189 (1927).

16. CSONKA, F. A., J. C. MURPHY and D. B. JONES: The Iso-Electric Points of Various Proteins. J. Amer. Chem. Soc. **48**, 763 (1926).

17. DEUTSCH, T.: Die N-endständigen Aminosäuren der Weizen- und Roggengliadine. Acta physiol. Acad. Sci. hung. **6**, 209 (1954).

18. — Die Bestimmung der N-endständigen Aminosäuren einiger Prolamine. Ung. Z. Chem. **61**, 135 (1955).

19. DUCLAUX, J. et A. DOBRY: Pression osmotique des protéines dans les solvants autres que l'eau. C. R. trav. Lab. Carlsberg **22**, 155 (1938).

20. ESCRIBANO, M. J., H. KEILOVA et P. GRABAR: Etude de la gliadine et de la glutenine après réduction ou oxydation. Biochim. Biophys. Acta **127**, 94 (1966).

21. FEILLET, P.: Contribution à l'étude des protéines du blé. Influence des facteurs génétiques, agronomiques et technologiques. Thèse, Univ. Paris. 1965.

22. FOSTER, J. F. and D. FRENCH: The Partial Specific Volumes of Zein and Gliadin. J. Amer. Chem. Soc. **67**, 687 (1945).

23. FOSTER, J. F., J. T. YANG and N. H. YUI: Extraction and Electrophoretic Analysis of the Proteins of Corn. Cereal Chem. **27**, 477 (1950).

24. GERPE, M. COMENGE and G. TABARES: Analytical Study of some Cereals Cultivated in Spain. Anales fis. y quim. (Madrid) **43**, 1033 (1947).

25. GORTNER, R. A. and R. T. McDONALD: Studies on the Fractionation of Zein. Cereal Chem. **21**, 324 (1944).

26. HAUROWITZ, F. and F. BURSA: The Linkage of Glutamic Acid in Protein Molecules. Biochem. J. **44**, 509 (1949).

27. HEINER, D. D., M. E. LAHEY, F. WILSON, G. GERRAYD, H. SWACHMAN and K. T. KHAW: Precipitins to Antigens of Wheat and Cow's Milk in Celiac Disease. J. Pediat. **61**, 813 (1962).

28. HOFFMAN, W. F.: An Alcohol-soluble Protein Isolated from Polished Rice. J. Biol. Chem. **66**, 501 (1925).

29. HOPKINS, F. G. and E. G. WILLCOCK: The Importance of Individual Amino-Acids in Metabolism. J. Physiol. (London) **35**, 88 (1906).

30. JONES, R. W., G. E. BABCOCK, N. W. TAYLOR and F. R. SENTI: Molecular Weights of Wheat Gluten Fractions. Arch. Biochem. Biophys. **94**, 483 (1961).

31. Jones, R. W., N. W. Taylor and F. R. Senti: Electrophoresis and Fractionation of Wheat Gluten. Arch. Biochem. Biophys. **84**, 363 (1959).
32. Kent-Jones, D. W. and A. J. Amos: Modern Cereal Chemistry, 4th ed. Liverpool: Northern Publ. Co. 1947.
33. Kling, H.: Über Hordein und Gliadin im unreifen und reifen Samenkorn. Brauwiss. **19**, 388 (1966).
34. Kloos, G.: Über Zusammensetzung und Anthocyanogengehalt des Hordeins. Brauwiss. **14**, 223 (1961).
35. — (unveröffentlicht).
36. Kovács, J., I. Kandel, M. Kandel und V. Bruckner: Über die Bindungsart der amidierten Reste der β-Aminodicarbonsäuren in Eiweißstoffen. Experientia **11**, 96 (1955).
37. Kühnau, J.: Über den Aminosäuregehalt der Getreidearten und der Sojabohne. Z. Lebensmittel-Unters. Forsch. **90**, 434 (1950).
38. Lamm, O. and A. Polson: The Determination of Diffusion Constants of Proteins by a Refractometric Method. Biochem. J. **30**, 528 (1936).
39. Lewicki, S.: Mém. inst. natl. polon. écon. rurale Pulawy **10**, 293 (1929).
40. Ljungdahl, L. und E. Sandegren: Über die Zusammensetzung der Kältetrübung im Bier. Proc. E. B. C. Congr. Baden-Baden. S. 98 (1955).
41. Luers, H. und M. Siegert: Zur Kenntnis der Proteine des Hafers. Biochem Z. **144**, 467 (1924).
42. Lundgren, H. P. and W. H. Ward: Molecular Size of Proteins. In: D. M. Greenberg (Edit.), Amino Acids and Proteins, p. 312. Springfield, Ill.: C. C. Thomas. 1951.
43. McCalla, A. G.: Fractionation of Nitrogen in Developing Wheat Kernels. Canad. J. Res. **16**, 263 (1938).
44. McCalla, A. G. and R. C. Rose: Fractionation of Gluten Dispersed in Sodium Salicylate Solution. Canad. J. Res. **12**, 346 (1935).
45. Neurath, H.: The Investigation of Proteins by Diffusion Measurements. Chem. Rev. **30**, 357 (1942).
46. Osborne, T. B.: The Proteids of Barley. J. Amer. Chem. Soc. **17**, 587 (1895).
47. — Die Pflanzenproteine. Erg. Physiol. **10**, 47 (1910).
48. — The Vegetable Proteins, 2nd ed. London: Longmans, Green and Co. 1924.
49. Osborne, T. B., D. D. Van Slyke, C. S. Leavenworth and M. Vinograd: Some Products of Hydrolysis of Gliadin. Lactalbumin, and the Protein of the Rice Kernel. J. Biol. Chem. **22**, 259 (1915).
50. Osborne, T. B. and C. G. Voerhees: The Proteids of Wheat Kernel. J. Amer. Chem. Soc. **15**, 392 (1893).
51. Pokorna, J. Sourek und J. Svejcar: Die Bedeutung der Präzipitine gegenüber Gliadin im Serum von Kindern mit Cöliakie. Helv. Paediat. Acta **18**, Fasc. 5, 393 (1963).
52. Pollock, J. R. A. and A. A. Pool: Hordein in Relation to Anthocyanogens (Leucoanthocyanins) in Ungerminated Barley. J. Inst. Brew. **65**, 483 (1959).
53. Pollock, J. R. A., A. A. Pool and T. Reynolds: Chemical Aspects of Malting. IX. Anthocyanogens in Barley and Other Cereals and their Fate During Malting. J. Inst. Brew. **66**, 389 (1960).
54. Préaux, G., M. P. Holemans et A. Monshouwer: Etude des protéines et des polypeptides de l'orge. Echo brasserie **20**, 617 (1964).
55. Quensel, O. and T. Svedberg: Studies of the Brewing Process by Means of Ultracentrifugal Sedimentation, Diffusion and Electrophoresis Measurements. C. R. trav. Lab. Carlsberg **22**, 441 (1938).

56. RAMACHANDRAN, L. K. and W. B. MCCONNELL: The Terminal Amino Acids of Wheat Gliadin. Canad. J. Chem. 33, 1463 (1955).

57. REES, E. D. and S. J. SINGER: A Preliminary Study of the Properties of Proteins in Some Nonaqueous Solvents. Arch. Biochem. Biophys. 63, 144 (1956).

58. REZNICHENKO, M. S, L. I. POLOTNOVA and E. V. TSVETKOVA: N-terminal Amino Acid Residues of Wheat Gliadin. Biokhimiya 23, 649 (1958). [Chem. Abstr. 53, 2305 (1959)].

59. SANDEGREN, E.: Proc. E. B. C. Congr. Scheveningen, S. 28 (1947).

60. SANGER, F.: The Free Amino Groups of Insulin. Biochem. J. 39, 507 (1945).

61. SCALLET, B. L.: Zein Solutions as Association-dissociation Systems. J. Amer. Chem. Soc. 69, 1602 (1947).

62. SCRIBAN, R.: Les protides de l'orge du malt et du moût. Thèse, Univ. Lille. 1951.

63. SCRIBAN, R. et G. BISERTE: Observations sur quelques prolamines. Bull. soc. botan. Nord 3, 95 (1950).

64. SELA, M., N. LUPU, A. YARON and A. BERGER: Water-soluble Polypeptidyl Gliadins. Biochim. Biophys. Acta 62, 594 (1962).

65. SVEDBERG, T. and K. O. PEDERSEN: The Ultracentrifuge. Oxford: Clarendon Press. 1940.

66. SWALLEN, L. G. and J. P. DANKY: In: J. ALEXANDER (Edit.), Colloid Chemistry, Vol. VI, p. 1140. New York: Reinhold Publ. 1946.

67. TAGUE, E. L.: The Iso-Electric Points of Gliadin and Glutenin. J. Amer. Chem. Soc. 47, 418 (1925).

68. TENG-YI, LO: J. Chinese Chem. Soc. 8, 170 (1941).

69. URION, E., G. LEJEUNE et M. FRONSACQ-COLLIN: Variations saisonnières dans la composition protéique des graines de céréales stockées. Bull. soc. chim. biol. (Paris) 33, 120 (1951).

70. URION, E. und M. METCHE: Isolierung und Struktur von Anthocyanosiden des Gerstenkorns. Brauwiss. 14, 227 (1961).

71. VERCOUTEREN, R. et R. LONTIE: La solubilisation du glutène de froment par la diméthylformamide. Arch. internat. physiol. 62, 579 (1954).

72. WALDSCHMIDT-LEITZ, E.: Zur chemischen Systematik der Getreideeiweiß-körper. Proc. E. B. C. Congr., S. 78 (1963).

73. — (unveröffentlicht).

74. WALDSCHMIDT-LEITZ, E. und H. BRUTSCHECK: Zur Analyse der Gersteneiweiß-körper. Brauwiss. 12, 278 (1955).

75. — — Über die Zusammensetzung der elektrophoretisch unterscheidbaren Komponenten des Hordeins. Z. physiol. Chem. 311, 1 (1958).

76. WALDSCHMIDT-LEITZ, E. und K. GAUSS: Zur Bestimmung der carboxylend-ständigen Bausteine in Proteinen. Z. physiol. Chem. 293, 10 (1953).

77. WALDSCHMIDT-LEITZ, E. und H. KLING: Über die Veränderlichkeit der Zu-sammensetzung des Hordeins. III. Brauwiss. 19, 17 (1966).

78. — — Zur Darstellung elektrophoretisch einheitlicher Komponenten von Pro-laminen. Z. physiol. Chem. 346, 17 (1966).

79. WALDSCHMIDT-LEITZ, E. und G. KLOOS: Über die Veränderlichkeit des Kom-ponentenverhältnisses im Hordein. Z. physiol. Chem. 314, 218 (1959).

80. — — Über die spezifische Zusammensetzung der Kältetrübung aus Bier. Z. physiol. Chem. 316, 88 (1959).

81. — — Über den Einfluß von Düngungs- und Standortbedingungen auf die Zusammensetzung des Hordeins. Z. physiol. Chem. 321, 114 (1960).

82. — — Über die Veränderlichkeit der Zusammensetzung des Hordeins. Brau-wiss. 13, 64 (1960).

83. Waldschmidt-Leitz, E. und G. Kloos: Über die Veränderlichkeit der Zusammensetzung des Hordeins. I. Braugerstenjahrbuch **1960—62**, 16.

84. — — Über die Veränderlichkeit der Zusammensetzung des Hordeins. II. Brauwiss. **16**, 459 (1963).

85. Waldschmidt-Leitz, E. und P. Metzner: Über die Prolamine aus Weizen, Roggen, Mais und Hirse. Z. physiol. Chem. **329**, 52 (1962).

86. Waldschmidt-Leitz, E. und E. Reicheneder: Über das Vorkommen von γ-Glutamyl-Bindungen in Getreideproteinen. Z. physiol. Chem. **323**, 124 (1961).

87. Waldschmidt-Leitz, E. und O. Zwisler: Über die Proteine des Hafers. Z. physiol. Chem. **332**, 216 (1963).

88. Watson, C. C., S. Arrhenius and J. W. Williams: Physical Chemistry of Zein. Nature **137**, 322 (1936).

89. Weissberger, A. (Edit.): Technique of Organic Chemistry, 3rd ed., Vol. I, p. 1102. New York: Interscience Publ. 1959.

90. Woychik, J. H., J. A. Boundy and R. J. Dimler: Starch Gel Electrophoresis of Wheat Gluten Proteins with Concentrated Urea. Arch. Biochem. Biophys. **94**, 477 (1961).

(Eingelaufen am 27. September 1966.)

Conformational Analysis of Some Alkaloids.

By **G. A. Morrison**, Leeds.

Contents. Page

I. Introduction.

The first clear statement of the principles of conformational analysis, as applied to cyclohexane derivatives, was made by BARTON in 1950 (*21*). Since then conformational arguments have been invoked to deal with questions of relative stereochemistry throughout the whole realm of natural products. A previous review in this Series has discussed the application of conformational analysis to steroids and related natural products (*24*) and recently the numerous uses of this approach in the field of carbohydrate chemistry have been surveyed (*61a*). The extension of conformational principles to nitrogen-containing molecules was an early development (*43*, *110*), and alkaloids now constitute a great area of natural products which has been extensively investigated by the methods of conformational analysis.

Conformational analysis has been employed in elucidating the configurations and in interpreting the reactions of practically every group of alkaloids where such an approach is valid. Within the compass of a short review an exhaustive coverage of the subject is not possible; accordingly, in this review, the application of conformational analysis to the study of alkaloids is illustrated by reference mainly to the elucidation of the configurations of certain yohimbinoid and *Amaryllidaceae* alkaloids.

II. Summary of the Principles of Conformational Analysis.

At the time of writing, a large number of review articles (e. g., *22, 24, 59, 60, 121, 132, 145*) and two textbooks (*61, 86*) are available which provide detailed accounts of conformational principles. Only a very brief summary of these principles will therefore be attempted here.

Conformational analysis is an analysis of the physical and chemical properties of a compound in terms of the conformation or conformations adopted by its constituent molecules. Since the conformations of a molecule are all those non-superimposable steric arrangements of its atoms which may arise by rotation about single bonds, it is clear that for a given configuration there will be an infinite number of possible conformations. The application of conformational analysis is, however, greatly simplified by the fact that, in the ground state, the conformation adopted by most of the molecules of a compound is the one of lowest energy, usually referred to as the "preferred conformation" of the molecule.

In the absence of hydrogen-bonding or electrostatic effects the preferred conformation of a molecule is that in which non-bonded interactions are at a minimum. Thus, the preferred conformation of ethane may be represented as (1) (*167*), in which there is maximum staggering of the C—H bonds. In general, in aliphatic hydrocarbons, the carbon skeleton adopts a zig-zag arrangement so as to minimise the non-bonded interactions.

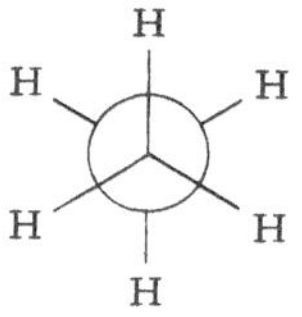

(1)

Ethane (Newman projection).

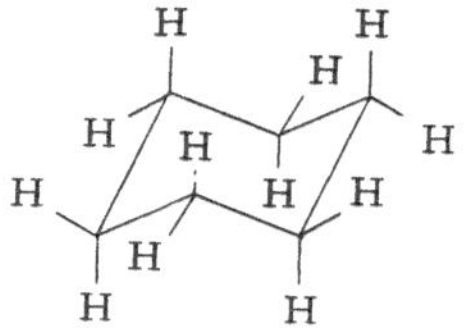

(2)

Cyclohexane (Chair conformation).

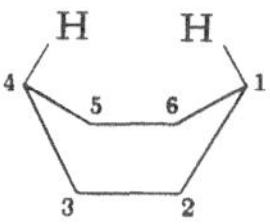

(3)

Cyclohexane (Boat conformation).

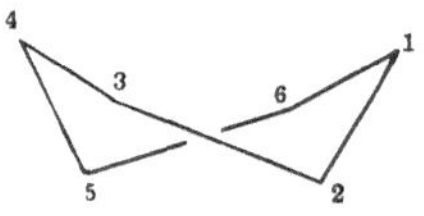

(4)

Cyclohexane (Skew – boat conformation).

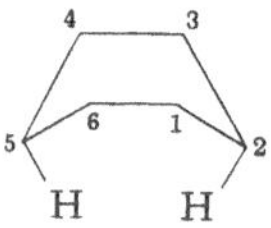

(5)

References, pp. 307—317.

Among natural products the cyclohexane ring is a more important structural unit. It can exist in two conformations which are free from angle strain, the rigid or chair conformation (2) and the flexible conformation.

The most stable flexible conformation of cyclohexane is (4) (90), described variously as the "skew", "twist", "skew-boat", "twist-boat" and "stretched" form. It is intermediate between the two classical boat conformations (3) and (5) (93), into either of which it may be converted simply by rotation about the ring bonds. The skew-boat does not possess the severe non-bonded interaction between the "flagpole" hydrogen atoms present in the boat form. Experimental determinations of the enthalpy of the flexible form (9, 14, 116, 138) have afforded results indicating that it is some 4.8–5.9 kcal./mole less stable than the chair. These results are in fair agreement with values of ΔH obtained by calculation (5, 39, 90, 92, 98, 148).

Only in the chair conformation are C—H bonds emanating from adjacent carbon atoms fully staggered*. Usually, therefore, wherever it is configurationally possible, the chair conformation of cyclohexane and its derivatives is preferred. In special circumstances, however, especially when one of the ring atoms is trigonal, the flexible conformation may be the more stable (18, 24, 61, 129). For a recent example of a saturated cyclohexane derivative which prefers the flexible conformation see (35).

Two distinct types of C—H bond may be discerned in the chair conformation of cyclohexane. Six "axial" (a) bonds lie parallel to the major axis of the ring (6); those bonds pointing out from the ring are designated "equatorial" (e) (7) (145). A very important property of the cyclohexane chair is that, by a flexing of bond angles, it can be converted through the flexible form, into an enantiomeric chair conformation in which all the C—H bonds which were originally axial have become equatorial and vice versa. Thus, a monosubstituted cyclohexane can exist in two chair conformations, with the substituent either axial (8) or equatorial (9). For cyclohexane itself, the free energy of activation is about 10.3 kcal./mole, as determined from NMR spectroscopic data (11).

---

* In the ideal chair conformation all the valence angles have the exact tetrahedral value of 109° 28', and the dihedral angle between adjacent equatorial C—H bonds and between an adjacent pair of axial and equatorial C—H bonds is 60°. Electron diffraction data (48, 118), however, indicate that a ring angle of 111.5° is more common and this has been shown by calculation to be of lower energy than the regular chair (38, 39). This has the important consequence that the trans-ee dihedral angle is approximately 65.5° and the cis-ea dihedral angle becomes approximately 54.5°. On the basis of this slightly flattened chair conformation it is possible to explain some of the more subtle features of the chemistry of cyclohexane and its derivatives (193).

At ordinary temperatures, therefore, the two possible chair conformations of cyclohexane are in rapid equilibrium. However, at very low temperatures the rate of interconversion is greatly reduced, and the isolation of a pure specimen of the equatorial conformer of chlorocyclohexane by fractional crystallisation at — 150° has recently been reported (*113*).

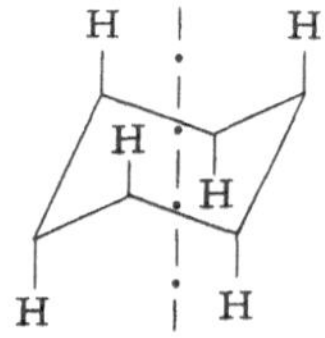

(6)

Axial C — H bonds in cyclohexane.

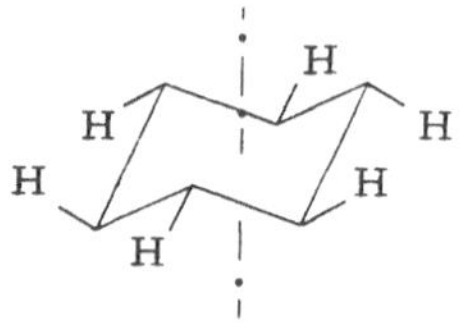

(7)

Equatorial C — H bonds in cyclohexane.

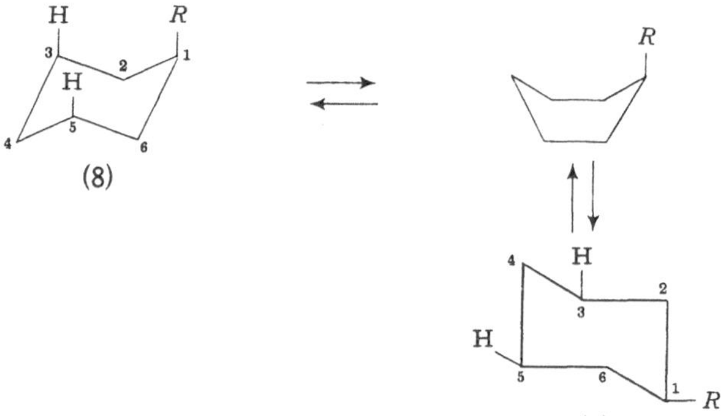

(8)

(9)

Since a substituent in an axial conformation is subject to greater non-bonded interactions than when it is equatorial (due mainly to the syn-axial interactions with axial hydrogen atoms on the same side of the ring), the preferred chair conformation of a simple cyclohexane derivative is that which results in the greater number of substituents with an equatorial orientation. For the same reason, in systems where it is impossible for the six-membered ring to undergo a conformational inversion (in *trans*-decalin, for example) an axial substituent attached to a configurationally labile carbon atom is easily epimerised. The free energy change associated with changing the conformation of a substituent on a cyclohexane ring has been determined for a large number of groups (*61b*). For exceptions to the general rule that substituents on six-membered rings are more stable when they are equatorially oriented the reader is referred to an earlier review (*24*).

In cyclohexene, $C_{(1)}$, $C_{(2)}$, $C_{(3)}$ and $C_{(6)}$ must of necessity be coplanar. The most stable conformation of the molecule is the "half-chair" form

(10) in which the remaining two carbon atoms are symmetrically disposed above and below the plane of the other four (27, 38, 39). The "half-boat" conformation (11), in which the two extraplanar carbon atoms lie on the same side of the $C_{(1)}$—$C_{(2)}$—$C_{(3)}$—$C_{(6)}$ plane and at the same distance from it, has been shown (38, 39) not to correspond to a minimum of conformational energy; it is, rather, to be regarded as one extreme in the vibration of the half-chair form. Like the chair conformation of cyclohexane, the cyclohexene half-chair is interconvertible with an enantiomeric conformation; the free energy of activation for this process as determined from NMR studies is 5.3 kcal./mole (12). Only $C_{(4)}$ and $C_{(5)}$ (10) carry true axial and equatorial bonds; those at $C_{(3)}$ and $C_{(6)}$ are referred to as pseudoaxial (a') and pseudoequatorial (e'). A similar half-chair conformation is adopted by ethylene oxide (146).

(10)
Cyclohexene
(Half - chair conformation).

(11)
Cyclohexene
(Half - boat conformation).

Many natural products consist of assemblages of fused six-membered rings. Generally, wherever it is configurationally possible, such fused systems have a preferred all-chair conformation. Thus *trans*-decalin

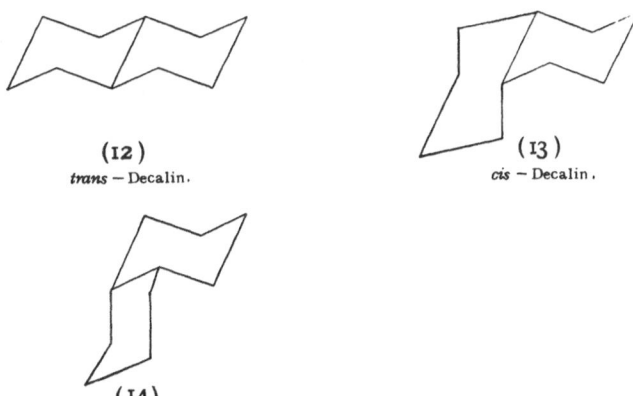

(12)
trans – Decalin.

(13)
cis – Decalin.

(14)

is to be represented by (12) and *cis*-decalin by (13) (25, 49; cf. 39, 82, 83). Because there are greater non-bonded interactions in the *cis* form its enthalpy is some 2.7 kcal./mole greater than that of the *trans*-isomer

(*8, 168*). Since it is impossible to fuse two six-membered rings through adjacent *trans*-diaxial bonds, neither of the rings of *trans*-decalin can undergo a chair inversion. In rigid molecules of this type there is a fixed relation between the configuration of any substituent and its conformation. *cis*-Decalin, where the rings are fused through adjacent axial and equatorial bonds, is subject to no such constraint. It can, like cyclohexane, adopt either of two enantiomeric two-chair conformations, (13) and (14). In its preferred conformation a substituted *cis*-decalin has the maximum number of substituents equatorially disposed.

The concept of axial and equatorial bonds has led to a number of very useful generalisations (*21, 22, 24, 61, 86*). An axial group is subjected to greater steric compression and is more hindered than it would be if it were equatorial. Consequently, equatorial carboxyl and hydroxyl groups are more easily esterified and the resulting esters are more readily hydrolysed. Exceptions to this rule occur when saponification of an axial ester grouping is assisted by participation of a suitably located hydroxyl group; for a discussion of this effect in certain indole alkaloids see (*3, 4*).

Since to some extent the strengths of acids and bases depend upon the degree to which the derived ions are solvated, it is usually the case that the less hindered equatorial epimer is stronger. In general, an axially disposed secondary hydroxyl group is oxidised with chromium trioxide faster than its epimer; for allylic hydroxyl groups, however, the reverse holds (*61c*). In S_N2 reactions an axial group is displaced faster, since the approach of the nucleophile from the rear is less hindered. Axial groups are also displaced faster in S_N1 reactions; in this case formation of a trigonal carbonium ion is the rate determining step, and hence there is greater release of steric compression if an axial leaving group is involved.

Certain reactions proceed through transition states which are stereochemically very demanding and these are of course very sensitive to conformational factors. Thus, for maximum ease of reaction, the four centres involved in an E2 elimination (the two groups to be eliminated and the atoms to which they are bonded) must be coplanar, with the groups to be eliminated disposed on opposite sides of the central bond. In conformational terms, this means that two groups bonded to adjacent carbon atoms of a cyclohexane ring must both be axial if their elimination is to proceed easily (see 15). (If an *anti*-coplanar conformation is not possible, the next most favoured is a *syn*-coplanar arrangement (*50, 61d, 96*); this is of importance when considering cyclopentane derivatives and cyclohexane boat conformations.)

Similarly, the formation of epoxides by base treatment of halohydrins requires an anticoplanar disposition of the centres involved, i. e. if the halogen and hydroxyl groups are substituents on a cyclohexane ring, it must be possible for both to assume an axial conformation (see 16) (*23*). The converse of this is also true; nucleophilic or electrophilic cleavage of a cyclohexanic epoxide usually proceeds in a diaxial fashion.

Halonium ions have a similar geometry to epoxide rings and it is not surprising therefore that the overall result of halogenation of a cyclohexane derivative is diaxial addition (17 → 18 → 19).

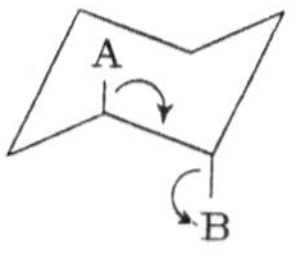

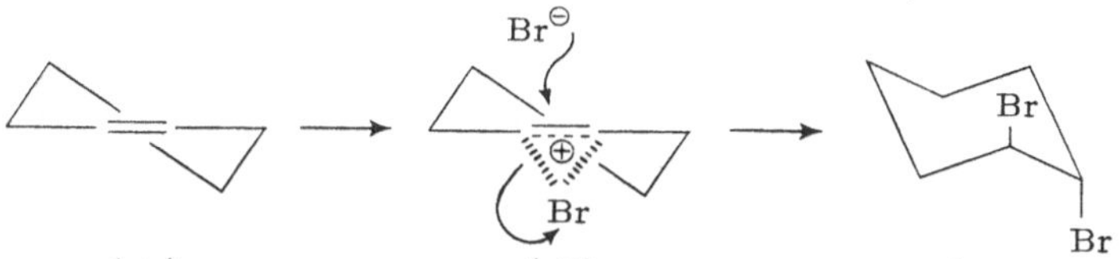

Halonium ions are also intermediates in the replacement reactions of halohydrins which proceed with retention of configuration. Hence, in neighbouring group participation of this type the importance of an anticoplanar arrangement of the four centres involved is again apparent. In cyclohexanic compounds such reactions proceed much more easily if the initial halohydrin can assume a *trans*-diaxial conformation (*10*).

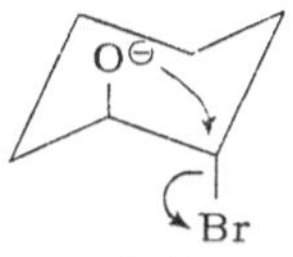

Many molecular rearrangements are highly stereoselective, and in such transformations the conformation of the system undergoing rearrangement is critical. Reactions involving a concerted 1,2-migration require an anticoplanar arrangement of the departing and migrating groups. In cyclohexane derivatives, if the rearrangement proceeds without change in ring size, this implies that both the departing and migrating groups have to be axial. In ring-contraction processes, on the other hand, the leaving group has to be equatorial in order to achieve the required anticoplanar relationship to the migrating C—C bond of the ring.

One of the most striking illustrations of how the course of a rearrangement may be affected by conformation is provided by the contrasting behaviour of the epimeric 17-amino-17 aβ-hydroxy-17 aα-methyl-D-homo steroids upon treatment with nitrous acid (*20* and *21*). In each case, the reaction follows the course imposed upon it by the geometry of the molecule (*47*). The stereochemistry of ring expansion reactions is controlled by similar conformational factors (see, for example, *189*).

18*

(20)

(21)

   The energy required to bring a pair of adjacent bonds into coplanarity
is less if they are *cis* (e, a) than if they are *trans* (e, e)*. As a consequence,
reactions which involve, either as an intermediate or in the transition
state, a 1,2-fusion of an approximately planar ring, proceed more easily
with *cis* compounds. Some reactions of this type are, the cleavage of
glycols with periodic acid or lead tetraacetate, the thermal elimination
of esters, the decarboxylation of $\beta\gamma$-unsaturated carboxylic acids, and the
pyrolysis of amine oxides (*24*, *61e*).
   The physical properties of a compound may also be interpreted in
terms of the conformations of its molecules. The relation between the
strengths of acids and bases and the conformations of their functional
groups has already been noted. The behaviour of alcohols in adsorption,
partition and gas-liquid chromatography is also affected by conformation. Generally speaking, axial hydroxyl groups, being less accessible,
are less strongly adsorbed than are their equatorial epimers. A number
of exceptions have been reported (see, for example, *44*, *131*, *165*), but
many of these may be attributed to the effect of hydrogen bonding (*91*)
or to preferential solvation of the less hindered equatorial epimer (*149*,
171).
   The most important physical methods for investigating conformation
are, however, spectroscopic. Infrared (*22*, *24*, *61*) and ultraviolet (*22*,
24, *61*, *160*) spectroscopy have long been employed in this way, and
within the past decade optical rotatory dispersion (*45*, *46*, *53*, *61*) and
circular dichroism (*45*, *46*, *176*) have been developed into valuable techniques for the study of conformational problems. A discussion of these
methods is outside the scope of the present brief summary; detailed
reviews of each of these topics are indicated.

---

   * This has been explained in terms of the non-bonded interactions set up in
each case (*22*, *63*); it may, however, be regarded as a natural consequence of the
fact that a smaller angular distortion is required in the latter case (*193*; see footnote
on p. 271).

References, pp. 307—317.

Perhaps the most powerful spectroscopic technique available for conformational studies is proton magnetic resonance (*29, 71, 108*). Generally, an equatorial proton is less shielded than its axial epimer and so resonates at a lower field*. The difference in signal positions is usually of the order of 0.5 p.p.m. of the applied field (*108*). Where both members of an epimeric pair are available for study, this difference in chemical shift may often permit configurations to be assigned. However, the interpretation of chemical shift differences must be made with caution, as differential shielding by a grouping elsewhere in the molecule may result in an inversion of the usual relationship between chemical shift and conformation. Thus, the 4α (e) protons in 4β-acetoxy-$5\alpha,6\alpha$-epoxycholestane (**22**) and its 3β-methoxy analogue (**23**) absorb at higher field than do the 4β (a) protons in their respective 4-epimers (*40*). These results may be ascribed to the anisotropy of the oxirane ring (cf. *56, 112, 172*).

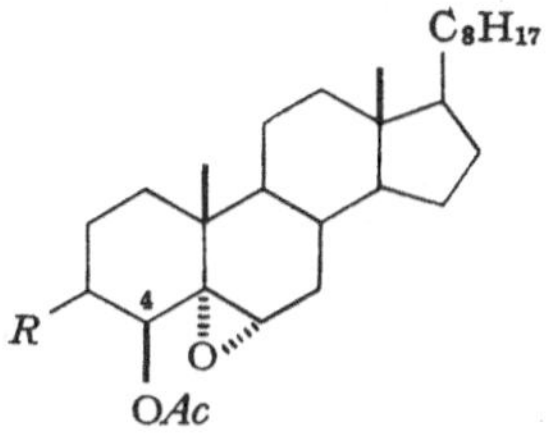

C_8H_{17}

R

OAc

(**22**) 4β-Acetoxy $- 5\alpha, 6\alpha -$ epoxycholestane
($R = H$).

(**23**) $3\beta -$ Methoxy $- 4\beta -$ acetoxy $- 5\alpha, 6\alpha -$
epoxycholestane ($R = OCH_3$)

From early NMR studies on acylated sugars and related compounds (*126*) another very valuable means of tackling conformational problems emerged. It was found that, whereas the coupling constant between adjacent equatorial protons or between an axial and an equatorial proton is 2–3.5 cps., that between adjacent axial protons is 2–3 times larger (cf. *188*). It was later shown by KARPLUS (*117*) that the coupling constant J_{HH} between hydrogen atoms bonded to adjacent carbon atoms in a chain depends upon the dihedral angle (φ) between them, according to the relationship

$$J_{HH} = a \cos^2 \varphi,$$

the value of a depending upon the system under consideration (*34, 103, 118, 190*). When φ is 0°, J_{HH} has a medium value which falls off progressively until it is almost zero when φ is 90°, and then increases to a maximum value at $\varphi = 180°$.

---

* For epimeric pairs of α-haloketones a reversal of this rule appears to hold for the protons bonded to the halogenated carbon atoms (*144, 180*).

For the most part, the foregoing discussion has dealt with cyclohexane and its congeners. Fortunately, most of the principles described
can be applied equally to six-membered rings in which one or more of
the carbon atoms has been replaced with nitrogen. This is a consequence
of the close similarity between the C—C and C—N bond lengths (1.54 Å
and 1.47 Å, respectively) and between the C—C—C and C—N—C bond
angles (both 109°).

The preferred conformation of piperidine (24; R = H) and of N-methyl-
piperidine (24; R = CH_3) is a chair form, with the lone pair on nitrogen
axially disposed (6, 7, 31). This implies, contrary to an earlier view (16),
that the steric requirement of the piperidine lone pair is less than that
of a hydrogen atom (see also 41, 62, 154). There is some evidence that
the effective bulk of the lone pair on nitrogen is increased when in a sol-
vating solvent (37, 116).

It is now established that in the tropane alkaloids the piperidine ring
adopts a chair conformation with the N-methyl group equatorial (30,
42, 58). Thus, ψ-tropine is to be represented as (25); in tropine itself (26)
there appears to be some distortion of the chair form of the ring (30).

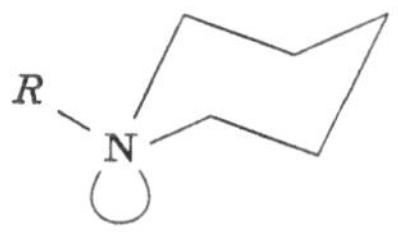

(24)
Piperidine (R = H).

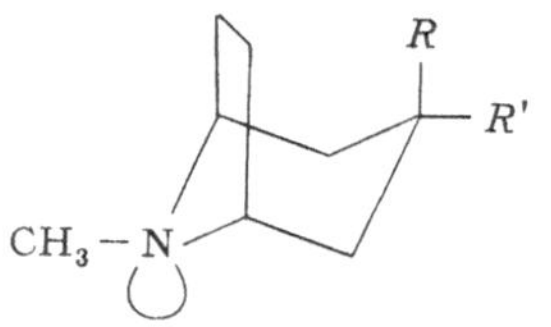

(25) ψ - Tropine (R = OH, R' = H).
(26) Tropine (R = H, R' = OH).

The inversion of substituents on trivalent nitrogen involves a very
low activation energy. Hence, when an N-alkylpiperidine derivative is
quaternised, it is not possible to deduce the conformation of the N-alkyl
group in the starting material from the conformation adopted by the
incoming group (42, 115, 141). In the tropane series, for example, the
incoming group in almost all cases assumes an equatorial conformation;
as a result it was at first concluded, erroneously, that the N-methyl
group in the tropane alkaloids was axially disposed (79, 80). In the
alkylation of simple piperidines, on the other hand, axial quaternisation
appears to be more general (26, 36, 97, 133–135; see however, 105).

Because of the stereochemical instability of trivalent nitrogen, quinolizidine differs from its carbocyclic analogue decalin, in that the trans- (27)
and cis- (28) forms are not separately isolable compounds but are in rapid
equilibrium at room temperature. For quinolizidine itself it has been
estimated (1, 116) that the trans- form (27) is favoured over the cis-

conformer (28) by a free energy difference of approximately 4.4–4.6 kcal./ mole. The preferred conformation of a substituted quinolizidine may involve either a *cis-* or *trans-* arrangement of the rings, depending upon the configurations of the substituents. It has been established that the 1-, 2-, 3- and 4-methylquinolizidines all adopt the *trans-* conformation, irrespective of the configuration of the methylated carbon atom relative to that of $C_{(10)}$ (*1, 115, 141*).

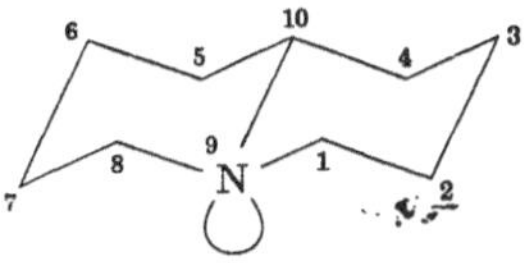

(27)
trans – Quinolizidine.

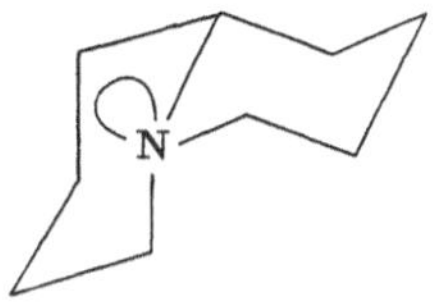

(28)
cis – Quinolizidine.

III. Yohimbine and Related Alkaloids.

The yohimbinoid alkaloids possess structures based upon the yohimbane skeleton, which has the gross structure (29). In nature, this nucleus occurs in four stereoisomeric variants, from which are derived compounds of the "normal" (3α, 15α, 20β-) series (30), the "pseudo-" (3β, 15α, 20β-) series (31), the "allo-" (3α, 15α, 20α-) series (32) and the "epiallo-" (3β, 15α, 20α-) series (33)*.

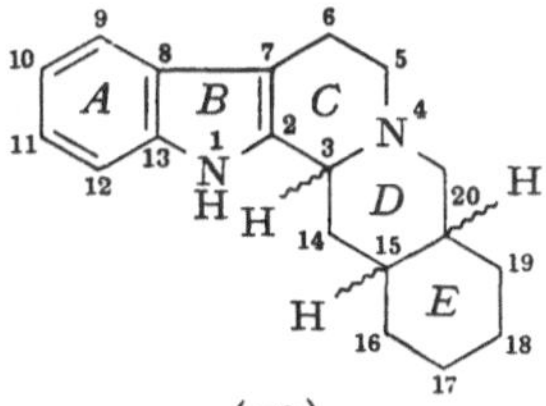

(29)

Because of the *trans*-fusion of rings D and E, alkaloids of the normal and pseudo series are conformationally rigid; the only possible all-chair conformations available to them are, respectively, (30a) and (31a). In conformation (30a) the indole ring system is equatorially disposed on

---

* All known naturally occurring yohimbinoid alkaloids have the 15α-configuration. The absolute configurations depicted here were established as a result of studies of molecular rotation differences (*122*) and of optical rotatory dispersion (*54*), and have since been confirmed by the application of PRELOG's asymmetric synthesis to yohimbine and to methyl resperate (*19*).

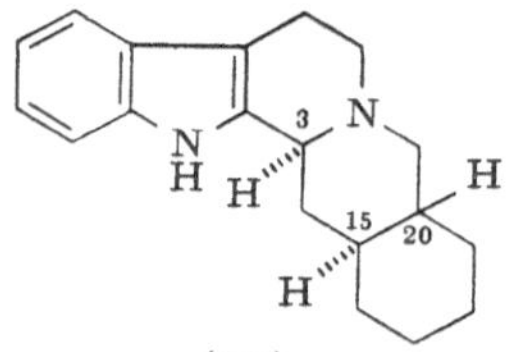

(30)
Yohimbane.

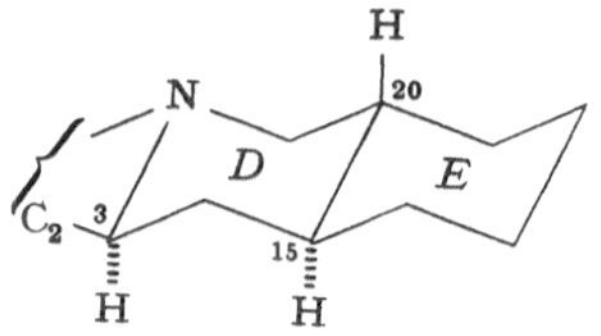

(30a)

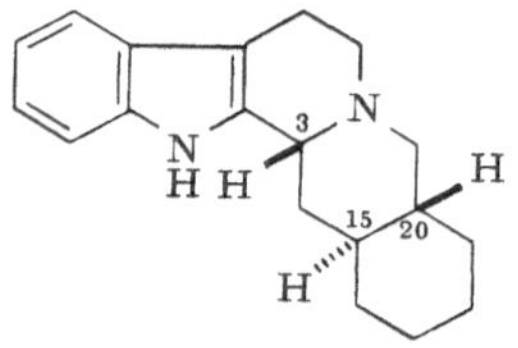

(31)
Pseudoyohimbane.

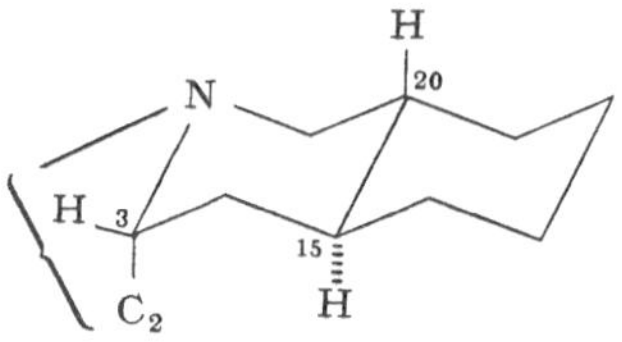

(31a)

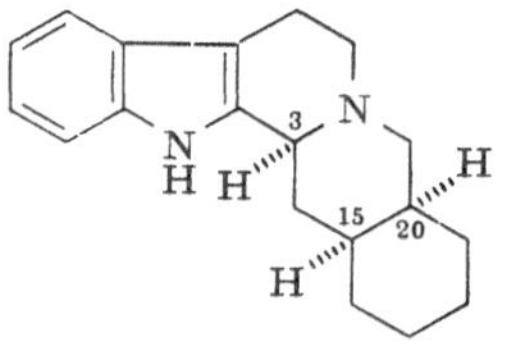

(32)
Alloyohimbane.

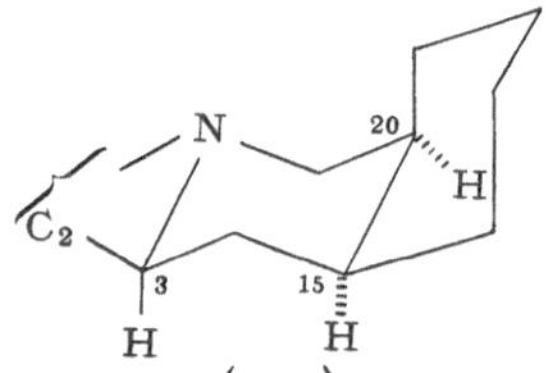

(32a)

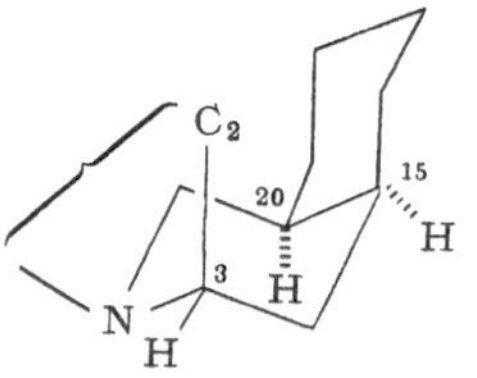

(32b)

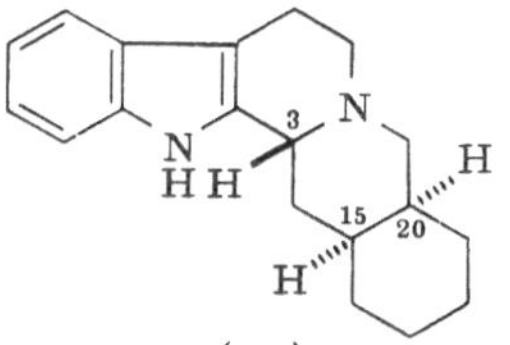

(33)
3-Epialloyohimbane.

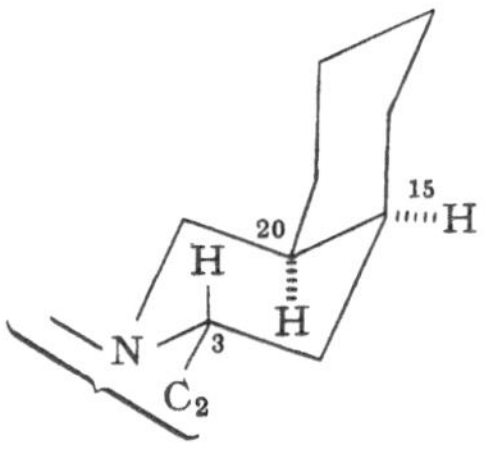

(33a)

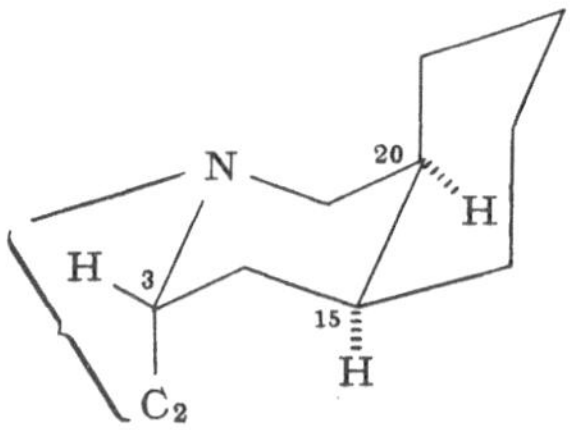

(33b)

ring *D*, while in (31 a) it is attached axially. As a consequence, a compound of the pseudo-series is always less stable than its 3-epimer of the normal series.

References, pp. 307—317.

Compounds of the allo- and epiallo- series have a *cis-D/E* ring junction and can exist in either of two possible all-chair conformations. For the allo- series these conformations are (32a) and (32b); and for the epiallo-series they are (33a) and (33b). For the parent hydrocarbons, allo-yohimbane (32) and epialloyohimbane (33), it may be assumed that the preferred conformations are (32a) and (33a), respectively, in which the indole ring is equatorially disposed. The results of acid-catalysed equi-libration of alloyohimbane and epialloyohimbane indicate that the latter is the thermodynamically more stable arrangement (*181*). However, when (32a) and (33a) are interconverted, all the axial C—H bonds in rings *D* and *E* become equatorial and vice versa. It follows, therefore, that the stability relationship between compounds of the allo- and epi-allo- series possessing substituents in rings *D* and *E* will depend upon the nature and configuration of the substituents.

The elucidation of the configurations of yohimbine (34) and of its 16-epimer corynanthine (35), which was achieved independently by COOKSON (*43*) and by JANOT, GOUTAREL, LE HIR, AMIN and PRELOG (*110*), represented an important early application of confor-mational analysis to the stereochemical problems presented by alkaloids. Rings *D* and *E* of yohimbine were known to be *trans*-fused from its degradation to (+)-N-methyl-*trans*-decahydroisoquinoline (40) (*192*); and since the configuration at $C_{(3)}$ could be shown to be the thermo-dynamically more stable one (*43*, *111*), yohimbine could be assigned to the normal (3α, 15α, 20β-) series (30).

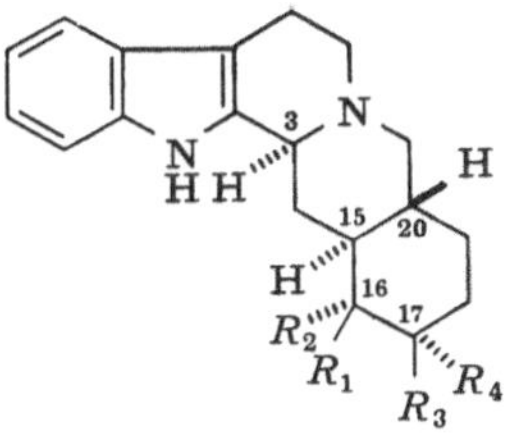

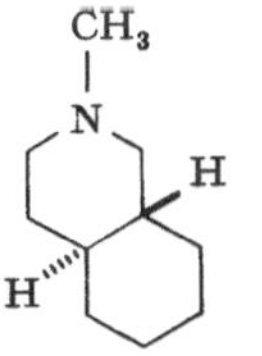

N – Methyl – *trans* – decahydroisoquinoline.

(34) Yohimbine ($R_1 = R_3 = H$; $R_2 = CO_2CH_3$; $R_4 = OH$).

(35) Corynanthine ($R_1 = CO_2CH_3$; $R_2 = R_3 = H$; $R_4 = OH$).

(36) Yohimbic acid ($R_1 = R_3 = H$; $R_2 = CO_2H$; $R_4 = OH$).

(37) Yohimbone ($R_1 = R_2 = H$; $R_3, R_4 = O$).

(38) *epi* – Yohimbol ($R_1 = R_2 = R_4 = H$; $R_3 = OH$).

(39) β – Yohimbine ($R_1 = R_4 = H$; $R_2 = CO_2CH_3$; $R_3 = OH$).

The $C_{(16)}$-epimeric relationship between yohimbine and corynanthine followed from the observation that upon alkaline hydrolysis both gave rise to yohimbic acid (36) (*109*). The less stable 16-methoxycarbonyl group of corynanthine could be assigned an axial conformation; reference to formula (30 a) will make it clear that in the rigid normal series this implies the β-configuration.

When yohimbone (37) is reduced by the Meerwein-Ponndorf method, the proportion of epi-yohimbol (38) in the mixture of epimeric alcohols produced becomes progressively greater as the reaction time is increased (191). From this it may be concluded that the hydroxyl group of epi-yohimbol is in the thermodynamically more stable equatorial conformation; it follows that yohimbine and corynanthine have a 17α(axial)-hydroxyl function. The reduction of yohimbone to epi-yohimbol with lithium aluminium hydride (184) provides confirmation of these assignments, since it is well established that hydride reduction of an unhindered ketone affords mainly the equatorial alcohol (24).

The effect of conformation on the course of a chemical reaction is strikingly illustrated by the differing behaviour of the sulphate esters derived from corynanthine and from yohimbine when each is treated with dilute alkali. In the corynanthine derivative (41) the carboxyl and sulphate groups bear an anti-coplanar relationship; as a result the elimination of sulphate occurs with concomitant decarboxylation and the product is apo-corynanthol (42) (109). The methoxycarbonyl group in the sulphate (43) derived from yohimbine is equatorial and so is not suitably disposed to react in the same way; in this case the product is apo-yohimbine (44) (20), arising by diaxial elimination of a molecule of sulphuric acid.

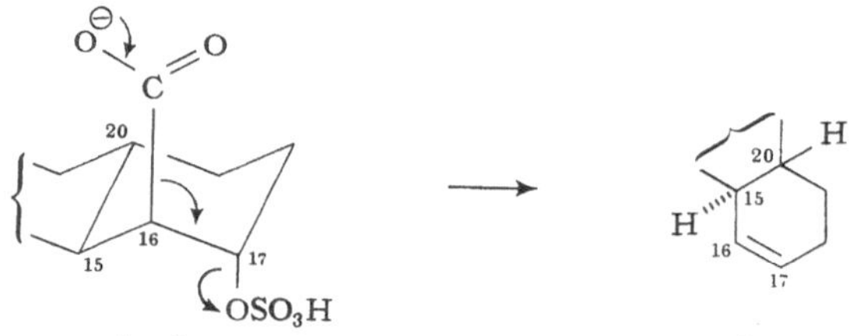

(41)   (42)

(43)   (44)

Yohimbine and corynanthine are two members of a group of seven naturally-occurring stereoisomers (for a recent review, see 150), all of which are listed in Table 1*.

---

* 19-Dehydroyohimbine also occurs in nature (15; cf. 163, 166).

Table 1. Naturally Occurring Isomers of Yohimbine

Skeletal type	Name	Configuration of nucleus	Configuration of substituents	
			CO_2CH_3	OH
Normal	Yohimbine (**34**)	3α, 15α, 20β	16α	17α
	Corynanthine (**35**)	3α, 15α, 20β	16β	17α
	β-Yohimbine (**39**)	3α, 15α, 20β	16α	17β
Pseudo	pseudo-Yohimbine (**45**)..	3β, 15α, 20β	16α	17α
Allo	allo-Yohimbine (**47**).....	3α, 15α, 20α	16β	17β
	α-Yohimbine (**50**)	3α, 15α, 20α	16β	17α
Epiallo	3-epi-α-Yohimbine (**49**)..	3β, 15α, 20α	16β	17α

In addition, several other stereoisomers have been prepared by epimerisations carried out in the laboratory.

β-Yohimbine (**39**) the third member of the normal series, does not undergo epimerisation upon base treatment (*125*) indicating that its methoxycarbonyl group, like that of yohimbine, is in the stable equatorial conformation. This assignment is supported by the closely similar pK values found for the carboxylic acids derived from the two alkaloids. Oppenauer oxidation of β-yohimbine gives yohimbone (**37**), thus establishing a 17-epimeric relationship between yohimbine and β-yohimbine. In agreement with the equatorial conformation assigned to the methoxycarbonyl group of β-yohimbine, it undergoes saponification with base more rapidly than that of corynanthine, which is axial (*125*). The much faster saponification reported for yohimbine (*125*) probably arises as a result of hydrogen bonding between the *cis*-disposed ester and hydroxyl groups in that compound.

Pseudo-yohimbine (**45**) undergoes dehydrogenation with lead tetraacetate to give tetradehydroyohimbine (**46**) which can be reduced by catalytic hydrogenation to yohimbine (**34**) (*110*). It follows therefore that yohimbine and pseudoyohimbine are $C_{(3)}$ epimers. Alkaloids of the pseudo and epiallo series may also be obtained from their 3-epimers by oxidation of the latter with, for example, mercuric acetate, to the corresponding Δ^3-dehydro compounds followed by reduction with zinc and hydrochloric acid. α-Yohimbine (**50**) has been transformed into 3-epi-α-yohimbine (**49**) in this way (*179*)*. Since 3-epialloyohimbane is a stronger base than alloyohimbane and affords an N-oxide under conditions which leave alloyohimbane untouched (*181*), it can be inferred that the quinolizidine nitrogen atom is more hindered in the latter compound. Thus, alloyohimbane may be assigned the 3α-configuration (**32**) and the

---

* For detailed reviews of isomerisations carried out in the yohimbine series see (*111*, *150*).

preferred conformations of alloyohimbane and its 3-epimer are then to be represented, respectively, as (32 a) and (33 a); only in (32 a) is $N_{(4)}$ seriously hindered.

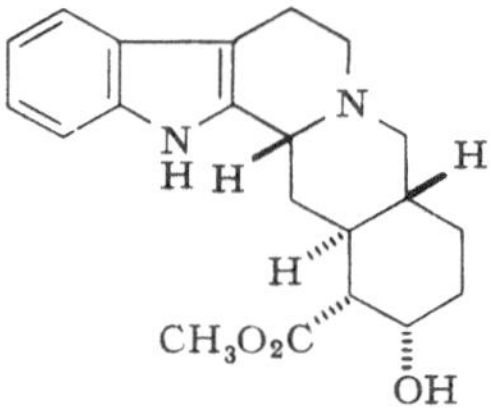

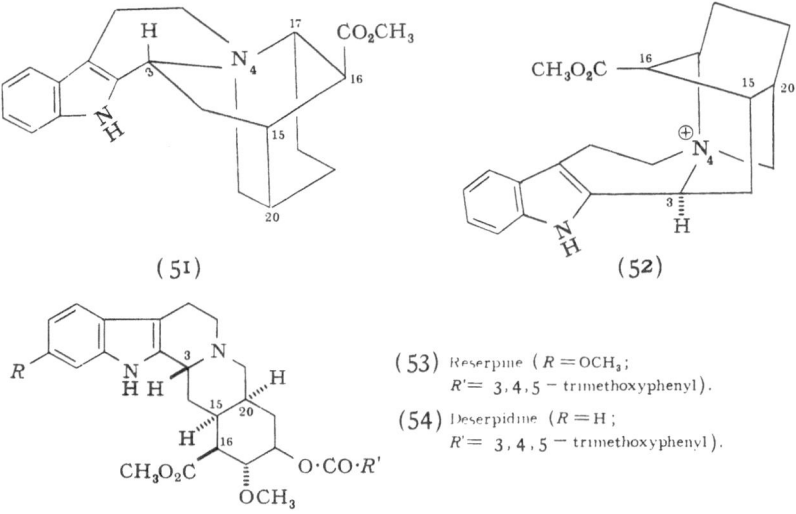

(45)
Pseudoyohimbine.

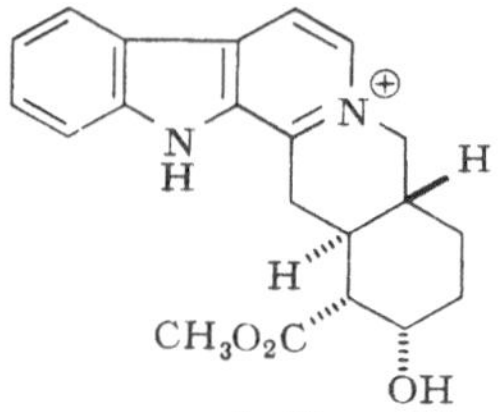

(46)
Tetradehydroyohimbine.

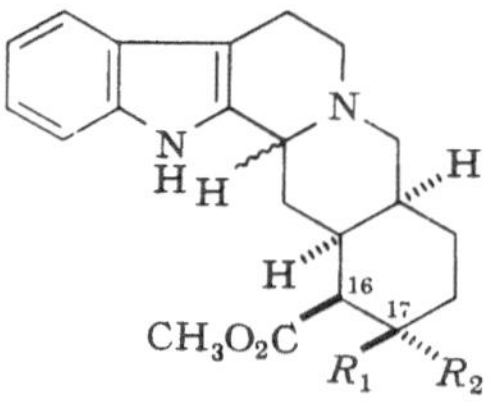

(47) Alloyohimbine ($R_1 = OH$; $R_2 = H$; $3\alpha - H$).

(48) 3 - Epialloyohimbine ($R_1 = OH$; $R_2 = H$; $3\beta - H$).

(49) 3 - epi - α - Yohimbine ($R_1 = H$; $R_2 = OH$; $3\beta - H$).

(50) α - Yohimbine ($R_1 = H$; $R_2 = OH$; $3\alpha - H$).

The *cis* D/E ring junction which exists in members of the allo- and epiallo- series permits the occurrence of internal quaternisation of the quinolizidine nitrogen atom in suitably substituted compounds. Thus, the 17-tosylate of 3-epi-α-yohimbine (49) is transformed under very mild conditions into the quaternary ammonium tosylate (51) (99). The

(51)

(52)

(53) Reserpine ($R = OCH_3$; $R' = 3,4,5 -$ trimethoxyphenyl).

(54) Deserpidine ($R = H$; $R' = 3,4,5 -$ trimethoxyphenyl).

great ease with which this reaction occurs indicates a direct intramolecular displacement of the tosyloxy group by $N_{(4)}$, from which it follows that 3-epi-α-yohimbine (49) [and hence also α-yohimbine (50)] possesses the 17α-configuration. The fact that α-yohimbine tosylate does not undergo a similar internal quaternisation may be attributed to the severe non-bonded interaction which would exist between the methoxycarbonyl group and the indole ring in the analogous structure (52).

Results obtained in the yohimbine series were of great importance in the elucidation of the configurations of the *Rauwolfia* alkaloids reserpine (53) and deserpidine (54). The complete stereochemistry of these compounds resulted from a series of investigations carried out by SCHLITTLER, VAN TAMELEN, WENKERT, WINTERSTEINER and their colleagues (2). Both of these alkaloids were known to possess the thermodynamically unstable configuration at $C_{(3)}$ (*101, 136, 137*); and since deserpidine could be converted into α-yohimbine (50) under conditions known to cause epimerisation at $C_{(3)}$, it

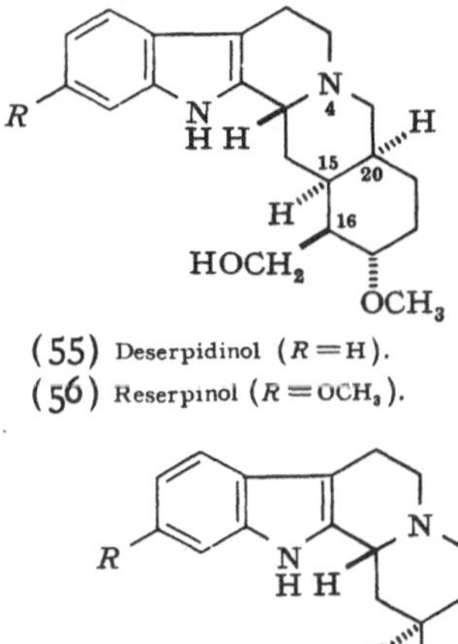

(55) Deserpidinol ($R=$H).
(56) Reserpinol ($R=$OCH$_3$).

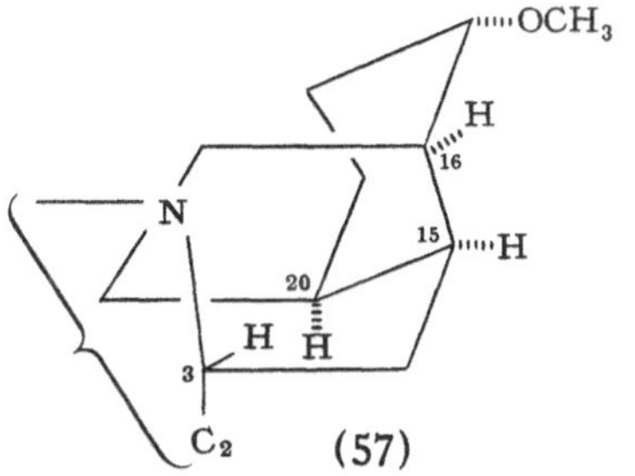

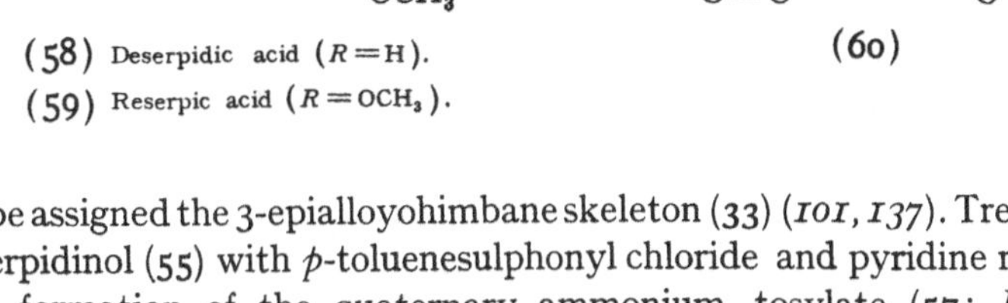

(57)

(58) Deserpidic acid ($R=$H).
(59) Reserpic acid ($R=$OCH$_3$).

(60)

could be assigned the 3-epialloyohimbane skeleton (33) (*101, 137*). Treatment of deserpidinol (55) with p-toluenesulphonyl chloride and pyridine resulted in the formation of the quaternary ammonium tosylate (57; $R =$ H) (*2*; cf. *85*). The obtainment of this compound established the 16β-configuration for deserpidine, and confirmed that rings D and E are *cis*-fused. Since deserpidine has the same configuration at $C_{(17)}$ as α-yohimbine (*137*), only the stereochemistry at $C_{(18)}$ remains for discussion. That there is a *cis*-relationship between the substituent at that position and the 16β-substituent follows from the ready formation of deserpidic acid lactone (60; $R = H$) by treatment of deserpidic acid (58) with acetic anhydride (*137*).

Reserpine (53) is the 11-methoxy analogue of deserpidine, and its stereochemistry was established largely by employment of arguments analogous to those already described for deserpidine (*2*). Thus, in the reserpine series, compounds possessing the structures (57; $R = OCH_3$) (*52, 102, 175*) and (60; $R = OCH_3$) (55) were obtained, from which the relative configurations at $C_{(15)}$, $C_{(16)}$, $C_{(18)}$ and $C_{(20)}$ could be inferred. The absolute configurations at $C_{(3)}$ and at $C_{(18)}$ were deduced from molecular rotation data (*51, 100*).

The configurational problem posed by $C_{(17)}$ of reserpine was resolved by consideration of the products formed by treatment of methyl reserpate tosylate (53; OTs instead of $O \cdot CO \cdot R'$) with collidine. In addition to methyl anhydroreserpate (61) the quaternary ammonium tosylate (62)

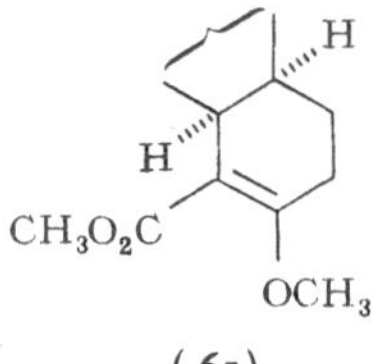

(61)

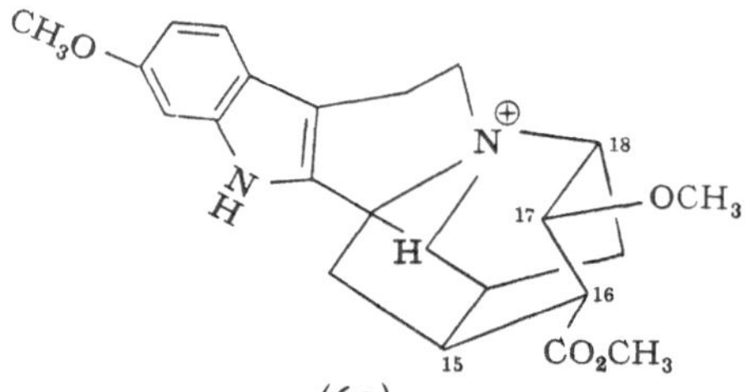

(62)

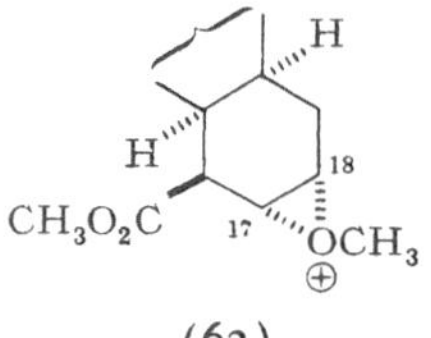

(63)

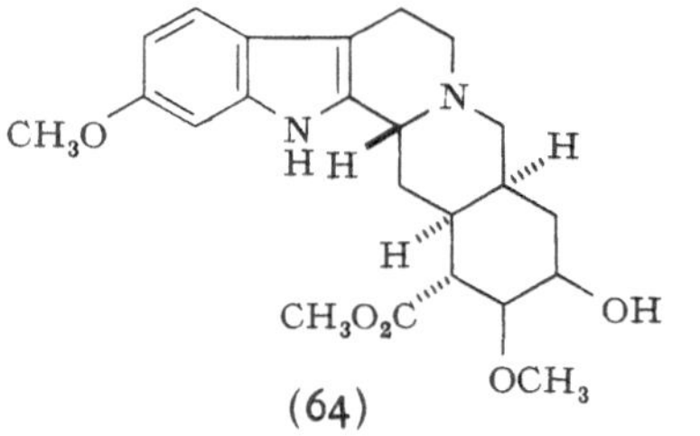

(64)
Methyl neoreserpate.

was also obtained (*51*). Clearly, since the 18-tosyloxy group of methyl reserpate tosylate is β-oriented, the quaternary ammonium tosylate (62) could not have arisen by direct intramolecular displacement; neighbouring group participation involving the 17-methoxyl group must be postulated (*175, 181*), possibly involving the intermediate oxonium ion (63) (*175*; but see also *2*). In order for such participation to occur, it must be possible for the molecule to assume a conformation in which the 17- and 18- substituents are both axial; it could therefore be concluded that the 17-methoxyl group of reserpine and its derivatives is α-oriented.

It should be noted that when the tosylate of methyl neoreserpate (64) is heated with collidine, no internal quaternisation occurs (*157*); in this case the 17- and 18- substituents are *cis* to one another and neighbouring group participation by the methoxyl is not possible.

The configurations of reserpine and deserpidine having been established, it became clear that certain of the properties of these compounds have a conformational basis. Thus, unlike the unsubstituted parent compound, 3-epialloyohimbane (33), reserpine and deserpidine have the unstable configuration at $C_{(3)}$. Doubtless, the fact that 3-isoreserpine and 3-isodeserpidine can adopt a two-chair conformation (65) for rings D and E in which the indole ring and all three ring E substituents are equatorially disposed, is responsible for the greater stability of these isomers. Of the two di-chair conformations available for reserpine and deserpidine themselves, one (based on 33a) has all three ring E substituents axially disposed, while the other (based on 33b) requires the indole ring to adopt the unfavourable axial conformation.

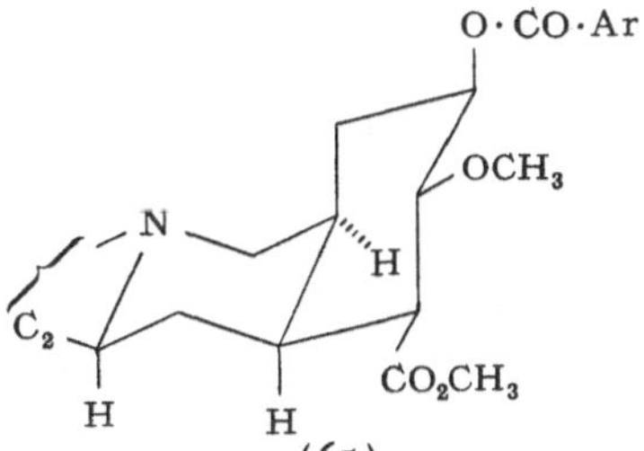

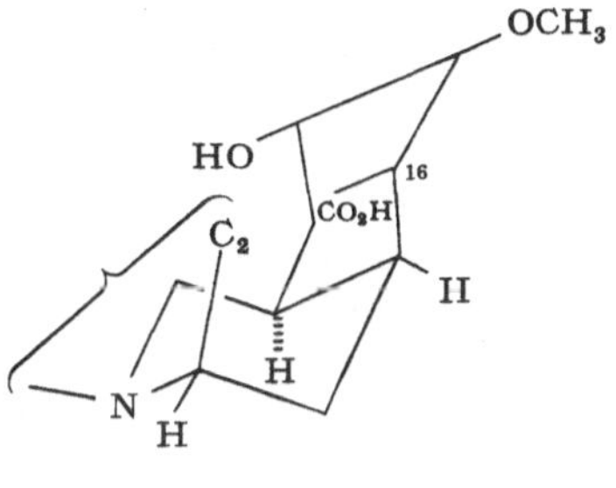

The marked resistance to lactonisation of 3-isoreserpic acid as compared with reserpic acid (59) itself (*100*) may also be ascribed to conformational factors. In order for lactonisation to occur 3-isoreserpic acid has to adopt conformation (66) in which there is an extremely severe non-bonded interaction between $C_{(2)}$ and the $C_{(16)}$-carboxyl group. Consequently, lactonisation proceeds only under forcing conditions (*194*). The conformation which reserpic acid must assume for lactonisation is based on (33a); here the 16β-carboxyl group interacts only with the hydrogen atom attached to $C_{(3)}$, and much milder conditions suffice for the reaction (*100*).

Investigations of the yohimbines and related indole alkaloids have led to a number of useful generalisations relating the reactivity and spectral properties of these compounds to the conformation of the 3-hydrogen atom. Thus, it has been suggested (*127, 128*) that the oxidation of quinolizidine derivatives by mercuric acetate involves an *anti*-coplanar elimination from an N-mercurated derivative (**67**). In agreement with this, yohimbinoid alkaloids of the normal- (**30**) and allo- (**32**) series for which a preferred conformation of this type might be expected, readily afford the corresponding 3-dehydro derivatives upon dehydrogenation with mercuric acetate (**70**, *179*, *182*), while pseudo-yohimbine (**45**), which cannot assume such a conformation, resists oxidation under the usual conditions (*179*).

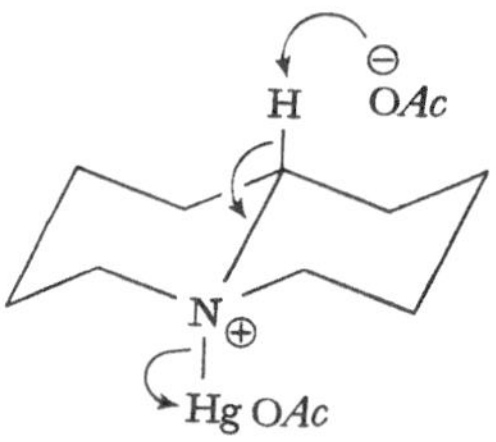

(**67**)

Results obtained in the epiallo series, however, suggest that caution should be exercised in the use of mercuric acetate oxidation as a diagnostic reaction. Thus, while the failure of reserpine (**53**), deserpidine (**54**), and methyl reserpate (**53**; OH instead of $O \cdot CO \cdot R'$) to react under the usual conditions (*179*, but see also **70**) may be ascribed to their preferred conformation being of type (**33b**), so as to accommodate all their ring E substituents equatorially, the resistance of *dl*-3-epialloyohimbane (**33**) to oxidation (*182*) is certainly anomalous. The preferred conformation of this latter compound is almost certainly (**33a**) in which the 3-hydrogen and the p-electrons of $N_{(4)}$ are disposed in an anticoplanar fashion.

Data obtained from mercuric acetate oxidations have been presented in support of configurational assignments for certain spiro-oxindole alkaloids related to the yohimbines and heteroyohimbines (see, for example, *114*); however, in this series also, it has been observed that compounds possessing the appropriate anticoplanar arrangement of the 3-hydrogen and the p-electrons of $N_{(4)}$ sometimes fail to react (*196*).

A relationship between the configuration at $C_{(3)}$ and the infrared absorption in the 2800 cm^{-1} region, first proposed by Wenkert and Roychaudhuri in 1956 (*183*) was later refined by Bohlmann (*32, 33*) who noted that in the quinolizidine series complex absorption between 2700 cm^{-1} and 2800 cm^{-1} is associated with the presence of two or more hydrogen atoms α to the nitrogen which are in a *trans*- diaxial relationship with the nitrogen lone pair. With reference to the yohimbine (155, 195) and heteroyohimbine (162) alkaloids the correlation may be expressed as follows: Only those compounds which, in their preferred conformation, possess an axial hydrogen atom at $C_{(3)}$ exhibit two or more peaks or distinct shoulders between $2700\ cm^{-1}$ and $2900\ cm^{-1}$, at least one of which appears below $2800\ cm^{-1}$. The value of the infrared criterion is well illustrated by its application to an investigation of the conformations of five of the eight possible ring E epimers of methyl reserpate (53; OH instead of $OCOR'$) and its derivatives (156).

Since the beginning of the present decade, nuclear magnetic resonance has assumed great importance in conformational studies. In the

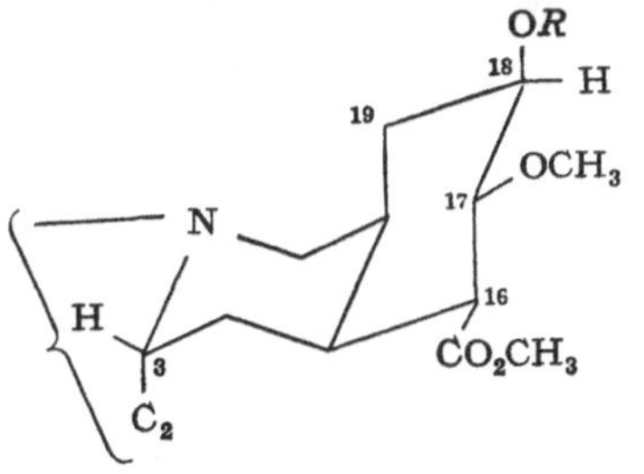

(68)

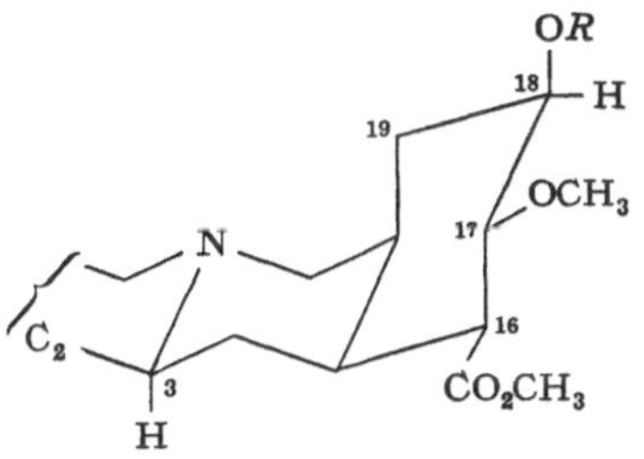

(69)

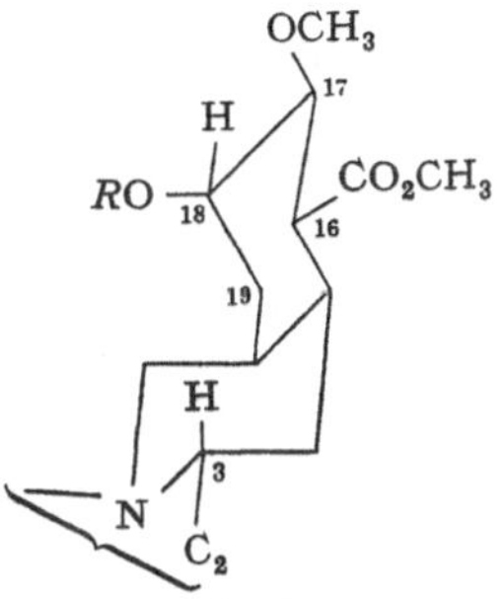

(70)

yohimbine series the conformation of the $C_{(3)}$—H may be inferred from its chemical shift. Thus, while methyl reserpate (53; OH instead of $OCOR'$) exhibits a signal at $5.57\ \tau$, the corresponding signals from methyl 3-isoreserpate (53; OH instead of $OCOR'$; 3α—H) and methyl neoreserpate (64) occur much higher upfield and are obscured by peaks due to other hydrogen atoms (159). The $C_{(3)}$—H atom in methyl reserpate may therefore be assigned an equatorial conformation while in the other two compounds it must be axially oriented. Hence, the preferred conformations of methyl reserpate, methyl 3-isoreserpate and methyl neoreserpate are to be represented as (68; $R = H$), (69; $R = H$), and (70; $R = H$), respectively.

Further support for these conformations is provided by the signal due to the $C_{(18)}$ proton in each of these compounds. In methyl neo-reserpate (70; $R = H$) it appears as a quartet ($J \sim 3$ cps) at 5.79 τ which is consistent with an equatorial proton equally coupled with three neighbouring protons (two axial and one equatorial). The $C_{(18)}$ protons in methyl reserpate (68; $R = H$) and in methyl 3-isoreserpate (69; $R = H$), being axial, occur at higher fields and can not be distinguished; however, in the derived trimethoxybenzoates, (68; $R = 3,4,5$-trimethoxybenzoyl) and (69; $R = 3,4,5$-trimethoxybenzoyl) the $C_{(18)}$-protons appear as broad signals centred on 4.92 τ. The breadth of the signals arises from their structure which is that of a triplet ($J \sim 10$ cps), due to the axial $C_{(18)}$-proton coupling equally with two adjacent axial protons, in which each line is further split ($J \sim 3$ cps) as a result of coupling with the equatorial proton at $C_{(19)}$. The $C_{(18)}$ proton of methyl neoreserpate is also shifted downfield (to 4.22 τ) in the aroyl derivative (70; $R = 3,4,5$-trimethoxybenzoyl), but retains its quartet structure.

Methyl neoreserpate (70; $R = H$) is obtained from methyl reserpate (68; $R = H$) by heating a solution of the latter compound under reflux with sodium methoxide (157; see also 139). The driving force for this rearrangement is clearly the reduction in non-bonded interactions achieved by accommodating the bulky indole ring (represented by $C_{(2)}$ in 70) in an equatorial conformation. In methyl reserpate this is possible only if the molecule adopts a conformation of the type (33a) in which all the ring E substituents would have to be axial. It has been established by tracer studies (158) that the rearrangement proceeds by reversible β-elimination of the 17-methoxyl group of methyl reserpate (68; $R = H$).

IV. The Heteroyohimbines and Related Oxindole Alkaloids.

The application of physical methods to conformational analysis is well illustrated in the elucidation of the configurations of the hetereyohimbine alkaloids. These compounds may all be represented by the gross structure (71) in which R and R' may be either hydrogen or methoxyl. As with the yohimbinoid alkaloids, normal-, pseudo-, allo- and epiallo-configurations are possible; since the 19-methyl group may be either α- or β-oriented, there are thus eight different stereoisomeric skeletons.

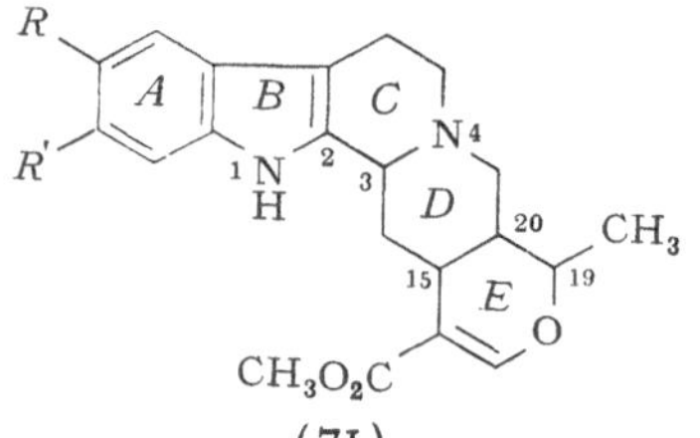

(71)

The spectroscopic properties of the heteroyohimbines are very sensitive to changes in stereochemistry. By studying the chemical shift of the $C_{(19)}$-methyl group, and the infrared absorption in the region of 1200 cm^{-1} (*143*), SHAMMA and RICHEY (*162*) were able to classify all the members of the series into eight stereochemical groups, A—H (*Table 2*).

Table 2. Stereochemical Classification of the Heteroyohimbine Alkaloids (*162*).

Stereochemical group	Skeleton	Configuration of 19-CH$_3$
A	Allo-	α
B	Epiallo-	α
C	Allo-	β
D	Epiallo-	β
E	Normal	α
F	Normal	β
G	Pseudo	α
H	Pseudo	β

The *cis*- or *trans*- nature of the C/D ring fusion in each group was inferred by examining the infrared absorption of appropriate compounds in the 2800 cm^{-1} region. The configurations of the D/E ring fusions in compounds belonging to groups A, B, E and G were deduced by measuring the pKa values and rates of methylation of appropriate compounds. It was found that the basicity of $N_{(4)}$ varied from one stereochemical group to another, as a result of the differing steric hindrance to which the lone pair was subject.

Parallel, but greater, differences were observed in the rates of methiodide formation. Compounds of group A (in which rings C and D were already known to be *trans*-fused), were found to methylate more slowly than any of the others, and were therefore assigned the allo- configuration

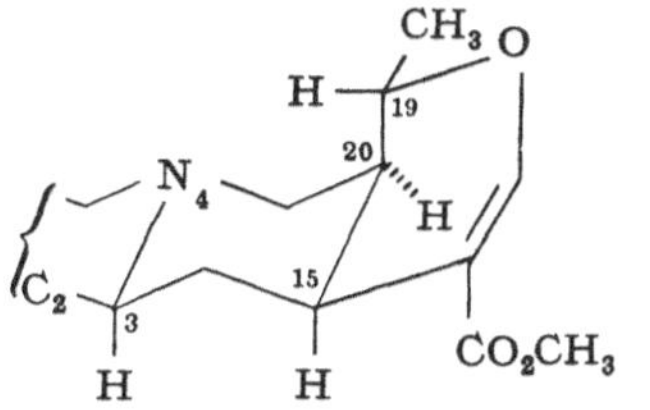

(72) 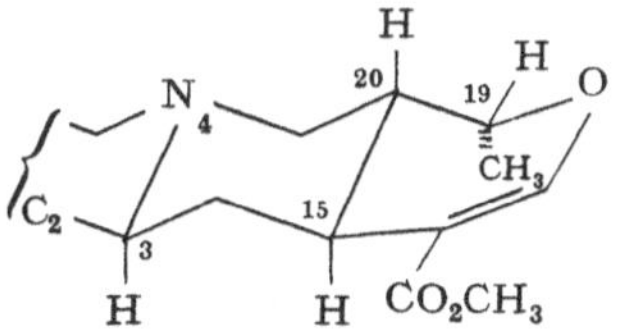   (73)

(72) in which the lone pair of $N_{(4)}$ enters into a *syn*-axial interaction with $C_{(19)}$. Similarly, the very fast rates recorded for the alkaloids of group E (C/D *trans*-fused) required that they be assigned the normal configuration (73) in which $N_{(4)}$ enters into no serious non-bonded interactions. From the known $C_{(3)}$ configurational relationships of certain alkaloids in these

two groups with those in groups B and G, the latter groups could be assigned, respectively, the epiallo- and pseudo- configurations.

Although the alkaloids of group C methylated somewhat more slowly than those of group D, the difference was not sufficiently clear cut to permit configurational assignments. The problem was solved by the

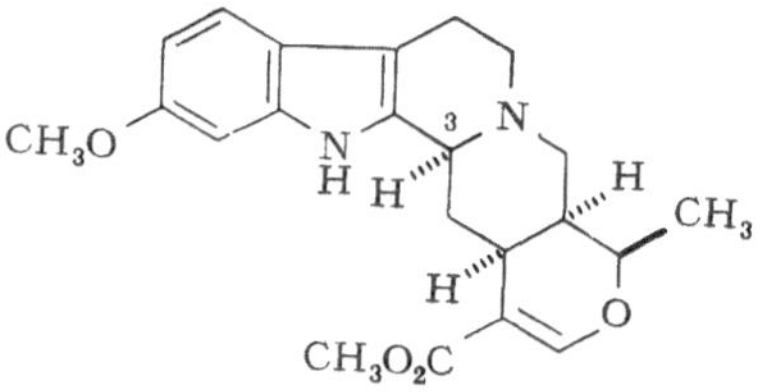

(74)

Raunitidine.

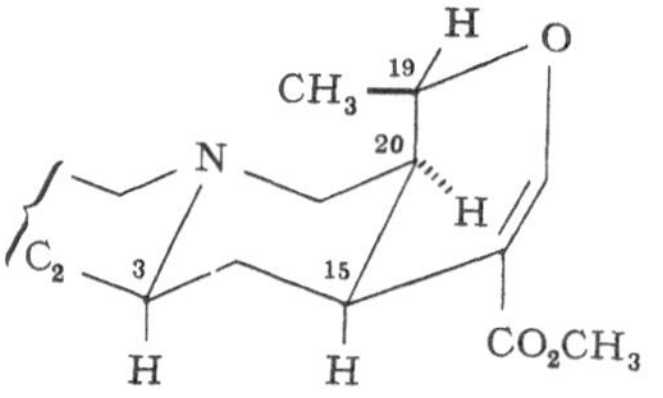

(75)

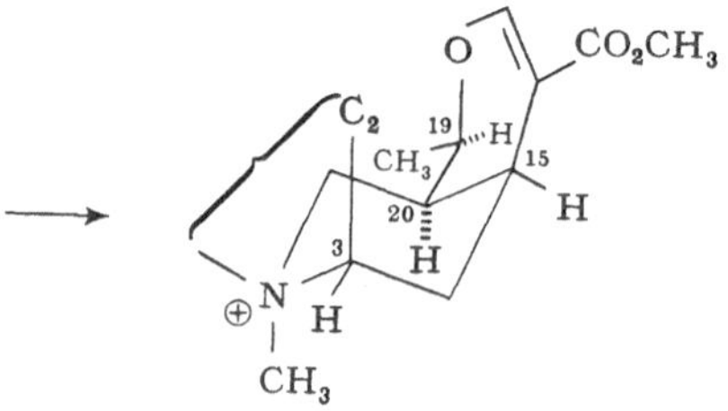

(76)

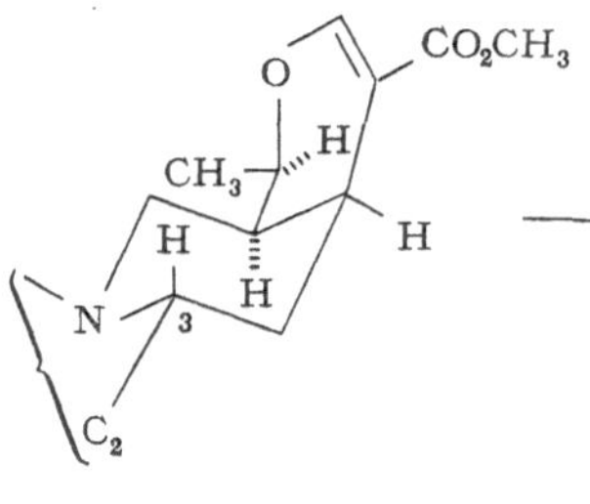

(77)

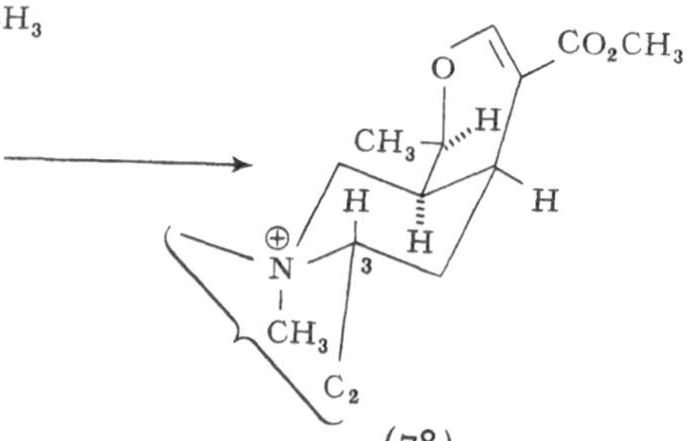

(78)

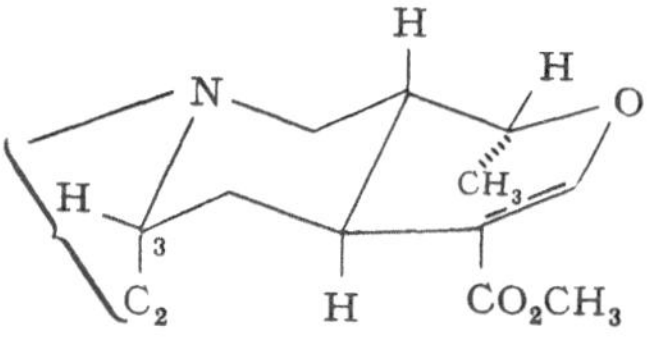

(79)

application of spectroscopic techniques to raunitidine (74) (a member of group C), its 3-epimer, isoraunitidine (group D) and to their methiodides. Both isomers exhibited complex infrared absorption in the 2800 cm⁻¹

region, implying an axial 3-hydrogen in each case (i. e. *trans*-fusion of rings C and D). It followed therefore that one of the alkaloids possessed the allo- configuration (75) and the other the epiallo- configuration (77). The final assignment of configuration was made on the basis of the NMR spectra of the derived methiodides. Raunitidine methiodide exhibited an $\overset{\oplus}{N}$—CH$_3$ peak at 6.51 τ while the corresponding signal in isoraunitidine methiodide appeared further upfield at 6.69 τ. In N-methylquinolizidinium cations, the $\overset{\oplus}{N}$—CH$_3$ protons are known to absorb at lower fields when the rings are *cis*-fused than when they are *trans*-fused (*141*). It followed therefore that raunitidine but not isoraunitidine had undergone a conformational change before methylation, with the result that in the derived methiodide rings C and D were *cis*-fused. Raunitidine could therefore be assigned the allo- configuration (75). Direct methylation being subject to severe hindrance by $C_{(19)}$ and its attached axial methyl group, the reaction proceeds to give the methiodide (76) derived from the alternative C/D-*cis*-conformation. Isoraunitidine, which may now be assigned the epiallo- configuration (77), is relatively unhindered and methylation takes place directly to afford (78).

These configurational assignments are, of course, in conformity with the observation noted above that raunitidine methylates more slowly than isoraunitidine.

Compounds of group F (Table 2) could be assigned a *trans-C/D* configuration from their infrared spectra. Hence, since groups A—D represented the four possible allo- and epiallo- arrangements, these could be assigned a normal configuration, like the compounds of group E. All the possible normal, allo- and epiallo- arrangements being accommodated in groups A—F, it followed that compounds of group H (for an example, see *151*) like those of group G, must possess the pseudo- configuration.

The skeletal configurations of compounds belonging to groups A, B, E, and G were also deduced independently by Wenkert and his colleagues (*186*) on the basis of chemical degradations.

The mass spectra and optical rotatory dispersion of the heteroyohimbine alkaloids also yield information concerning their skeletal configurations. It is possible to distinguish by mass spectroscopy between a *cis*- and *trans D/E* ring junction (*13*), while a recent ORD study has revealed that the Cotton Effect exhibited by the heteroyohimbine alkaloids at 235–255 mμ is negative only for those possessing the allo- configuration (77).

The last remaining stereochemical problem presented by the heteroyohimbine alkaloids (namely, the configuration of the 19-methyl in each stereochemical group) was solved mainly by studying the NMR spectra of representative compounds (*162, 185, 186*). The large coupling constant (10.3 c. p. s.) between the $C_{(19)}$ and $C_{(20)}$ protons of tetrahydro-alstonine (group A) implied a *trans*-diaxial relationship (*72*); from this it followed that the 19-methyl group had the α-configuration (*186*). The much smaller coupling constants observed for ajmalicine (group E, $J = 2.7$ c. p. s.) (*186*), 3-isoajmalicine (group G, $J = 1.8$ c. p. s.) (*186*), and mayumbine (group D, $J = 2.6$ c. p. s.) (*185*) indicated that in these compounds the 19- and 20-hydrogen atoms were *cis*- disposed. As a consequence, the 19-methyl group is α- in groups E (*73*) and G (*79*) and β- in group D (*77*). Since groups B, C, F and H are, respectively, 19-epi-mers of groups D, A, E and G, their stereochemistry follows as a conse-quence of the above assignments. The same $C_{(19)}$ configurations were also deduced from the observed rates of methylation of various hetero-yohimbine alkaloids and from the chemical shifts of the 19-methyl groups in the free alkaloids and in the derived methiodides (*162*).

The elucidation of the configurations of the heteroyohimbine alka-loids led to a rapid increase in knowledge of the stereochemistry of the structurally related oxindole alkaloids. From a comparison of the NMR

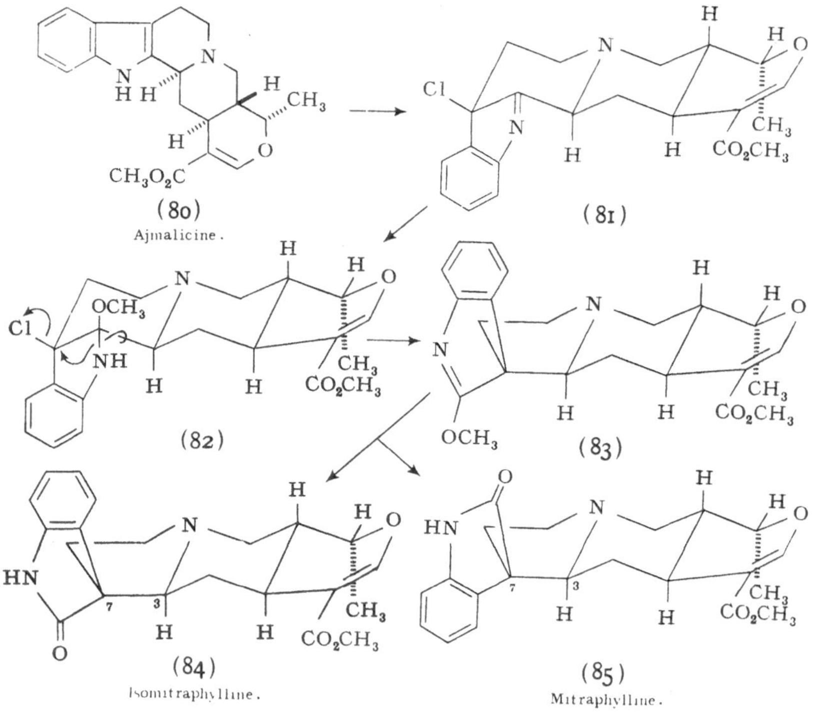

Scheme 1. Partial Synthesis of Mitraphylline and Isomitraphylline.

spectra of mitraphylline (85) and its $C_{(7)}$-epimer, isomitraphylline (84), with that of their heteroyohimbine analogue, ajmalicine (80), WENKERT and his colleagues (185) concluded that all three alkaloids possessed the same configuration at $C_{(15)}$, $C_{(19)}$ and $C_{(20)}$. The 3α—H configuration depicted in (84) and in (85) was assigned (185) in order to account for the observation that the mitraphyllines, which are readily equilibrated by a process affecting the $C_{(3)}$ and $C_{(7)}$ asymmetric centres only (see 86 $\rightleftarrows$ 87), are of comparable thermodynamic stability [K (mitraphylline/isomitraphylline) = 4 (161)]. If either of the isomers had possessed an equatorial (3β—H)-configuration, it would be expected to be isomerised almost completely into its 3-epimer in order to relieve the severe non-bonded interactions associated with the presence of an axial quaternary $C_{(3)}$ substituent (see 88)*.

These deductions were strikingly confirmed by the achievement of a partial synthesis of mitraphylline (85) and isomitraphylline (84) from ajmalicine (80) (75, 76, 164, 196) (Scheme 1). Oxidation of ajmalicine with tert-butyl hypochlorite gave the chloroindolenine (81), methanolysis of which afforded the imidoether (83) presumably via the intermediate (82). Acid hydrolysis of the imidoether resulted in the formation of the equilibrium mixture of mitraphylline and isomitraphylline. These transformations conclusively established the configurational identity of ajmalicine and the mitraphyllines at $C_{(15)}$, $C_{(19)}$ and $C_{(20)}$. However, because the conditions of hydrolysis of the imidoether (83) were such as to cause interconversion of the initially formed oxindole with its $C_{(7)}$-epimer, by the mechanism indicated in (86) and (87), they did not immediately permit configurational assignments to be made for $C_{(3)}$ and for $C_{(7)}$. These remaining stereochemical ambiguities were resolved as a result of experiments carried out on yohimbine (34) (76).

---

* By similar reasoning formosanine and isoformosanine were assigned the same configurations at $C_{(3)}$, $C_{(15)}$, $C_{(19)}$, and $C_{(20)}$ as their heteroyohimbine analogue mayumbine (partial structure 77) (185). Interpretation of equilibration data for oxindole alkaloids possessing a cis-junction of rings D and E is made difficult, however, by the fact that in such compounds two chair conformations are available for ring D so that, irrespective of the $C_{(3)}$-configuration, it is possible for the molecule to adopt a conformation such that the $C_{(7)}$ quaternary grouping is equatorially disposed. It has recently been demonstrated that uncarines C, D, E, and F ($C_{(3)}$ and $C_{(7)}$ epimers of 89) can be equilibrated in pyridine solution (87); and NMR studies have revealed that in all four isomers $C_{(7)}$ has an equatorial conformation with respect to ring D. Confirmation of the $C_{(3)}$-configuration in formosanine and isoformosanine would therefore be desirable. For the mitraphyllines, the trans D/E ring junction permits only one chair conformation for ring D and equilibration data can be interpreted with greater confidence.

Added in Proof. Since the foregoing note was set in print the stereochemistry of a large number of the pentacyclic oxindole alkaloids has been clarified, and the configurations of formosanine and isoformosanine have been revised (163 a).

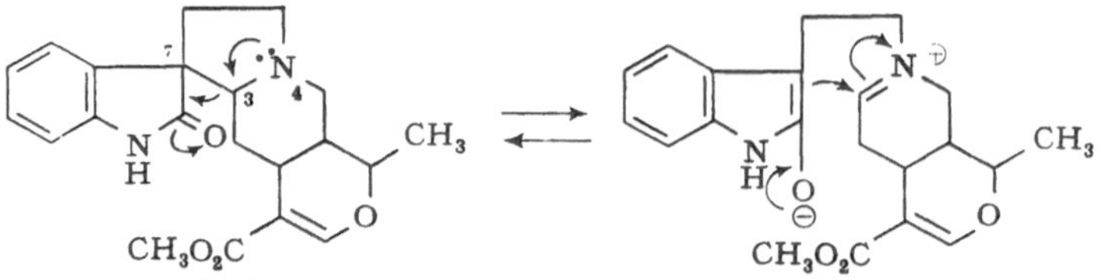

(86)   (87)

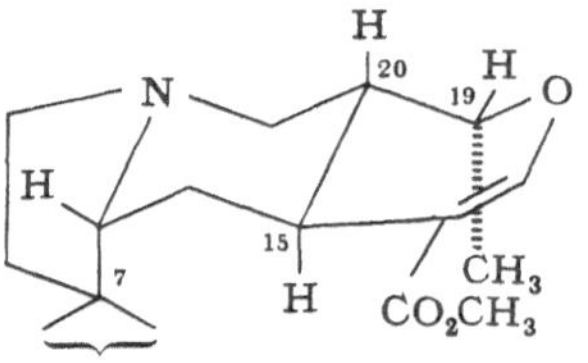

(88) 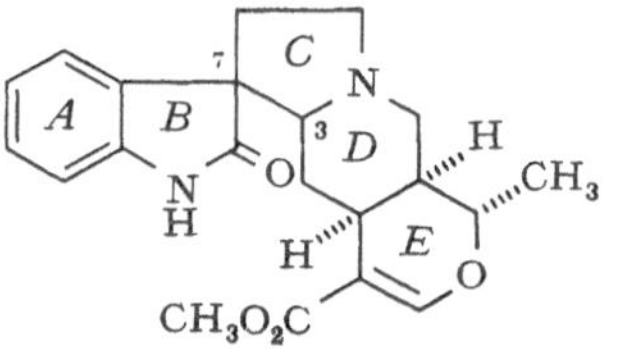   (89)

It was concluded that rearrangement of the chloroindolenine derived from yohimbine to the intermediate imidoether (90) proceeds with almost complete retention of configuration at $C_{(3)}$ (as expected by consideration of the stereoelectronic requirements of the reaction) when it was found that pseudoyohimbine (45), the 3-epimer of yohimbine, gave only a minute yield of the same imidoether. As in the case of the ajmalicine derivative, acid hydrolysis of the imidoether (90) gave two isomeric oxindoles, designated A and B. However, when the methiodide of (90) was rearranged, the methiodide of yohimbine oxindole A was obtained exclusively. Since quaternisation of $N_{(4)}$ prohibits an equilibration of the type (86 ⇄ 87), yohimbine oxindole A could be formulated as (91); NMR and pKa data indicated that yohimbine oxindole B is its 7-epimer (76). These assignments are in accord with the observation that the infrared and NMR spectra of both oxindoles suggest an axial 3α-hydrogen (196).

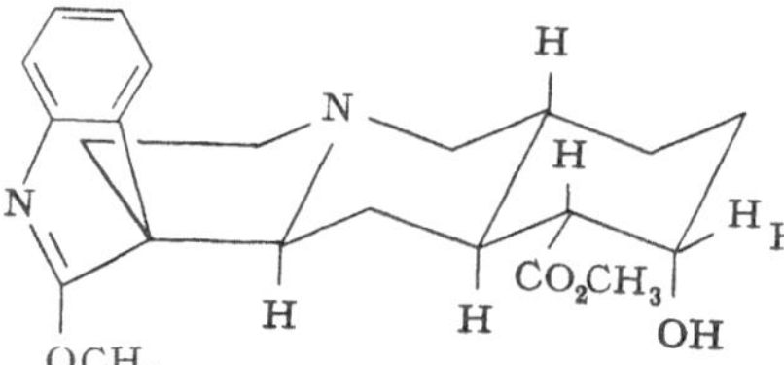

(90) 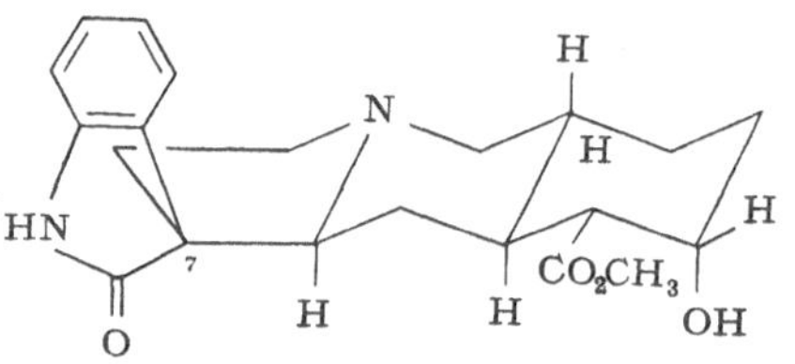   (91)
                                 Yohimbine oxindole A.

From the results obtained in the yohimbane series it could be inferred that the reaction sequence (80 → 83) was stereospecific. Thus, when a thin-layer chromatographic examination of the hydrolysis of (83) revealed that the initially formed product was isomitraphylline, the latter could with confidence be assigned the configuration implied in
References, pp. 307—317.

(84). Further evidence for the configurations of mitraphylline (85) and isomitraphylline (84) was obtained by comparing their NMR spectra with those of yohimbine oxindole *A* (91) and its 7-epimer, yohimbine oxindole B (76). Recently, it has been shown that circular dichroism can be employed to investigate the configurations of oxindole alkaloids at $C_{(3)}$ and $C_{(7)}$ (28, 153).

Conversion of yohimbinoid and heteroyohimbine alkaloids into their oxindole analogues via chloroindolenine and imidoether intermediates is satisfactory only for compounds possessing a *trans-D/E* ring junction (76, 196). In the *cis-D/E* series the same overall transformation can be achieved by rearrangement under acid conditions of the 7-acyloxindolenine derivative (72, 73). Thus, isoreserpiline (92; 3α—H), upon oxidation with lead tetraacetate affords the 7-acetoxyindolenine (93; $R = CH_3$, 3α—H)* which, by treatment with acid, is rearranged to carapanaubine (94) (78), an alkaloid of natural occurrence (84) *(Scheme 2)*. The course of the reaction is insensitive to the stereochemistry at $C_{(3)}$; similar treat-

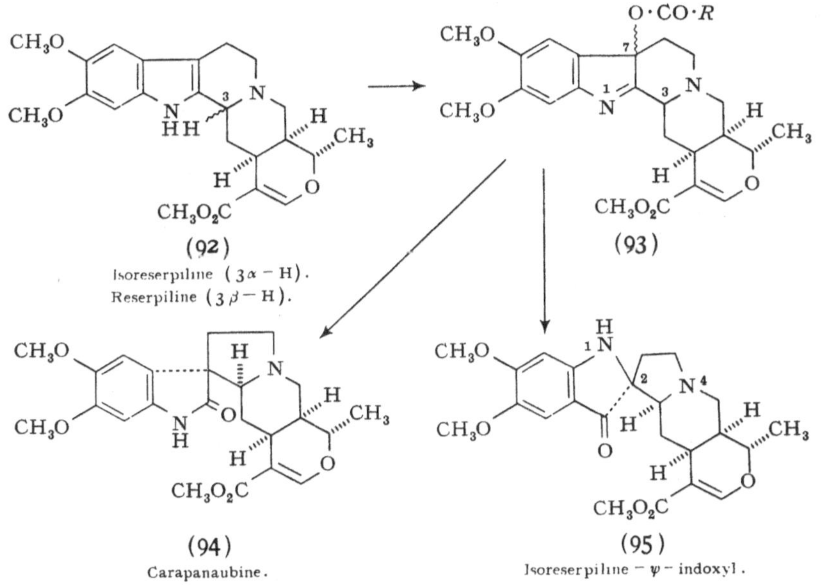

Scheme 2. Conversions of Reserpiline and Isoreserpiline.

---

* It appears, from the results of investigations employing X-ray diffraction and ORD techniques (72), that in such oxidations the newly-introduced 7-acetate group adopts the same configuration as the 3-hydrogen atom of the starting material. On this basis, the acetoxyindolenine derived from isoreserpiline is to be assigned the 7α-configuration, and that from reserpiline the 7β-configuration.

ment of reserpiline (92; 3β—H) also affords carapanaubine (73, 152), together with rauvoxine and rauvoxinine ($C_{(3)}$ epimers of carapanaubine differing in configuration at $C_{(7)}$) (152). As a result of the degree of flexibility conferred on these molecules by the cis-fusion of rings D and E it is probable that in all three isomers the $C_{(7)}$ quaternary centre has an equatorial conformation.

Such an arrangement has been demonstrated for the $C_{(3)}$ and $C_{(7)}$ epimeric uncarines C, D, E, and F (89), all of which are obtainable from tetrahydroalstonine (partial structure 72) by rearrangement of the corresponding acetoxyindolenine (87).

7-Acyloxyindolenines with a trans-D/E ring junction are stable under the rearrangement conditions, and it has been suggested (73) that this may follow from the necessity for protonation of $N_{(1)}$ during the rearrangement sequence (see 96). The greater flexibility of ring E in compounds with a cis-D/E ring fusion may render $N_{(1)}$ more accessible to protonation (Scheme 3).

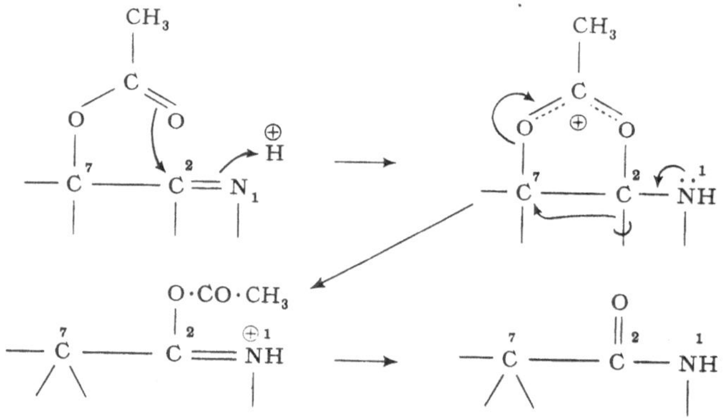

(96)

Scheme 3. Postulated Mechanism for Acid catalysed Rearrangement of 7-Acyloxyindolenines (73).

Under base catalysis the 7-m-bromobenzoyloxyindolenine (93; R = 3-bromophenyl, 3α—H) derived from isoreserpiline is rearranged to the pseudoindoxyl (95) (72, 78; see also 74), which is of natural occurrence. The configuration at $C_{(2)}$ of the pseudoindoxyl derivative (95) was assigned by analogy with that of the corresponding derivative of yohimbine. The latter was deduced from the results of isomerisation studies, and from the observation that the basicity of $N_{(4)}$ is reduced when $N_{(1)}$ is methylated, an effect which implies that the configuration of the $N_{(1)}$-methyl group is such as to produce steric hindrance to the protonation of $N_{(4)}$.

V. Alkaloids of the Amaryllidaceae.

For an excellent review of this class of alkaloids see (*187*).

Most of the known members of the Amaryllidaceae family of alkaloids have been discovered within the past fifteen years, and the first reports concerning the configurations of these compounds appeared in 1956. From the first, therefore, conformational principles were available to workers in this field. In the space available some indication will be given of the way in which these principles have been applied to each of the main classes of Amaryllidaceae alkaloids.

I. Tazettine.

The constitution of tazettine (97) with the exception of the configuration of the hemiketal grouping, was first proposed by Taylor and Uyeo and their colleagues (*104*) on the basis of an elegant series of degradations.

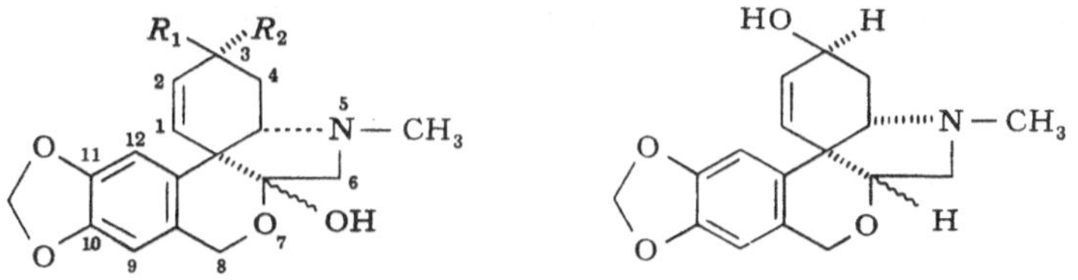

(97) Tazettine ($R_1 = OCH_3$; $R_2 = H$).
(98) Criwelline ($R_1 = H$; $R_2 = OCH_3$).

(99)
Deoxytazettinol.

The relative configuration of the methoxyl group of tazettine with respect to the basic nitrogen atom was deduced from the pKa values of deoxytazettinol (99) and its 3-epimer. The latter compound being a stronger base by 1.6 pKa units, it was argued that the proton of the conjugate acid must enter into hydrogen bonding with the 3-hydroxyl group. Since the groups involved are in a 1,3-relationship on a six-membered ring, this is possible only if each can assume an axial or pseudo-axial conformation. It follows that in 3-isotazettinol the 3-hydroxyl group and the nitrogen atom are *cis*-oriented on the cyclohexene ring; and that in tazettine itself the corresponding functions bear a *trans* relationship.

That the nitrogen function of tazettine, and those of its transformation products where such a consideration is appropriate, can assume an axial conformation is also indicated by the ease with which the derived quaternary ammonium salts undergo Hofmann elimination to afford degradation products containing a Δ^4-double bond. Since fusion of a five-membered ring to a six-membered ring through adjacent *trans*-diaxial bonds is not possible, it must be postulated that the pyrrolidine ring of tazettine is *cis*-fused to the cyclohexene ring.

Confirmation of these deductions, which are summarised in the expression (97), has been provided by synthetic studies which have established that the methoxyl and phenyl groups of tazettine are *cis*-related (*107*).

UYEO and his coworkers (*173*) have proposed that the configuration of the hemiketal grouping in tazettine is to be represented as in (100). The experimental evidence quoted in support of this assertion is that when the pyrrolidine ring of tazettine is reformed by cyclisation of Hofmann degradation products of the type (101), the nitrogen atom affixes itself on the side of ring B to which it was attached in the parent compound. However, inspection of Dreiding models reveals that if the hemiketal moiety of tazettine is opposite in configuration to that of (100), then α-attachment of the dimethylamino side-chain in the cyclisation is equally favoured; indeed, in such a case the configuration of the molecule would permit *only* α-cyclisation. Hence, on this basis, the configuration of the hemiketal group can not be regarded as firmly established.

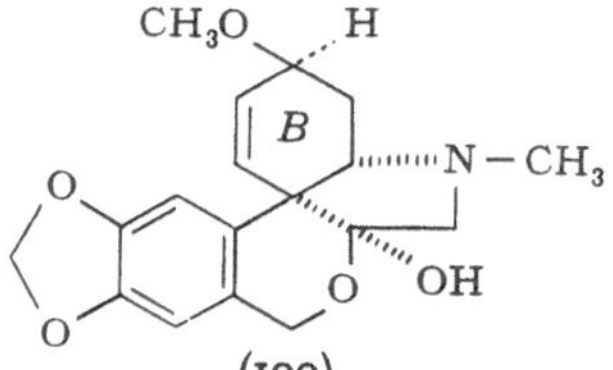

(100) 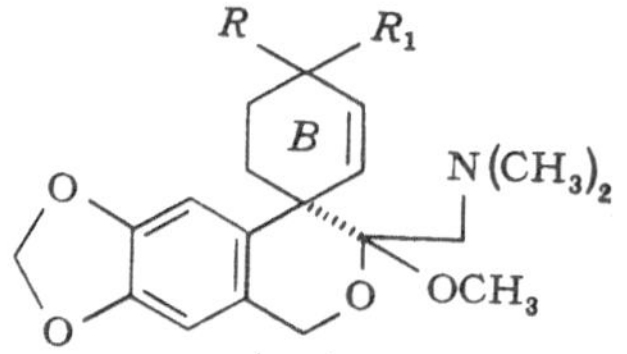   (101)

The NMR spectra of tazettine and that of its 3-epimer, criwelline have been shown to be consistent with the stereochemistry indicated in (97) and (98), respectively (*88*). These data do not, however, distinguish between the two possible configurations of the hemiketal group.

Although they differ in configuration only at $C_{(3)}$, tazettine (97) and criwelline (98) afford quite different mass spectra. It has been suggested (*57*) that the dominant ion (104, m/e = 247) in the tazettine spectrum arises by hydrogen transfer in the molecular ion (102) followed by cleavage of the resultant ion (103) *(Scheme 4)*. In the spectrum of criwelline (98) on the other hand there is a prominent ion at m/e 301, thought to be (106) arising by loss of formaldehyde from the molecular ion (105). Subsequent cleavage (106 → 104) then gives rise to a weak line in the spectrum corresponding to m/e 247. These differences in the mass spectra of tazettine and criwelline may be directly attributable to a conformational effect, since in both (102) and (105) it is the pseudo-axial substituent on $C_{(3)}$ which is involved in the initial process (cf. *88*).

References, pp. 307—317.

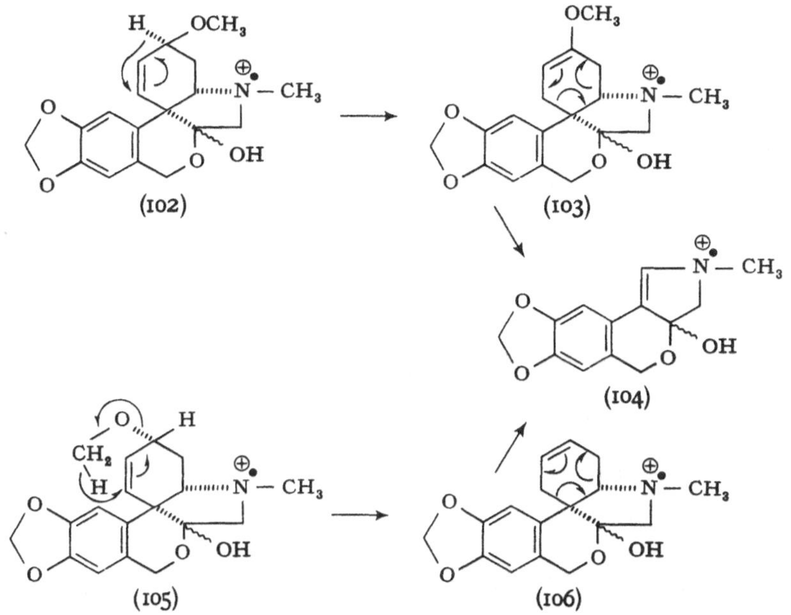

H OCH₃

(102)

OCH₃

(103)

⊕N—CH₃
OH
(104)

CH₂
H
H
(105)

(106)

Scheme 4. Mass Spectral Fragmentation of Tazettine and Criwelline.

2. Lycorine and Related Compounds.

The stereochemistry of lycorine (107) was derived by application of conformational analysis to the chemistry of dihydrolycorine (108 ≡ 112) (*169*).

The *trans*-diaxial nature of the two hydroxyl groups was established by acid cleavage of the epoxide (113), formed by base treatment of the monotosylate (109), when dihydrolycorine (108) was re-formed. The obtainment of 2-epilycorine by borohydride reduction and saponification of the ketone derived from the 1-monoacetate of lycorine (107) (*142*) supports this formulation, since hydride reduction of an unhindered ketone generally gives a preponderance of equatorial alcohol (*24*). 2-Epilycorine, being a *cis*-diol, readily affords an isopropylidene derivative (*142*); no corresponding derivative can be obtained from lycorine itself.

Hydride reduction of the epoxide (113) gave, following the usual pattern of diaxial epoxide cleavage, α-dihydrocaranine (110) (*169*). Dehydration with phosphorus oxychloride proceeded smoothly, yielding the styrene derivative (114) from which the *trans*-diaxial relationship of the 1-hydroxyl group and the 11b-hydrogen atom was inferred (*169*). von Braun degradation of dihydrolycorine diacetate afforded (115) which could be converted into (116), thus fixing the 2-hydroxyl and 4-methylene

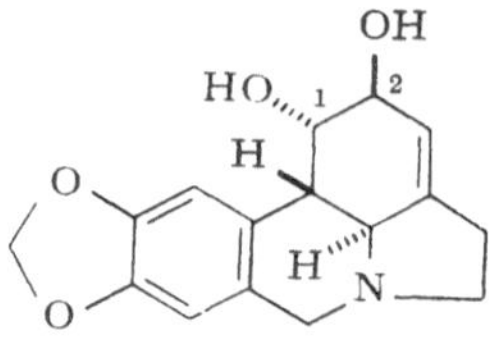

(107)
Lycorine.

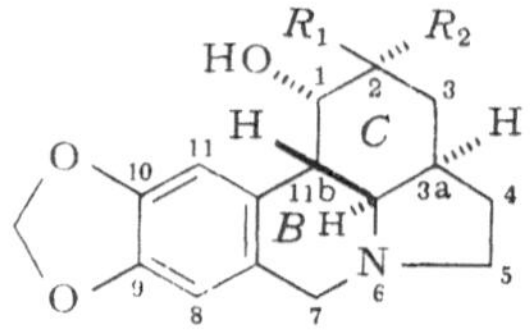

(108) Dihydrolycorine ($R_1 = OH$, $R_2 = H$).
(109) Dihydrolycorine monotosylate
($R_1 = OTs$, $R_2 = H$).
(110) α-Dihydrocaranine ($R_1 = R_2 = H$).
(111) Zephyranthine ($R_1 = H$, $R_2 = OH$).

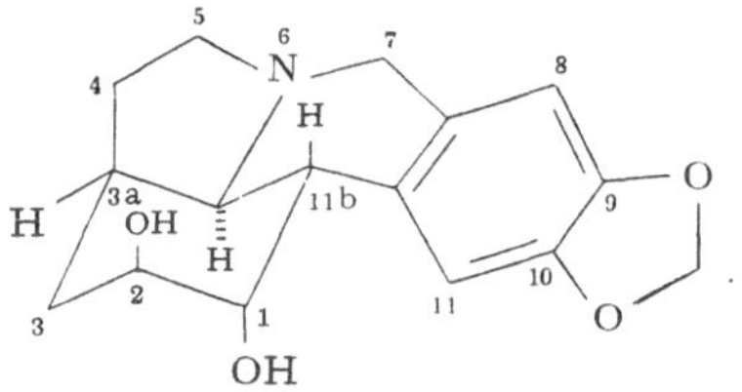

(112)
Dihydrolycorine.

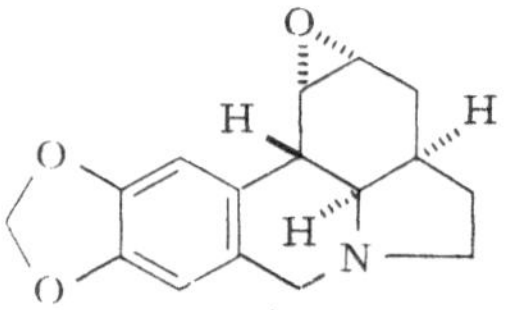

(113)

groups as *cis*-oriented (*169*). Further, since the 2-hydroxyl group of
(108) is axial, the $C_{(4)}$ methylene group must also be axial, if ring C
exists in a chair conformation. Consequently, the pyrrolidine ring must
be *cis*-fused to ring C, and the configurations of lycorine and of dihydrolycorine are correctly represented by (107) and (108), respectively. The
trans-B/C ring junction and the steric relationship of the hydroxyl groups
in ring C have been confirmed by NMR studies involving the technique
of spin-decoupling (*124*).

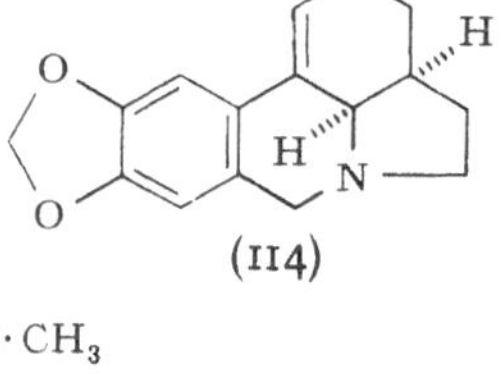

(114)

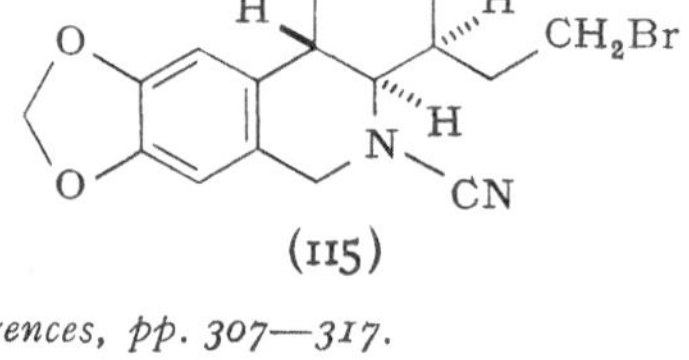

(115)

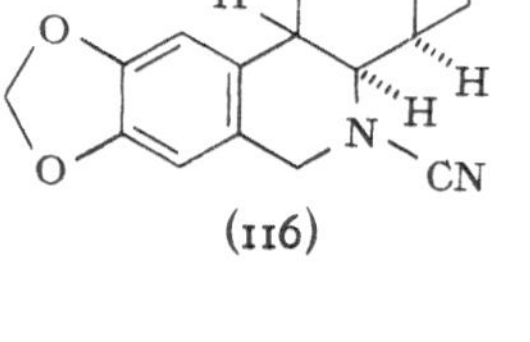

(116)

The experiments described above, since they interrelated lycorine (107) and α-dihydrocaranine (110), also established the stereochemistry of caranine (117) (*169, 177*), which has since been confirmed by the partial synthesis of this alkaloid from lycorine (*170*). A stereospecific synthesis of racemic β-lycorane (119), derivable from either caranine or lycorine (*123*), has also been reported (*95*).

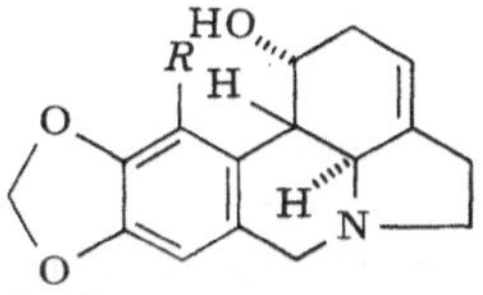

(117) Caranine (R = H).
(118) Falcatine (R = OCH₃).

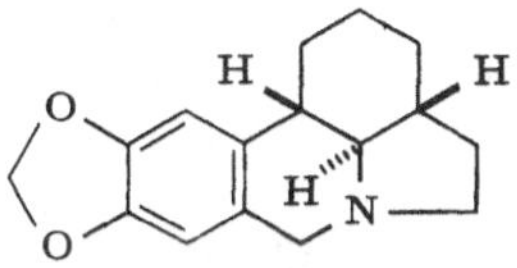

(119)
β – Lycorane.

The stereochemistry of zephyranthine (111), another member of this group, has recently been elucidated (*147*). The *cis*-relationship of the hydroxyl groups followed from the observation that the compound reacted faster with periodic acid than did dihydrolycorine (108 ≡ 112), and from the ready formation of an isopropylidene derivative. Selective tosylation of the equatorial 2α-hydroxyl group of zephyranthine (cf. 112; 2α—OH in place of 2β—OH), followed by hydride reduction, afforded α-dihydrocaranine (110).

Correlations established between lycorine (107), caranine (117), pluviine (120) and falcatine (118) and those alkaloids of the Amaryllidaceae which have structures based on the [2]-benzopyrano-[3,4-g]-

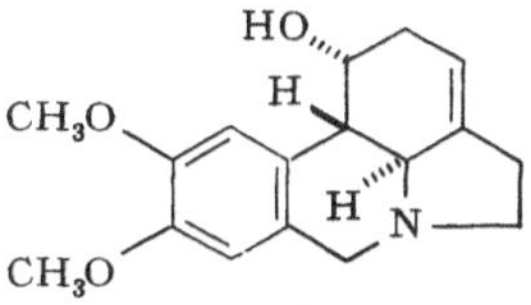

(120)
Pluviine.

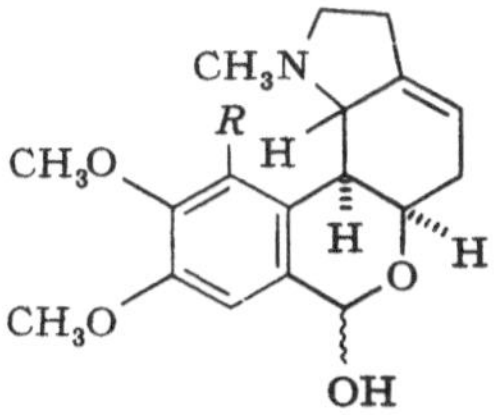

(121) Lycorenine (R = H).
(122) Krigenamine (R = OCH₃; – OCH₂O – instead of di – OCH₃).

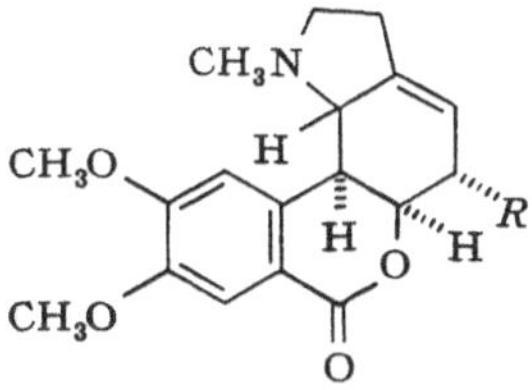

(123) Homolycorine (R = H).
(124) Hippeastrine (R = OH; – OCH₂O – instead of di – OCH₃).

indole nucleus, have made it possible to deduce the stereochemistry of lycorenine (**121**), homolycorine (**123**), hippeastrine (**124**), and krigenamine (**122**) (*81, 120, 140, 174*).

Configurational and conformational aspects of lycorenine and related alkaloids have also been investigated by means of NMR spectroscopy and by measuring their rates of methylation (*89*).

3. Alkaloids Possessing the 5,10 b-Ethanophenanthridine Skeleton.

Alkaloids in this group form two enantiomorphic series, based respectively on (—)-crinane (**125**) and on (+)-crinane (**126**) (*66, 120*). The relative configurations of compounds in both these series have been determined largely as a result of the work done by WILDMAN and his collaborators (*64–68, 94, 106, 178, 187*). The arguments employed in elucidating the stereochemistry of undulatine (**127**) (*130, 178*) typify the approach employed throughout their investigations.

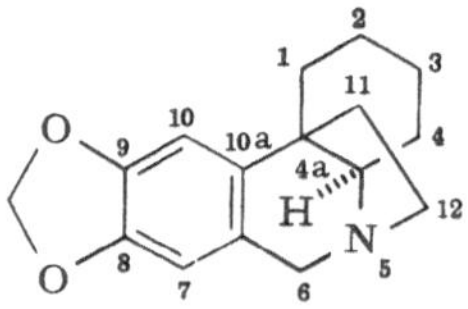

(**125**)
(-) — Crinane.

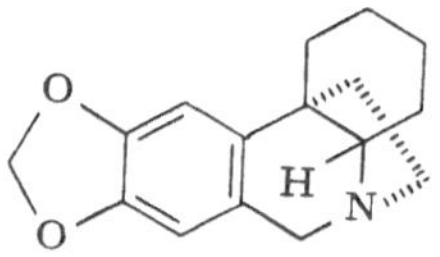

(**126**)
(+) — Crinane.

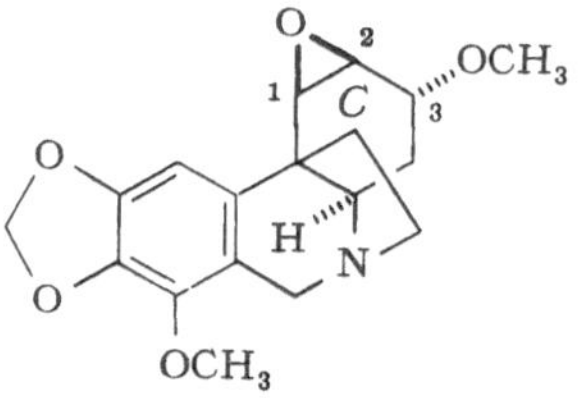

(**127**)
Undulatine.

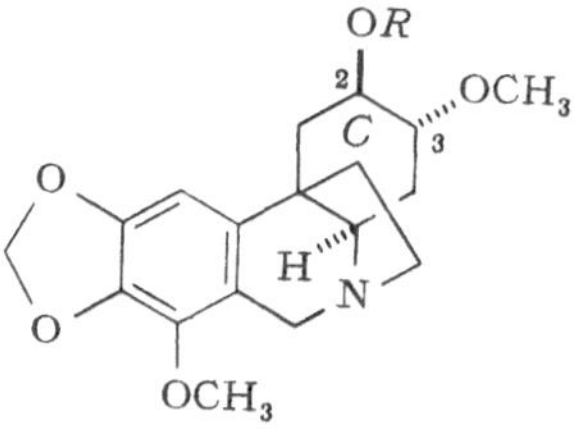

(**128**)

An axial conformation was assigned to the 3-methoxyl group of undulatine (**127**), since this function was found to be readily epimerised when the ketone derived from dihydroundulatine (**128**; $R = H$) was treated under basic conditions. Dihydroundulatine having been obtained as the major product of hydride reduction of the epoxide ring of undulatine, it could reasonably be inferred that the 2-hydroxyl group of the former was also axial, and hence *trans* to the methoxyl group. The correctness of this surmise was confirmed when it was shown that (**128**; $R = SO_2CH_3$) was converted into (**128**; $R = CO \cdot CH_3$) exclusively by treatment with acetic anhydride and acetic acid under reflux.

References, pp. 307—317.

The retention of configuration observed in this reaction is a clear example of neighbouring group participation by the methoxyl group, a process which occurs most readily if both the functions involved are axially disposed. Further evidence for the axial nature of the 2-hydroxyl group of dihydroundulatine (128; $R = H$) was provided by the observation that this compound is oxidised by chromium trioxide 2.8 times faster than its epimer.

Assuming a chair conformation for ring C of dihydroundulatine (which is reasonable since four of the ring atoms are free to adopt the most stable conformation), only two possible configurations, namely (129) and (130), are possible for this molecule. The correctness of (129 $\equiv$ 128) was established by converting (131), the von Braun degradation product of dihydroundulatine, into (132), the formation of which is possible only if the ethano-bridge and the hydroxyl group bear a *cis*-relationship. The configuration expressed in (127) follows for undulatine itself.

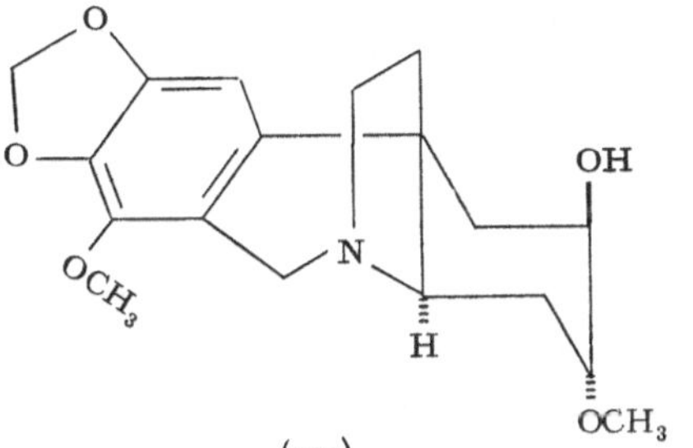

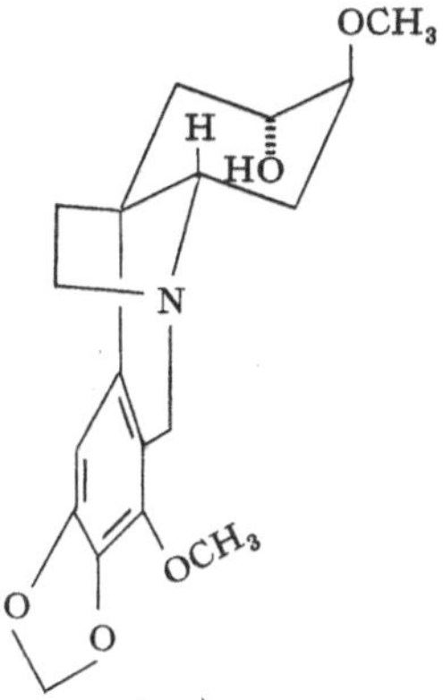

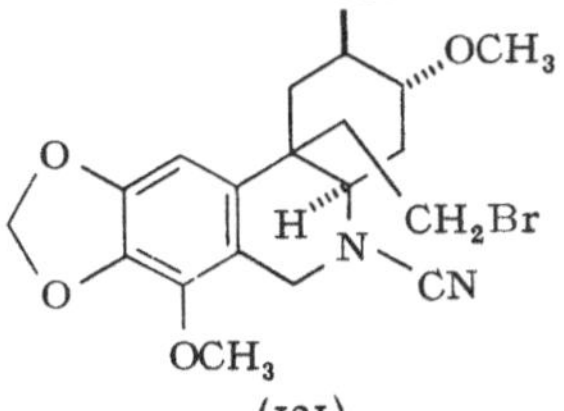

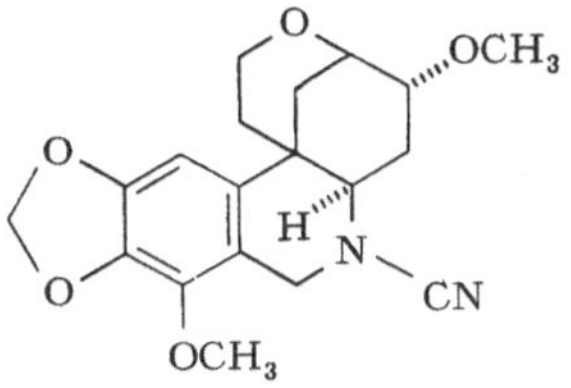

Haemanthamine (133) and crinamine (134), both of which have structures based on the skeleton of (+)-crinane (126), undergo an interesting skeletal rearrangement when treated with methanesulphonyl chloride and pyridine (*106*). Haemanthamine yields isohaemanthamine (137), while from crinamine a mixture of α-isocrinamine (138) and β-isocrinamine (139) was obtained.

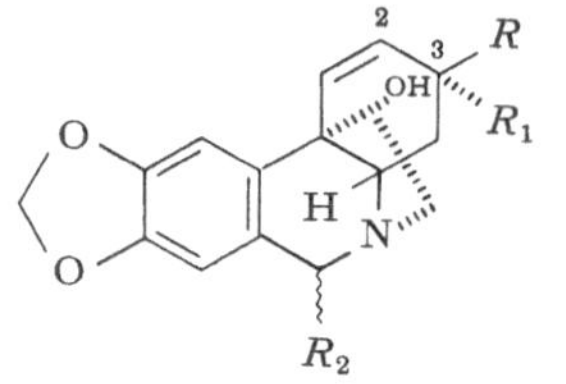

(133) Haemanthamine ($R = OCH_3$; $R_1 = R_2 = H$).
(134) Crinamine ($R = R_2 = H$; $R_1 = OCH_3$).
(135) 6 – Hydroxycrinamine ($R = H$; $R_1 = OCH_3$; $R_2 = OH$).
(136) Haemanthidine ($R = OCH_3$; $R_1 = H$; $R_2 = OH$).

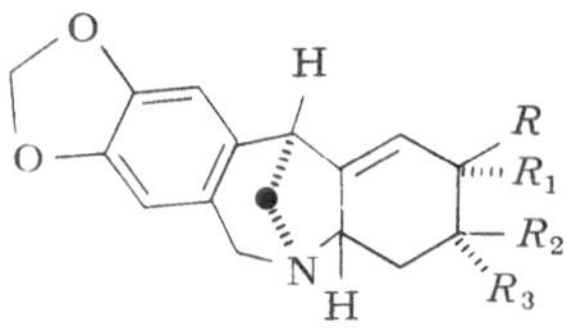

(137) Isohaemanthamine ($R = R_3 = H$; $R_1 = OH$; $R_2 = OCH_3$).
(138) α – Isocrinamine ($R = OH$; $R_1 = R_2 = H$; $R_3 = OCH_3$).
(139) β – Isocrinamine ($R = R_2 = H$; $R_1 = OH$; $R_3 = OCH_3$).
(140) Coccinine ($R = OCH_3$; $R_1 = R_3 = H$; $R_2 = OH$).
(141) Montanine ($R = R_3 = H$; $R_1 = OCH_3$; $R_2 = OH$).
(142) Manthine ($R = R_3 = H$; $R_1 = R_2 = OCH_3$).

The stereochemistry of these rearrangement products, save for the configuration of the 2-hydroxyl group, follows from the mechanism of the reaction, summarised, in the case of haemanthamine, by the expression (143 → 144) *(Scheme 5)*. Clearly, this process is an example of a 1,2-rearrangement in which the four centres involved occupy an *anti*-coplanar conformation.

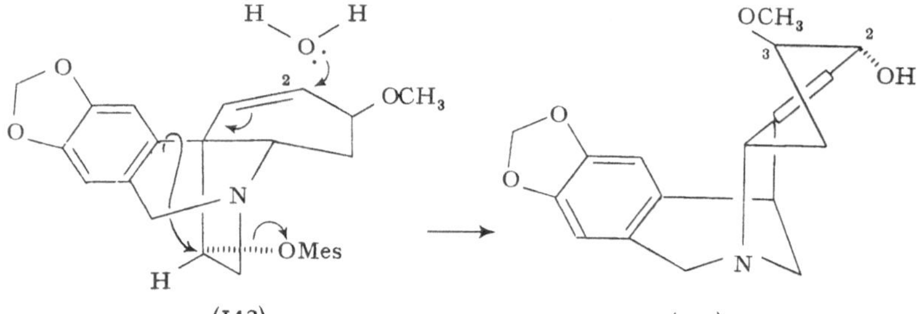

(143)   (144)

Scheme 5. Rearrangement of Haemanthamine Mesylate to Isohaemanthamine.

Configurations at $C_{(2)}$ of the products were assigned on the basis of studies of their infrared absorption in the O—H region (cf. *69*). Thus, for example, the *trans*-diaxial relationship of the 2- and 3-oxygen functions of (**144**) was inferred from the absence of hydrogen bonding as judged by the infrared spectrum.

The above rearrangements acquired a special significance when it was found that the skeleton of isohaemanthamine (and also of the α- and β-isocrinamines) occurs in nature in the Amaryllidaceae alkaloids coccinine (**140**), montanine (**141**) and manthine (**142**). The stereochemistry of these compounds was established by direct correlation with isohaemanthamine and the α- and β-isocrinamines.

An investigation by NMR spectroscopy of a number of Amaryllidaceae alkaloids has confirmed that in solution the preferred conformation of those based on the 5,10b-ethanophenanthridine ring system is as indicated in (**143**) (*88*). Interestingly, it has been shown by NMR spectroscopy that 6-hydroxycrinamine (**135**) and haemanthidine (**136**), which are both of natural occurrence, exist in solution as mixtures of $C_{(6)}$-epimers (*119*).

References.

1. AARON, H. S.: Conformational Analysis of Quinolizidine and Indolizidine Systems. Chem. and Ind. **1965**, 1338.

2. ALDRICH, P. E., P. A. DIASSI, D. F. DICKEL, C. M. DYLION, P. D. HANCE, C. F. HUEBNER, B. KORZUN, M. E. KUEHNE, L. H. LIU, H. B. MACPHILLAMY, E. W. ROBB, D. K. ROYCHAUDHURI, E. SCHLITTLER, A. F. ST. ANDRÉ, E. E. VAN TAMELEN, F. L. WEISENBORN, E. WENKERT and O. WINTERSTEINER: The Stereochemistry of Reserpine, Deserpidine and Related Alkaloids. J. Amer. Chem. Soc. **81**, 2481 (1959).

3. ALLEN, M. J.: A Comparative Study of Hydrolysis Rates of Some Indole Alkaloids. J. Chem. Soc. (London) **1960**, 4904.

4. — Suppression of 1,3-Diaxial Interaction in Methyl Reserpate during Alkaline Hydrolysis. J. Chem. Soc. (London) **1961**, 4252.

5. ALLINGER, N. L.: Conformational Analysis. III. Applications to Some Medium Ring Compounds. J. Amer. Chem. Soc. **81**, 5727 (1959).

6. ALLINGER, N. L., J. G. D. CARPENTER and F. M. KARKOWSKI: The Effective Size of the Lone Pair on Nitrogen. Tetrahedron Letters **1964**, 3345.

7. — — — Conformational Analysis. XLII. Experimental Approaches to the Problem of the Size of the Lone Pair on Nitrogen. J. Amer. Chem. Soc. **87**, 1232 (1965).

8. ALLINGER, N. L. and J. L. COKE: The Relative Stabilities of *cis* and *trans* Isomers. VI. The Decalins. J. Amer. Chem. Soc. **81**, 4080 (1959).

9. ALLINGER, N. L. and L. A. FREIBERG: Conformational Analysis. X. The Energy of the Boat Form of the Cyclohexane Ring. J. Amer. Chem. Soc. **82**, 2393 (1960).

10. ALT, G. H. and D. H. R. BARTON: Some Conformational Aspects of Neighbouring-group Participation. J. Chem. Soc. (London) **1954**, 4284.

11. ANDERSON, J. E.: The Study of Ring Inversions by Nuclear Magnetic Resonance Spectroscopy. Quart. Rev. (Chem. Soc. London) **1965**, 426.

12. ANET, F. A. L. and M. Z. HAQ: Ring Inversion in Cyclohexene. J. Amer. Chem. Soc. **87**, 3147 (1965).

13. ANTONACCIO, L. D., N. A. PEREIRA, B. GILBERT, H. VORBRUEGGEN, H. BUDZIKIEWICZ, J. M. WILSON, L. J. DURHAM and C. DJERASSI: Alkaloid Studies. XXXIII. Mass Spectrometry in Structural and Stereochemical Problems.

VI. Polyneuridine, A New Alkaloid from *Aspidosperma polyneuron* and Some Observations on Mass Spectra of Indole Alkaloids. J. Amer. Chem. Soc. **84**, 2161 (1962).

14. Armitage, B. J., G. W. Kenner and M. J. T. Robinson: Conformational Effects in Compounds with Six-membered Rings. II. Conformational Equilibra in Monosubstituted and *cis*-1,3-Disubstituted Cyclohexanes. Tetrahedron **20**, 747 (1964).

15. Arndt, R. R. and C. Djerassi: Alkaloid Studies. LV. 19-Dehydroyohimbine, a Novel Alkaloid from *Aspidosperma pyricollum*. Experientia **21**, 566 (1965).

16. Aroney, M. and R. J. W. Le Fèvre: Molecular Polarisability. Application of the N—H and N—C Link Polarisabilities to the Conformations of Tertiary Amines, Piperidine, and Morpholine. J. Chem. Soc. (London) **1958**, 3002.

17. Atkinson, V. A.: The Conformational Analysis of Chloro*cyclo*hexane by Electron Diffraction. Acta Chem. Scand. **15**, 599 (1961).

18. Balasubramanian, M.: The Boat Form in Six-membered Rings. Chem. Rev. **62**, 591 (1962).

19. Ban, Y. and O. Yonemitsu: Absolute Stereochemistry of Yohimbine and Reserpine. Tetrahedron **20**, 2877 (1964).

20. Barger, G. and E. Field: Yohimbine (Quebrachine). II. *apo*-Yohimbine and Deoxy-yohimbine. J. Chem. Soc. (London) **123**, 1038, (1923).

21. Barton, D. H. R.: The Conformation of the Steroid Nucleus. Experientia **6**, 316 (1950).

22. Barton, D. H. R. and R. C. Cookson: The Principles of Conformational Analysis. Quart. Rev. Chem. Soc., London **10**, 44 (1956).

23. Barton, D. H. R., D. A. Lewis and J. F. McGhie: Conformational Anomalies in Some Triterpenoid Bromo-ketones. J. Chem. Soc. (London) **1957**, 2907.

24. Barton, D. H. R. and G. A. Morrison: Conformational Analysis of Steroids and Related Natural Products. Fortschr. Chem. organ. Naturstoffe **19**, 165 (1961).

25. Bastiansen, O. and O. Hassel: Structure of the So-called *cis*-Decalin. Nature **157**, 765 (1946).

26. Becconsall, J. K., R. A. Y. Jones and J. McKenna: Stereoisomeric Pairs of Cyclic Quaternary Ammonium Salts. II. Configurational Analysis by Proton Magnetic Resonance Spectroscopy. J. Chem. Soc. (London) **1965**, 1726.

27. Beckett, C. W., N. K. Freeman and K. S. Pitzer: The Thermodynamic Properties and Molecular Structure of Cyclopentene and Cyclohexene. J. Amer. Chem. Soc. **70**, 4227 (1948).

28. Beecham, A. F., N. K. Hart, S. R. Johns and J. A. Lamberton: A Study of the C3/C7 Stereochemistry of Uncarines C, D, E and F by Circular Dichroism. Tetrahedron Letters **1967**, 991.

29. Bhacca, N. S. and D. H. Williams: Applications of NMR Spectroscopy in Organic Chemistry. San Francisco: Holden-Day. 1964.

30. Bishop, R. J., G. Fodor, A. R. Katritzky, F. Soti, L. E. Sutton and F. J. Swinbourne: The Conformations of Tropanes. J. Chem. Soc. (London) C **1966**, 74.

31. Bishop, R. J., L. E. Sutton, D. Dineen, R. A. Y. Jones and A. R. Katritzky: Conformational Analysis of Heterocycles: Steric Requirements of the Piperidine Lone-pair. J. Chem. Soc. (London) **1964**, 257.

32. Bohlmann, F.: Zur Konfigurationsbestimmung von Chinolizidin-Derivaten. Chem. Ber. **91**, 2157 (1958).

33. — Stereochemie der 3-[Piperidyl-(2)]-chinolizidine; zugleich ein Beitrag zur Dehydrierung des Sparteins. Chem. Ber. **92**, 1798 (1959).

34. Booth, H.: The Variation of Vicinal Proton-Proton Coupling Constants with Orientation of Electronegative Substituents. Tetrahedron Letters **1965**, 411.

35. Booth, H. and G. C. Gidley: N. M. R. Evidence for the Twist Conformations of Some *cis*-1,4-Disubstituted Cyclohexanes. Tetrahedron Letters **1964**, 1449.

36. Brown, D. R., B. G. Hutley, J. McKenna and J. M. McKenna: The Preferred Steric Course of Quaternisation of Certain 1-Alkylpiperidines. Chem. Commun. **1966**, 719.

37. Brown, K., A. R. Katritzky and A. J. Waring: Conformational Analysis of Heterocycles: Steric Requirements of Solvated and Free Lone-pairs. Proc. Chem. Soc. (London) **1964**, 257.

38. Bucourt, R. et D. Hainaut: Analyse conformationnelle du cyclohexène. C. R. hebd. séances Acad. Sci. **258**, 3305 (1964).

39. — — L'emploi des angles dièdres en analyse conformationnelle. Calcul des géométries et des énergies conformationnelles du cyclohexane, du cyclohexène et de quelques systèmes bicycliques. Bull. soc. chim. France **1965**, 1366.

40. Campion, T. H. and G. A. Morrison: Unpublished.

41. Claxton, T. A.: On the Use of Lone Pairs in Conformational Analysis. Chem. and Ind. **1964**, 1713.

42. Closs, G. L.: The Configurational Equilibrium of the N-Methyl Group in Some Tropane Deuteriohalides. J. Amer. Chem. Soc. **81**, 5456 (1959).

43. Cookson, R. C.: The Stereochemistry of Alkaloids. Chem. and Ind. **1953**, 337.

44. Corey, E. J.: The Stereochemistry of α-Haloketones. IV. The Stable Orientation of Bromine in 2-Bromocholestane-3-one. J. Amer. Chem. Soc. **75**, 4832 (1953).

45. Crabbé, P.: Etude comparative de la dispersion rotatoire optique et du dichroïsme circulaire en chimie organique. Tetrahedron **20**, 1211 (1964).

46. — Optical Rotatory Dispersion and Circular Dichroism in Organic Chemistry. San Francisco: Holden-Day. 1965.

47. Cremlyn, R. J. W., D. L. Garmaise and C. W. Shoppee: Steroids and Walden Inversion. X. The Reconversion of D-Homosteroids into Steroids. J. Chem. Soc. (London) **1953**, 1847.

48. Davis, M. and O. Hassel: Electron Diffraction Investigation of Molecules Containing a Cyclohexane Type Six-membered Ring. Acta Chem. Scand. **17**, 1181 (1963).

49. — — Conformational Analyses of some Fused Alicyclics by Means of Gas Electron Diffraction. Acta Chem. Scand. **18**, 813 (1964).

50. DePuy, C. H., R. D. Thurn and G. F. Morris: Concerted Bimolecular Eliminations and Some Comments on the Effect of Dihedral Angle on E2 Reactions. J. Amer. Chem. Soc. **84**, 1314 (1962).

51. Diassi, P. A., F. L. Weisenborn, C. M. Dylion and O. Wintersteiner: The Stereochemistry of Reserpine. J. Amer. Chem. Soc. **77**, 2028 (1955).

52. — — — — On The Stereochemistry of Reserpine. J. Amer. Chem. Soc. **77**, 4687 (1955).

53. Djerassi, C.: Optical Rotatory Dispersion: Applications to Organic Chemistry. New York: McGraw Hill. 1960.

54. Djerassi, C., R. Riniker and B. Riniker: Optical Rotatory Dispersion Studies. VII. Application to Problems of Absolute Configurations. J. Amer. Chem. Soc. **78**, 6362 (1956).

55. Dorfman, L., A. Furlenmeier, C. F. Huebner, R. Lucas, H. B. MacPhillamy, J. M. Mueller, E. Schlittler, R. Schwyzer und A. F. St. André: Die Konstitution des Reserpins. Helv. Chim. Acta **37**, 59 (1954).

56. Dreyer, D. L.: Citrus Bitter Principles. II. Application of N. M. R. to Structural and Stereochemical Problems. Tetrahedron **21**, 75 (1965).
57. Duffield, A. M., R. T. Aplin, H. Budzikiewicz, C. Djerassi, C. F. Murphy and W. C. Wildman: Mass Spectrometry in Structural and Stereochemical Problems. LXXXII. A Study of the Fragmentation of Some Amaryllidaceae Alkaloids. J. Amer. Chem. Soc. **87**, 4902 (1965).
58. Eckert, J. M. and R. J. W. Le Fèvre: Molecular Polarisability: The Conformations of Tropinone and 3-Halogenotropanes as Solutes in Benzene. J. Chem. Soc. (London) **1962**, 3991.
59. Eliel, E. L.: Conformational Analysis in Mobile Systems. J. Chem. Education **37**, 126 (1960).
60. — Conformational Analysis in Mobile Cyclohexane Systems. Angew. Chem., Internat. Ed. **4**, 761 (1965) [German Ed. **77**, 784 (1965)].
61. Eliel, E. L., N. L. Allinger, S. J. Angyal and G. A. Morrison: Conformational Analysis. New York: Wiley-Interscience. 1965. (a) Chapter 6; (b) pp. 436–442; (c) p. 272; (d) pp. 482–484; (e) p. 305.
62. Eliel, E. L. and M. C. Knoeber: The "Size" of a Lone Pair of Electrons. Evidence for an Axial *t*-Butyl Group. J. Amer. Chem. Soc. **88**, 5347 (1966).
63. Eliel, E. L., L. A. Pilato and J. C. Richer: Oxidation of Monocyclic Cyclohexanols with Chromic Acid. Chem. and Ind. **1961**, 2007.
64. Fales, H. M., D. H. S. Horn and W. C. Wildman: Structure of Criwelline. Chem. and Ind. **1959**, 1415.
65. Fales, H. M. and W. C. Wildman: The Structures of Haemanthamine and Crinamine. J. Amer. Chem. Soc. **82**, 197 (1960).
66. — — The Stereochemistry of Amaryllidaceae Alkaloids Derived from 5,10b-Ethanophenanthridine. J. Amer. Chem. Soc. **82**, 3368 (1960).
67. — — Alkaloids of the Amaryllidaceae. XIX. On the Structures of Crinamidine, Flexinine and Nerbowdine. J. Organ. Chem. (USA) **26**, 181 (1961).
68. — — Structure and Stereochemistry of Buphanamine. J. Organ. Chem. (USA) **26**, 881 (1961).
69. — — Intramolecular Hydrogen Bonding Studies with Semi-rigid Molecules. I. Derivatives of 5,10b-Ethanophenanthridine. J. Amer. Chem. Soc. **85**, 784 (1963).
70. Farkas, E., E. R. Lavagnino and R. T. Rapala: Preparation of 3-Dehydroreserpic Acid Lactone and Its Conversion to Reserpic Acid Lactone. J. Organ. Chem. (USA) **22**, 1261 (1957).
71. Feltkamp, H. and N. C. Franklin: Conformational Analysis of Cyclohexane Derivatives by Nuclear Magnetic Resonance Spectroscopy. Angew. Chem., Internat. Ed. **4**, 774 (1965) [German Ed. **77**, 798 (1965)].
72. Finch, N., C. W. Gemenden, I. H.-C. Hsu, A. Kerr, G. A. Sim and W. I. Taylor: Oxidative Transformations of Indole Alkaloids. III. Pseudoindoxyls from Yohimbinoid Alkaloids and Their Conversion to "Invert" Alkaloids. J. Amer. Chem. Soc. **87**, 2229 (1965).
73. Finch, N., C. W. Gemenden, I. H.-C. Hsu and W. I. Taylor: Oxidative Transformations of Indole Alkaloids. II. The Preparation of Oxindoles from *cis*-DE-Yohimbinoid Alkaloids. The Partial Synthesis of Carapanaubine. J. Amer. Chem. Soc. **85**, 1520 (1963).
74. Finch, N., I. H.-C. Hsu, W. I. Taylor, H. Budzikiewicz and C. Djerassi: Mass Spectrometry in Structural and Stereochemical Problems. XLVII. Some Observations on Mass Spectra of Pseudoindoxyl Alkaloids. J. Amer. Chem. Soc. **86**, 2620 (1964).

75. FINCH, N. and W. I. TAYLOR: The Conversion of Tetrahydro-β-carboline Alkaloids into Oxindoles. The Structures and Partial Syntheses of Mitraphylline and Rhyncophylline. J. Amer. Chem. Soc. **84**, 1318 (1962).

76. — — Oxidative Transformations of Indole Alkaloids. I. The Preparation of Oxindoles from Yohimbine; the Structures and Partial Syntheses of Mitraphylline, Rhyncophylline and Corynoxeine. J. Amer. Chem. Soc. **84**, 3871 (1962).

77. FINCH, N., W. I. TAYLOR, T. R. EMERSON, W. KLYNE and R. J. SWAN: Optical Rotatory Dispersion Curves of Heteroyohimbine Alkaloids. Tetrahedron **22**, 1327 (1966).

78. FINCH, N., W. I. TAYLOR and P. R. ULSHAFER: Rauwolfia Alkaloids. XLVII. Isoreserpiline-ψ-Indoxyl, its Isolation, Synthesis and Structure. Experientia **19**, 296 (1963).

79. FODOR, G.: Neuere Ergebnisse über die Raumstruktur der Tropanalkaloide. Experientia **11**, 129 (1955).

80. — Recent Developments in the Synthesis and Stereochemistry of Tropane Alkaloids. Tetrahedron **1**, 87 (1957).

81. GARBUTT, D. F. C., P. W. JEFFS and F. L. WARREN: The Alkaloids of the Amaryllidaceae. XI. The Alkaloids of *Nerine krigeii* and the Structure of Krigenamine. J. Chem. Soc. (London) **1962**, 5010.

82. GENESTE, P. et G. LAMATY: Analyse conformationnelle de systèmes mobiles. II. *trans*-Décaline. Tetrahedron Letters **1964**, 3545.

83. — — Analyse conformationnelle des systèmes mobiles. I. *Cis*-décaline. Bull. soc. chim. France **1964**, 2439.

84. GILBERT, B., J. A. BRISSOLESE, N. FINCH, W. I. TAYLOR, H. BUDZIKIEWICZ, J. M. WILSON and C. DJERASSI: Mass Spectrometry in Structural and Stereochemical Problems. XX. Carapanaubine, a New Alkaloid from *Aspidosperma carapanauba* and Some Observations on Mass Spectra of Oxindole Alkaloids. J. Amer. Chem. Soc. **85**, 1523 (1963).

85. GOVINDACHARI, T. R., N. VISWANATHAN, B. R. PAI and T. S. SAVITRI: Chemical Constituents of *Alstonia venenata* R. Br. Tetrahedron **21**, 2951 (1965).

86. HANACK, M.: Conformation Theory. New York: Academic Press. 1965.

87. HART, N. K., S. R. JOHNS and J. A. LAMBERTON: Uncarine C, D (Speciophylline), E, and F: C-3 and C-7 Epimeric Oxindoles Related to Tetrahydroalstonine. Chem. Commun. **1967**, 87.

88. HAUGWITZ, R. D., P. W. JEFFS and E. WENKERT: Proton Magnetic Resonance Spectral Studies of Some Amaryllidaceae Alkaloids of the 5,10b-Ethanophenanthridine Series and of Criwelline and Tazettine. J. Chem. Soc. (London) **1965**, 2001.

89. HAWKSWORTH, W. A., P. W. JEFFS, B. K. TIDD and T. P. TOUBE: The Alkaloids of the Amaryllidaceae. XII. The Aromatic Oxygenation Patterns and Stereochemistry of Some Trioxyaryl Alkaloids of the Hemiacetal and Lactone Series. J. Chem. Soc. (London) **1965**, 1991.

90. HAZEBROEK, P. and L. J. OOSTERHOFF: The Isomers of Cyclohexane. Discuss. Faraday Soc. **10**, 87 (1951).

91. HENBEST, H. B. and J. McENTEE: Aspects of Stereochemistry. XVIII. Synthesis of the 10-Methyldecalin-2,9-diols. J. Chem. Soc. (London) **1961**, 4478.

92. HENDRICKSON, J. B.: Molecular Geometry. I. Machine Computation of the Common Rings. J. Amer. Chem. Soc. **83**, 4537 (1961).

93. — The Twist-Boat Form of Cyclohexane. J. Organ. Chem. (USA) **29**, 991 (1964) et loc. cit.

94. HIGHET, P. F. and W. C. WILDMAN: Hofmann Degradation of 3a-(3,4-Methylene-dioxyphenyl)-1-methyloctahydroindole. J. Organ. Chem. (USA) 25, 287 (1960).
95. HILL, R. K., J. A. JOULE and L. J. LOEFFLER: Stereoselective Syntheses of d,l-α- and β-Lycoranes. J. Amer. Chem. Soc. 84, 4951 (1962).
96. HINE, J.: Application of the Principle of Least Motion to the Stereochemistry of Elimination Reactions. J. Amer. Chem. Soc. 88, 5525 (1966).
97. HOUSE, H. O., B. A. TEFERTILLER and C. G. PITT: Stereochemistry of the N-Alkylation of 4-t-Butylpiperidine Derivatives. J. Organ. Chem. (USA) 31, 1073 (1966).
98. HOWLETT, K. E.: Intramolecular Non-bonded Effects involving Hydrogen Atoms. J. Chem. Soc. (London) 1957, 4353.
99. HUEBNER, C. F. and D. F. DICKEL: Rauwolfia Alkaloids. XXVI. Stereochemistry at C-17. Experientia 12, 250 (1956).
100. HUEBNER, C. F., H. B. MACPHILLAMY, E. SCHLITTLER and A. F. ST. ANDRÉ: Rauwolfia Alkaloids. XXI. The Stereochemistry of Reserpine and Deserpidine. Experientia 11, 303 (1955).
101. HUEBNER, C. F., A. F. ST. ANDRÉ, E. SCHLITTLER and A. UFFER: Rauwolfia Alkaloids. XX. 11-Methoxyalloyohimbane from Reserpine. J. Amer. Chem. Soc. 77, 5725 (1955).
102. HUEBNER, C. F. and E. WENKERT: Rauwolfia Alkaloids. XXII. Further Observations of the Stereochemistry of Reserpine. J. Amer. Chem. Soc. 77, 4180 (1955).
103. HUITRIC, A. C., J. B. CARR, W. F. TRAGER and B. J. NIST: Configurational and Conformational Analysis. Axial-axial and Axial-equatorial Coupling Constants in Six-membered Ring Compounds. Tetrahedron 19, 2145 (1963).
104. IKEDA, T., W. I. TAYLOR, Y. TSUDA, S. UYEO and H. YAJIMA: The Structure and Stereochemistry of Tazettine. J. Chem. Soc. (London) 1956, 4749.
105. IMBACH, J.-L., A. R. KATRITZKY and R. A. KOLINSKI: The Conformational Analysis of Heterocyclic Systems. VIII. Kinetics of Quaternisation of N-Alkyl-piperidines and N-Alkylpiperazines in Acetonitrile. J. Chem. Soc. (London) B 1966, 556.
106. INUBUSHI, Y., H. M. FALES, E. W. WARNHOFF and W. C. WILDMAN: Structures of Montanine, Coccinine and Manthine. J. Organ. Chem. (USA) 25, 2153 (1960).
107. IRIE, H., Y. TSUDA and S. UYEO: The Structure of Tazettine. A Synthesis of the Enide Degradation Product derived from Tazettamide. J. Chem. Soc. (London) 1959, 1446.
108. JACKMAN, L. M.: Applications of Nuclear Magnetic Resonance Spectroscopy in Organic Chemistry. New York: Pergamon Press. 1959.
109. JANOT, M.-M. et R. GOUTAREL: La corynanthine: ses rapports de constitution avec la yohimbine. Bull. soc. chim. France 1949, 509.
110. JANOT, M.-M., R. GOUTAREL, A. LE HIR, M. AMIN et V. PRELOG: Stéréochimie de la pseudo-yohimbine, de la yohimbine et de la corynanthine. Bull. soc. chim. France 1952, 1085.
111. JANOT, M.-M., R. GOUTAREL, E. W. WARNHOFF et A. LE HIR: Stéréoisomères de la yohimbine. IV. Nouvelles isomérisations et obtention de l'épi-3 cory-nanthine et de l'épi-3 allo-yohimbine. Bull. soc. chim. France 1961, 637.
112. JEFFERIES, P. R., R. B. ROSICH and D. E. WHITE: Long Range Shielding by the Epoxide Ring. Tetrahedron Letters 1963, 1853.
113. JENSEN, F. R. and C. H. BUSHWELLER: Separation of Conformers. I. Axial and Equatorial Isomers of Monosubstituted Cyclohexanes. J. Amer. Chem. Soc. 88, 4279 (1966).

114. Johns, S. R. and J. A. Lamberton: *Uncaria* Alkaloids: Two Stereoisomers ot Mitraphylline from *Uncaria bernaysii* F. v. Muell. and *U. ferrea* D. C. Tetrahedron Letters **1966**, 4883.

115. Johnson, C. D., R. A. Y. Jones, A. R. Katritzky, C. R. Palmer, K. Schofield and R. J. Wells: A Re-examination of the Stereochemistry of Quinolizidine and the Methylquinolizidines through Measurement of their Rates of Quaternisation and those of the Hexahydrojulolidines. J. Chem. Soc. (London) **1965**, 6797.

116. Johnson, W. S., V. J. Bauer, J. L. Margrave, M. A. Frisch, L. H. Dreger and W. N. Hubbard: The Energy Difference between the Chair and Boat Forms of Cyclohexane. The Twist Conformation of Cyclohexane. J. Amer. Chem. Soc. **83**, 606 (1961).

117. Karplus, M.: Contact Electron-Spin Coupling of Nuclear Magnetic Moments. J. Chem. Phys. **30**, 11 (1959).

118. — Vicinal Proton Coupling in Nuclear Magnetic Resonance. J. Amer. Chem. Soc. **85**, 2870 (1963).

119. King, R. W., C. F. Murphy and W. C. Wildman: 6-Hydroxycrinamine and Haemanthidine. J. Amer. Chem. Soc. **87**, 4912 (1965).

120. Kitagawa, T., S. Uyeo and N. Yokoyama: Stereochemistry of Lycorenine, Homolycorine, Pluviine, and their Hydrogenation Products. J. Chem. Soc. (London) **1959**, 3741.

121. Klyne, W.: The Conformations of Six-membered Ring Systems. Progr. Stereochem. **1**, 36 (1954).

122. — Stereochemical Correlation of the Yohimbine Alkaloids with the Steroids. Chem. and Ind. **1953**, 1032.

123. Kotera, K.: The Stereochemistry of Lycorane. II. (—)-β-Lycorane. Tetrahedron **12**, 240 (1961).

124. Kotera, K., Y. Hamada, K. Tori, K. Aono and K. Kuriyama: Absolute Configurations and the C-Ring Conformations of Lycorine and Related Compounds Evidenced by N. M. R and CD Spectroscopies. Tetrahedron Letters **1966**, 2009.

125. Le Hir, A. et R. Goutarel: Stéréoisomères de la yohimbine. II. β-Yohimbine. Bull. soc. chim. France **1953**, 1023.

126. Lemieux, R. U., R. K. Kullnig, H. J. Bernstein and W. G. Schneider: Configurational Effects on the Proton Magnetic Resonance Spectra of Six-membered Ring Compounds. J. Amer. Chem. Soc. **80**, 6098 (1958).

127. Leonard, N. J., A. S. Hay, R. W. Fulmer and V. W. Gash: Unsaturated Amines. III. Introduction of α,β-Unsaturation by Means of Mercuric Acetate: Δ1(10)-Dehydroquinolizidine. J. Amer. Chem. Soc. **77**, 439 (1955).

128. Leonard, N. J. and D. F. Morrow: Unsaturated Amines. XII. Steric Requirements of Mercuric Acetate Oxidation of Tertiary Amines. J. Amer. Chem. Soc. **80**, 371 (1958).

129. Levisalles, J.: La forme bateau dans les équilibres et les réactions des cycles hexatomiques. Bull. soc. chim. France **1960**, 551.

130. Lloyd, H. A., E. A. Kielar, R. J. Highet, S. Uyeo, H. M. Fales and W. C. Wildman: Position of the Aromatic Methoxyl in Alkaloids Related to Powelline. J. Organ. Chem. (USA) **27**, 373 (1962).

131. McAleer, W. J. and M. A. Kozlowski: Application of Zaffaroni-type Partition Systems to the Paper Chromatography of Steroidal Sapogenins. Arch. Biochem. Biophys. **66**, 120 (1957).

132. McKenna, J.: Conformational Analysis of Organic Compounds. Roy. Inst. Chem. (London), Lectures **1966**, No. 1.

133. McKenna, J., B. G. Hutley and J. White: Stereoisomeric Pairs of Cyclic Quaternary Ammonium Salts. III. Examination of Differential Reactivity of Certain N-Ethyl-N-methyl Pairs in Nucleophilic Substitution Reactions. J. Chem. Soc. (London) **1965**, 1729.

134. McKenna, J., J. M. McKenna, A. T. Tulley and J. White: Stereoisomeric Pairs of Cyclic Quaternary Ammonium Salts. I. Stereospecificity in Quaternisations of N-Alkylcamphidines, 2-Methylpyrrolidines, 2-Methyl- and 4-Phenylpiperidines, *trans*-Decahydroquinolines, and Tropanes, and Configurations of the Diastereoisomeric Salts. J. Chem. Soc. (London) **1965**, 1711.

135. McKenna, J., J. M. McKenna and J. White: Stereoisomeric Pairs of Cyclic Quaternary Ammonium Salts. IV. Interconversion of N-Benzyl-N-methyl Quaternary Iodides in Hot Chloroform, and General Theoretical Discussion of Reactions Involving a Change Between Co-ordination Numbers 3 and 4 at a Six-ring Atom. J. Chem. Soc. (London) **1965**, 1733.

136. MacPhillamy, H. B., L. Dorfman, C. F. Huebner, E. Schlittler and A. F. St. André: Rauwolfia Alkaloids. XVIII. On the Constitution of Deserpidine and Reserpine. J. Amer. Chem. Soc. **77**, 1071 (1955).

137. MacPhillamy, H. B., C. F. Huebner, E. Schlittler, A. F. St. André and P. R. Ulshafer: Rauwolfia Alkaloids. XIX. The Constitution of Deserpidine and Reserpine. J. Amer. Chem. Soc. **77**, 4335 (1955).

138. Margrave, J. L., M. A. Frisch, R. G. Bautista, R. L. Clarke and W. S. Johnson: Further Studies on the Energy Difference between the Chair and Twist Forms of Cyclohexane. J. Amer. Chem. Soc. **85**, 546 (1963).

139. Mitscher, L. A., J. K. Paul and L. Goldman: Methyl 11-Methoxy-18-oxo-3-epialloyohimban-16α-carboxylate, a New Keto Ester Derived from Reserpine. Experientia **19**, 195 (1963).

140. Mizukami, S.: Stereochemical Relations Between Lycorine, Caranine, Pluviine, and Lycorenine. Tetrahedron **11**, 89 (1960).

141. Moynehan, T. M., K. Schofield, R. A. Y. Jones and A. R. Katritzky: The Synthesis and Stereochemistry of Quinolizidine and the Monomethylquinolizidines, and of their Salts and Quaternary Salts. J. Chem. Soc. (London) **1962**, 2637.

142. Nakagawa, Y. and S. Uyeo: Stereochemistry of Reduction Products of 1-Acetyl-lycorin-2-one. J. Chem. Soc. (London) **1959**, 3736.

143. Neuss, N. and H. E. Boaz: Rauwolfia Alkaloids. V. Stereochemical Correlation of Some Indole Alkaloids from the Infrared Spectra. J. Organ. Chem. (USA) **22**, 1001 (1957).

144. Nickon, A., M. A. Castle, R. Harada, C. E. Berkoff and R. O. Williams: Chemical Shifts of Axial and Equatorial α-Protons in the N. M. R. of Steroidal α-Haloketones. J. Amer. Chem. Soc. **85**, 2185 (1963).

145. Orloff, H. D.: The Stereoisomerism of Cyclohexane Derivatives. Chem. Rev. **54**, 347 (1954).

146. Ottar, B.: The Structure of 1,2-Epoxy-cyclohexane. An Electron Diffraction Investigation Based on the Rotating Sector Method. Acta Chem. Scand. **1**, 283 (1947).

147. Ozeki, S.: Alkaloids of *Zephyranthes candida* Herb. II. Structure of Zephyranthine. J. Pharmac. Soc. Japan **85**, 200 (1965).

148. Pauncz, R. and D. Ginsburg: Conformational Analysis of Alicyclic Compounds. I. Considerations of Molecular Geometry and Energy in Medium and Large Rings. Tetrahedron **9**, 40 (1960).

149. Petrowitz, H.-J.: Zur Kieselgelschicht-Chromatographie der stereoisomeren Menthole. Angew. Chem. **72**, 921 (1960).

150. POISSON, J.: Recherches récentes sur les alcaloïdes du Pseudocinchona et du Yohimbe. Ann. chimie [13] **9**, 99 (1964).

151. POISSON, J., R. BERGOEING, N. CHAUVEAU, M. SHAMMA et R. GOUTAREL: Alcaloïdes des Rauwolfia: Structure de la raumitorine et relations avec celle de la rauvanine. Bull. soc. chim. France **1964**, 2853.

152. POUSSET, J.-L. et J. POISSON: Rauvoxine et rauvoxinine, alcaloïdes oxindoliques des feuilles du *Rauwolfia vomitoria* Afz. C. R. hebd. séances Acad. Sci. **259**, 597 (1964).

153. POUSSET, J.-L., J. POISSON et M. LEGRAND: Application des spectres de dichroïsme circulaire a l'étude des alcaloïdes oxindoliques. Tetrahedron Letters **1966**, 6283.

154. PUMPHREY, N. W. J. and M. J. T. ROBINSON: The Volume Requirement of the Unshared Pair of Electrons in Amines. Chem. and Ind. **1963**, 1903.

155. ROSEN, W. E.: Rauwolfia Alkaloids. XLII. Methyl Neoreserpate, an Isomer of Methyl Reserpate. 4. Infrared Spectra and Configuration at C-3. Tetrahedron Letters **1961**, 481.

156. — Rauwolfia Alkaloids. XLIX. Ring E Isomers in the Epiallo Series. J. Organ. Chem. (USA) **30**, 2044 (1965).

157. ROSEN, W. E. and J. M. O'CONNOR: Rauwolfia Alkaloids. XXXVII. Methyl *neo*-Reserpate, an Isomer of Methyl Reserpate. J. Organ. Chem. (USA) **26**, 3051 (1961).

158. ROSEN, W. E. and H. SHEPPARD: Rauwolfia Alkaloids. XXXIX. Methyl Neoreserpate, an Isomer of Methyl Reserpate. 2. Mechanism of Formation. J. Amer. Chem. Soc. **83**, 4240 (1961).

159. ROSEN, W. E. and J. N. SHOOLERY: Rauwolfia Alkaloids. XLI. Methyl Neoreserpate, an Isomer of Methyl Reserpate. 3. Conformations and N. M. R. Spectra. J. Amer. Chem. Soc. **83**, 4816 (1961).

160. SCOTT, A. I.: Interpretation of the Ultraviolet Spectra of Natural Products. London: Pergamon Press. 1964.

161. SEATON, J. C., M. D. NAIR, O. E. EDWARDS and L. MARION: The Structure and Stereoisomerism of Three Mitragyna Alkaloids. Canad. J. Chem. **38**, 1035 (1960).

162. SHAMMA, M. and J. M. RICHEY: The Stereochemistry of the Heteroyohimbine Alkaloids. J. Amer. Chem. Soc. **85**, 2507 (1963).

163. SHAMMA, M. and R. J. SHINE: The Structure and Stereochemistry of Raujemidine. Tetrahedron Letters **1964**, 2277.

163 a. SHAMMA, M., R. J. SHINE, I. KOMPIŠ, T. STICZAY, F. MORSINGH, J. POISSON and J. L. POUSSET: The Stereochemistry of the Pentacyclic Oxindole Alkaloids. J. Amer. Soc. **89**, 1739 (1967).

164. SHAVEL, J., Jr. and H. ZINNES: Oxindole Alkaloids. I. Oxidative Rearrangement of Indole Alkaloids to their Oxindole Analogs. J. Amer. Chem. Soc. **84**, 1320 (1962).

165. SHOPPEE, C. W., D. N. JONES and G. H. R. SUMMERS: Steroids and Walden Inversion. XXXVII. The Epimeric Cholestane-2 : 3-diols. J. Chem. Soc. (London) **1957**, 3100.

166. SMITH, E., R. S. JARET, M. SHAMMA and R. J. SHINE: Deserpideine, a New Yohimbinoid Type Alkaloid. J. Amer. Chem. Soc. **86**, 2083 (1964); see also Lloydia **27**, 440 (1964).

167. SMITH, L. G.: The Infra-Red Spectrum of C_2H_6. J. Chem. Physics **17**, 139 (1949).

168. SPEROS, D. M. and F. D. ROSSINI: Heats of Combustion and Formation of Naphthalene, the Two Methylnaphthalenes, *cis* and *trans*-Decahydronaphthalene, and Related Compounds. J. Physic. Chem. **64**, 1723 (1960).

169. TAKEDA, K. and K. KOTERA: The Stereochemistry of Dihydrolycorine. Pharmac. Bull. (Japan) 5, 234 (1957); Chem. and Ind. 1956, 347.

170. TAKEDA, K., K. KOTERA and S. MIZUKAMI: A Partial Synthesis of Caranine. J. Amer. Chem. Soc. 80, 2562 (1958).

171. TAKEUCHI, M.: Systematic Analysis of Steroids. I. Systematic Analysis of Steroid Hormones by Thin Layer Chromatography. Chem. and Pharmac. Bull. (Japan) 11, 1183 (1963).

172. TORI, K., K. AONO, K. KITAHONOKI, R. MUNEYUKI, Y. TAKANO, H. TANIDA and T. TSUJI: N. M. R. Studies of Bridged Ring Systems. X. Long-range Anisotropic Shielding Effects of an Epoxide and an Aziridine Ring. Tetrahedron Letters 1966, 2921 et loc. cit.

173. TSUDA, Y. and S. UYEO: Stereochemistry of the Hemiketal Moiety of Tazettine. J. Chem. Soc. (London) 1961, 2485.

174. UYEO, S., T. KITAGAWA and Y. YAMAMOTO: Oxidation of the Hydrogenation Products of Lycorenine and Homolycorine. Chem. and Pharmac. Bull. (Tokyo) 12, 408 (1964).

175. VAN TAMELEN, E. E. and P. D. HANCE: The Stereochemical Formulation of Reserpine. J. Amer. Chem. Soc. 77, 4692 (1955).

176. VELLUZ, L., M. LEGRAND and M. GROSJEAN: Optical Circular Dichroism: Principles, Measurements, and Applications. Weinheim: Verlag Chemie, and New York and London: Academic Press. 1965.

177. WARNHOFF, E. W. and W. C. WILDMAN: Alkaloids of the Amaryllidaceae. X. The Structure of Caranine. J. Amer. Chem. Soc. 79, 2192 (1957).

178. — — The Structure and Stereochemistry of Undulatine. J. Amer. Chem. Soc. 82, 1472 (1960).

179. WEISENBORN, F. L. and P. A. DIASSI: The Reaction of Rauwolfia Alkaloids with Mercuric Acetate. Conversion of 3-Isoreserpine to Reserpine. J. Amer. Chem. Soc. 78, 2022 (1956).

180. WELLMAN, K. M. and F. G. BORDWELL: Magnetic Shielding of Alpha Protons by the Carbonyl Group in Cyclohexanones. Tetrahedron Letters 1963, 1703.

181. WENKERT, E. and L. H. LIU: The Constitution of the Alloyohimbanes. Experientia 11, 302 (1955).

182. WENKERT, E. and D. K. ROYCHAUDHURI: 3-Dehydro Derivatives of Some Indole Alkaloids. J. Organ. Chem. (USA) 21, 1315 (1956).

183. — — The C-3 Configuration of Certain Indole Alkaloids. J. Amer. Chem. Soc. 78, 6417 (1956).

184. — — Oxidation-Reduction Studies in the Realm of Indole Alkaloids. J. Amer. Chem. Soc. 80, 1613 (1958).

185. WENKERT, E., B. WICKBERG and C. LEICHT: The Stereochemistry of Mayumbine and Structurally Related Oxindole Alkaloids. Tetrahedron Letters 1961, 822.

186. — — — The Stereochemistry of Ajmalicine and Tetrahydroalstonine. J. Amer. Chem. Soc. 83, 5037 (1961).

187. WILDMAN, W. C.: Alkaloids of the Amaryllidaceae. In: R. H. F. MANSKE (Edit.), The Alkaloids, Vol. VI, p. 289. New York: Academic Press. 1960.

188. WILLIAMS, D. H. and N. S. BHACCA: Dependency of Vicinal Coupling Constants on the Configuration of Electronegative Substituents. J. Amer. Chem. Soc. 86, 2742 (1964).

189. WILLIAMS, K. I. H., M. SMULOWITZ and D. K. FUKUSHIMA: D-Homoannulation of Pregnane 17,20-Glycols. J. Organ. Chem. (USA) 30, 1447 (1965).

190. WILLIAMSON, K. L.: Substituent Effects on Nuclear Magnetic Resonance Coupling Constants and Chemical Shifts in a Saturated System: Hexachlorobicyclo[2.2.1]heptenes. J. Amer. Chem. Soc. 85, 516 (1963).

191. WITKOP, B.: Zur Konstitution des Yohimbins und seiner Abbauprodukte. Liebigs Ann. Chem. **554**, 83 (1943).

192. — On The Stereochemistry of Yohimbine. J. Amer. Chem. Soc. **71**, 2559 (1949).

193. WOHL, R. A.: The Flattened Cyclohexane Ring. Chimia **18**, 219 (1964).

194. WOODWARD, R. B., F. E. BADER, H. BICKEL, A. J. FREY and R. W. KIERSTEAD: The Total Synthesis of Reserpine. Tetrahedron **2**, 1 (1958).

195. YAMAZAKI, F.: The Conformation of Reserpines. J. Chem. Soc. Japan, Pure Chem. Sect. **82**, 72 (1961).

196. ZINNES, H. and J. SHAVEL, Jr.: Yohimbane Derivatives. III. The Oxidative Rearrangement of Indole Alkaloids to Their Spirooxindole Analogs. J. Organ. Chem. (USA) **31**, 1765 (1966).

(Received, March 20, 1967.)

Namenverzeichnis. Index of Names. Index des Auteurs.

Kursiv gedruckte Seitenzahlen beziehen sich auf Literaturverzeichnisse.

Page numbers printed in *italic* refer to References.

Les chiffres en *italique* indiquent les pages de bibliographie.

Sachverzeichnis. Index of Subjects. Index des Matières.

Phalloidin, Edman degradation- →phenyl-
thiohydantoin deriv. 224.
—, → desthio- 222.
—, enzymatic conversion (liver) 245.
—, hydrolyzate, electrophoresis 223.
—, inhibition of incorporation of amino
acids (liver) 245.
—, and 2-S-methylindolyl acetic acid,
comparative UV spectra 222.
—, mol. weight 222, 224.
—, properties 220.
—, radioactive, adsorption (liver) 245.
—, similarity with 2-S-methylindolyl
acetic acid 222.
—, spectrum 222, 237.
—, structure 225.
—, thioamide structure 222.
—, toxic action, mechanism 245.
—, → toxic dithiolane with [35]S 229.
— deriv., toxic, labeled 228, 245.
—, desthio- 222.
—, —, structure 225.
—, desthio-seco- 224, 228.
—, epoxy-, structure 229.
—, keto-, structure 229.
—, keto-seco-, cyclization 244.
—, seco compounds 224, 228, 243.
—, δ-tosyl-, structure 229.
Phalloin 217, 225.
—, amino acids 225.
—, properties 220.
—, structure 225.
—, desmethyl-, structure 229.
—, seco compounds 243.
Phenylacetylene, o-hydroxy-benzoyl-,
→ Benzalcumaranon-Deriv. 164.
Phenylpolyine, Biogenese (durch Verfütterung) 40, 41.
—, oxydative Entfernung der endständigen Methylgruppe in der Pflanze 42,
48.
C_{18}-Phenylpolyine, Biogenese 42.
Phloracylphenones, tetra-(3-methyl-2-
enyl)-, deriv. 82.
Phlorisovalerophenone, tetraisoprenyl-,
structure 82, 83.
Photo-isohumulone 71.
Piperidine, conformation 278.
—, N-alkyl-, quaternized, conformation,
278.
—, N-methyl-, preferred conformation
278.

Pluviine, structure 303.
Polyine, s. auch Acetylenverbindungen.
—, aus *Fistulina hepatica* 44.
—, markierte, Synthese 7.
—, Methylmercaptan-Anlagerung 10.
—, natürliche, mit Ketogruppe 28.
—, [14]C- oder tritium-markiert 1.
C_{13}- und C_{14}-Polyine, Biogenese 23, 24.
—, O-reichere Derivate, Biogenese 24.
C_{16}-Polyine, Biogenese aus C_{18}-Verbin-
dungen 22.
C_{17}-Polyine in Compositen und Umbelliferen 18.
—, toxische 19.
Polyine, phenyl-, Biogenese (durch Verfütterung) 40, 41.
C_{15}-C_{17}-Polyin-aldehyde *(Centaurea)* 22,
23.
C_{13}-Polyinen-Kohlenwasserstoffe als Vorstufen 8.
C_{14}-Polyinsäureester, natürlicher 26.
Posthumulone 65.
Prehumulone 65.
Prolamine 251.
—, Bausteinanalyse 254.
—, Beziehung zu Anthocyanogenen 262.
—, und Cöliakie 264.
—, elektrophoretische Analyse, Komponenten 254.
—, endständige Aminosäuren 257, 258.
—, ernährungsphysiologische Bedeutung
263.
—, im Gersteneiweiß 251.
—, im Getreidesamen, Gehalt 252.
—, günstig für die Ernährung des pflanzlichen Embryos 264.
—, isoelektrischer Punkt 256.
—, Komponenten, mit extremer Ladung
256.
—, —, Veränderlichkeit der Verhältnisse
258.
—, Löslichkeit, Isolierung 253.
—, Mol.-Gew. 253.
—, Peptidketten 258.
—, Vorkommen 252.
—, aus Weizen und Roggen, Unterschied
256.
L-Prolin, allo-hydroxy-, synthesis 239.
—, hydroxy-, structure 230.
Proton magnetic resonance and conformational studies 277.
Pseudoguaianolides 90.

Fortschritte der Chemie organischer Naturstoffe. Progress in the Chemistry of Organic Natural Products. Progrès dans la chimie des substances organiques naturelles. Herausgegeben von **L. Zechmeister,** California Institute of Technology, Pasadena, California, U. S. A.

Springer-Verlag / Wien · New York

Bisher erschienen:

Erster Band: Mit 41 Abbildungen im Text. VI, 371 Seiten. Gr.-8°. 1938. Ganzleinen S 348.—, DM 72.25, $ 17.20

Zweiter Band: Mit 24 Abbildungen im Text. VII, 366 Seiten. Gr.-8°. 1939. Ganzleinen S 348.—, DM 72.25, $ 17.20

Dritter Band: Mit 10 Abbildungen im Text. VI, 252 Seiten. Gr.-8°. 1939. Ganzleinen S 264.—, DM 55.45, $ 13.20

Vierter Band: Mit 47 Abbildungen im Text. VIII, 499 Seiten. Gr.-8°. 1945. Ganzleinen S 474.—, DM 99.10, $ 23.60

Fünfter Band: Mit 34 Abbildungen. VIII, 417 Seiten. Gr.-8°. 1948. Ganzleinen S 305.—, DM 50.40, $ 12.—

Sechster Band: Mit 32 Abbildungen. VIII, 392 Seiten. Gr.-8°. 1950. Ganzleinen S 338.—, DM 55.80, $ 13.30

Siebenter Band: Mit 12 Abbildungen. VII, 330 Seiten. Gr.-8°. 1950. Ganzleinen S 325.—, DM 53.70, $ 12.80

Achter Band: Mit 47 Abbildungen. XI, 400 Seiten. Gr.-8°. 1951. Ganzleinen S 427.—, DM 70.50, $ 16.80

Neunter Band: Mit 20 Abbildungen. XI, 535 Seiten. Gr.-8°. 1952. Ganzleinen S 498.—, DM 82.50, $ 19.60

Zehnter Band: Mit 19 Abbildungen. IX, 529 Seiten. Gr.-8°. 1953. Ganzleinen S 498.—, DM 83.—, $ 19.80

Elfter Band: Mit 67 Abbildungen. VIII, 457 Seiten. Gr.-8°. 1954. Ganzleinen S 448.—, DM 74.80, $ 18.—

Zwölfter Band: Mit 15 Abbildungen. X, 550 Seiten. Gr.-8°. 1955. Ganzleinen S 497.—, DM 82.80, $ 19.80

Dreizehnter Band: Mit 48 Abbildungen. XII, 624 Seiten. Gr.-8°. 1956. Ganzleinen S 645.—, DM 107.50, $ 25.60

Vierzehnter Band: Mit 38 Abbildungen. VIII, 377 Seiten. Gr.-8°. 1957. Ganzleinen S 450.—, DM 75.—, $ 17.85

Weitere Bände siehe nächste Seite!

Zu beziehen durch Ihre Buchhandlung
Auslieferung für die U. S. A. und Canada: Springer-Verlag New York Inc., 175 Fifth Avenue, New York N. Y. 10010

Springer-Verlag/ Wien · New York

Fortsetzung von vorhergehender Seite

Fünfzehnter Band: Mit 81 Abbildungen. VI, 244 Seiten. Gr.-8°. 1958.
Ganzleinen S 246.—, DM 41.—, $ 9.75

Sechzehnter Band: Mit 27 Abbildungen. VI, 226 Seiten. Gr.-8°. 1958.
Ganzleinen S 240.—, DM 40.—, $ 9.50

Siebzehnter Band: Mit 57 Abbildungen. X, 515 Seiten. Gr.-8°. 1959.
Ganzleinen S 498.60, DM 83.10, $ 19.80

Achtzehnter Band: Mit 65 Abbildungen. X, 600 Seiten. Gr.-8°. 1960.
Ganzleinen S 618.—, DM 103.—, $ 24.50

Neunzehnter Band: Mit 16 Abbildungen. VIII, 420 Seiten. Gr.-8°. 1961.
Ganzleinen S 490.—, DM 78.—, $ 19.50

Zwanzigster Band: Mit 33 Abbildungen. XIII, 509 Seiten. Gr.-8°. 1962.
Ganzleinen S 604.—, DM 96.—, $ 24.—

Einundzwanzigster Band: Mit 14 Abbildungen. VII, 362 Seiten. Gr.-8°. 1963.
Ganzleinen S 479.—, DM 76.—, $ 19.—

Über den Inhalt der Bände gibt der Verlag bereitwilligst Auskunft

Generalregister / Cumulative Index / Index Général I—XX. 1938—1962.
XVI, 369 Seiten. Gr.-8°. 1964. Ganzleinen S 378.—, DM 60. —, $ 15. —

Zweiundzwanzigster Band: Mit 8 Abbildungen. VII, 370 Seiten. Gr.-8°.
1964. Ganzleinen S 554.—, DM 88. —, $ 22.—

Inhalt: **Schaffner, K.** Photochemische Umwandlungen ausgewählter Naturstoffe. — **Billek, G.** Stilbene im Pflanzenreich. — **Halsall, T. G.,** and **R. T. Aplin.** A Pattern of Development in the Chemistry of Pentacyclic Triterpenes. — **Grove, J. F.** Griseofulvin and Some Analogues. — **Scheuer, P. J.** The Chemistry of Toxins Isolated from Some Marine Organisms. — **Keller-Schierlein, W., V. Prelog** und **H. Zähner.** Siderochrome.

Dreiundzwanzigster Band: Mit 58 Abbildungen. VIII, 397 Seiten. Gr.-8°.
1965. Ganzleinen S 590.—, DM 93.60, $ 23.40

Inhalt: **Peat, S.,** and **J. R. Turvey.** Polysaccharides of Marine Algae. — **Schlubach, H. H.** Der Kohlenhydratstoffwechsel in Gerste, Hafer und Rispenhirse. — **Schlenk, F.** The Chemistry of Biological Sulfonium Compounds. — **Schroeder, W. A.,** and **R. T. Jones.** Some Aspects of the Chemistry and Function of Human and Animal Hemoglobins. — **Grassmann, W.** Kollagen. — **Jackman, L. M.** Some Applications of Nuclear Magnetic Resonance Spectroscopy in Natural Products Chemistry.

Vierundzwanzigster Band: Mit 25 Abbildungen. VIII, 475 Seiten. Gr.-8°.
1966. Ganzleinen S 700.—, DM 111.—, $ 27.80

Inhalt: **Biemann, K.** Mass Spectrometry of Selected Natural Products. — **Tschesche, R.** Pflanzliche Steroide mit 21 Kohlenstoffatomen. — **Kindl, H.,** und **O. Hoffmann-Ostenhof.** Cyclite: Biosynthese, Stoffwechsel und Vorkommen. — **Erdtman, H.,** and **T. Norin.** The Chemistry of the Order Cupressales. — **Turner, A. B.** Quinone Methides in Nature. — **Warren, F. L.** The Pyrrolizidine Alkaloids II. — **Fraenkel-Conrat, H.** Some Aspects of Virus Chemistry.

Subskribenten auf die künftig erscheinenden Bände erhalten diese mit einem Nachlaß von 10% auf den Ladenpreis.

Bei Bezug der Serie Band 1—20, inklusive Generalregister für die Bände 1—20: 20% Nachlaß auf den Ladenpreis.

Zu beziehen durch Ihre Buchhandlung
Auslieferung für die U. S. A. und Canada: Springer-Verlag New York Inc.,
175 Fifth Avenue, New York N. Y. 10010